GROUP THEORY AND PHYSICS

GROUP THEORY AND PHYSICS

S. STERNBERG

GEORGE PUTNAM PROFESSOR
OF PURE AND APPLIED MATHEMATICS,
HARVARD UNIVERSITY
AND PERMANENT SACKLER FELLOW,
UNIVERSITY OF TEL AVIV.

CAMBRIDGE
UNIVERSITY PRESS

PUBLISHED BY THE PRESS SYNDICATE OF THE UNIVERSITY OF CAMBRIDGE
The Pitt Building, Trumpington Street, Cambridge, United Kingdom

CAMBRIDGE UNIVERSITY PRESS
The Edinburgh Building, Cambridge CB2 2RU, UK   http://www.cup.cam.ac.uk
40 West 20th Street, New York, NY 10011–4211, USA   http://www.cup.org
10 Stamford Road, Oakleigh, Melbourne 3166, Australia

First published 1994
Reprinted 1995
First paperback edition 1995
Reprinted 1997, 1999

A catalogue record for this book is available from the British Library

ISBN 0 521 24870 1 hardback
ISBN 0 521 55885 9 paperback

Transferred to digital printing 2003

CONTENTS

PREFACE

Group theory is one of the great achievements of 19th century mathematics. It emerged as a unifying idea drawing on four different sources: number theory, the theory of equations, geometry, and crystallography. The early motivation from number theory stemmed from the work of Euler, Legendre and Gauss on power residues. In the theory of equations, the study of various permutation groups became increasingly important through the work of Lagrange, Ruffini, Gauss, Abel, Cauchy, and especially Galois. The discovery of new types of geometries – including non-Euclidean, affine, projective etc. – led, eventually, to the famous Erlangen program of Klein, which proposed that the true study of any geometry lies in an analysis of its group of motions. In crystallography, the possible symmetries of the internal structure of a crystal were enumerated long before there was any possibility of its physical determination (by X-ray analysis).

The definition of an abstract group was proposed by Cayley in two remarkable papers in 1854, reflecting perhaps some influence of Boole (for abstract formulation) and Hamilton's quaternions (for the existence of algebras with noncommutative multiplication). This definition was not immediately appreciated by the mathematical community. In 1870, Kronecker (independently of Cayley) introduced the axioms for an abstract commutative group. When Cayley reiterated his definition in 1878, the reception was much warmer. In the period from 1870 to 1900, enormous progress was made in group theory. For example, the idea of a continuous group was introduced and studied by Lie; this led to a wealth of applications to geometry and differential equations, culminating in the classification by Killing and Cartan of the simple finite dimensional Lie groups. The theory of finite groups was greatly advanced through the work of Jordan, Hölder and Burnside.

The theory of group representations was created by Frobenius, Schur and Burnside in the last decade of the 19th century, although some of the ideas were anticipated in Jordan's monumental *Traité des substitutions* of 1870. Their theory for finite groups was extended to compact groups and brought into fruitful contact with Lie theory in a series of fundamental papers by Hermann Weyl in the 1920s. Almost all the key mathematical ideas presented in this book were developed during this period; thus, some 70 to 120 years ago. (The one principal exception is our description of Wigner's seminal paper, extending Frobenius's method to obtain the representations of the Poincaré group. This paper appeared around 50 years ago.) Of course, there has been huge progress in the last half century. I have tried to present this classical material from a geometric viewpoint, which will, hopefully, help the reader to enter the realm of the more recent advances.

It is more difficult to trace the early sources of the applications of group theory to physics. Symmetry considerations entered into the solutions of physical problems at the very beginning of mathematical physics. Mathematical crystallography, a major success of 19th century physics, is essentially group theoretical, but it had developed before the abstract language of group theory had been accepted. We explain some of the more elementary ideas of this subject in Chapter 1, and go into somewhat more detail in Appendix A. The spirit of Klein's Erlangen program pervades Poincaré's *La Science et l'Hypothese*, and other philosophical writings, and through them influenced the development of special relativity. The culmination of this group theoretical approach to relativity is Wigner's paper mentioned above, where the physical characteristics, mass and spin, arise as parameters in the description of irreducible representations. One of the goals of our method of presentation is to reach this central result.

The explicit recognition of the importance of group representation theory in physics started very soon after the discovery of quantum mechanics, with the path-breaking work of Weyl, Wigner, and others. In fact, Weyl's classic book of 1928, *Gruppentheorie und Quantenmechanik*, makes instructive and inspiring reading even today. (In his book, Weyl adopts the pedagogic strategy of segregating the mathematics and the physics into separate chapters. There is much to be said for this strategy, especially from the point of view of logical coherence. But it had the unintended effect that physicists and mathematicians would read alternate chapters. I have taken the risk of going to the opposite extreme here, trying to use the physics to motivate the mathematics and vice versa, mixing the two.) The uses of group theory in quantum mechanics extended from chemistry and spectroscopy in the 1920s and 1930s, to nuclear and particle physics in the 1930s and 1940s, and then to high energy physics and the discovery of the theory of colored quarks in the 1960s and 1970s. It is this story of the interweaving of mathematics and physics that I try to tell in this book.

It should not be supposed that there was a warm reception in the physics community to the introduction of group theoretical methods. In fact, the contrary was true. To get a feeling for a typical early reaction, let me quote at length from the autobiography of John Slater, who was a leading American physicist and head of the MIT Physics Department for many years. The following quotes are taken from pages 60–2 of his autobiography:

> It was at this point that Wigner, Hund, Heitler, and Weyl entered the picture with their "Gruppenpest": the pest of the group theory.... The authors of the "Gruppenpest" wrote papers which were incomprehensible to those like me who had not studied group theory, in which they applied these theoretical results to the study of the many electron problem. The practical consequences appeared to be negligible, but everyone felt that to be in the mainstream one had to learn about it. Yet there were no good texts from which one could learn group theory. It was a frustrating experience, worthy of the name of a pest.

> I had what I can only describe as a feeling of outrage at the turn which the subject had taken....
>
> As soon as this [Slater's] paper became known, it was obvious that a great many other physicists were as disgusted as I had been with the group-theoretical approach to the problem. As I heard later, there were remarks made such as "Slater has slain the 'Gruppenpest'". I believe that no other piece of work I have done was so universally popular.

Outrage, disgust, the characterization of group theory as a plague or as a dragon to be slain – this is not an atypical physicist's reaction in the 1930s–50s to the use of group theory in physics. It is, however, amazing to consider that this autobiography was published in 1975, after the major triumphs of group theory in elementary particle physics.

When I was a student in the early 1950s, the basic facts of abstract group theory were part of the algebra course, but the theory of group representations was not included in the standard mathematics curriculum. My introduction to representation theory and its physical applications was at the hands of Prof. George W. Mackey. He gave a wonderful and justly famous course of lectures at the University of Chicago in the summer of 1955, where I visited as a special summer student. From the time that I joined the Harvard faculty in 1959, George has given me access to his voluminous handwritten notes on mathematical physics, and has, on occasion, written me long letters explaining various points. Much of his influence can be felt in the first half of this book.

In 1962, I was invited by Prof. Yuval Ne'eman to give a series of lectures on the topology of Lie groups at his seminar, then held at Nahal Soreq. This was after the prediction of the existence of the Ω^- particle (by Gell-Mann and by Ne'eman on the basis of $SU(3)$ symmetry in 1961), but before its momentous discovery at Brookhaven National Laboratory in 1964. This series of lectures developed into a lifelong collaboration. Much of my own work in physics has been in collaboration with Prof. Ne'eman, or an outgrowth of the seminars we have held together over the past 32 years.

Let me now describe the contents. The key idea in Chapter 1 is an action of a group on a set, with the concomitant notions of fixed point sets and stabilizer subgroups. We use these notions to clarify the notion of form and habit in a crystal, and to classify the finite subgroups of $O(3)$. Along the way we show that $Sl(2, C)$ is the double cover of the connected component of the Lorentz group, and hence that $SU(2)$ is a double cover of the three-dimensional rotation group, facts that are central to the understanding of the concept of spin. We conclude with a discussion of icosahedral symmetry in conjunction with the newly discovered carbon molecules, the buckyballs.

Chapter 2 presents the basic facts in the representation theory of finite groups. The central unifying theme is that of character formulas as fixed point formulas, both in this chapter and the next. Of course, we present these formulas in the purely finite context. But they represent the finite prototypes of the more powerful fixed point formulas in modern analysis, such as the Atiyah–Bott theorem. A partial transition to these formulas is given in Appendix G using differential geometric methods and generalized functions. At the end of the chapter we make a first pass at the representation theory of the symmetric groups. I return to this topic in Chapter 5 and in Appendix C.

Chapter 3 discusses induced representations from the point of view of vector bundles. The motivating example is the study of the vibrational spectrum of a molecule, both in classical and quantum mechanics. The main physical idea is the use of Schur's lemma to determine the number of possible vibrational modes and also to derive the quantum mechanical selection rules that determine which transitions are forbidden. In this latter connection, one needs to use tensor products and what are known in the physics literature as 'tensor operators'. I give the Frobenius theory of the representations of a semidirect product, and describe Wigner's use of this method to obtain the irreducible representations of the symmetry group of special relativity. The chapter includes a careful mathematical discussion of the question of the discrete symmetries of space time such as parity and time reversal. The chapter concludes with the Mackey theorems on induced representations and Mackey's approach to exchange forces.

Chapter 4 makes the transition from finite to compact groups. The Peter–Weyl theorem is stated, but its proof is deferred to Appendix E. The irreducible representations of $SU(2)$ and $SO(3)$ are derived, with the concomitant theory of spherical harmonics. Applications include a discussion of the hydrogen spectrum and the role of the representation theory of the rotation group in the periodic table and the magic numbers of nuclear physics. I discuss the role of the Clebsch–Gordan coefficients in isospin, in particular in pion–nucleon scattering experiments, and show how the Klein–Gordan equation, the Dirac equation, Weyl's neutrino equation and Maxwell's equation are related to the appropriate irreducible representations of the Poincaré group in Wigner's list. A brief introduction to Lie algebra methods is included.

Chapter 5 is devoted to the Schur–Weyl duality between representations of the symmetric groups and the general linear groups. The representations of the special unitary groups are derived from this duality in the standard fashion. The results are applied to the study of quarks. A typical application is the derivation of the nucleon magnetic moments from the quark theory. The physics in this chapter represents discoveries up to the early 1970s. I do not include a discussion of electroweak unification, but end, somewhat out of context, with a discussion of the differential geometry of the Higgs mechanism. I have not included any of the material on grand unified or supersymmetric models. My feeling is that no one of these models has won the day, and that the fundamental problems, such as confinement, the mass spectrum, divergences, the source of the Higgs field, etc., must be regarded as open questions. In the meantime, the attention of much of the theoretical physics community has turned elsewhere.

There are seven appendices. Appendix A takes the study of mathematical crystallography a bit further than does the treatment in Chapter 1. It does not go through the detailed classification of the crystallographic groups, but does give a description of how this classification proceeds. In particular, it gives a precise definition of the Bravais lattices and their classification. My feeling is that this material is of general cultural importance (and is, of course, central to solid state physics) but too technical to be included in the introductory first chapter.

Appendix B provides the necessary background on tensor products, as this material is not always included in the standard linear algebra course.

Appendix C provides proofs of some of the more technical aspects of the representation theory of the symmetric group. I chose to follow the beautiful 1977 paper of James. This method illustrates the power of the Gel'fand approach to integral geometry. Alternative approaches to this theory, such as via Hopf algebras or combinatorics, have their own individual merits, and are available elsewhere.

Appendix D gives the proof (following Bargmann) of Wigner's theorem on the symmetries of quantum logic. This theorem lies at the heart of the application of group theory to quantum mechanics. It is the quantum mechanical version of the fundamental theorem of projective geometry. As it is not usually included in the standard texts, I thought it important to include it here.

Appendix E gives the proofs of the basic facts about the representations of compact groups. The treatment is concise and standard, and practically no hard theorems in functional analysis are used. Nevertheless, I felt that it would be too much of a distraction from the main storyline to include this material in the main text.

Appendix F includes no mathematics at all. It is devoted to a history of 19th century spectroscopy. Many quantum mechanics texts start with a little of the prehistory of the subject, usually beginning with the Bohr atom. But it took a century of research to reach Bohr's epoch making paper, and during most of that period the existence of atoms was in dispute, not to say the existence of subatomic constituents. My feeling is that we are in a similar state today with regards to quarks and their possible constituent components. So a look back at how the science of spectroscopy actually progressed might be a source of comfort and amusement during our present period of groping towards an understanding of the deeper components of matter. My guides to the original literature were early Encyclopedia Britannica articles, all written by key players (that is how it was in those days) and the excellent unpublished Ph.D. thesis by Clifford Lawrence Meier entitled 'The role of spectroscopy in the acceptance of an internally structured atom 1860–1920', submitted to the University of Wisconsin in 1964. Of course, I bear the responsibility for the judgment calls in the shaping of the story.

Appendix G is taken from my joint book *Geometric Asymptotics*, written with Victor Guillemin. I try to give a taste of how the fixed point theorems given in the text in the finitistic setting can be formulated and proved in the framework of differential topology.

A word about prerequisites. I have tried to make the demands on the mathematical background of the reader as modest as possible. A course in multivariate calculus and linear algebra, together with an elementary physics course, should suffice. Especially if the reader will forgive my occasional lapses into more advanced material.

תושלב"ע

1

BASIC DEFINITIONS

AND EXAMPLES

1.1 Groups: definition and examples

In this chapter we will introduce the basic mathematical concepts associated with symmetry: the notion of a group and the action of a group on a set. A group G is a set on which we are given a binary operation which behaves much like ordinary multiplication; that is, we are given a map of $G \times G \to G$ sending the pair (p, q) into pq, satisfying the associative law, the existence of an identity element e, and the existence of an inverse. That is, we assume that

- $(pq)r = p(qr)$ for any three elements p, q, r in G;
- there exists an element, e, in G such that $ep = pe = p$ for all p in G; and
- for every p in G there is a p^{-1} in G such that $pp^{-1} = p^{-1}p = e$.

Example 1

(a) Let $\mathbb{Z}_4$ denote the additive group of the integers modulo 4. The elements of this group are equivalence classes which we shall call e, a, b and c:

$$
\begin{aligned}
e &= \{0, \quad 4, \quad -4, \quad 8, \quad -8, \quad \ldots\} \\
a &= \{1, \quad 5, \quad -3, \quad 9, \quad -7, \quad \ldots\} \\
b &= \{2, \quad 6, \quad -2, \quad 10, \quad -6, \quad \ldots\} \\
c &= \{3, \quad 7, \quad -1, \quad 11, \quad -5, \quad \ldots\}.
\end{aligned}
$$

The binary operation is addition modulo 4; for example, since $1 + 3 = 4$, which equals 0 modulo 4, we have $ac = e$. The identity element is e. Since $ac = e$, $a^{-1} = c$ and $c^{-1} = a$; since $bb = e$, $b^{-1} = b$.

(b) Let G denote the following set of four 2×2 real matrices:

$$
e = \begin{pmatrix} 1 & 0 \\ 0 & 1 \end{pmatrix} \quad a = \begin{pmatrix} 0 & -1 \\ 1 & 0 \end{pmatrix} \quad b = \begin{pmatrix} -1 & 0 \\ 0 & -1 \end{pmatrix} \quad c = \begin{pmatrix} 0 & 1 \\ -1 & 0 \end{pmatrix}.
$$

The binary operation is matrix multiplication; for example,

$$
ac = \begin{pmatrix} 0 & -1 \\ 1 & 0 \end{pmatrix}\begin{pmatrix} 0 & 1 \\ -1 & 0 \end{pmatrix} = \begin{pmatrix} 1 & 0 \\ 0 & 1 \end{pmatrix} = e.
$$

The identity element is the identity matrix

$$\begin{pmatrix} 1 & 0 \\ 0 & 1 \end{pmatrix},$$

and the inverse of each element is its matrix inverse; for example,

$$a^{-1} = \begin{pmatrix} 0 & -1 \\ 1 & 0 \end{pmatrix}^{-1} = \begin{pmatrix} 0 & 1 \\ -1 & 0 \end{pmatrix} = c.$$

(c) Let C_4 denote the group of rotational symmetries of the square, as follows:

e = identity (rotation through 0)
a = counterclockwise rotation through $\pi/2$
b = counterclockwise rotation through π
c = counterclockwise rotation through $3\pi/2$ (clockwise rotation through $\pi/2$).

Now the group operation is composition of transformations. Clearly the 'multiplication table' is the same as in the preceding two examples; we have considered three different realizations of the same abstract group, the so-called 'cyclic group of four elements'. It is a simple example of a *finite* group.

Example 2
We turn now to an example of a group which has an infinite number of elements. Let $SL(2, \mathbb{C})$ denote the set of 2×2 matrices of determinant 1 with complex entries. Thus, an element A of $SL(2, \mathbb{C})$ is given as

$$A = \begin{pmatrix} a & b \\ c & d \end{pmatrix},$$

where a, b, c and d are complex numbers satisfying

$$ad - bc = 1.$$

Multiplication is the ordinary multiplication of matrices. Since the determinant of the product of two matrices is the product of their determinants, we see that if A and B are elements of $SL(2, \mathbb{C})$, then so is their product AB. If A is an element of $SL(2, \mathbb{C})$, so that $\det A = 1$, then A is invertible and $\det A^{-1} = 1$, so that A^{-1} exists and lies in $SL(2, \mathbb{C})$. The identity element of the group is the identity matrix, i.e.

$$e = \begin{pmatrix} 1 & 0 \\ 0 & 1 \end{pmatrix}.$$

The associative law holds for matrix multiplication and thus $SL(2, \mathbb{C})$ is indeed a group. Notice that the commutative law does not hold in general for this group.

More generally, we can consider $n \times n$ matrices with either real or complex entries. The collection of real invertible $n \times n$ matrices is denoted by $GL(n, \mathbb{R})$. (Notice that here the condition of invertibility has to be added as a supplemental hypothesis. Not all

2×2 or $n \times n$ matrices are invertible, but those that are invertible form a group.) The group $GL(n, \mathbb{R})$ is called the *real general linear* group in n variables. We can also consider the group $SL(n, \mathbb{R})$ consisting of the $n \times n$ real matrices of determinant 1. It is called the *real special linear* group in n variables. Similarly, we can consider the group $GL(n, \mathbb{C})$ of all invertible complex $n \times n$ matrices or the group $SL(n, \mathbb{C})$ of all $n \times n$ complex matrices of determinant 1.

Example 3

As a third example of a group we can consider the group, $O(3)$, of all orthogonal transformations in Euclidean three-dimensional space. This is the group of all linear transformations of three-dimensional space which preserve the Euclidean distance; that is, those transformations, A, which satisfy

$$\| A\mathbf{v} \| = \| \mathbf{v} \|$$

for all vectors $\mathbf{v}$ in ordinary three-dimensional space. If we choose an orthonormal basis for three-dimensional space so that every A becomes identified with a matrix, then A is an orthogonal transformation if and only if

$$A A^t = e,$$

where e denotes the identity matrix in three dimensions. Notice that this equation is the same as $A^t = A^{-1}$. We see immediately that the product of any two orthogonal transformations is again orthogonal and that the inverse of any orthogonal transformation exists and is orthogonal. Thus, the collection of all orthogonal transformations does indeed form a group. Since $\det A = \det A^t$, it follows from $A A^t = e$ that $(\det A)^2 = 1$. Thus, for any orthogonal transformation A we have $\det A = \pm 1$. The collection of those matrices which are orthogonal, and which satisfy the further condition that $\det A = +1$, forms a subcollection of $O(3)$, which in itself is a group and which we will denote by $SO(3)$. We say that $SO(3)$ is a *subgroup* of $O(3)$. $SO(3)$ is called the special orthogonal group in three variables. (Similarly, $SL(n, \mathbb{C})$ is a subgroup of $GL(n, \mathbb{C})$, and $SL(n, \mathbb{R})$ is a subgroup of $GL(n, \mathbb{R})$.) More generally, if we put the standard Euclidean scalar product on the n-dimensional space $\mathbb{R}^n$, we can consider the orthogonal group $O(n)$ of all orthogonal $n \times n$ matrices and the corresponding subgroup $SO(n)$ of those orthogonal matrices with determinant 1.

Example 4

Let $\mathbb{C}^n$ denote the n-dimensional complex vector space of all complex n-tuples with its standard Hermitian scalar product, so that

$$(\mathbf{z}, \mathbf{w}) = z_1 \bar{w}_1 + \cdots + z_n \bar{w}_n,$$

where

$$\mathbf{z} = \begin{pmatrix} z_1 \\ \vdots \\ z_n \end{pmatrix} \quad \text{and} \quad \mathbf{w} = \begin{pmatrix} w_1 \\ \vdots \\ w_n \end{pmatrix}.$$

Table 1.

	1	−1
1	1	−1
−1	−1	1

Table 2.

	1	ω	ω^2
1	1	ω	ω^2
ω	ω	ω^2	1
ω^2	ω^2	1	ω

A complex matrix A is *unitary* if

$$(A\mathbf{z}, A\mathbf{w}) = (\mathbf{z}, \mathbf{w})$$

for all $\mathbf{z}$ and $\mathbf{w}$ in $\mathbb{C}^n$. If we denote $\overline{A^t}$ (the complex conjugate transpose of A) by A^*, we may say that A is unitary only if $AA^* = e$. The product of two unitary matrices is unitary, and the inverse of a unitary matrix is unitary; so the collection of unitary $n \times n$ matrices forms a group which we denote by $U(n)$. Since $\det A^* = \overline{\det A}$, we see that $|\det A| = 1$ for A in $U(n)$. The subgroup of $U(n)$ consisting of those matrices which in addition satisfy $\det A = 1$ is denoted by $SU(n)$.

Thus, for example, the group $SU(2)$ consists of all 2×2 matrices of the form

$$\begin{pmatrix} a & b \\ -\bar{b} & \bar{a} \end{pmatrix}, \quad \text{where} \quad |a|^2 + |b|^2 = 1.$$

Example 5

We can generalize Examples 1(*a*), (*b*) and (*c*) by replacing the number 4 by any positive integer. For instance, we can consider the group C_2 consisting of two elements with the 'multiplication table' as in Table 1, which is isomorphic to the additive group of the integers modulo 2. Similarly, we can think of the three-element group, C_3, with elements $1, \omega, \omega^2$ where $\omega = \exp 2\pi i/3$ which obey the 'multiplication table' shown in Table 2.

The group C_3 can be thought of as the additive group of the integers modulo 3, or as the group of all rotations in the plane which preserve an equilateral triangle centered at the origin. Thus, ω represents rotation through $2\pi/3 = 120°$.

We have already considered the group C_4 of all rotations preserving a square. It

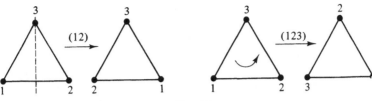

Fig. 1.1

contains four elements, consisting of the identity, rotation through $\pi/2$, rotation through π, and rotation through $3\pi/2$. We can now recognize that C_4 is a subgroup of $SO(2)$, the group of all rotations in the plane. More generally, we can consider C_n as the group of all rotations which preserve a regular polygon with n sides. It will consist of the identity and all rotations through angles of the form $2\pi k/n$.

Example 6

Let us go back to the equilateral triangle. We can consider the group of *all* symmetries of the triangle, not only the rotations. That is, we can allow reflection about perpendicular bisectors as well. This group has six elements; we will denote it by S_3. Notice that we can find some element in S_3 which has the effect of making any desired permutation of the vertices of the triangle. Let us denote the vertices of the triangle by 1, 2 and 3 (Fig. 1.1). Suppose, for example, that (12) denotes the permutation which interchanges the vertices 1 and 2 but leaves a third vertex, 3, fixed. This permutation can be achieved by a reflection about the perpendicular bisector of the edge joining 1 to 2.

Similarly, let (123) denote the permutation that sends 1 into 2, 2 into 3, and 3 into 1. This can be achieved by rotating the triangle through 120°. The permutation (132), which sends 1 into 3, 3 into 2, and 2 into 1, is achieved by rotating the triangle through 240°. From this we see that the group of symmetries of an equilateral triangle is the same as the group of all permutations on three symbols.

Suppose we consider four symbols 1, 2, 3, 4, instead of three. Let s be a one-to-one map of this four-element set onto itself. Thus, s is a permutation of this four-element set. There are four possibilities for $s(1)$: it can be any of the numbers 1, 2, 3, 4. Once we know what $s(1)$ is, then there are three remaining possibilities for $s(2)$, then two remaining possibilities for $s(3)$. Finally, $s(4)$ will be completely determined by being the last remaining number. Thus, there are $4 \cdot 3 \cdot 2 \cdot 1 = 4! = 24$ permutations on four letters. The group S_4 is the group of all these permutations. Similarly, we define the group S_n to be the group of all permutations; that is, all one-to-one transformations on a set with n elements.

Example 7

As a final example, we consider the group of all symmetries of the square, denoted by D_4. D_4 contains eight elements: four rotations, together with four reflections – the reflections about the two diagonals, and the reflections about the two perpendicular bisectors (see Fig. 1.2).

Each element of D_4 permutes the vertices 1, 2, 3, 4 of the square. Thus, we may regard

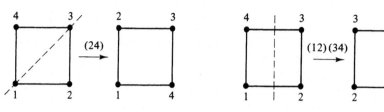

Fig. 1.2

D_4 as a subgroup of S_4, but not every element of S_4 (which has 24 elements all together) lies in D_4, which has only eight elements. Similarly, the group D_n, the group of symmetries of the regular polygon with n sides, is a subgroup of the group S_n of permutations of n symbols. The reader should check that D_n contains $2n$ elements.

1.2 Homomorphisms: the relation between $SL(2, \mathbb{C})$ and the Lorentz group

Let G_1 and G_2 be groups. Let ϕ be a map from G_1 to G_2. We say that ϕ is a *homomorphism* if

$$\phi(ab) = \phi(a)\phi(b) \quad \text{for all } a \text{ and } b \text{ in } G_1.$$

The notion of homomorphism is central to the study of groups and so we give some examples. Take $G_1 = \mathbb{Z}$ to be the integers and $G_2 = C_2$. Define the map ϕ by

$$\phi(n) = (-1)^n.$$

Recall that group 'multiplication' is ordinary addition in $\mathbb{Z}$ so that the condition that ϕ be a homomorphism reduces to the assertion that

$$\phi(a + b) = \phi(a)\phi(b),$$

i.e that

$$(-1)^{a+b} = (-1)^a(-1)^b$$

which is clearly true. More generally, we can define a homomorphism from $\mathbb{Z}$ to C_k by

$$\phi(a) = \exp 2\pi i a/k = \omega^a, \quad \text{where } \omega \text{ equals } \exp 2\pi i/k.$$

This generalizes the construction of Example 1 of the preceding section. Basically, what the homomorphism ϕ is telling us is that we can regard multiplication in C_k as 'addition modulo k' in the integers.

We now want to describe another homomorphism which has many important physical applications and which will recur frequently in the rest of this book. For this we need to introduce still another group, the Lorentz group. Let M denote the four-dimensional space $M = \mathbb{R}^4$, with the 'Lorentz metric'

$$\|\mathbf{x}\|^2 = x_0^2 - x_1^2 - x_2^2 - x_3^2, \quad \text{where } \mathbf{x} = \begin{pmatrix} x_0 \\ x_1 \\ x_2 \\ x_3 \end{pmatrix}.$$

Thus, M is the ordinary Minkowski space of special relativity, where we have chosen units in which the speed of light is unity. A Lorentz transformation, B, is a linear transformation of M into itself which preserves the Lorentz metric, i.e. which satisfies

$$\| B\mathbf{x} \|^2 = \| \mathbf{x} \|^2, \quad \text{for all } \mathbf{x} \text{ in } M.$$

We let L denote the group of all Lorentz transformations; L is called the Lorentz group.

We now describe a homomorphism from the group $SL(2, \mathbb{C})$ to the group L. For this purpose we shall identify every point $\mathbf{x}$ in M with a two-by-two self-adjoint matrix, as follows:

$$x = \begin{pmatrix} x_0 + x_3 & x_1 - ix_2 \\ x_1 + ix_2 & x_0 - x_3 \end{pmatrix} \quad \text{represents } \mathbf{x} = \begin{pmatrix} x_0 \\ x_1 \\ x_2 \\ x_3 \end{pmatrix}.$$

Notice that

$$x^* = \overline{x^{\mathrm{t}}} = x,$$

and that

$$\det x = \| \mathbf{x} \|^2 = x_0^2 - x_1^2 - x_2^2 - x_3^2.$$

Indeed, the most general self-adjoint 2×2 matrix can be written in this form: if $x = x^*$ is a self-adjoint matrix, then its diagonal entries must be real. We can let $x_0 = \frac{1}{2}\operatorname{tr} x = \frac{1}{2}$ (the sum of the diagonal entries of x) and similarly $x_3 = \frac{1}{2}$ (the difference of the diagonal entries of x). Also, we can write the entry in the lower left-hand corner of x as $x_1 + ix_2$. Then the entry in the upper right-hand corner will be $x_1 - ix_2$. In effect, what we have done is to note that the collection of 2×2 self-adjoint matrices is a four-dimensional real vector space, for which a convenient basis consists of the identity matrix $I = \begin{pmatrix} 1 & 0 \\ 0 & 1 \end{pmatrix}$ and the three so-called 'Pauli matrices'

$$\sigma_1 = \begin{pmatrix} 0 & 1 \\ 1 & 0 \end{pmatrix} \quad \sigma_2 = \begin{pmatrix} 0 & -i \\ i & 0 \end{pmatrix} \quad \sigma_3 = \begin{pmatrix} 1 & 0 \\ 0 & -1 \end{pmatrix}.$$

We have identified the vector

$$\mathbf{x} = \begin{pmatrix} x_0 \\ x_1 \\ x_2 \\ x_3 \end{pmatrix}$$

with the matrix

$$x = x_0 e + x_1 \sigma_1 + x_2 \sigma_2 + x_3 \sigma_3.$$

Now let A be any 2×2 matrix. We define the action of the matrix A on the self-adjoint matrix x by

$$x \to AxA^*$$

and we denote the corresponding action on the vector $\mathbf{x}$ by $\phi(A)\mathbf{x}$. Notice that $(AxA^*)^* = A^{**}x^*A^* = AxA^*$, so that AxA^* is again self-adjoint. Notice also that

$$\det(AxA^*) = |\det A|^2 \det x.$$

Therefore, if A is in $SL(2, \mathbb{C})$, then

$$\| \phi(A)\mathbf{x} \|^2 = \| \mathbf{x} \|^2,$$

so that, if A is in $SL(2, \mathbb{C})$, $\phi(A)$ represents a Lorentz transformation. Notice also that

$$ABx(AB)^* = ABxB^*A^* = A(BxB^*)A^*$$

so that

$$\phi(AB)\mathbf{x} = \phi(A)\phi(B)\mathbf{x}.$$

Thus, $\phi(AB) = \phi(A)\phi(B)$, so that ϕ is a homomorphism. Notice, however, that $\phi(-A) = \phi(A)$ so that ϕ is not one-to-one. The matrices A and $-A$ correspond to the same Lorentz transformation.

Suppose now that A belongs to the subgroup $SU(2)$ of $SL(2, \mathbb{C})$. This means that A is a unitary matrix, satisfying

$$AA^* = I; \quad \text{i.e.} \quad AIA^* = I.$$

Therefore, if $\mathbf{e}_0$ denotes the vector

$$\begin{pmatrix} 1 \\ 0 \\ 0 \\ 0 \end{pmatrix},$$

which is represented by the 2×2 identity matrix I, then

$$\phi(A)\mathbf{e}_0 = \mathbf{e}_0.$$

If a Lorentz transformation C satisfies $C\mathbf{e}_0 = \mathbf{e}_0$, then C also carries the three-dimensional space $e_0^\perp$, consisting of vectors

$$\begin{pmatrix} 0 \\ x_1 \\ x_2 \\ x_3 \end{pmatrix},$$

into itself and C is an orthogonal transformation on that three-dimensional space. Put another way, we can regard $O(3)$ as the subgroup of L consisting precisely of those Lorentz transformations which satisfy $C\mathbf{e}_0 = \mathbf{e}_0$. Thus, the mapping ϕ, when restricted to $SU(2)$, maps $SU(2)$ into $O(3)$.

For example, let us consider the diagonal matrix

$$U_\theta = \begin{pmatrix} e^{-i\theta} & 0 \\ 0 & e^{i\theta} \end{pmatrix}.$$

The $\phi(U_\theta)\mathbf{x}$ may be computed by matrix multiplication as follows:

$$U_\theta x U_{-\theta} = \begin{pmatrix} e^{-i\theta} & 0 \\ 0 & e^{i\theta} \end{pmatrix} \begin{pmatrix} x_0 + x_3 & x_1 - ix_2 \\ x_1 + ix_2 & x_0 - x_3 \end{pmatrix} \begin{pmatrix} e^{i\theta} & 0 \\ 0 & e^{-i\theta} \end{pmatrix}$$

$$= \begin{pmatrix} x_0 + x_3 & e^{-2i\theta}(x_1 - ix_2) \\ e^{2i\theta}(x_1 + ix_2) & x_0 - x_3 \end{pmatrix}.$$

Thus, $\phi(U_\theta)$ leaves x_0 and x_3 unchanged and hence is a rotation about the x_3 axis. Since it sends $x_1 + ix_2$ into $e^{2i\theta}(x_1 + ix_2)$, we see that it is a rotation through angle 2θ in the x_1, x_2 plane. We have thus shown that

$$\phi(U_\theta) \text{ is rotation through angle } 2\theta \text{ about the } x_3 \text{ axis.}$$

Notice that as θ ranges from 0 to π the corresponding rotation goes from 0 to 2π, making a complete circuit. As θ ranges from 0 to 2π, the corresponding rotation goes through *two* complete circuits. This is a reflection of the fact that $\phi(-A) = \phi(A)$.

Similarly, consider the action of the unitary matrix

$$V_\alpha = \begin{pmatrix} \cos\alpha & -\sin\alpha \\ \sin\alpha & \cos\alpha \end{pmatrix} \quad \text{for which} \quad V_\alpha^* = V_{-\alpha}.$$

We calculate $\phi(V_\alpha)\mathbf{x}$ by matrix multiplication as follows:

$$V_\alpha x V_{-\alpha} = \begin{pmatrix} \cos\alpha & -\sin\alpha \\ \sin\alpha & \cos\alpha \end{pmatrix} \begin{pmatrix} x_0 + x_3 & x_1 - ix_2 \\ x_1 + ix_2 & x_0 - x_3 \end{pmatrix} \begin{pmatrix} \cos\alpha & \sin\alpha \\ -\sin\alpha & \cos\alpha \end{pmatrix}.$$

We can easily determine the action of $\phi(V_\alpha)$ on the vector

$$\mathbf{e}_2 = \begin{pmatrix} 0 \\ 0 \\ 1 \\ 0 \end{pmatrix}$$

by taking $x_0 = x_1 = x_3 = 0$, so that

$$x = \sigma_2 = \begin{pmatrix} 0 & -i \\ i & 0 \end{pmatrix}.$$

We find that $\phi(V_\alpha)\mathbf{e}_2 = \mathbf{e}_2$, so that $\phi(V_\alpha)$ must be a rotation about the x_2 axis.

We now determine the action of $\phi(V_\alpha)$ on the basis vector

$$\mathbf{e}_3 = \begin{pmatrix} 0 \\ 0 \\ 0 \\ 1 \end{pmatrix}$$

by taking $x = \sigma_3 = \begin{pmatrix} 1 & 0 \\ 0 & -1 \end{pmatrix}$. We find

$$V_\alpha \sigma_3 V_{-\alpha} = \begin{pmatrix} \cos\alpha & -\sin\alpha \\ \sin\alpha & \cos\alpha \end{pmatrix} \begin{pmatrix} 1 & 0 \\ 0 & -1 \end{pmatrix} \begin{pmatrix} \cos\alpha & \sin\alpha \\ -\sin\alpha & \cos\alpha \end{pmatrix} = \begin{pmatrix} \cos 2\alpha & \sin 2\alpha \\ \sin 2\alpha & -\cos 2\alpha \end{pmatrix}$$

which corresponds to the vector

$$\begin{pmatrix} 0 \\ \sin 2\alpha \\ 0 \\ \cos 2\alpha \end{pmatrix}.$$

We conclude that

$$\phi(V_\alpha)\mathbf{e}_3 = \mathbf{e}_3 \cos 2\alpha + \mathbf{e}_1 \sin 2\alpha$$

so that V_α represents rotation through angle 2α about the x_2 axis.

As a third example, consider the diagonal matrix with real entries

$$M_r = \begin{pmatrix} r & 0 \\ 0 & r^{-1} \end{pmatrix}.$$

Since $M_r = M_r^*$, we have

$$M_r \times M_r^* = \begin{pmatrix} r & 0 \\ 0 & r^{-1} \end{pmatrix} \begin{pmatrix} x_0 + x_3 & x_1 + ix_2 \\ x_1 - ix_2 & x_0 - x_3 \end{pmatrix} \begin{pmatrix} r & 0 \\ 0 & r^{-1} \end{pmatrix}$$

$$= \begin{pmatrix} r^2(x_0 + x_3) & x_1 + ix_2 \\ x_1 - ix_2 & r^{-2}(x_0 - x_3) \end{pmatrix}.$$

Thus the Lorentz transformation $\phi(M_r)$ leaves x_1 and x_2 alone whereas the x_0 and the x_3 coordinates are transformed into

$$x_0' = \tfrac{1}{2}(r^2 + r^{-2})x_0 + \tfrac{1}{2}(r^2 - r^{-2})x_3 \quad \text{and} \quad x_3' = \tfrac{1}{2}(r^2 - r^{-2})x_0 + \tfrac{1}{2}(r^2 + r^{-2})x_3.$$

We recall the definition of the hyperbolic functions:

$$\cosh u = \tfrac{1}{2}(e^u + e^{-u}) \quad \text{and} \quad \sinh u = \tfrac{1}{2}(e^u - e^{-u}).$$

The *Lorentz boost* in the z direction with parameter t, denoted by L_t, is defined as the transformation given by

$$x_1' = x_1, \; x_2' = x_2, \; x_0' = (\cosh t)x_0 + (\sinh t)x_3 \quad \text{and} \quad x_3' = (\sinh t)x_0 + (\cosh t)x_3.$$

In other words, L_t^z is the Lorentz transformation given by the matrix

$$L_t^z = \begin{pmatrix} \cosh t & 0 & 0 & \sinh t \\ 0 & 1 & 0 & 0 \\ 0 & 0 & 1 & 0 \\ \sinh t & 0 & 0 & \cosh t \end{pmatrix}.$$

If we set $r = e^t$ then our preceding computation shows that

$$\phi(M_{e^t}) = L_{2t}^z.$$

To summarize: let R_θ^z denote rotation through angle θ about the z axis, let R_θ^y denote rotation through angle θ about the y axis. We have shown that

$$\phi(U_\theta) = R_{2\theta}^z, \; \phi(V_\theta) = R_{2\theta}^y \quad \text{and} \quad \phi(M_{e^t}) = L_{2t}^z.$$

We might now ask what is the range of ϕ: that is, which elements C in the Lorentz group L are actually of the form $\phi(A)$ for some A in $SL(2, \mathbb{C})$? We first show that every A in $SL(2, \mathbb{C})$ can be continuously joined to the identity by a curve A_t of matrices which all lie in $SL(2, \mathbb{C})$. By a standard theorem of linear algebra we know that any A is conjugate to an upper triangular matrix, that is

$$A = B\begin{pmatrix} a & b \\ 0 & a^{-1} \end{pmatrix} B^{-1}.$$

Now simply let a_t be any curve of non-zero complex numbers with $a_0 = 1, a_1 = a$, and let b_t be any curve of complex numbers with $b_0 = 0$ and $b_1 = b$. Then

$$A = B\begin{pmatrix} a_t & b_t \\ 0 & a_t^{-1} \end{pmatrix} B^{-1}$$

is a curve of matrices in $SL(2, \mathbb{C})$, with $A_0 = I$ and $A_1 = A$. This demonstrates explicitly that any A in $SL(2, \mathbb{C})$ can be joined to the identity by a continuous curve.

However, not every element of L can be joined to the identity element by a continuous curve. For instance, there exist elements in L which are 4×4 matrices with negative determinant. As the determinant is a continuous function, there can be no curve joining an element with a negative determinant to the identity. Furthermore, every element of L preserves the set of time-like vectors, that is those vectors with $\|\mathbf{x}\|^2 > 0$. The set of time-like vectors falls into two components according to whether x_0 is positive or negative. An element of L can interchange these two components, but obviously any element of L which can be continuously joined to the identity must preserve each component. It follows that there are elements in L which cannot lie in the range of ϕ.

We shall denote by L^0 the *proper* Lorentz group, the subgroup of L consisting of those transformations which have positive determinant and which preserve the forward light cone, that is, which send each component of the set of time-like vectors into itself. It will emerge from discussions in the course of the next few sections that the mapping ϕ sends $SL(2, \mathbb{C})$ onto L^0 and sends the subgroup $SU(2)$ onto the group $SO(3)$ of rotations in three-space.

The proof goes as follows:

Lemma

Every proper Lorentz transformation, B, can be written as

$$B = R_1 L_u^z R_2,$$

where R_1 and R_2 are rotations, and L_u^z is a suitable Lorentz boost in the z direction.

Proof We can write

$$Be_0 = \begin{pmatrix} x_0 \\ \mathbf{x} \end{pmatrix},$$

where $\mathbf{x} = x_1 e_1 + x_2 e_2 + x_3 e_3$ and $x_0^2 - \|\mathbf{x}\|^2 = 1$. We can find a rotation S which rotates the vector $\mathbf{x}$ to the positive z axis, so

$$SBe_0 = \begin{pmatrix} x_0 \\ 0 \\ 0 \\ \|\mathbf{x}\| \end{pmatrix}.$$

The self-adjoint matrix that corresponds to SBe_0 is thus

$$\begin{pmatrix} x_0 + \|\mathbf{x}\| & 0 \\ 0 & x_0 - \|\mathbf{x}\| \end{pmatrix}.$$

Now choose r so that $r^2 = (x_0 + \|\mathbf{x}\|)^{-1} = x_0 - \|\mathbf{x}\|$. (Remember that

$$(x_0 + \|\mathbf{x}\|)(x_0 - \|\mathbf{x}\|) = x_0^2 - \|\mathbf{x}\|^2 = 1.)$$

Then applying M_r gives $\phi(M_r)SBe_0 = e_0$. Thus $\phi(M_r)SB$ is a rotation; call it R_2. We thus have

$$\phi(M_r)SB = R_2$$

or

$$B = S^{-1}[\phi(M_r)]^{-1}R_1.$$

This is the desired decomposition with $R_1 = S^{-1}$ and $L_u^z = [\phi(M_r)]^{-1}$.

In Section 1.6 we will prove a theorem due to Euler which asserts that every rotation R in three-dimensional space can be written as a product

$$R = R_\theta^z R_\phi^y R_\psi^z$$

that is, as a rotation about the z axis, followed by a rotation about the y axis, followed by a rotation about the z axis again. (The angles θ, ϕ, ψ are called the *Euler angles* of the rotation R.) Combined with the above lemma, we conclude that every element of the proper Lorentz group can be written as a product of elements of the form L_u^z, R_θ^z and R_ϕ^y. But each of these is in the image of ϕ. So, granted Euler's theorem, we conclude that $\phi(SL(2, \mathbb{C}))$ is all of the proper Lorentz group.

1.3 The action of a group on a set

We now return to some general definitions. Let G be a group and let M be a set. We say that we have an *action* of G on M if we are given a mapping of $G \times M \to M$ sending (a, m) into am which satisfies the associative law

$$a(bm) = (ab)m$$

and

$$em = m \quad \text{for any } m \text{ in } M,$$

where e is the identity element of the group. Thus, an action of a group G on a set M is nothing other than a homomorphism from G into the group of all one-to-one

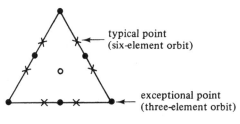

typical point
(six-element orbit)

exceptional point
(three-element orbit)

Fig. 1.3

transformations of M. Examples from the preceding section include the action of $SL(2, \mathbb{C})$ (a group) on Minkowski space (a set) and the action of the permutation group S_3 on the set of vertices of the equilateral triangle.

Let G act on M and let m be a point of M. We can consider the subset of M consisting of all points of the form am as a ranges over all elements of G. This subset of M is called the *orbit* of the point m under the action of G on M and is denoted by $G \cdot m$. Thus, $G \cdot m = \{am \,|\, a \in G\}$. Thus, for example, in Fig. 1.3 we have marked some orbits of various points in the triangle under the action of S_3. Notice in this example that a 'typical' point x has six points in its orbit, but there exist certain exceptional points $\bullet$ and $\times$ with three-element orbits and one point $\circ$ whose orbit consists of itself alone.

As another example, we can consider the group $SO(3)$, whose elements act as rotations in three-dimensional Euclidean space. The orbit of any point which is not the origin is the sphere centered at the origin and passing through that point, but the orbit of the origin is just one point, the origin itself.

Given any point m in M we can consider the subset of G consisting of those a in G which satisfy $am = m$. It is clear that if $am = m$, then $a^{-1}m = m$. Furthermore, if $am = m$ and $bm = m$, then $(ab)m = m$. Thus, the set of such a forms a subgroup of G, which is called the *isotropy group* of m and is denoted by G_m. Thus, for example, in the case of $SO(3)$ acting on three-dimensional space, the isotropy group of any point $m \neq 0$ will consist of those rotations which preserve the point in question, that is the subgroup of rotations about the axis passing through the point. Thus for each non-zero point m the group G_m is isomorphic to $SO(2)$ and consists of rotations in the plane perpendicular to m. Notice that any two such isotropy subgroups G_m are isomorphic as abstract groups, but are different subgroups of $SO(3)$. The isotropy group of the origin is the entire group $SO(3)$.

In the case of S_3 acting on the triangle, the isotropy group of a typical point x will consist of the identity alone. The isotropy group of a point $\bullet$ with a three-element orbit will consist of the identity and one reflection. Finally, the isotropy group of the center point $\circ$ is the entire group S_3. Notice that in all cases the product of the number of elements in the orbit by the number of elements in the isotropy subgroup is six, which is the number of elements in S_3.

This fact is true in general. Suppose that G is a finite group, and let $\#G$ denote the number of elements in G. Let m be a point of M, and consider the orbit of m under G, which contains $\#(G \cdot m)$ elements. If n is an element of this orbit, then $n = am$ for some a in G. If $n = bm$ as well, then $a^{-1}b$ must lie in G_m. This means that to each element n there are exactly $\#G_m$ group elements which map m into n. Therefore, since every element of G

carries m into some orbit element, we conclude that

$$\#G = \#(G \cdot m) \cdot \#G_m.$$

If M consists of a single orbit, we say that G acts *transitively* on M. Any time that we have a transitive group action of G on M, we can, by choosing a point m in M, determine a partition of G into classes, called *cosets*, of the form

$$aG_m = \{ab \quad \text{for all } b \text{ in } G_m\}.$$

Each coset contains the same number of elements as the subgroup G_m. The coset aG_m, which consists of all group elements that carry the point m into the point am, may be identified with the point am. Since G acts transitively on M, there is, for each element n of M, at least one element b of G such that $n = bm$. We have therefore identified each point of M with a coset of G. By introducing the notation G/G_m for the set of cosets, we may identify

$$M \quad \text{with} \quad G/G_m.$$

As an example of this identification, let M be the set of vertices of an equilateral triangle and let G be the group S_3, which permutes these vertices. We partition G into cosets by selecting vertex 1 (though 2 or 3 would have served equally well). Then the coset $\{e, (23)\}$, which is the isotropy group G_1 itself, is identified with vertex 1. The coset $\{(123), (12)\}$, consisting of elements which carry vertex 1 into vertex 2, is identified with vertex 2, and the coset $\{(132), (13)\}$ is similarly identified with vertex 3.

This identification of elements of M with cosets of G provides a useful way of constructing a set M on which G acts transitively. Suppose that H is a subgroup of G. We can form the cosets aH, each containing $\#H$ elements, so that there are $\#G/\#H$ cosets in all. We define the action of a group element g on a coset aH by $g(aH) = (ga)H$. This construction yields the same result no matter which element of the coset we choose. If, instead of a, we had chosen a different element $a' = ah$ (where $h \in H$, so that a' and a are in the same coset), then $g(a'H) = (ga')H = (gah)H = (ga)(hH) = (ga)H$ since hH is just the subgroup H. Thus we have a rule by which G acts on the space of cosets $M = G/H$, and we may label each coset in M by the name of any element of G which lies in that coset.

As an example of the construction, we construct a two-element set M on which the group S_3 acts transitively. We start with the subgroup $H = \{e, (123), (132)\}$, which corresponds to *rotational* symmetries of an equilateral triangle. The set M consists of two cosets: $\{e, (123), (132)\}$, which we call simply $[e]$, and $\{(12), (13), (23)\}$, which we call $[(12)]$. The elements of H carry each point of M into itself; the other three elements of S_3 carry $[e]$ into $[(12)]$ and $[(12)]$ into $[e]$.

1.4 Conjugation and conjugacy classes

One set M on which a group G can act is the group G itself. When G acts on itself by left multiplication, so that a transforms b into ab, then the action is always transitive. For any group elements c and b there is an element $a = cb^{-1}$ such that $ab = cb^{-1}b = c$.

Correspondingly, the isotropy subgroup of each group element, G_b, consists of the identity alone, since $ab = b$ implies that $a = e$.

There is another natural way for a group to act on itself which leads to less trivial orbits and isotropy subgroups. We say that G acts on itself by *conjugation* if a acting on b is aba^{-1}. Conjugation always defines a group action, since the action of ac on b yields

$$(ac)b(ac)^{-1} = acbc^{-1}a^{-1} = a(cbc^{-1})a^{-1},$$

which is the same as the action (by conjugation) of c followed by the action of a.

The orbits of a group under conjugation are called the *conjugacy classes* of the group. Two elements b and c belong to the same conjugacy class if there exists an element a such that

$$aba^{-1} = c.$$

From this equation follow two straightforward consequences:

(1) The identity is always a one-element conjugacy class. *Proof*: $aea^{-1} = e$ for any a.

(2) For an Abelian (commutative) group, every element is in a conjugacy class by itself. *Proof*: $aba^{-1} = aa^{-1}b = b$ in this case.

The isotropy group G_b of an element b under conjugation consists of those elements a for which $aba^{-1} = b$. One such element is $a = e$, another is b (if $b \neq e$), and a third is b^{-1} (if $b^{-1} \neq b$). The number of elements in the subgroup G_b of course obeys the general rule $\#G = \#(G \cdot b) \cdot \#G_b$ for any finite group. It follows that the number of elements in any conjugacy class is always a divisor of the number of elements of the group.

For any group of matrices, conjugate group elements B and C satisfy

$$ABA^{-1} = C.$$

It follows immediately that matrices B and C have the same eigenvalues, and hence the same determinant and trace.

As an example, we list the conjugacy classes of the group S_3. One class consists of the identity element e alone. The second class consists of the two elements (123) and (132) (note that $(23)(123)(23)^{-1} = (132)$.) Both these elements represent 120° rotations of an equilateral triangle. The third conjugacy class consists of the remaining three elements, (12), (13) and (23), which represent reflections of an equilateral triangle in the perpendicular bisector of a side.

More generally, for any permutation group, one can show that elements with the same 'cycle structure' always belong to the same conjugacy class*. For example, the group S_4 has five conjugacy classes as follows:

- the identity e (one element);
- (123), (132), (124), etc. (eight elements);
- (12), (13), etc. (six elements);
- (12)(34), (13)(24), (14)(23) (three elements);
- (1234), (1243), etc. (six elements).

* See Section 8, Chapter 2.

1.5 Applications to crystallography

The concept of an orbit of a group action is useful in clarifying the notion of 'form' in crystallography. Ordinary table salt, NaCl, if allowed to crystallize under carefully controlled conditions, forms cubic crystals. Let C be the cube in $\mathbb{R}^3$ whose vertices are the points $(\pm 1, \pm 1, \pm 1)$. The group of all orthogonal transformations which carry C into itself is denoted in crystallographic notation by O_h. It is clear that any linear transformation sending (x, y, z) into (u, v, w), where the u, v, w are permutations of $\pm x$, $\pm y, \pm z$, carries the cube into itself, and that these are the only linear transformations which preserve the cube. There are eight choices of $\pm$ and six possible permutations of x, y, z. The group O_h thus contains 48 elements.

NaCl can, however, crystallize in other forms. If a small amount of urea is added, then small equilateral triangles replace the corners of the cube. These triangles are congruent, so the crystal is still invariant under O_h. As more urea is added, the triangles open up into hexagons. Finally, the cube faces disappear altogether and we have an octahedron. (See Figs. 1.4(a)–(d).)

The group preserving the cube is thus the same as the group of the octahedron (which is the reason for the notation O_h).

The crystal whose appearance is as in the figure is said to show faces of two different *forms*; the eight triangular faces constitute one form whereas the six octagonal faces (modifications of the original cube faces) constitute the other form. The relative size of the various forms is called the *habit* of the crystal.

Crystals that are grown under careful conditions will develop into cubes or octahedra or other regular shapes. But if you go and pick up a crystal in the field, it will most likely have a rather irregular shape. Nevertheless, it does exhibit a symmetry, but in a more subtle sense. The first major discovery, by Nicolas Steno (1669) and Christian Huyghens in the field of crystallography is the celebrated 'law of corresponding angles' that says that the angles between 'corresponding faces' on all crystals of the same substance are equal. Put another way, suppose that we draw the normals to each of the faces, so as to get a set of points, on the unit sphere. Then (up to a rotation of the sphere) this set of points is the same for crystals of the same substance, although some of the points may be missing in a given crystal (i.e. some faces do not make their appearance). Furthermore, if a substance does (under controlled conditions)

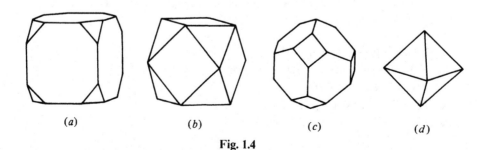

(a)   (b)   (c)   (d)

Fig. 1.4

crystallize into a regular body, then the set of normals of *any* crystal of that substance is invariant under the group of symmetries of the body. For example, suppose a substance crystallizes on occasion as a cube. Then the set of normals of any crystal of this substance will be invariant under O_h. Steno's 'law' appears in a caption to a figure in the appendix to his book entitled *De Solido intra Solidum Naturaliter Contento* (Florence, 1669).

The concept of an orbit of a group action allows us to explain in more detail what the crystallographers mean by the concept of the 'form' of a crystal. As we have already indicated, the observed symmetry of a crystal is best expressed in terms of directions. The law of the constancy of corresponding angles says that it is the collection of normal directions to the faces that is invariant (up to an overall rotation). These directions are the same for all crystals of the same substance, even though the outward appearance of the crystal might be quite irregular. We can consider these directions as points on the unit sphere. The group of symmetries of the crystal acts on this set of points. An orbit of the symmetry group acting on this set of directions is what is called a form of the crystal. To quote from a standard textbook (Phillips, *An Introduction to Crystallography*, p. 9): 'A rigid definition of a form is "the assemblage of faces necessitated by the symmetry when one face is given."'

Let us describe the possible types of orbits for the cubic group, $G = O_h$. We know that the number of elements in an orbit must divide the order of the group, which in this case is 48. It is clear that if we pick a generic point (x, y, z), with no two coordinates equal in absolute value and no coordinate zero, then the isotropy group $G_{(x,y,z)}$ consists only of the identity, and thus the orbit of this point contains the full 48 elements. The isotropy group of a point of the form (x, x, z), $|x| \neq |z|$, $x \neq 0$, $z \neq 0$, consists of the two-element permutation group – the group of transformations which permute the first two variables. The orbit through (x, x, z) thus contains 24 elements. Similarly, the orbit through $(x, x, 0)$ contains 12 elements. The orbit through (x, x, x) contains eight elements and the orbit through $(1, 0, 0)$ six elements. This last orbit corresponds to the set of faces of a cube – it is called the cubic form of the crystal. The eight-element orbit corresponds to the set of faces of the octahedron – it is called the octahedral form.

The crystallographers use the notation $\bar{x}$ for $-x$. Thus, the cubic form consists of the six elements

$$(1, 0, 0), \quad (\bar{1}, 0, 0), \quad (0, 1, 0), \quad (0, \bar{1}, 0), \quad (0, 0, 1), \quad (0, 0, \bar{1}).$$

The octahedral orbit is the orbit of the point $(1, 1, 1)$. (We write $(1, 1, 1)$ instead of $(1/3^{\frac{1}{2}}, 1/3^{\frac{1}{2}}, 1/3^{\frac{1}{2}})$, since we are really interested in the direction. We can always remember to normalize if we want the points to lie on the unit sphere.) The octahedral form thus consists of the eight points

$$(1, 1, 1), \quad (\bar{1}, 1, 1), \quad (1, \bar{1}, 1), \quad (1, 1, \bar{1}), \quad (\bar{1}, \bar{1}, 1), \quad (1, \bar{1}, \bar{1}) \quad (\bar{1}, 1, \bar{1}), \quad (\bar{1}, \bar{1}, \bar{1}).$$

It is convenient to have a way of representing points of the sphere on the plane of the paper. A standard convention is to use stereographic projections. If we project from the south pole, as shown in Fig. 1.5, all points on the upper hemisphere are mapped into the

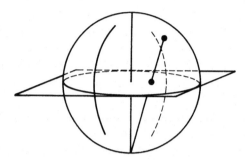

Fig. 1.5

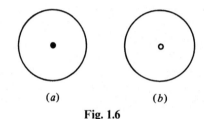

(*a*)   (*b*)

Fig. 1.6

interior of the circle. We use a solid dot in the interior (or boundary) of the circle to represent a point coming from the upper hemisphere (or equator). Thus, Fig. 1.6(a) represents the north pole, $(0, 0, 1)$. For points on the lower hemisphere we project from the north pole, and represent the corresponding points (which are now also inside the circle) by open dots. Thus, Fig. 1.6(*b*) represents the south pole, $(0, 0, \bar{1})$. If points from the upper and lower hemisphere project onto the same point, we use a circle with a center dot. The $+x$ direction runs down and the $+y$ direction runs to the right. We can now enumerate the different forms of the group O_h, corresponding to the different possible orbits. For each form it is worthwhile to try to visualize the isotropy group of a point in the orbit (see Fig. 1.7).

There are, of course, many different forms corresponding to (x, x, z), to $(x, y, 0)$ and (x, y, z). It turns out to be convenient to distinguish, among the (x, x, z) forms, between the cases $x > z$ and $x < z$. Representative crystals of the first case have triangular faces, while in the second case the faces are quadrilaterals (Figs. 1.7(*a*),(*c*),(*e*)). These are the 'pure forms' of a crystal of cubic symmetry. The substance may also crystallize in a combination of forms. That is, the set of normals to the faces may constitute more than one orbit.

The relative preponderance of the constituent forms is called the habit of the crystal. If we examine the forms that actually arise for a substance possessing cubic symmetry, we discover a remarkable fact: although, in principle, the values of the unrestricted variables x, y or z in forms of the last four types can be arbitrary, the observed values are always small integers. (Actually, the (x, y, z) is only determined up to scalar multiple; it is the set of ratios such as x/y that are the invariants. What is observed is that these values are always rational numbers, so, multiplying by a suitable factor, we can arrange that

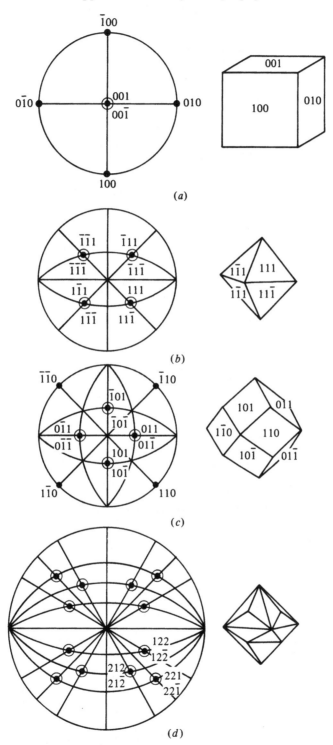

Fig. 1.7

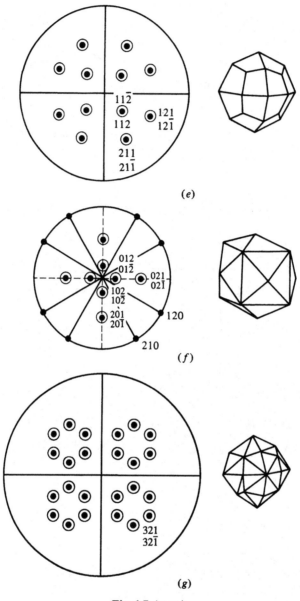

Fig. 1.7 (cont.)

the (x, y, z) are integers; and these turn out to be relatively small integers.) This is a special case of the celebrated law of rational indices, discovered by Haüy in 1784. We shall discuss this law later on.

We can now pose the following problem: what are the possible observed symmetry groups of crystals, and for each group what are the possible forms? Since a crystal has a finite number of faces, we can clearly tackle the first part of our problem by looking for all possible finite subgroups of the group of Euclidean motions in $\mathbb{R}^3$. We shall turn to this problem in Section 1.9.

1.6 The topology of $SU(2)$ and $SO(3)$

An important result about isotropy subgroups will enable us to learn more about the structure of the group $SO(3)$. Let m be a point of M and let a be an element of G. We claim that

$$G_{am} = aG_m a^{-1}. \tag{6.1}$$

By this we mean that a group element c lies in G_{am} if and only if $c = aba^{-1}$, where b is in G_m. Indeed, if b is in G_m so that $bm = m$, then $c(am) = (aba^{-1})\, am = abm = am$, so that c is in G_{am}. Conversely, if $cam = am$, then $a^{-1}ca = b$ lies in G_m, so $c = aba^{-1}$.

We will have many occasions to use this important formula. As our first application, we will explain the concept of Euler angles and use them to prove that the mapping ϕ of Section 1.2 sends $SU(2)$ onto all of $SO(3)$. Euler's description of an arbitrary element R of $SO(3)$ is that any such element can be written as the product of three rotations

$$R = R_\phi^z R_\theta^y R_\psi^z.$$

Here R_ψ^z denotes a rotation through angle ψ about the z axis, R_θ^y denotes rotation through angle θ about the y axis, and R_ϕ^z is another rotation through angle ϕ about the z axis. The angles ψ, θ and ϕ are known as the Euler angles of R. Observe that once we know that any R has such a decomposition, then, in fact, we know that R can be represented by a matrix in $SU(2)$. Indeed, we have already constructed, in Section 1.2, matrices which map onto any rotation about the z axis and any rotation about the y axis, so we can construct a matrix in $SU(2)$ which maps onto any product of such rotations. To prove Euler's theorem, let n be the north pole of the unit sphere (Fig. 1.8). The rotation R is completely determined by a knowledge of the image Rn of n, and of the image of any unit tangent vector to the sphere passing through n.

If B is some other rotation with $Bn = Rn$, then $R = CB$, where C is some rotation with $CBn = Bn$. In other words, C is in the isotropy group $SO(3)_{Rn}$ and thus is a rotation about the axis through Rn. Now we can get from n to any point on the unit sphere (in particular to $p = Rn$) by first applying a rotation about the y axis to move n to a point on the same latitude as p, then rotating about the z axis to move this point into p (Figs. 1.9(a)–(c)). Taking $p = Rn$, we thus see that we can write $Rn = R_\phi^z R_\theta^y n$.

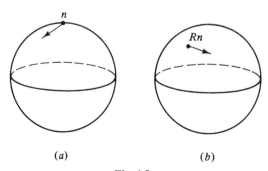

(a)   (b)

Fig. 1.8

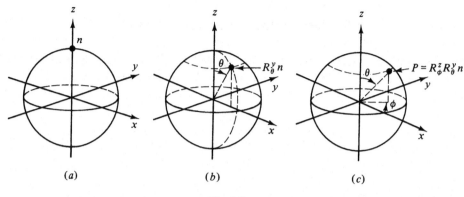

Fig. 1.9

That is, $Rn = Bn$ with $B = R_\phi^z R_\theta^y$. But this means that $R = CB$, where C is in $SO(3)_{Bn}$, so $C = BDB^{-1}$, where D is in $SO(3)_n$. Thus, D is a rotation about the z axis, i.e. $D = R_\psi^z$ for some angle ψ. Thus,

$$R = BDB^{-1}B$$
$$= BD = R_\phi^z R_\theta^y R_\psi^z$$

proving Euler's theorem.

We have shown that the mapping ϕ from $SU(2)$ to $SO(3)$ is surjective and in fact two-to-one. From this fact, we shall deduce an interesting topological property of the rotation group $SO(3)$. Recall that every element of $SU(2)$ can be written as

$$\begin{pmatrix} a & b \\ -\bar{b} & \bar{a} \end{pmatrix}, \quad \text{where} \quad |a|^2 + |b|^2 = 1.$$

Let us write $a = y_1 - iy_2$ and $b = y_3 - iy_4$. Then the parameters y_1, y_2, y_3, y_4 describe the points of $SU(2)$ and are subject to the constraint

$$y_1^2 + y_2^2 + y_3^2 + y_4^2 = 1.$$

In other words, relative to this parametrization, the group $SU(2)$ looks like a three-dimensional unit sphere in the four-dimensional y space. The identity element e corresponds to the point $(1, 0, 0, 0)$ and $-e$ corresponds to the antipodal point $(-1, 0, 0, 0)$. Now the sphere in any positive dimension has the property that any closed curve can be shrunk to a point. In more detail: let $\gamma(t)$ be a continuous curve with $\gamma(0) = \gamma(1) = e$. By moving γ a little we may assume that γ does not pass through the point $(-1, 0, 0, 0)$ for any value of t. We can then continuously deform γ to the constant map by pushing each point $\gamma(t)$ along the great circle from the 'south pole' $(-1, 0, 0, 0)$ to the 'north pole' $(1, 0, 0, 0)$. We say that $SU(2)$ is *simply connected*.

Now let us examine the situation for $SO(3)$. Consider a rotation through angle t about the z axis in $SO(3)$. As t ranges from zero to 2π, the rotation R_t^z describes a closed curve in $SO(3)$. We will prove that this closed curve cannot be continuously deformed to the constant curve in $SO(3)$. We can consider the pre-image of this curve in $SU(2)$. That is, we can consider the curve $C(t)$ in $SU(2)$ with the property that $C(0) = e$, that C is continuous, and that $\phi(C(t)) = R_t^z$. Remember that ϕ is two-to-one, and that the two points in $SU(2)$ which map into the same point in $SO(3)$ are antipodal points on the three-dimensional sphere. Therefore the requirement of continuity fixes C completely and it is in fact the semicircle in the $y_3 = y_4 = 0$ plane given by $y_1(t) = \cos t/2$, $y_2(t) = \sin t/2$. Thus, $C(2\pi) = -e$. By continuity, the same must hold true for any nearby curve; that is, if $B(s, t)$, a curve in $SO(3)$, depends continuously on s and t with $B(s, 0) = B(s, 2\pi) =$ the identity element in $SO(3)$, and if $B(0, t) = R_t^z$, then we can find a curve $C(s, t)$ in $SU(2)$ so that C depends continuously on s and t and has values in $SU(2)$ with $\phi(C(s, t)) = B(s, t)$. Then the same argument shows that $C(s, 2\pi) = -e$. But this implies that there is no way of shrinking the curve R_t^z to the constant curve, i.e. there cannot be any continuous B with $B(1, t)$ identically equal to the identity element in $SO(3)$.

Now suppose we go around the circle in $SO(3)$ twice, that is we let t range from zero to 4π. Then the pre-image curve in $SU(2)$ will be a complete circle, which we *can* shrink to a point in $SU(2)$. Hence, by looking at the image under ϕ of the various deformed curves, we can shrink our original double circuit to a point in $SO(3)$. To summarize, we have shown that the curve consisting of a family of rotations about an axis traversed once from 0 to 2π cannot be deformed to a point, while this same curve when traversed twice *can* be shrunk to a point! It is easy to see that this same argument can be generalized to prove the following: take any curve in $SO(3)$ which is closed. Either it can be deformed to a constant curve, or it cannot be, but going around it twice gives a curve which can be deformed to the identity.

This observation is of importance in the theory of elementary particles and is related to the notion of spin. Rotating an electron 360° about an axis does not bring the electron back to the same state, but rotating it 720° does. This property is crucial to the Dirac theory of electrons. Dirac also suggested the following physical experiment to indicate the difference between a 360° rotation and a 720° rotation. Suppose we take a wrench-shaped object and attach two rubber bands to each of the prongs and one to the handle, fixing the ends of the bands by attaching them to the wall as in Fig. 1.10. If the wrench is rotated through 360°, the bands will become entangled and it can be shown that there is no way by stretching or looping to disentangle them. However, if we rotate the wrench through 720°, the bands can be disentangled without moving the wrench. For details, we refer the reader to a very clear exposition given by Bolker (1973), *Am.*

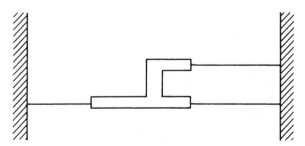

Fig. 1.10

Math Mthly, Vol. 80, No. 9. A simpler, but less convincing, experiment is the following: hold a cup of water in your hand. Try, without changing your grip on the cup or spilling the water, to rotate the cup (do not move your legs). You will find that you can make a $720°$ rotation but not $360°$.

1.7 Morphisms

Let the group G act on two different sets M_1 and M_2. Suppose that we are given a mapping f from M_1 to M_2. We say that f is *equivariant* with respect to the actions of G, or that f is a *G-morphism* if, for all a in G and all $m_1 \in M_1$, we have

$$f(am_1) = af(m_1).$$

In other words, it does not matter whether we first apply a group element and then the mapping f, or first apply f and then the group element. The notion of morphism will be very important in all that follows.

As a first example, suppose that the action of G on M_2 is trivial. That is, we suppose that $m_2 = am_2$ for all a in G and m_2 in M_2. Then the requirement of equivariance is that $f(am_1) = f(m_1)$ for all m_1 in M_1. In other words, f must be constant on the various orbits in M_1. This trivial example is of great importance in practical applications where we use measurements to deduce the existence of a group action. For example, suppose we measure the values of some function on a set with five elements. For a function 'chosen at random' we would expect to get five different values. Suppose we consistently observe only two values. We are then led to believe that there is a group acting on a set with two orbits and the function f is equivariant (for the trivial action, this is called *invariant*) under the action of G.

As a second example, let M_1 be an arbitrary set on which G acts. Let $M_2 = Subg(G)$ be the set of all subgroups of G. We let G act on M_2 by conjugation: that is, if H is a subgroup of G, then a sends H into aHa^{-1}. It is easy to check that this does indeed define an action of G on $Subg(G)$. Now define the mapping, f, from M_1 into M_2 by $f(m) = G_m$. In other words, f assigns to each point of M its isotropy subgroup G_m. The assertion that f be a morphism reads

$$f(am) = G_{am} = aG_m a^{-1}$$

which we know to be true.

Here is a closely related example: let $G \times M$ denote the Cartesian product of G and M, so that $G \times M$ consists of all pairs (b, m), where b is in G and m is in M. Let G act on $G \times M$ by

$$a(b, m) = (aba^{-1}, am).$$

Relative to this action, we can define two simple morphisms

$$G \times M \xrightarrow{\ \theta\ } M$$

$$\tau \downarrow$$

$$G$$

$$\theta(b, m) = m$$

$$\tau(b, m) = b.$$

We let G act on itself by conjugation:

$$a \text{ acting on } b \text{ is } aba^{-1}.$$

Then we have indeed described an action of G on $G \times M$ and the maps θ and τ are G-morphisms.

Let $Z \subset G \times M$ be the subset consisting of all pairs (b, m) such that $bm = m$:

$$Z = \{(b, m) \mid bm = m\}.$$

Notice that Z is carried into itself, as a set, by any $a \in G$:

$$\text{if} \quad bm = m, \quad \text{then } (aba^{-1}) \, am = am$$

so

$$\text{if} \quad (b, m) \in Z, \quad \text{then } a(b, m) \in Z.$$

We write this as $aZ \subset Z$. (Of course, the same holds for a^{-1}, so $aZ = Z$ as a set.)

Let $\rho: Z \to M$ denote the restriction of θ to Z and let $\sigma: Z \to G$ denote the restriction of τ to Z. So we have the diagram of maps:

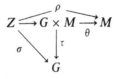

$$Z \xrightarrow{\ \ } G \times M \xrightarrow{\ \ } M$$

or simply

$$Z \xrightarrow{\ \rho\ } M \qquad\qquad (*)$$

$$\sigma \downarrow$$

$$G$$

For a fixed $m \in M$, its inverse image, $\rho^{-1}(m)$, is precisely the set of all (a, m), where a belongs to the isotropy group of m:

$$\rho^{-1}(m) = G_m \times \{m\} = \{(a, m) \mid a \in G_m\}.$$

For an a in G, its *fixed point set* consists of all m in M left fixed by a, so $FP(a) = \{m \mid am = m\}$. Then the inverse image $\sigma^{-1}(a)$ of an a in G is the set of all (a, m), where m

belongs to the fixed point set of a:

$$\sigma^{-1}(a) = \{a\} \times FP(a) = \{(a, m) \mid m \in FP(a)\}.$$

We can thus think of our diagram (∗) as saying that Z is 'fibered' over M, the fiber over an individual point m being its isotropy group, and also is fibered over G, the fiber over each group element being its fixed point set.

Notice that $e \in G_m$ for every m, that is $\{e\} \times M \subset Z$. For an application in the next section, it is convenient to remove this subset from Z. Let

$$Y = Z - \{e\} \times M = \{(a, m) \mid am = m, a \neq e\}.$$

We let ρ_1 and σ_1 denote the restrictions of ρ and σ to Y. So $\rho_1(a, m) = m$ and $\sigma_1(a, m) = a$. As before, $\sigma_1^{-1}(a) = \{a\} \times FP(a)$, while

$$\rho_1^{-1}(m) = (G_m - \{e\}) \times \{m\} = \{(a, m) \mid am = m, a \neq e\}.$$

As a simple example, let G be the group of symmetries of the equilateral triangle; let M be the set of three vertices of the triangle. Then Z consists of six elements:

$$(e, 1), ((23), 1)$$
$$(e, 2), ((13), 2)$$
$$(e, 3), ((12), 3).$$

The mapping ρ^{-1} assigns to each vertex of the triangle the pairs of elements involving that vertex: e.g.

$$\rho^{-1}(1) = \{(e, 1), ((23), 1)\}.$$

The mapping σ^{-1} assigns to each group element its fixed point set, so

$$\sigma^{-1}(e) = \{(e, 1), (e, 2), (e, 3)\}$$

$$\sigma^{-1}((23)) = \{((23), 1)\}, \text{etc.}$$

$\sigma^{-1}((123))$ and $\sigma^{-1}((132))$ are the empty set, since these rotations have no fixed points.

To construct Y in these examples, we delete $(e, 1)$, $(e, 2)$ and $(e, 3)$ from Z, leaving just three elements. Then $\rho_1^{-1}(1)$ is the single element $((23), 1)$, for example, and $\sigma_1^{-1}(e)$ is empty.

Suppose that G is a *finite* group, and that each $a \neq e$ has only finitely many fixed points. (For example, we might take G to be a finite subgroup of the group $SO(3)$ and take M to be the unit sphere S^2 in $\mathbb{R}^3$. If a point $m \in S^2$ is fixed by $a \in G$, then $s = -m$ is also left fixed by a. But if any other point is left fixed, then a leaves two linearly independent vectors fixed, and hence must leave the entire plane through these vectors pointwise fixed. Since a is orthogonal, it must carry the line perpendicular to this plane into itself, and since the determinant of a is 1, it must be the identity on this line, i.e. $a = e$. Thus, any $a \neq e$ in G can have at most two fixed points. (In the next section we shall see that each $a \neq e$ in $SO(3)$ has exactly two fixed points on M.))

In any event, under our hypothesis it is clear that Y is a *finite* set. We can count the number of elements of Y in two ways. Using the fibration over G we see that

$$\# Y = \sum_{a \neq e} \# FP(a).$$

On the other hand, let P denote the subset of M consisting of those $m \in M$ which are left fixed by some $a \neq e$ in G. Then using the fibration over M, we see that

$$\# Y = \sum_{m \in P} (\# G_m - 1).$$

(The -1 occurs because we removed e.) We can simplify this last expression as follows: suppose $m \in P$ and $a \in G$. Since

$$G_{am} = aG_m a^{-1},$$

we conclude that $am \in P$ and that

$$\# G_{am} = \# G_m.$$

Thus we can decompose P into orbits under G:

$$P = P_1 \cup \cdots \cup P_r.$$

On each orbit the value $\# G_m$ is constant. The number of elements in each orbit is $\# G / \# G_m$, so

$$\# Y = \sum_{\text{orbits}} \frac{\# G}{\# G_m} (\# G_m - 1).$$

Thus,

$$\sum_{a \neq e} \# FP(a) = \sum_{\text{orbits}} \frac{\# G}{\# G_m} (\# G_m - 1).$$

1.8 The classification of the finite subgroups of $SO(3)$

The determination of the possible symmetry groups of crystals and of molecules depends on the classification of the finite subgroups of $SO(3)$.

We begin with another theorem of Euler which asserts that every $a \neq e$ in $SO(3)$ is a rotation about some axis. We must prove that there is some vector $\mathbf{v}$ in $\mathbb{R}^3$ such that $a\mathbf{v} = \mathbf{v}$. Then a will carry the plane perpendicular to $\mathbf{v}$ into itself and, since $\det a = 1$, we see that a is a rotation in this plane, i.e. a is a rotation about the axis through $\mathbf{v}$.

To prove the existence of such a $\mathbf{v}$, we must prove that $a - I$ has a non-trivial kernel, i.e. that

$$\det (a - I) = 0.$$

We first note that

$$a - I = a(I - a^{-1}) = a(I - a^{t})$$

since $aa^t = I$. Also,

$$\det(I - a^t) = \det(I - a)^t = \det(I - a).$$

Furthermore, $\det a = 1$ by hypothesis. Thus,

$$\det(a - I) = \det(I - a) = \det(-I)\det(a - I).$$

But $\det(-I) = -1$, so $\det(a - I) = -\det(a - I)$ or $\det(a - I) = 0$. In n dimensions we have $\det(-I) = (-1)^n$ so the above proof works whenever n is odd: any a in $SO(n)$ with n odd leaves invariant at least one non-zero vector. (Rotations in the plane show that this is not true when n is even.)

Now let G be a finite subgroup of $SO(3)$. Each $a \neq e$ in G leaves precisely one line of vectors pointwise fixed, and hence has precisely two fixed points on the unit sphere, M. Thus, the formula $\#Y = \sum_{a \neq e} \#FP(a)$ simplifies to $\#Y = 2(\#G - 1)$.

Let

$$n = \#G$$

$$r = \#(\text{orbits of } G \text{ on } P)$$

$$n_i = \#G_m, \text{ where } m \in i\text{th orbit}.$$

Then

$$2(n - 1) = \sum_{i=1}^{r} \frac{n}{n_i}(n_i - 1)$$

or, dividing by n,

$$2 - \frac{2}{n} = r - \sum_{1}^{r} \frac{1}{n_i}.$$

We shall see that this equation imposes severe restrictions on r, n and n_i.

Since P consists of points which are held fixed by at least one element of G apart from the identity, we can be sure that $G_m \neq \{e\}$ for $m \in P$; hence $\#G_m \geq 2$. Thus all the n_i are ≥ 2, and the right-hand side of the above expression is not less than $r - (r/2)$. It follows that $r/2 < 2$, and so $r < 4$. There are at most three orbits. But $r = 1$ is also excluded since $n_i \leq n$ and thus

$$2 - \frac{2}{n} = 1 - \frac{1}{n_1}$$

is impossible; thus $r = 2$ or $r = 3$. We consider the two cases separately.

Suppose that $r = 2$. Then

$$\frac{2}{n} = \frac{1}{n_1} + \frac{1}{n_2}.$$

Since $n_i \leq n$, this can only happen if $n = n_1 = n_2$. This implies that $G_m = G$ for each pole. Thus all rotations are about a fixed axis. The group G thus consists of all rotations through angles $2\pi/n$ about fixed axis. As an abstract group, it is the cyclic group of order n. There are two systems of notation in use to describe the various finite subgroups of the orthogonal group. The older notation, due to Schoenflies, describes the above group as C_n. The Hermann–Mauguin (H–M) notation describes it simply by n.

Now suppose $r = 3$. Then

$$\frac{1}{n_1} + \frac{1}{n_2} + \frac{1}{n_3} = 1 + \frac{2}{n}.$$

We may assume that we have chosen our notation so that $n_1 \leqslant n_2 \leqslant n_3$. We must have $n_1 = 2$, for otherwise the left-hand side of the above equation would be $\leqslant 1$. Similarly, $n_2 \geqslant 4$ is impossible. If $n_2 = 3$, then $1/n_3 = 1/6 + 2/n = (1 + (12/n))/6$ and so $n_3 < 6$. Thus the possibilities for (n_1, n_2, n_3) and n are

$$(2, 2, k) \quad n = 2k, k \geqslant 2, \text{ arbitrary}$$

$$(2, 3, 3) \quad n = 12$$

$$(2, 3, 4) \quad n = 24$$

$$(2, 3, 5) \quad n = 60.$$

Each of these possiblities actually occurs and represents exactly one group. We proceed down the list.

(2, 2, k)   Since $n_3 = n/2$, the orbit M_3 consists of two elements, and thus $G_{p_3} = C_k$. For any q belonging to M_1 or M_2, the isotropy group G_q is of order two, and thus consists of the identity and a rotation through 180°. The subgroup C_k rotates the various poles of each orbit amongst themselves, and acts transitively on each orbit, since $\#C_k = k = \#M_1 = \#M_2$, and only the identity in C_k fixes any pole in M_1 or M_2. All the points of M_1 lie in a plane (and on a circle) and are rotated into one another by rotations of angles $2\pi/k$. They therefore form the vertices of a regular k-polygon. If we consider this as a polygon in space having two sides, then we can consider the 180° rotations also as symmetries, which have the effect of turning the figure over. We can consider such a two-sided polygon as a (degenerate) polyhedron having two sides, a 'dihedron'. Its group of symmetries does have all the desired properties and is known as the dihedral group of order k. (The case $k = 2$ is still further degenerate; in this case the group consists of the 180° rotations about each of the three coordinate axes.) There is a slight difference between the case k even and the case k odd. For k even there are two families of axes – those which pass through the vertices and those which bisect the sides. For k odd, there is one family of axes, but two families of poles. Cases $k = 3$ and $k = 4$ are illustrated in Fig. 1.11.

The Schoenflies notation for the dihedral groups is D_k. The H–M notation for D_2 is 222 since there are three two-fold axes of rotation. The group D_k (for $k > 2$) is denoted by $k2$ since there is a k-fold principal axis of rotation plus two-fold axes.

(2, 3, 3) with n = 12   Then $\#M_1 = 6$ and $\#M_2 = \#M_3 = 4$. There are thus six poles corresponding to three axes of 180° rotation, i.e. diad axes and two systems of triad axes. If we examine the group of rotations which preserve the regular tetrahedron, we see that

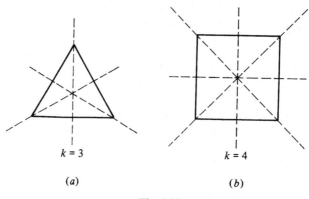

$$k = 3$$

$$k = 4$$

$$(a) \qquad\qquad (b)$$

Fig. 1.11

the lines which join the midpoints of non-adjacent edges form diad axes. The points of intersection of these axes with the sphere form one orbit. The lines joining a vertex to the midpoint of the opposite face form triad axes. The vertices themselves form one orbit and their opposite points form another. Thus the group of rotations of a regular tetrahedron is of the type $(2, 3, 3)$. The Schoenflies notation for this group is T, the H–M notation is 23 since there are diad and triad axes, and to distinguish it from the dihedral group, $32 = D_3$. Conversely, T is the only group of type $(2, 3, 3)$. To prove this, it suffices to prove that the points of the orbit M_2 form the vertices of a regular tetrahedron, and that G contains all permutations of these points which give a linear transformation with positive determinant. Let m_1, m_2, m_3 and m_4 be the poles in M_2. Then G_{m_4} contains a rotation through $120°$ and hence G_{m_4} acts transitively on m_1, m_2, m_3. Thus the segments joining m_4 to m_1, m_2, and m_3 all have the same length. But since G acts transitively on M_2, we may conclude that all the segments joining the m_i have equal length, and so they form the vertices of a tetrahedron. Thus G is a subgroup of T. Since it has order 12, $G = T$.

$(2, 3, 4)$ *with* $n = 24$ Then $\#M_1 = 12, \#M_2 = 8$ and $\#M_3 = 6$. The group of rotations of the cube has this set of poles. Conversely, looking at the orbit M_3, we can argue as above to conclude that the segments joining two adjacent points in M_3 all have the same length, and hence form the vertices of a regular octahedron.

As we know, the group of the octahedron and the group of the cube are the same. The Schoenflies notation for this group is O. The H–M notation is 432 since it contains fourth-order, third-order and second-order axes.

$(2,3,5)$ *with* $n = 60$ Then $\#M_1 = 30, \#M_2 = 20$ and $\#M_3 = 12$ Choose some m_{11} in M_3 and let m_{12} be the opposite pole of the axis through m_{11}. We may take these to be the north and south poles of our sphere. The remaining ten poles in M_3 cannot all lie on the equator, because if they did there would be some five-fold axis lying in the equatorial plane, and it would have to rotate the remaining poles out of the plane. Thus five of the ten remaining poles lie in the upper hemisphere, and are all equidistant from m_{11}, and the remaining five lie in the lower hemisphere and are equidistant from m_{12}. This

Table 3. *Finite rotation groups*

Type	#G	Name
(n, n)	n	C_n
$(2, 2, k)$	$2k$	D_k, $k \geqslant 2$
$(2, 3, 3)$	12	T
$(2, 3, 4)$	24	O
$(2, 3, 5)$	60	I

implies that each pole in M_3 has five nearest neighbors and the distances between nearest neighbors is constant. Thus the poles of M_3 lie at the vertices of a regular icosahedron. It is easy to check that the rotation group of the icosahedron, denoted by I, does indeed have type $(2, 3, 5)$. We can summarize our results so far by Table 3.

We now point out a remarkable fact of far-reaching consequence – no rotation of order five or of order greater than six ever occurs in the symmetry group of any known natural crystal! In particular, in the above list, the values of n in the first row are restricted to $n = 1, 2, 3, 4$ and 6; the values of k in the second row are restricted to $k = 2, 3, 4$ and 6; and the group I does not occur as the group of rotational symmetries of any crystal.

What is so special about the values $1, 2, 3, 4$ and 6? Well, 1 and 2 are clearly special values – 1 corresponds to no rotational symmetry at all, and 2 to a rotational plane of symmetry, and so our question really is, what is so special about the values 3, 4 and 6? If we examine regular polygons in the plane, there is indeed something special about triangles, squares and hexagons: these are the only polygons that can fit together to cover the plane without overlap and without leaving any empty spaces. (Indeed, the interior angle of k-sided regular polygon is $\pi - (2\pi/k)$, and this must be an integral subunit of 2π. Thus the condition on k is that $2/(1 - 2/k)$ be an integer, and this holds only for $k = 3, 4$ and 6. This fact was already known to Kepler.) This result suggests that perhaps there is an analogous (and more general) result in three dimensions. Suppose that the crystals were made up of discrete repetitive subunits, and the overall symmetry of the crystal has to preserve the repetitive pattern. Let us make this assumption more precise: suppose that there are three linearly independent vectors, $\mathbf{u}, \mathbf{v}$ and $\mathbf{w}$, such that the interior arrangement of the crystal looks the same when viewed from any point r as when viewed from $r + n_1\mathbf{u} + n_2\mathbf{v} + n_3\mathbf{w}$, where the n_i are *integers*. (Here we think of the units as being so small relative to the physical size of the crystal, that the n_i can take on extremely large values, whose bound we can ignore for the relevant purposes of the succeeding argument.) Our discreteness assumption is that it is possible to find such $\mathbf{u}, \mathbf{v}$ and $\mathbf{w}$ with the property that if some translation $r \to r + \mathbf{t}$ is a symmetry of the pattern, then $\mathbf{t}$ is a linear combination of $\mathbf{u}, \mathbf{v}$ and $\mathbf{w}$ with integer coefficients. The set of linear combinations of $\mathbf{u}, \mathbf{v}$ and $\mathbf{w}$ with integer coefficients is called the *lattice* generated by $\mathbf{u}, \mathbf{v}$ and $\mathbf{w}$. A *basis* of a lattice is a set of linearly independent elements in the lattice such that every lattice element can be expressed as an integral combination of them. It is true that every lattice possesses a basis. At the moment we will take this for granted, or assume it as part of our definition of a lattice. In any event, we make the assumption that the pattern of the crystal has a three-dimensional lattice as its group of translational symmetries. Now let a be a rotation which also preserves the pattern of the crystal. For any $\mathbf{t}$ in the lattice, the transformations

$$r \to a^{-1}r$$

$$r \to a^{-1}r + \mathbf{t}$$

$$r \to a(a^{-1}r + \mathbf{t}) = r + a\mathbf{t}$$

all preserve the lattice. In particular, $a\mathbf{t}$ again belongs to the lattice. This means that $a\mathbf{u}$, $a\mathbf{v}$ and $a\mathbf{w}$ are all in the lattice, and hence can be expressed as linear combinations of $\mathbf{u}, \mathbf{v}$ and $\mathbf{w}$ with integer coefficients. Put another way, this says that the matrix of a, when expressed in terms of the basis $\mathbf{u}, \mathbf{v}$ and $\mathbf{w}$, has all integer entries. In particular,

$$\text{tr}\, a \text{ is an integer.}$$

On the other hand, if we introduce orthogonal coordinates x, y, z, so that a is a rotation through angle ϕ about the z axis, then the matrix of a takes the form

$$\begin{pmatrix} \cos\phi & -\sin\phi & 0 \\ \sin\phi & \cos\phi & 0 \\ 0 & 0 & 1 \end{pmatrix}$$

and we see that

$$\text{tr}\, a = 1 + 2\cos\phi.$$

Thus

$$2\cos\phi \text{ is an integer,}$$

which implies that

$$\phi = 2\pi/k \quad \text{with} \quad k = 1, 2, 3, 4 \text{ or } 6.$$

Thus, from the observation that only these angles occur in rotational symmetries of crystals, we are led to the *atomic hypothesis*, that a solid is not a continuum, but is rather built up from discrete subunits in a regular repetitive pattern.

The actual historical development of this idea followed similar, but somewhat different, lines. In 1784, the Abbe Haüy published a work entitled *Essai d'une théorie sur la structure des crystaux appliquée à plusieurs genres de substances cristallisées*. In this work he presented various ideas about crystals, to which he had been led by observing the cleavage of certain minerals. By an accident during the examination of some crystals in the mineralogical collection of an amateur friend, one of the larger crystals was dropped and it broke. When it broke along a rather regular line Haüy pursued the study of cleavages of other crystals. By repeated experiments on many different substances, he was led to propound the view that continued cleavage would ultimately lead to a basic unit, and that the whole crystal is built up of a repetition of this unit. If the unit is repeated the same number of times along each of its faces, the crystal as a whole would have a shape similar to that of the unit. But if rows of units were omitted regularly, then faces of other slopes could be formed. For example, suppose that the basic unit is a cube. Then by stacking the same number of cubes in all three directions, we would obtain a large cube. By regularly omitting cubes at the corners, we could form an octahedron. This theory led Haüy to formulate the *law of rational indices*, which says the following: suppose that we choose coordinate axes along edges of the crystal. Suppose that we then choose some plane, parallel to a crystal face but not parallel to any of the three axes we have chosen. The intercept of this plane determines three vectors, one on each of the above crystallographic axes. The law of rational indices asserts that if we choose any other faces of the crystal, the ratio of its intercepts to those we have already chosen are simple rational numbers (or ∞ if the plane is parallel to an axis).

Haüy was led to formulate this law by considering the consequence of the simple stacking model he proposed. Of course, we no longer believe that the fundamental units of the crystal are little blocks, as proposed by Haüy. But the law of rational indices is true, and has been verified over and over again. Our conception of the nature of the underlying unit has changed, but the fundamental lesson that a crystal is composed of a discrete pattern remains with us.

It should be pointed out that one can argue directly from the law of rational indices to exclude all angles of rotation other than $2\pi/k$ with $k = 1, 2, 3, 4$ and 6. Any group preserving a polyhedron which satisfies the law of rational indices can only have the above angles as its angles of rotation. Such a proof, for example, can be found in Hamermesh, *Group Theory*, pp. 45–8. But, in a sense, this misses the main point. The existence of discrete subunits implies the law of rational indices for Haüy, and they imply the non-existence of other angles of rotation, by the argument we gave above; and it is the existence of the discrete subunits which is the most important deduction.

By the restriction on the angle of rotation we have narrowed our list of finite rotation groups down to 11:

$$C_1, C_2, C_3, C_4, C_6, D_2, D_3, D_4, D_6, T, O.$$

1.9 The classification of the finite subgroups of $O(3)$

Let G be a finite subgroup of $O(3)$. The map det: $G \to C_2 = \{\pm 1\}$ sending $a \in G$ into det a is a homomorphism: $\det(ab) = (\det a)(\det b)$. The set H of all $a \in G$ with det $a = 1$ is a subgroup. One possibility is that this subgroup is all of G, i.e. $G \subset SO(3)$. We have already studied this case. The other possibility is that H is a proper subgroup of G. In this case, if $a \notin H$, then $ab \notin H$ for any $b \in H$ and $G = H \cup aH$. There will be as many $c \in G$ with det $c = +1$ as with det $c = -1$.

For example, consider the group, T_d, of all symmetries of the tetrahedron. It certainly contains reflections; for instance, reflections through the plane passing through an edge and bisecting the opposite face (Fig. 1.12). Since T is the subgroup of T_d consisting of rotations, and T contains 12 elements, we see that T_d contains 24 elements. Notice

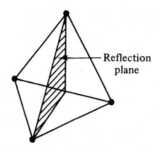

Reflection plane

Fig. 1.12

that any $a \in T_d$ must permute the four vertices of the tetrahedron, and only the identity leaves all the vertices fixed. So we have an isomorphism of T_d with a subgroup of S_4. Since $\#S_4 = 24 = \#T_d$, we see that T_d is isomorphic to S_4.

We can thus say that we have a homomorphism of $S_4 \rightarrow O(3)$ which identifies S_4 with T_d, the group of all symmetries of the tetrahedron.

Notice that $-I$ does not lie in T_d. On the other hand, we can consider our tetrahedron as situated inside a cube, with its vertices at the diagonally opposite vertices of each face of the cube. For a given cube there are two such tetrahedra, as shown in Fig. 1.13. One tetrahedron is carried into the other by the inversion, $-I$. Now $T \subset O$. Suppose $A \in O$ and $A \notin T$. Then A must carry one tetrahedron into the other and hence $-A \in T_d$. Conversely, if $B \in T_d$ and $\det B = -1$, then B preserves the cube and hence so does $-B$, so $-B \in O$ and $-B \notin T$.

We conclude that we can decompose

$$S_4 = T_d = T \cup T'$$

where T is a subgroup and T' is the set of A with $\det A = -1$. We can then map S_4 into O by sending

$$A \rightarrow A \qquad \text{for } A \in T$$

$$B \rightarrow -B \quad \text{for } B \in T'.$$

This is a homomorphism:

$$A_1 A_2 \rightarrow A_1 A_2$$

$$A_1 B_2 \in T' \quad \text{so} \quad A_1 B_2 \rightarrow -(A_1 B_2) = A_1(-B_2)$$

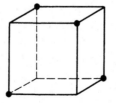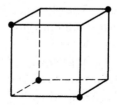

Fig. 1.13

and

$$B_1 B_2 \in T \quad \text{so} \quad B_1 B_2 \to B_1 B_2 = (-B_1)(-B_2).$$

Since O also has 24 elements, we see that O is also isomorphic to S_4. Thus O and T_d are isomorphic as abstract groups, and they are both isomorphic to S_4. However, they are *not* conjugate in $\mathbb{R}^3$: there cannot exist any matrix $R \in GL(3, \mathbb{R})$ with

$$ROR^{-1} = T_d$$

since all elements of O, and hence of ROR^{-1}, have positive determinant, while T_d contains some elements with negative determinant. We say that we have two *inequivalent representations* of S_4 on $\mathbb{R}^3$: as all symmetries of the tetrahedron and as all rotational symmetries of the cube. The theory of representations will be developed in the next chapter.

The preceding analysis applies more generally: for any G containing an element a with $\det a = -1$ there are two possibilities:

(a) the matrix $-I \in G$ (where I is the identity matrix); or

(b) $-I \notin G$.

Let G_+ denote the subgroup consisting of a with $\det a = +1$ and let G_- denote the set of elements with $\det a = -1$. In case (a) we have $G_- = (-I)G_+$ and so G is completely determined once we know G_+. In case (b) consider the set

$$G^\vee = G_+ \cup (-I)G_-.$$

We claim that $G^\vee$ is a group. Indeed, the product of any two elements of G_+ is in G_+. The product of an element, a, of G_+ and $(-I)b$ of $(-I)G_-$ equals $(-I)ab \in (-I)G_-$, and $(-I)b(-I)c = bc$ is in G_+ for b and c in G_-. Thus, $G^\vee$ is a group, and clearly all the elements of $G^\vee$ have positive determinant. Furthermore, G_+ is a normal subgroup of index* two in $G^\vee$. (To say a subgroup H is normal means that $aHa^{-1} = H$ in all a in G.) Given $G^\vee$ and its normal subgroup, G_+, we can recover G: we write $G^\vee = G_+ \cup aG_+$, where $a \notin G_+$, and we set $G_- = (-I)aG_+$. Since we know all finite groups of rotations (i.e. of orthogonal matrices with determinant $+1$), then we can apply the preceding analysis to obtain the list of all finite subgroups of $O(3)$. In this section, we shall restrict our attention to the groups arising in crystallography. In Section 1.10 we shall make a special study of the icosahedral group.

From the 11 rotation groups listed at the end of the preceding section, we obtain 11 non-rotation groups by including $-I$. In addition, we must check for the normal subgroups of index two in each rotation group in order to construct non-rotation groups which do not include $-I$. We now carry out this procedure explicitly for each rotation group.

Starting with C_1, which consists of the identity alone, we obtain, by including $-I$, the two-element group called C_i. This is isomorphic, as an abstract group, to C_2, but C_i and C_2 are not conjugate subgroups of $O(3)$.

The next rotation group, C_2, consists of the identity I and a $180°$ rotation R_π. The identity alone constitutes a normal subgroup of index 2, which we use as G_+; the $180°$

* The *index* of a subgroup H of a group G is the number of elements in G/H; index $H = \#(G/H)$.

rotation constitutes $(-I)G_-$. We multiply the 180° rotation,

$$R_\pi = \begin{pmatrix} -1 & 0 & 0 \\ 0 & -1 & 0 \\ 0 & 0 & 1 \end{pmatrix}$$

by $-I$, obtaining

$$\bar{R}_\pi = \begin{pmatrix} 1 & 0 & 0 \\ 0 & 1 & 0 \\ 0 & 0 & -1 \end{pmatrix},$$

which of course represents a reflection in the xy plane. The resulting group $\{I, \bar{R}_\pi\}$ is called C_s.

The groups C_2 and C_s are of course isomorphic as abstract groups, but they are different subgroups of $O(3)$ in the sense that one cannot be obtained from the other by conjugation with any element of $O(3)$.

From C_2 we obtain also a four-element group by including $-I$. This group, $\{I, -I, R_\pi, \bar{R}_\pi\}$, is called C_{2h} because it contains both a two-fold rotation axis and a 'horizontal' reflection plane perpendicular to that axis. As an abstract group C_{2h} is sometimes called V, the 'four-group'. Its four elements $\{e, a, b, c\}$ satisfy $a^2 = b^2 = c^2 = e$, $ab = c$, $bc = a$, $ac = b$. This group is Abelian (commutative), but it is not isomorphic to the cyclic group C_4.

The group C_3, which contains an odd number of elements, of course cannot include a normal subgroup of index 2. We can, however, obtain a six-element non-rotation group from C_3 by multiplying each element by the inversion $-I$. Since $-I$ is the product of a 180° rotation and a reflection in the xy plane, the three new elements are rotations through 180°, 120° + 180° = 300°, and $-120° + 180° = 60°$ all followed by reflection in the xy plane. This group, denoted S^6 or $\bar{3}$, is isomorphic as an abstract group to C_6, but the 'odd' rotations, through 60°, 180° and 300°, are multiplied by a reflection in the **xy** plane. Fig. 1.14(*a*), where $\otimes$ denotes a vector into the page, $\odot$ a vector out of the page, has S^6 as its symmetry group. Note that it has no plane of reflection symmetry and no axis of rotational symmetry in the plane of the page.

By including the inversion $-I$ along with C_4, we obtain the group C_{4h}, whose elements consist of rotations through 0°, 90°, 180° and 270°, plus the same four rotations multiplied by reflection in the xy plane. Fig. 1.14(*b*) has C_{4h} symmetry.

Since C_4 contains a normal subgroup C_2 of index 2 consisting of the identity I and a 180° rotation R, we can also obtain from it a four-element non-rotation group by multiplying the 90° and 270° rotations by $-I$. The resulting group is called S^4; it is isomorphic as an abstract group to C^4, but the 90° and 270° rotations are multiplied by a reflection in the xy plane. Fig. 1.14(*c*) has S^4 symmetry.

By combining the inversion $-I$ with C_6, we obtain the 12-element group C_{6h}. Since C_6 contains C_3 as a normal subgroup of index 2, we can also create a six-element non-rotation group from C_6 by multiplying the 60°, 180° and 300° rotations, which are not elements of C_3, by $-I$. This yields rotations of 240°, 0° and 120°, respectively, all multiplied by a reflection in the xy plane. The resulting group is C_{3h}; it consists of

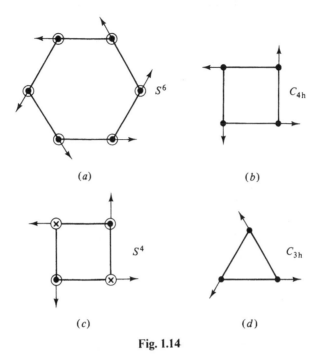

Fig. 1.14

rotations through $0°$, $120°$ and $240°$, each one with or without reflection in a horizontal plane. Fig. 1.14(d) has C_{3h} symmetry.

We now construct non-rotation subgroups of $O(3)$ which include the dihedral groups as subgroups. We begin with D_2, whose four elements consist of the identity and rotations about the x, y and z axes. Multiplying each of these elements by the inversion $-I$, we obtain, in addition to $-I$, reflections in the three coordinate planes. The resulting eight-element group is called D_{2h}; it is the complete symmetry group of a rectangular 'dihedron'. Note that there are two elements of D_{2h} which interchange vertices 1 and 3 and interchange vertices 2 and 4; one a rotation about the dotted line shown in Fig. 1.15(a), the other a reflection in a vertical plane containing this line.

The group D_2 contains three equivalent normal subgroups of index 2. We choose as G_+ the one consisting of I and the $180°$ rotation about the vertical (z) axis. Multiplying the $180°$ rotations about the x and y axes by $-I$, we obtain reflections in the xz and yz planes. The resulting four-element non-rotation group is called C_{2v}, where the subscript v indicates the presence of vertical reflection planes. The rectangle shown in Fig. 1.15(b), which has a vector into the page attached to each corner, has C_{2v} as its symmetry group.

From D_3 we may similarly obtain two non-rotation groups. One is found by multiplying each element of D_3 by $-I$. From the $0°$, $120°$ and $240°$ rotations we obtain elements which are most easily described as rotations through $60°$, $180°$ and $300°$ multiplied by reflection in the horizontal plane. Each rotation about an axis in the plane of the page, when multiplied by $-I$, gives a reflection in a vertical plane perpendicular to that axis. The resulting group is called D_{3d}; it is the symmetry group of

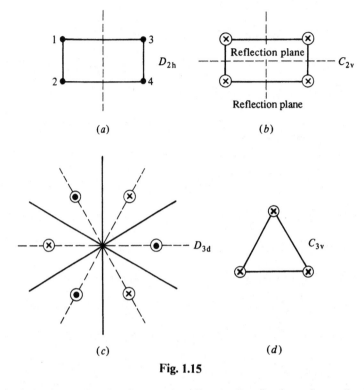

Fig. 1.15

Fig. 1.15(c). The subscript d refers to the 'diagonal' reflection planes (broken lines) which bisect the angles between the rotation axes (solid lines).

Since D_3 contains C_3 as a normal subgroup of index 2, we can also obtain a six-element non-rotation group from it by multiplying the three 180° rotations of D_3 by $-I$ to obtain three reflections in vertical planes. The resulting group, called C_{3v}, is the symmetry of Fig. 1.15(d).

By multiplying the elements of D_4 and D_6 by $-I$, we obtain groups D_{4h} (16 elements) and D_{6h} (24 elements), which are the complete symmetry groups (including reflections) of the square and the hexagon, respectively (Fig. 1.16). Using the normal subgroups C_4 and C_6, and multiplying all the 180° rotations by $-I$, we obtain the groups C_{4v} and C_{6v}, which include reflections in vertical planes (broken lines). There is another way of converting D_4 to a non-rotation group. We choose the normal subgroup D_2, which is not isomorphic to C_4. Multiplying the four elements of D_4 which are not in D_2 by $-I$, we obtain the eight-element group D_{2d}, which includes two reflections in vertical planes at 45° to the coordinate axes and two rotations, through 90° and 270°, which are followed by reflections in a horizontal plane. D_{2d} is the symmetry group of Fig. 1.17. In a similar manner, starting with D_3 as a normal subgroup D_6, we obtain the 12-element group D_{3h}, which is the complete symmetry group (including reflections) of the equilateral triangle.

By multiplying each element of the group T by $-I$, we add 12 more elements to complete the group T_h. This may be interpreted as the symmetry group of a cube which has 'right-hand' objects attached to one tetrahedron of four vertices, 'left-handed'

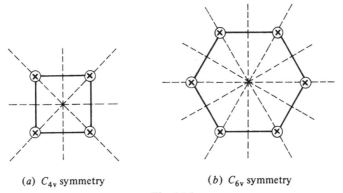

(a) C_{4v} symmetry   (b) C_{6v} symmetry

Fig. 1.16

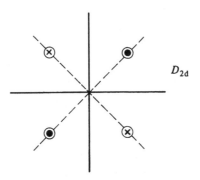

D_{2d}

Fig. 1.17

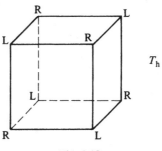

T_h

Fig. 1.18

objects attached to the other four vertices (Fig. 1.18). The elements of T_h with determinant $+1$ preserve each tetrahedron; those with determinant -1 interchange the two tetrahedra but also turn 'R' into 'L', thus preserving the figure. The notation T_h indicates the presence of a horizontal reflection plane.

By multiplying each element of the group O by $-I$, we obtain O_h, the group of all symmetries of the cube, with 48 elements. By choosing T as a normal subgroup of O, and multiplying each of the 12 elements of O which is not in T by $-I$, we obtain T_d, the group of all symmetries (including reflections) of the tetrahedron. We have already encountered this group, which is isomorphic to the permutation group S_4.

Table 4.

Rotation groups		Non-rotation groups containing $-I$		Non-rotation groups not containing $-I$	
S	H–M	S	H–M	S	H–M
C_1	1	C_i	$\bar{1}$		
C_2	2	C_{2h}	$2/m$	C_s	m
C_3	3	S^6	$\bar{3}$		
C_4	4	C_{4h}	$4/m$	S^4	$\bar{4}$
C_6	6	C_{6h}	$6/m$	C_{3h}	$3/m$
D_2	222	D_{2h}	mmm	C_{2v}	mm2
D_3	32	D_{3d}	$\bar{3}m$	C_{3v}	3m
D_4	422	D_{4h}	$4/mmm$	C_{4v}	4mm
				D_{2d}	$\bar{4}2m$
D_6	62	D_{6h}	$6/mmm$	C_{6v}	6mm
				D_{3h}	$\bar{6}2m$
T	23	T_h	m3		
O	432	O_h	m3m	T_d	$\bar{4}3m$

This exhausts the list of finite subgroups of $O(3)$ which contain only two-fold, three-fold, four-fold and six-fold axes of rotation. We can summarize our results in Table 4, where we list all 32 groups both in Schoenflies (S) and in H–M notation. In the Schoenflies notation, a principal rotation axis is taken as vertical. If a horizontal reflection plane is also present, this is denoted by the subscript h and a vertical reflection plane by the subscript v. Diagonal vertical planes added to D_n are denoted by D_{nd} and similarly for T. The presence of a vertical mirror plane is denoted in the H–M notation by m and a horizontal mirror plane by /m.

Perhaps some words are in order about procedures that allow the determination of the actual symmetry groups of a given substance. Since the discovery of the X-ray diffraction technique by von Laue in 1912, X-ray methods have been developed which not only allow the determination of the symmetry group of the crystal as a whole, but also determine the underlying pattern. We refer the reader to the standard texts on X-ray crystallography for a description of these methods. Here we discuss methods available to 19th century crystallographers. For details, we refer to Phillips. If a substance crystallizes in the general form of a given group, then we know that its symmetry group is indeed the full group. Thus, for instance, if a substance sometimes crystallizes as a hexoctahedron, then we know that its symmetry group is all of O_h. But if it always crystallizes as a cube, then we cannot tell from this fact alone which of the groups T, T_h, O, O_h or T_d is the correct symmetry group. One method is to attack the crystal with an acid, and to study the etch figures induced on the faces. The symmetry pattern of the etch figures may allow the determination of the correct group. The presence or absence of $-I$ in the group may sometimes be deduced from piezoelectricity: certain crystals, in the presence of mechanical stresses, react by having a separation of charges on their surface. This is possible only if $-I$ is not in the group. Another test is rotatory optical activity: some substances have the property of rotating the plane of polarization of plane polarized light passing through it, either to the right or to the left. This can only happen if the group respects the difference between right and left, i.e. is a subgroup of the rotation group. This effect was made famous by a celebrated result of Pasteur: in 1811, Arago had discovered that quartz crystals have the power of rotating the polarization plane of a beam of polarized light passing through it. Some time later it was found that certain liquids are also optically active. It was then found that there were two kinds of tartaric acid naturally arising from wine. One was optically active and the other not. Pasteur precipitated a salt of the optically inactive tartaric acid. Superficially its crystals appeared the same as the salts of the active variety, but under closer examination he discovered that half of them were indeed the same, but the other half of the crystals were their mirror images. He separated the two different types of crystals and dissolved them separately. Each solution was indeed optically active, but in opposite sense. The optically inactive tartaric acid was really a mixture of two active varieties which cancelled each other out.

The mathematical proof of the existence of 32 geometrical crystal classes was achieved by Hessel in 1830, although his work was ignored for close to 60 years. The classification of substances into these classes has been steadily going on, and now there are over 20 000 substances listed and described. (Of these most fall into the classes of lower symmetry.) As we mentioned before, although X-rays are now used to analyse the structure of crystals, the assignment into classes was available, before the advent of X-rays, by the measurement of interfacial angles, examination of various habits, the use of etch figures, etc.

The study of the theoretical possibilities for the internal structure – the mathematical analysis of the classification of discrete repetitive patterns – began in the 1840s with the work of Frankenheim and Bravais. It culminated in 1890 with the classification, by Fedorov, of the 230 so-called crystallographic groups, or space groups. This result was obtained 'independently' by Schoenflies in 1891 and by Barlow in 1894. Fedorov was a Russian crystallographer, Schoenflies was a German mathematician, and Barlow was an English businessman. Of course, at the time, there was no means of detecting the actual internal structure. This came with von Laue and the Braggs who developed the X-ray techniques. So, the theoretical possibilities were known some 20 years before there was any means of observation! The 32 finite subgroups of $O(3)$ are portrayed in Fig. 1.19. The conjugacy classes of each group are included. The symbol r_4^2 means the square of a fourth-order rotation (a rotation through 90°) that occurs in the group. The symbol $3r_4^2$ means that there are three such elements in the conjugacy class. The symbol τ denotes a reflection. The symbol s_3 denotes $-I \cdot r_3$. For each group we draw a crystal whose symmetry is the given group (if such a crystal is known to exist). We will not give a complete classification of the crystallographic groups – the actual groups of internal symmetries of crystals – in this book. A very careful and readable account is available in the excellent book by Burckhardt (1966). In Appendix A we shall carry the analysis of crystals a bit further; roughly as far as Bravais did in his work of 1849, reprinted in his *Etudes cristallographiques*, Gauthier-Villars, Paris, 1866.

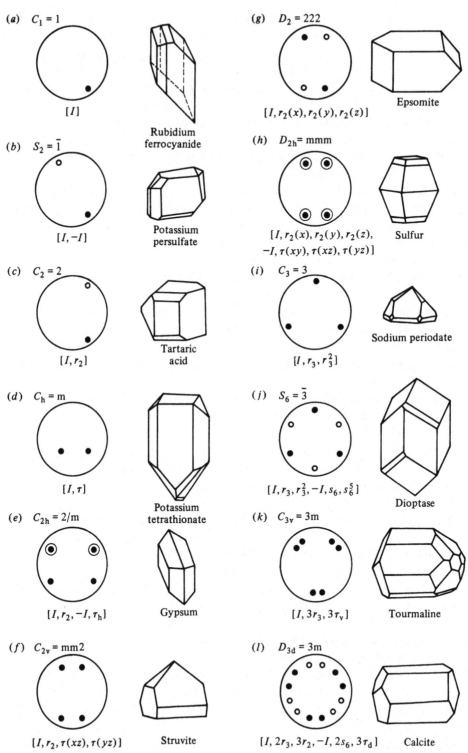

(a) $C_1 = 1$

$[I]$

Rubidium
ferrocyanide

(b) $S_2 = \bar{1}$

$[I, -I]$

Potassium
persulfate

(c) $C_2 = 2$

$[I, r_2]$

Tartaric
acid

(d) $C_h = m$

$[I, \tau]$

Potassium
tetrathionate

(e) $C_{2h} = 2/m$

$[I, r_2, -I, \tau_h]$ Gypsum

(f) $C_{2v} = mm2$

$[I, r_2, \tau(xz), \tau(yz)]$ Struvite

(g) $D_2 = 222$

$[I, r_2(x), r_2(y), r_2(z)]$

Epsomite

(h) $D_{2h} = mmm$

$[I, r_2(x), r_2(y), r_2(z),$
$-I, \tau(xy), \tau(xz), \tau(yz)]$ Sulfur

(i) $C_3 = 3$

$[I, r_3, r_3^2]$ Sodium periodate

(j) $S_6 = \bar{3}$

$[I, r_3, r_3^2, -I, s_6, s_6^5]$

Dioptase

(k) $C_{3v} = 3m$

$[I, 3r_3, 3\tau_v]$ Tourmaline

(l) $D_{3d} = \bar{3}m$

$[I, 2r_3, 3r_2, -I, 2s_6, 3\tau_d]$ Calcite

Fig. 1.19

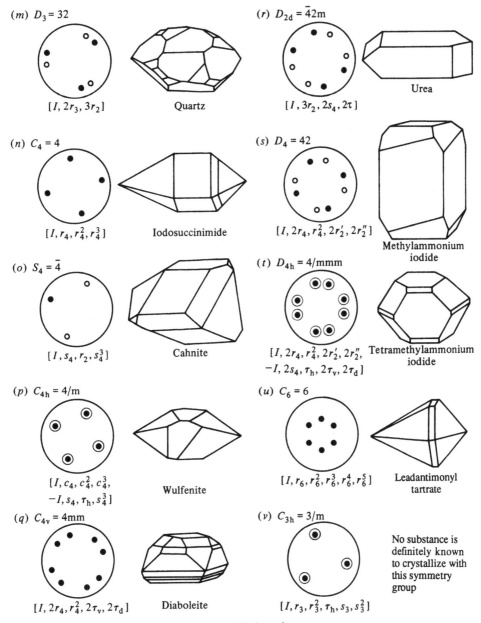

(m) $D_3 = 32$

$[I, 2r_3, 3r_2]$ Quartz

(n) $C_4 = 4$

$[I, r_4, r_4^2, r_4^3]$ Iodosuccinimide

(o) $S_4 = \bar{4}$

$[I, s_4, r_2, s_4^3]$ Cahnite

(p) $C_{4h} = 4/m$

$[I, c_4, c_4^2, c_4^3,$
$-I, s_4, \tau_h, s_4^3]$ Wulfenite

(q) $C_{4v} = 4mm$

$[I, 2r_4, r_4^2, 2\tau_v, 2\tau_d]$ Diaboleite

(r) $D_{2d} = \bar{4}2m$

Urea

$[I, 3r_2, 2s_4, 2\tau]$

(s) $D_4 = 42$

$[I, 2r_4, r_4^2, 2r_2', 2r_2'']$ Methylammonium
iodide

(t) $D_{4h} = 4/mmm$

$[I, 2r_4, r_4^2, 2r_2', 2r_2'',$ Tetramethylammonium
$-I, 2s_4, \tau_h, 2\tau_v, 2\tau_d]$ iodide

(u) $C_6 = 6$

$[I, r_6, r_6^2, r_6^3, r_6^4, r_6^5]$ Leadantimonyl
tartrate

(v) $C_{3h} = 3/m$

No substance is
definitely known
to crystallize with
this symmetry
group

$[I, r_3, r_3^2, \tau_h, s_3, s_3^2]$

Fig. 1.19 (cont.)

1.10 The icosahedral group and the fullerenes

One of the remarkable discoveries of the past few years has been the existence of new, stable forms of carbon consisting of large molecules built up out of hexagons and pentagons, with three 'bonds' emanating from each carbon atom.

All the known fullerenes are made of hexagons and pentagons. Whereas the number of hexagons varies from one type of fullerene to another, every fullerene has exactly

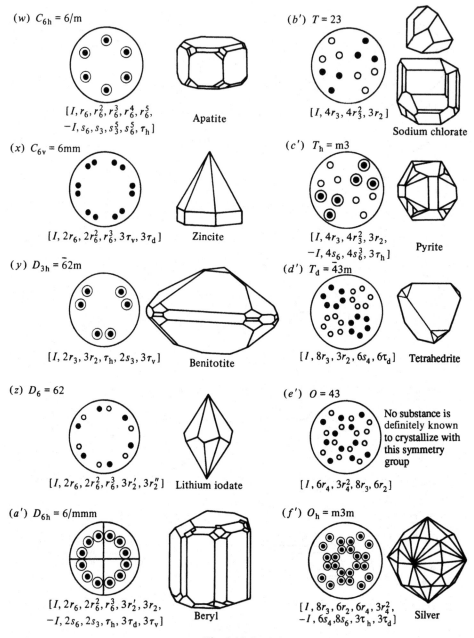

(w) $C_{6h} = 6/m$

$[I, r_6, r_6^2, r_6^3, r_6^4, r_6^5,$
$-I, s_6, s_3, s_3^5, s_6^5, \tau_h]$ Apatite

(x) $C_{6v} = 6mm$

$[I, 2r_6, 2r_6^2, r_6^3, 3\tau_v, 3\tau_d]$ Zincite

(y) $D_{3h} = \bar{6}2m$

$[I, 2r_3, 3r_2, \tau_h, 2s_3, 3\tau_v]$ Benitotite

(z) $D_6 = 62$

$[I, 2r_6, 2r_6^2, r_6^3, 3r_2', 3r_2'']$ Lithium iodate

(a') $D_{6h} = 6/mmm$

$[I, 2r_6, 2r_6^2, r_6^3, 3r_2', 3r_2,$
$-I, 2s_6, 2s_3, \tau_h, 3\tau_d, 3\tau_v]$ Beryl

(b') $T = 23$

$[I, 4r_3, 4r_3^2, 3r_2]$ Sodium chlorate

(c') $T_h = m3$

$[I, 4r_3, 4r_3^2, 3r_2,$
$-I, 4s_6, 4s_6^3, 3\tau_h]$ Pyrite

(d') $T_d = \bar{4}3m$

$[I, 8r_3, 3r_2, 6s_4, 6\tau_d]$ Tetrahedrite

(e') $O = 43$

No substance is
definitely known
to crystallize with
this symmetry
group

$[I, 6r_4, 3r_4^2, 8r_3, 6r_2]$

(f') $O_h = m3m$

$[I, 8r_3, 6r_2, 6r_4, 3r_4^2,$
$-I, 6s_4, 8s_6, 3\tau_h, 3\tau_d]$ Silver

Fig. 1.19 (cont.)

12 pentagons. This is *not* an accident. It is a consequence of a theorem of Euler. A famous formula of Euler, perhaps the first formula in topology, says the following: let us call a polyhedral surface *simple* if it can be mapped in a one-to-one bicontinuous fashion onto the surface of a sphere. The surface itself is polyhedral; that is, it is built up of polygons. So it has (two-dimensional) *faces*, (one-dimensional) *edges*, and (zero-

dimensional) *vertices*. Let f denote the number of faces, e the number of edges, and v the number of vertices. Then Euler's formula says

$$f - e + v = 2.$$

For example, the surface of a tetrahedron has four (triangular) faces, six edges (with three emanating from each vertex) and four vertices. And $4 - 6 + 4 = 2$ in accordance with Euler's formula. The surface of a cube has six (square) faces, 12 edges, and eight vertices. Again $6 - 12 + 8 = 2$. The regular icosahedron has 12 vertices and 20 (triangular) faces. Five edges emanate from each vertex, but each edge impinges on two vertices. So there are 30 edges ($5 \times 12/2$). Once again, $20 - 30 + 12 = 2$. Suppose we truncate the icosahedron at one of its vertices. This has the effect of replacing the vertex by a hexagonal face. We have added one face, deleted one vertex, and added five new edges and five new vertices. This clearly does not change the value of $f - e + v$, so it remains two. If we do this in a symmetric fashion at all the vertices, we obtain the *buckyball*. It has 32 faces (12 pentagons and 20 hexagons), 90 edges and 60 vertices: $32 - 90 + 60 = 2$.

A proof of Euler's formula can be found in every elementary text on topology or on graph theory. But Euler also derived a consequence of his formula if we make the additional assumption that exactly three edges emanate from each vertex.

Theorem (Euler)

For any simple polyhedral surface with three edges emanating from each vertex we have

$$\sum (6 - n) f_n = 12,$$

where f_n denotes the number of n-gons.

Proof The condition on the edges says that $v = \frac{2}{3} e$ since each edge impinges on two vertices and three edges emanate from each vertex. On the other hand, since each edge is on the boundary of two polygons,

$$e = \frac{1}{2} \sum n f_n.$$

Since $f = \sum f_n$ we have

$$2 = f - e + v = f - \frac{1}{3} e = \sum f_n - \frac{1}{6} \sum n f_n. \quad \text{QED}$$

For example, if all the polygons are to have four sides, the theorem says that there must be exactly six of them as in the cube. The tetrahedron, with only triangles, has exactly four of them. The dodecahedron with only pentagons has exactly 12. For fullerenes, built entirely out of pentagons and hexagons, Euler's theorem places no restriction on the number of hexagons, and says that there are exactly 12 pentagons. For a beautiful discussion of the implementation of Euler's theorem in biological forms, see Chapters 8 and 9 of the classic *On growth and form* by D'Arcy Thompson. For the rest of this section we shall focus on the buckyball C_{60}.

Recall from Section 1.8 that the icosahedral group, I, contains 60 elements. It acts as symmetries of the icosahedron, and hence as symmetries of the (symmetrically) truncated icosahedron, which is our buckyball. The buckyball has 60 vertices. We claim that the 60-element group, I, acts transitively on these vertices. To see this, observe that every vertex of our original icosahedron corresponds to a pentagon of the buckyball. Consider two vertices, X and Y, of the buckyball. The vertex X lies on some pentagon, U, of the buckyball, which was obtained by truncating our original icosahedron at some vertex, u. Similarly, Y lies on some pentagon, W, corresponding to a vertex, w, of the original icosahedron. Now we can find a rotation, s, in I which carries u to w, and hence r acts on the buckyball by carrying the pentagon U to W. But then, by a further rotation about the axis joining w to its opposite, we can arrange to move X to Y. We have proved that, given any two vertices X and Y of the buckyball, there is an element, g, of I which rotates X into Y. But there are 60 elements of I and 60 vertices of the buckyball. Hence there is a *unique* element of I which rotates X into Y.

Although all the vertices of the buckyball are alike, in the sense that there is a symmetry carrying any one into any other, this is not true of the edges. For example, if we truncate the icosahedron close to the vertices, the edges of the pentagons will be shorter than the edges which are left-over portions of the original triangular edges. In one chemical model of the buckyball the pentagonal edges are thought of as single bonds, and the left-over triangular edges are thought of as double bonds. In more sophisticated models the fourth valence electron of each carbon atom is considered as delocalized; we shall, however, adopt the convenient terminology of single and double bonds. So emanating from each vertex there are two single bonds, lying on a pentagon, and one double bond, lying on a hexagon. All pentagonal edges are single bonds, whereas every other edge on a hexagon is double. All double bonds of the buckyball are alike, as are all the single bonds.

These bond structures give us some insight into the nature of the group I. For example, if we pick a double bond, b, there will be a unique double bond, b', opposite it, in the sense that b and b' lie on a plane bisecting the buckyball. If we draw the line joining the midpoints of b and b', then a 180° rotation having this line as axis is a symmetry of the buckyball. There are 30 double bonds, and hence 15 such pairs. Thus there are 15 elements of order two (i.e. satisfying $r^2 = e$, where e denotes the identity element) in I. There are ten pairs of opposite hexagons. If we join the midpoints of these hexagons to form an axis, rotations through angles 120° and 240° about this axis are elements of I. Thus I contains 20 elements satisfying $t^3 = e$. There are six pairs of opposite pentagons; each non-trivial rotation about the axis joining the midpoints of such a pair is an element, p, of I satisfying $p^5 = e$, and so there are 24 such elements. Now $15 + 20 + 24 = 59$. Together with the identity element, e, we have accounted for all 60 elements of I.

We have seen that the group O is isomorphic to S_4, the group of all permutations of four objects, and that the group T is isomorphic to the group A_4 of *even* permutations of four objects. In the case of T, the four objects could be taken to be the vertices of the tetrahedron. In the case of the cube, the four objects can be taken to be the four diagonals passing through the center (and a pair of opposite vertices) of the cube. We shall now show that the group I is isomorphic to A_5, the group of even permutations of five objects. So our first task is to find five objects which are permuted by the elements of I. Let us go back to the plane passing through two opposite double bonds. Among all planes perpendicular to this plane and passing through the center of the buckyball, there will be two which also pass through pairs of double bonds, and these planes will be perpendicular to each other. In other words, we have a configuration of six double bonds consisting of three opposite pairs spanning three mutually perpendicular planes. The collection of 30 double bonds breaks up into five such configurations. These five configurations must be permuted amongst themselves by any element of I. A direct check, best visualized by actually performing rotations on a model, will show that only the identity element can fix all configurations. So we have identified I as a subgroup of the group S_5 of all permutations of five objects. But the only subgroup of S_5 containing 60 elements is the alternating group A_5. Under our identification of I with A_5, the elements of order five correspond to five cycles, permutations of the form $a \rightarrow b \rightarrow c \rightarrow d \rightarrow e \rightarrow a$, as these are the only elements of order five in S_5. Similarly, the elements of order three correspond to three cycles, $a \rightarrow b \rightarrow c \rightarrow a$, with d and e left fixed. The elements of order two correspond to permutations of the form $a \rightarrow b \rightarrow a$, $c \rightarrow d \rightarrow c$, with e left fixed.

For applications to spectroscopy, as we shall explain in Chapter 3, we must consider not only rotational symmetries of the buckyball but also allow reflections. The chemists call this larger group I_h. From the mathematical point of view, we can obtain this larger group by adjoining the inversion operator, P, where, in terms of a Cartesian coordinate system based at the center of the buckyball, $P(x, y, z) = (-x, -y, -z)$. The physicists call P the *parity* operator. Notice that $P^2 = e$ and that $Pg = gP$ for all elements, g, of I. The group I_h thus contains 120 elements. (But it is *not* isomorphic to the group S_5. Under the identification given in the preceding paragraph, the parity operator, P, preserves all configurations.) Suppose we concentrate on some definite vertex, X, of the buckyball. Since there are 120 elements in I_h and only 60 vertices, there must be some element of I_h besides the identity which keeps X fixed. It is easy to see what this element is. Consider the double bond emanating from X, the opposite double bond, and the plane passing through them. Reflection in this plane is a symmetry of the buckyball and keeps X fixed, since X lies in the plane. For future reference, we will denote this reflection by r_X.

2

REPRESENTATION THEORY

OF FINITE GROUPS

In this chapter we present the classical theory of representations of finite groups as linear transformations on vector spaces over the complex numbers. As we have seen in Chapter 1, geometrical problems give rise to the study of the representation theory of groups over the real numbers and also over the integers. These kinds of questions are far more subtle than the theory of representations over the complex numbers. However, the complex representation theory has a beautiful structure and a highly developed theory, has an enormous range of applications in mathematics, physics, and chemistry, and is the starting point for any of the other representation theories.

In all that follows in the present chapter, unless otherwise specified, all vector spaces will be finite-dimensional complex vector spaces and all groups will be finite. (In later chapters we will have something to say about infinite-dimensional vector spaces and certain kinds of non-finite groups.)

2.1 Definitions, examples, irreducibility

We recall that a representation of a group G on a vector space V is an action of G on V, in which each $a \in G$ acts as a linear transformation: the map

$$r(a): V \to V \quad \mathbf{v} \rightsquigarrow r(a)\mathbf{v}$$

is a linear transformation for each a. When we denote a representation by r, the symbol r is meant to specify the group, G, the vector space, V, and the particular action of G on V. For reasons of convenience, we will frequently write $a\mathbf{v}$ for the action of a on $\mathbf{v}$ in the representation r, so that

$$r(a)\mathbf{v} = a\mathbf{v}.$$

If we choose a basis of V, then each linear transformation, $r(a)$, has a matrix relative to the basis. We will denote the matrix associated to $r(a)$ by $(r_{ij}(a))$. For example,

$$(r_{ij}(e)) = (\delta_{ij})$$

is the identity matrix, where e is the identity element of the group. It is apparent that

$$(r_{ij}(a^{-1})) = (r_{ij}(a))^{-1}$$

and

$$r_{ij}(ab) = \sum_k r_{ik}(a)r_{kj}(b).$$

The matrices r_{ij} depend on the choice of basis for V. They are determined by the representation only up to conjugacy.

Let r and r' be representations of the same group, G, on vector spaces V and V'. We say that r and r' are *similar*, or *equivalent*, and write $r \sim r'$ if there exists a non-singular linear transformation, T, mapping V onto V' such that

$$r'(a)T = Tr(a) \quad \text{for all } a \in G.$$

In other words, T is a morphism for the group G and is an isomorphism of vector spaces.

Let r be a representation of a group G on a vector space V. A subspace W of V is said to be *invariant* if $r(a)W \subset W$ for all a in G. This means that for every $\mathbf{w}$ in W and every a in G the vector $r(a)\mathbf{w}$ is again in W. Applying $r(a^{-1}) = r(a)^{-1}$ we see that W is invariant if and only if $r(a)W = W$ for all a in G. If W is an invariant subspace, then the restriction of $r(a)$ to W, i.e. the operator $r(a)$ applied only to elements of W, is a linear transformation of W into itself, which we may denote by $r(a)|_W$. Since $r(a)|_W r(b)|_W = (r(a)r(b))|_W = r(ab)|_W$, we see that the map $r|_W$ sending $a \in G$ into $r(a)|_W$ is a representation of G on W. It is sometimes called a *subrepresentation* of the representation r.

For any r, the vector space V itself and the subspace $\{0\}$ are clearly (trivial) invariant subspaces. A representation r is called *irreducible* if these are the only invariant subspaces.

Examples

Any representation of G on a one-dimensional complex vector space is automatically irreducible, since there is no room for any subspace between $\{0\}$ and V. A linear transformation on a one-dimensional vector space is just multiplication by a complex number. So a linear representation of G on a one-dimensional vector space is a rule which assigns to each a in G a complex number, $\kappa(a)$; i.e. a representation on a one-dimensional vector space is just a complex-valued function, κ, on the group, subject to the conditions $\kappa(ab) = \kappa(a)\kappa(b)$ and $\kappa(e) = 1$. For any finite group, $a^{\#G} = e$, so $\kappa(a)^{\#G} = 1$. Thus $\kappa(a)$ is a root of unity, whose order is some divisor of $\#G$. Any two linear transformations on a one-dimensional space are equivalent if and only if they are multiplication by the same complex number. Thus the equivalence class of the one-dimensional representation is also completely determined by the function κ.

For example, consider the group C_n, and let b be a generator of C_n, i.e. an element whose powers include all elements of C_n. Then for each k with $0 \leqslant k < n$ we can define κ_k by

$$\kappa_k(b) = \exp 2\pi ik/n \quad \text{and hence} \quad \kappa_k(b^s) = \exp 2\pi isk/n.$$

These give n distinct irreducible representations of C_n. We shall see later on that it follows from the formula

$$1^2 + 1^2 + \cdots + 1^2 = n = \#C_n, \quad n \text{ summands,}$$

that these are all the irreducible representations of C_n. (Here the 1's represent the dimension of the irreducible representations, and we are summing the squares of the dimensions.)

Let us prove that if G is a finite *commutative* group, then any finite-dimensional irreducible representation of G over the complex numbers must, in fact, be one dimensional: Choose any $a \in G$. The operator $r(a)$, being a linear transformation on a finite-dimensional complex vector space, has at least one eigenvector $\mathbf{v}$ with eigenvalue λ_a. (Here is where we use the fact that we are over the complex numbers and that the vector space is finite dimensional.) Let W_1 be the subspace consisting of *all* $\mathbf{w}$ satisfying $r(a) = \lambda_a \mathbf{w}$. Then $W_1 \neq \{0\}$ since $\mathbf{v} \in W_1$. We claim that W_1 is invariant: that is, if $\mathbf{w} \in W_1$, $r(b)\mathbf{w} \in W_1$ for any b in G. The proof is straightforward:

$$r(a)(r(b)\mathbf{w}) = [r(a)r(b)]\mathbf{w}$$
$$= r(ab)\mathbf{w}$$
$$= r(ba)\mathbf{w} \quad \text{(here is where we use the commutativity of } G\text{)}$$
$$= r(b)r(a)\mathbf{w}$$
$$= r(b)(\lambda_a\mathbf{w}) = \lambda_a(r(b)\mathbf{w}).$$

Now choose some other element c in G. Inside W_1 we can find some eigenvector, $\mathbf{u}$, of $r(c)$, with eigenvalue λ_c. Let W_2 be the subspace of W_1 consisting of all vectors $\mathbf{z}$ satisfying $r(c)\mathbf{z} = \lambda_c\mathbf{z}$. Then W_2 is a non-zero invariant subspace of W_1 and every vector in W_2 satisfies

$$r(a)z = \lambda_a z \quad \text{and} \quad r(c)z = \lambda_c z.$$

Continuing this way, using all the group elements, we find a subspace $W \neq \{0\}$ and eigenvalues λ_a for every group element a such that every vector $\mathbf{v}$ in W satisfies $r(b)\mathbf{v} = \lambda_b\mathbf{v}$ for every b in G. This clearly implies that *any* one-dimensional subspace of W is invariant under G. If our original vector space V were irreducible, this one-dimensional subspace would have to be V itself. Thus, any irreducible representation of a commutative group is one dimensional.

If the group G is not commutative, it is rather hard for it to have many one-dimensional representations. In fact, we shall prove later on that if all the irreducible representations of G are one dimensional, then G must be commutative. For the moment, let us indicate partial results in this direction. If κ is any function on G which satisfies $\kappa(ab) = \kappa(a)\kappa(b)$, then

$$\kappa(bab^{-1}a^{-1}) = 1$$

for any a and b in G. Let us use this identity to determine the possible one-dimensional representations of S_3. We think of S_3 as symmetries of the triangle. If b is a reflection, then $b^2 = e$ so that $\kappa(b)^2 = 1$, i.e.

$$\kappa(b) = \pm 1 \quad \text{for reflections.}$$

If a is a rotation and b is a reflection, then $bab^{-1} = a^{-1}$ so $bab^{-1}a^{-1} = a^{-2}$. But every rotation in S_3 is of the form $c = a^{-2}$ so

$$\kappa(c) = 1 \quad \text{for rotations.}$$

Thus there are two possibilities:

$$\kappa \equiv 1 \quad \text{or} \quad \kappa(a) = \text{sgn}(a) = \begin{array}{l} + 1 \text{ for rotations} \\ - 1 \text{ for reflections.} \end{array}$$

(We shall see later that $\kappa \equiv 1$ and $\kappa(a) = \text{sgn}(a)$ are the only one-dimensional representations of S_n for any n.)

The group S_3 also has a natural two-dimensional representation. If we think of S_3 as symmetries of the triangle, this determines an action of S_3 on the plane $\mathbb{R}^2$, i.e. a homomorphism of S_3 into $GL(2, \mathbb{R})$. We can think of $GL(2, \mathbb{R})$ as a subgroup of $GL(2, \mathbb{C})$ so the action of S_3 as symmetries of the triangle gives rise to a two-dimensional representation of S_3. (More generally, the same argument shows that any represent-ation of any group on a real vector space gives rise to a representation of a complex vector space: in terms of a basis, just think of the real matrices $(r_{ij}(a))$ as being complex matrices.) The two-dimensional representation of S_3 that we have just constructed is irreducible. Indeed, let a be rotation through $120°$. As a complex matrix $r(a)$ has two distinct eigenvalues, $\omega = \exp(2\pi i/3)$ and $\bar{\omega} = \exp(-2\pi i/3)$ with corresponding eigen-vectors $\mathbf{v}$ and $\bar{\mathbf{v}}$. Since ω and $\bar{\omega}$ are distinct, the only lines in $\mathbb{C}^2$ which are left invariant by $r(a)$ are the lines through $\mathbf{v}$ and through $\bar{\mathbf{v}}$. Any non-trivial invariant subspace of $\mathbb{C}^2$ would have to be one dimensional and hence coincide with one of these lines. If b is a reflection, then $bab^{-1} = a^{-1}$, and hence $r(a)r(b^{-1})\mathbf{v} = r(b^{-1})r(a^{-1})\mathbf{v} = r(b^{-1})(\bar{\omega}\mathbf{v}) = \bar{\omega}r(b^{-1})\mathbf{v}$. In other words, $r(b)$ or $r(b^{-1})$ interchanges the lines through $\mathbf{v}$ and through $\bar{\mathbf{v}}$. Hence, neither of these lines can be invariant under S_3.

As we shall prove later, it follows from the equation

$$1^2 + 1^2 + 2^2 = 6 = \#S_3$$

(the sum of the squares
of the dimensions)

that the three representations, the two one-dimensional ones and the two-dimensional one, are, up to equivalence, the only irreducible representations of S_3.

In the preceding chapter we described two inequivalent real representations of S_4, both three dimensional – the representation r_T of S_4 as all symmetries of the tetrahedron, and the representation r_O of S_4 as all rotational symmetries of the cube. Each of these gives rise to a complex three-dimensional representation, and the proof given in Chapter 1 still works to prove that these two representations are not equivalent over the complex numbers. It is not hard to show directly, and will follow from machinery to be developed later, that these representations are both irreducible.

To discover a two-dimensional representation of S_4, we note that there is a homomorphism from S_4 to S_3. One way of defining it is as follows: think of S_4 as the rotational symmetries of the cube. There are three lines which pass through the centers of opposite faces; if the cube is the one with the vertices $(\pm 1, \pm 1, \pm 1)$, then these lines are the coordinate axes. Any symmetry of the cube must permute these three lines. This gives a homomorphism ϕ from S_4 to S_3. We have already constructed a two-dimensional representation, r, of S_3. Then $r \circ \phi$, sending each a in S_4 to $r(\phi(a))$, is an irreducible two-dimensional representation of S_4. We also have the two one-dimensional representations given by $\kappa \equiv 1$ and $\kappa(a) = \operatorname{sgn}(a)$. From

$$3^2 + 3^2 + 2^2 + 1^2 + 1^2 = 24 = \#S_4$$

we will be able to conclude that these are all the irreducible representations of S_4.

2.2 Complete reducibility

Let r^1 and r^2 be representations of the group G on the vector spaces V_1 and V_2. We obtain a representation of G on their direct sum, $V_1 \oplus V_2$ by setting

$$r(a)(\mathbf{v}_1 + \mathbf{v}_2) = r^1(a)\mathbf{v}_1 + r^2(a)\mathbf{v}_2 \quad \mathbf{v}_1 \in V_1, \ \mathbf{v}_2 \in V_2$$

for all $a \in G$. We denote the representation by $r^1 \oplus r^2$. With respect to a direct sum basis, the matrices of $r^1 \oplus r^2$ clearly have the form

$$\begin{pmatrix} (r^1_{ij}(a)) & (0) \\ (0) & (r^2_{kl}(a)) \end{pmatrix}.$$

Let r be a representation of G on V and suppose that there is an invariant subspace, W, of V. Then, by restricting each $r(a)$ to a linear transformation on W, we obtain a representation of G on W, which we have denoted by $r|_W$. We claim that there always exists a complementary subspace, W', invariant under G, so that $r \sim r_W \oplus r_{W'}$. To prove this claim, we assume that there is an Hermitian scalar product $(\ ,\ )$ defined on V which is *invariant* under G, i.e. a scalar product satisfying $(r(a)\mathbf{u}, r(a)\mathbf{v}) = (\mathbf{u}, \mathbf{v})$ for all $\mathbf{u}$ and $\mathbf{v}$ in V and all $a \in G$. Then let W' be the subspace orthogonal to W with respect to this scalar product. We claim that W' is invariant. Indeed, if $\mathbf{u} \in W'$, then $(r(a)\mathbf{u}, \mathbf{v}) = (\mathbf{u}, r(a^{-1})\mathbf{v})$. Taking $\mathbf{v}$ in W shows that $r(a)\mathbf{u} \in W'$, which proves that W' is invariant. Thus, the existence of an invariant complement follows from the existence of an invariant scalar product. But we can always construct an invariant scalar product by the process of 'averaging over the group', as follows: we can always choose some positive definite scalar product, $(\ ,\ )_0$, on V, which need not be invariant. (Just choose some basis for V and declare it to be orthonormal.) We then define $(\ ,\ )$ by averaging over the group:

$$(\mathbf{u}, \mathbf{v}) = (1/\#G) \sum_{b \in g} (r(b)\mathbf{u}, r(b)\mathbf{v})_0.$$

then

$$(r(a)\mathbf{u}, r(a)\mathbf{v}) = (1/\#G) \sum_{b \in G} (r(ba)\mathbf{u}, r(ba)\mathbf{v})_0$$
$$= (1/\#G) \sum_{c \in G} (r(c)u, r(c)\mathbf{v})_0 = (\mathbf{u}, \mathbf{v})$$

since, as b ranges over the group, so does $c = ba$. This proves the existence of the invariant scalar product. Notice that we have used the fact that the group is finite since we summed over all group elements.

This property of representations of finite groups is known as *complete reducibility*. It need not hold for representations of infinite groups. For instance, the map $t \rightsquigarrow \begin{pmatrix} 1 & t \\ 0 & 1 \end{pmatrix}$

is a representation of the additive group $\mathbb{R}$ by 2×2 matrices. The subspace $y = 0$ is invariant but has no invariant complement.

Recall that a representation r of G on V is said to be *irreducible* if there are no non-trivial subspaces invariant under G: every representation can be written as a direct sum of irreducibles. Suppose r is not irreducible. Then there is at least one proper invariant subspace, W. If there is more than one proper invariant subspace, we choose W to have the smallest possible dimension, so that it must be irreducible. The complement, W', has strictly smaller dimension than V. Either W' is irreducible, or we may again identify an irreducible subspace and take its complement. We continue the process. Eventually the complement will be irreducible, and we have then expressed r as a direct sum of irreducible representations.

Suppose that V is a vector space with a given positive definite scalar product, $(\ ,\ )$. A representation of G on V is said to be *unitary* if $(\ ,\ )$ is invariant under G. We can summarize the results of this section as saying that every unitary (finite-dimensional) representation of any group is completely reducible, and V can be decomposed into a direct sum of irreducible subspaces. We also proved that an arbitrary finite-dimensional representation by a finite group is equivalent to a unitary representation, and hence has the same complete reducibility properties.

As an example of a reducible representation, we construct the three-dimensional 'permutation representation' of S_3. We let S_3 act on the basis elements

$$\begin{pmatrix} 1 \\ 0 \\ 0 \end{pmatrix}, \begin{pmatrix} 0 \\ 1 \\ 0 \end{pmatrix}, \quad \text{and} \quad \begin{pmatrix} 0 \\ 0 \\ 1 \end{pmatrix}$$

of $\mathbb{R}^3$ as a three-element set. Thus, for example, the permutation (123) carries

$$\begin{pmatrix} 1 \\ 0 \\ 0 \end{pmatrix} \text{ into } \begin{pmatrix} 0 \\ 1 \\ 0 \end{pmatrix}, \begin{pmatrix} 0 \\ 1 \\ 0 \end{pmatrix} \text{ into } \begin{pmatrix} 0 \\ 0 \\ 1 \end{pmatrix}, \text{ and } \begin{pmatrix} 0 \\ 0 \\ 1 \end{pmatrix} \text{ into } \begin{pmatrix} 1 \\ 0 \\ 0 \end{pmatrix};$$

so it is represented by

$$\begin{pmatrix} 0 & 0 & 1 \\ 1 & 0 & 0 \\ 0 & 1 & 0 \end{pmatrix}.$$

Similarly, (23) is represented by

$$\begin{pmatrix} 1 & 0 & 0 \\ 0 & 0 & 1 \\ 0 & 1 & 0 \end{pmatrix}.$$

Clearly, the subspace W spanned by

$$\begin{pmatrix} 1 \\ 1 \\ 1 \end{pmatrix}$$

is invariant under all six 3×3 matrices in this representation, and on this subspace we have the identity representation of S_3. To find another invariant subspace, we note that every 3×3 matrix in the representation belongs to $O(3)$ and hence preserves the ordinary Euclidean scalar product. Therefore the subspace W' orthogonal to

$$\begin{pmatrix} 1 \\ 1 \\ 1 \end{pmatrix}$$

is also invariant. A convenient orthonormal basis for $\mathbb{R}^3$ is

$$\mathbf{e}_1 = \frac{1}{3^{\frac{1}{2}}} \begin{pmatrix} 1 \\ 1 \\ 1 \end{pmatrix}, \mathbf{e}_2 = \frac{1}{2^{\frac{1}{2}}} \begin{pmatrix} 1 \\ 0 \\ -1 \end{pmatrix}, \mathbf{e}_3 = \frac{1}{6^{\frac{1}{2}}} \begin{pmatrix} -1 \\ 2 \\ -1 \end{pmatrix}.$$

Here $\mathbf{e}_1$ spans W, while $\mathbf{e}_2$ and $\mathbf{e}_3$ span W'. With respect to this basis, each 3×3 matrix is reduced to a 1×1 matrix and a 2×2 matrix. For example, the matrix which represents (123) is

$$\begin{pmatrix} \frac{1}{3^{\frac{1}{2}}} & \frac{1}{3^{\frac{1}{2}}} & \frac{1}{3^{\frac{1}{2}}} \\ \frac{1}{2^{\frac{1}{2}}} & 0 & -\frac{1}{2^{\frac{1}{2}}} \\ -\frac{1}{6^{\frac{1}{2}}} & \frac{2}{6^{\frac{1}{2}}} & -\frac{1}{6^{\frac{1}{2}}} \end{pmatrix} \begin{pmatrix} 0 & 0 & 1 \\ 1 & 0 & 0 \\ 0 & 1 & 0 \end{pmatrix} \begin{pmatrix} \frac{1}{3^{\frac{1}{2}}} & \frac{1}{2^{\frac{1}{2}}} & -\frac{1}{6^{\frac{1}{2}}} \\ \frac{1}{3^{\frac{1}{2}}} & 0 & \frac{2}{6^{\frac{1}{2}}} \\ \frac{1}{3^{\frac{1}{2}}} & -\frac{1}{2^{\frac{1}{2}}} & -\frac{1}{6^{\frac{1}{2}}} \end{pmatrix} = \left(\begin{array}{c|cc} 1 & 0 & 0 \\ \hline 0 & -\frac{1}{2} & -\frac{3^{\frac{1}{2}}}{2} \\ 0 & \frac{3^{\frac{1}{2}}}{2} & -\frac{1}{2} \end{array} \right)$$

while the matrix which represents (23) is

$$\begin{pmatrix} \frac{1}{3^{\frac{1}{2}}} & \frac{1}{3^{\frac{1}{2}}} & \frac{1}{3^{\frac{1}{2}}} \\ \frac{1}{2^{\frac{1}{2}}} & 0 & -\frac{1}{2^{\frac{1}{2}}} \\ -\frac{1}{6^{\frac{1}{2}}} & \frac{2}{6^{\frac{1}{2}}} & -\frac{1}{6^{\frac{1}{2}}} \end{pmatrix} \begin{pmatrix} 1 & 0 & 0 \\ 0 & 0 & 1 \\ 0 & 1 & 0 \end{pmatrix} \begin{pmatrix} \frac{1}{3^{\frac{1}{2}}} & \frac{1}{2^{\frac{1}{2}}} & -\frac{1}{6^{\frac{1}{2}}} \\ \frac{1}{3^{\frac{1}{2}}} & 0 & \frac{2}{6^{\frac{1}{2}}} \\ \frac{1}{3^{\frac{1}{2}}} & -\frac{1}{2^{\frac{1}{2}}} & -\frac{1}{6^{\frac{1}{2}}} \end{pmatrix} = \left(\begin{array}{c|cc} 1 & 0 & 0 \\ 0 & \frac{1}{2} & -\frac{3^{\frac{1}{2}}}{2} \\ 0 & -\frac{3^{\frac{1}{2}}}{2} & -\frac{1}{2} \end{array} \right)$$

and the matrix which represents (12) is

$$\begin{pmatrix} \frac{1}{3^{\frac{1}{2}}} & \frac{1}{3^{\frac{1}{2}}} & \frac{1}{3^{\frac{1}{2}}} \\ \frac{1}{2^{\frac{1}{2}}} & 0 & -\frac{1}{2^{\frac{1}{2}}} \\ -\frac{1}{6^{\frac{1}{2}}} & \frac{2}{6^{\frac{1}{2}}} & -\frac{1}{6^{\frac{1}{2}}} \end{pmatrix} \begin{pmatrix} 0 & 1 & 0 \\ 1 & 0 & 0 \\ 0 & 0 & 1 \end{pmatrix} \begin{pmatrix} \frac{1}{3^{\frac{1}{2}}} & \frac{1}{2^{\frac{1}{2}}} & -\frac{1}{6^{\frac{1}{2}}} \\ \frac{1}{3^{\frac{1}{2}}} & 0 & \frac{2}{6^{\frac{1}{2}}} \\ \frac{1}{3^{\frac{1}{2}}} & -\frac{1}{2^{\frac{1}{2}}} & -\frac{1}{6^{\frac{1}{2}}} \end{pmatrix} = \left(\begin{array}{c|cc} 1 & 0 & 0 \\ \hline 0 & \frac{1}{2} & \frac{3^{\frac{1}{2}}}{2} \\ 0 & \frac{3^{\frac{1}{2}}}{2} & -\frac{1}{2} \end{array} \right)$$

We have explicitly decomposed the three-dimensional permutation representation into the direct sum of the one-dimensional identity representation (on the subspace spanned by $\mathbf{e}_1$) and a two-dimensional irreducible representation on the subspace spanned by $\mathbf{e}_2$

and $\mathbf{e}_3$. In fact, with the choice of basis which we made, the two-dimensional representation matrices are exactly the ones which would have been obtained by regarding S_3 as the symmetry group of the equilateral triangle and writing down the usual 2×2 matrices to represent these rotations and reflections of the plane.

2.3 Schur's lemma

Let V_1 and V_2 be vector spaces. The space $\mathrm{Hom}\,(V_1, V_2)$ denotes the (vector) space of all linear maps from V_1 to V_2. Suppose that we have a representation r_1 of a group G on V_1 and another representation, r_2, of G on V_2. We let $\mathrm{Hom}_G(V_1, V_2)$ denote the space of linear G morphisms from V_1 to V_2. Thus, $\mathrm{Hom}_G(V_1, V_2) \subset \mathrm{Hom}\,(V_1, V_2)$ consists of those linear maps T which satisfy $r_2(a)T = Tr_1(a)$ for all a in G. We can now state the main result.

Schur's lemma

Let r^1 and r^2 be irreducible representations of the group G on the vector spaces V_1 and V_2. Let T be an element of $\mathrm{Hom}_G(V_1, V_2)$, i.e. T is a linear map from V_1 to V_2 such that

$$r^2(a)T = Tr^1(a) \quad \text{for all} \quad a \in G.$$

then
   (i) If $r^1 \not\sim r^2$, then $T = 0$, and
   (ii) If $r^1 = r^2$ (so that $V_1 = V_2$), then T is a scalar operator i.e. $T = cI$ for some scalar c, where I is the identity operator.

Proof The subspace, ker T, of V_1 is invariant, and, hence, since V_1 is irreducible, the only alternatives are ker $T = V_1$ or ker $T = \{0\}$. In the first case, $T = 0$, and there is nothing further to prove. If ker $T = \{0\}$, then $T(V_1)$ is a non-trivial invariant subspace of V_2, hence coincides with V_2, showing that T is an isomorphism. This establishes (i). If $V_1 = V_2$, we can apply (i) to the operator $T - cI$. Since we are over the complex numbers, the operator T has at least one eigenvalue. Taking c to be this eigenvalue, the operator $T - cI$ is singular, and hence must be zero, by (i). This establishes (ii).

The group G acts on $\mathrm{Hom}\,(V_1, V_2)$ by sending $S \in \mathrm{Hom}\,(V_1, V_2)$ into $r^2(a)Sr^1(a)^{-1}$. It is easy to check that this defines a linear representation of G on $\mathrm{Hom}\,(V_1, V_2)$ which we shall denote by $\mathrm{Hom}\,(r^1, r^2)$. Thus

$$[\mathrm{Hom}\,(r^1, r^2)](a)S = r^2(a)Sr^1(a)^{-1}.$$

An element T in $\mathrm{Hom}\,(V_1, V_2)$ belongs to $\mathrm{Hom}_G(V_1, V_2)$, i.e. is a G morphism if and only if $r^2(a)Tr^1(a)^{-1} = T$, that is

$$[\mathrm{Hom}\,(r^1, r^2)](a)T = T$$

for all $a \in G$.

We now combine Schur's lemma with the method of averaging over the group:

Proposition 3.1

Let r^1 and r^2 be irreducible representations of G on V_1 and V_2. Let S_0 be any element of Hom (V_1, V_2) and set

$$S = (1/\#G) \sum_{a \in G} r^2(a) S_0 r^1(a)^{-1}$$

$$= (1/\#G) \sum_{a \in G} [\text{Hom}(r^1, r^2)](a) S_0.$$

Then

(i) If $r^1 \not\sim r^2$, then $S = 0$.

(ii) If $r^1 = r^2$ (so that $V_1 = V_2$), then $S = cI$, where $c = (1/n) \text{tr } S_0$ with $n = \dim V_1$.

Proof The element, S, is invariant, and hence (i) follows from part (i) of Schur's lemma, while (ii) follows from part (ii) of Schur's lemma except for the evaluation of the constant. To evaluate c, observe that $\text{tr } S = \text{tr } S_0$. But $\text{tr } cI = nc$, finishing the proof of (ii).

If we write out the assertions of Proposition 3.1 in matrix form, we obtain the so-called orthogonality relations for the matrix elements: choose bases in V_1 and V_2 so that $r^1(a)$ has the matrix representation, $(r^1_{ij}(a))$ and $r^2(a)$ has the matrix representation $(r^2_{kl}(a))$. Suppose that S_0 has the matrix representation (s_{jk}). Then the matrix of S has, as its entries,

$$(1/\#G) \sum_{\substack{a \in G \\ l,i}} r^2_{kl}(a) s_{li} r^1_{ij}(a^{-1}).$$

In case $r^1 \not\sim r^2$, this expression is to vanish identically, no matter what the values of the s_{li} actually are. In other words, the coefficients of the s_{li} must vanish identically. Thus,

$$\text{if} \quad r^1 \not\sim r^2, \quad \text{then } (1/\#G) \sum_{a \in G} r^2_{kl}(a) r^1_{ij}(a^{-1}) = 0 \text{ for all } k, l, i, j. \tag{3.1}$$

In case $r^1 = r^2$, let us denote the matrices by r_{ij} so that (ii) of Proposition 3.1 says that

$$(1/\#G) \sum_{l,i} \sum_{a \in G} r_{kl}(a) s_{li} r_{ij}(a^{-1}) = (1/n) \left(\sum s_{ii}\right) \delta_{kj}.$$

Comparing the coefficients of s_{li} on both sides, we get

$$(1/\#G) \sum_{a \in G} r_{kl}(a) r_{ij}(a^{-1}) = (1/n) \delta_{li} \delta_{kj} = \begin{cases} (1/n) \text{ if } l = i, k = j \\ 0 \text{ otherwise.} \end{cases} \tag{3.2}$$

Equations (3.1) and (3.2) take on a somewhat more pleasant form if we restrict attention to unitary representations, for which

$$r_{ij}(a^{-1}) = \overline{r_{ji}(a)}.$$

For any two functions, f_1 and f_2 on G, let us define their scalar product by the formula

$$(f_1, f_2) = (1/\#G) \sum_{a \in G} f_1(a) \overline{f_2(a)}.$$

Then, if we assume that we have chosen bases so that the matrices (r^1_{ij}) and (r^2_{kl}) are

Table 5.

Element	Identity	Sign	Two-dimensional
e	1	1	$\begin{pmatrix} 1 & 0 \\ 0 & 1 \end{pmatrix}$
(123)	1	1	$\begin{pmatrix} -\dfrac{1}{2} & -\dfrac{3^{\frac{1}{2}}}{2} \\ \dfrac{3^{\frac{1}{2}}}{2} & -\dfrac{1}{2} \end{pmatrix}$
(132)	1	1	$\begin{pmatrix} -\dfrac{1}{2} & \dfrac{3^{\frac{1}{2}}}{2} \\ -\dfrac{3^{\frac{1}{2}}}{2} & -\dfrac{1}{2} \end{pmatrix}$
(12)	1	-1	$\begin{pmatrix} 1 & 0 \\ 0 & -1 \end{pmatrix}$
(13)	1	-1	$\begin{pmatrix} -\dfrac{1}{2} & \dfrac{3^{\frac{1}{2}}}{2} \\ \dfrac{3^{\frac{1}{2}}}{2} & \dfrac{1}{2} \end{pmatrix}$
(23)	1	-1	$\begin{pmatrix} -\dfrac{1}{2} & -\dfrac{3^{\frac{1}{2}}}{2} \\ -\dfrac{3^{\frac{1}{2}}}{2} & \dfrac{1}{2} \end{pmatrix}$

unitary, we can write (3.1) as:

$$\text{If } r^1 \not\sim r^2, \text{ then } (r^2_{kl}, r^1_{ji}) = 0 \text{ for all } i, j, k \text{ and } l. \tag{3.3}$$

In other words, matrix entries from two inequivalent representations are orthogonal. We can write (3.2) as

$$(r^2_{kl}, r^1_{ij}) = (1/n)\delta_{ki}\delta_{lj} \tag{3.4}$$

so that distinct matrix entries from the same irreducible unitary representation are orthogonal, and each matrix entry has length $1/n^{\frac{1}{2}}$.

As an explicit example of these orthogonality relations, we write out the three representations of S_3: the identity representation, the 'sign' representation with $+1$ for even permutations, -1 for odd, and the 2×2 representations as rotations and reflections of an equilateral triangle (see Table 5).

You can check that the identity and sign representations are orthogonal to the entries in each position of the 2×2 matrices, and that these matrix entries are orthogonal to one another.

2.4   Characters and their orthogonality relations

Let r be a representation of the group G on the vector space V. The dimension of V is called the *degree* of the representation, r. The *character* of the representation r is the function χ^r defined on G by the formula

$$\chi^r(a) = \operatorname{tr} r(a) = \sum_i r_{ii}(a). \tag{4.1}$$

If we take $a = e$, so that $r(e)$ is the identity operator, whose trace is dim V, we see that

$$\chi^r(e) = \dim V. \tag{4.2}$$

For any two linear transformations we have $\operatorname{tr} AB = \operatorname{tr} BA$; so, if B is non-singular, $\operatorname{tr} BAB^{-1} = \operatorname{tr} A$. Thus,

$$\chi(bab^{-1}) = \chi(a) \tag{4.3}$$

if χ is the character of any representation. In other words, χ is a function which is constant on conjugacy classes. Such a function is called a *central* function.

For any representation r, we can introduce a Hermitian scalar product which is invariant under $r(a)$ for all $a \in G$. This means that if we take adjoints with respect to this scalar product, we have $r(a)^* = r(a^{-1})$. But $\operatorname{tr} r(a)^*$ is the complex conjugate of $\operatorname{tr} r(a)$, so

$$\chi(a^{-1}) = \overline{\chi(a)}. \tag{4.4}$$

Let r^1 and r^2 be representations of G. Then it follows from the matrix form of $r^1 \oplus r^2$ that

$$\chi^{r^1 \oplus r^2} = \chi^{r^1} + \chi^{r^2}. \tag{4.5}$$

The character χ^1 of the representation r^1 is given by

$$\chi^1 = \sum_i r_{ii}^1$$

and similarly for the character χ^2 of r^2. It now follows from (3.3) and (3.2) that

$$\text{if } r^1 \not\sim r^2, \quad \text{then } (\chi^1, \chi^2) = 0, \tag{4.6}$$

and

$$(\chi, \chi) = 1, \text{ if } \chi \text{ is the character of an irreducible representation.} \tag{4.7}$$

Now let r be a representation of G on a vector space, V, which is not necessarily irreducible, and let

$$r = r^1 \oplus \cdots \oplus r^k$$

be a decomposition of r into irreducible representations. Let ϕ be the character of r, and let χ_i be the character of r^i, so that

$$\phi = \chi_1 + \cdots + \chi_k.$$

Let s be some particular irreducible representation of G and let χ be its character. Then

$$(\phi, \chi) = (\chi_1, \chi) + \cdots + (\chi_k, \chi).$$

The terms on the right are all zero or one, according as $r^i \nsim r$ or $r^i \sim r$. Thus,

> (ϕ, χ) is the number of terms in the decomposition of r which are   (4.8)
> isomorphic to s. In particular, this number does not depend on the
> particular choice of decomposition.

From (4.8) it follows that *two representations with the same character are equivalent.* Indeed, by taking scalar products with the characters of all the irreducible representations, we can determine how many times each irreducible occurs in a decomposition of the given representation.

Notice that any character ϕ can be written as

$$\phi = m_1 \chi_1 + \cdots + m_p \chi_p,$$

where the χ_i are orthogonal characters and the m_i are non-negative integers. It follows that

$$(\phi, \phi) = m_1^2 + \cdots + m_p^2 \tag{4.9}$$

and, in particular,

$$\phi \text{ is irreducible if and only if } (\phi, \phi) = 1.$$

Let ϕ be the character of a representation of G on a vector space W, and let χ be the character of an irreducible representation of G on the vector space V. If we decompose

$$W = W_1 \oplus \cdots \oplus W_k$$

into irreducibles, we see that

$$\mathrm{Hom}_G(W, V) = \mathrm{Hom}_G(W_1, V) \oplus \cdots \oplus \mathrm{Hom}_G(W_k, V).$$

By Schur's lemma, each of these spaces is either one dimensional or zero dimensional according to whether the representation of G on W_i is or is not equivalent to the representation of G on V. Combining this with (4.8) we see that

$$(\phi, \chi) = \dim \mathrm{Hom}_G(W, V). \tag{4.10}$$

Now let r_u and r_v be representations of G on U and V. We do not assume that r_u and r_v are irreducible. We wish to compute $\dim \mathrm{Hom}_G(U, V)$.

Let us first consider a special case. Suppose $U = V = W \oplus W$, where W is irreducible. We can write any vector in U as $(\mathbf{w}_1, \mathbf{w}_2)$, where $\mathbf{w}_1$ and $\mathbf{w}_2$ are in W. Thus, for any $T \in \mathrm{Hom}(V, V)$ we have

$$T(\mathbf{w}_1, \mathbf{w}_2) = (T_{11}\mathbf{w}_1 + T_{12}\mathbf{w}_2, T_{21}\mathbf{w}_1 + T_{22}\mathbf{w}_2)$$

where $T_{ij} \in \mathrm{Hom}\,(W, W)$. So

$$T \circ r_{W \oplus W}(a)(\mathbf{w}_1, \mathbf{w}_2) = T(r_W(a)\mathbf{w}_1, r_W(a)\mathbf{w}_2)$$
$$= (T_{11} r_W(a)\mathbf{w}_1 + T_{12} r_W(a)\mathbf{w}_2, T_{21} r_W(a)\mathbf{w}_1 + T_{22} r_W(a)\mathbf{w}_2)$$

while

$$r_{W \oplus W}(a) T(\mathbf{w}_1, \mathbf{w}_2)$$
$$= (r_W(a)(T_{11}\mathbf{w}_1 + T_{12}\mathbf{w}_2), r_W(a)(T_{21}\mathbf{w}_1 + T_{22}\mathbf{w}_2))$$
$$= (r_W(a) T_{11}\mathbf{w}_1 + r_W(a) T_{12}\mathbf{w}_2, r_W(a) T_{21}\mathbf{w}_1 + r_W(a) T_{22}\mathbf{w}_2).$$

So $T \in \mathrm{Hom}_G(V, V)$ if and only if each $T_{ij} \in \mathrm{Hom}_G(W, W)$. By Schur's lemma, each T_{ij} ranges over a one-dimensional space, hence $\dim \mathrm{Hom}_G(W \oplus W, W \oplus W) = 4 = 2 \times 2$. For any representation, we may make the decomposition

$$U = (U_1 \oplus \cdots \oplus U_{p_1}) \oplus (U_{p_1 + 1} \oplus \cdots \oplus U_{p_1 + p_2}) \oplus \cdots (\cdots U_{p_1 + \cdots p_k})$$

where the first p_1 spaces are all equivalent to the irreducible representation W_1, the next p_2 spaces are all equivalent to the irreducible representation W_2 etc., and $W_1, \ldots, W_k$ are *inequivalent* irreducible representations of G. We may make the same decomposition

$$V = (V_1 \oplus \cdots \oplus V_{q_1}) \oplus (V_{q_1 + 1} \oplus \cdots \oplus V_{q_1 + q_2}) \oplus \cdots$$

for V. By Schur's lemma, any $T \in \mathrm{Hom}_G(U, V)$ when applied to any $\mathbf{u} \in U_1 \oplus \cdots \oplus U_{p_1}$ must give $T\mathbf{u}$ lying in $V_1 \oplus \cdots \oplus V_{q_1}$. Then the same argument as in the special case shows that

$$\dim \mathrm{Hom}_G(U, V) = p_1 q_1 + p_2 q_2 + \cdots + p_k q_k. \tag{4.11}$$

In particular, if $U = V$,

$$\dim \mathrm{Hom}_G(V, V) = p_1^2 + \cdots + p_k^2,$$

where p_i is the number of times that the ith irreducible representation occurs in V.

2.5   Action on function spaces

Suppose that we are given an action of the group G on the set M. Let $\mathscr{F}(M)$ denote the vector space of all complex-valued functions on M. Define an action of G on $\mathscr{F}(M)$ by

$$(af)(x) = f(a^{-1}x).$$

Put another way, we define af by

$$af = f \circ a^{-1},$$

where, on the right, we are considering the element a^{-1} as a transformation of M into itself. To verify that this is indeed a group action, observe that

$$bf = f \circ b^{-1}$$
$$\text{so } a(bf) = (bf) \circ a^{-1} = (f \circ b^{-1}) \circ a^{-1} = f \circ (b^{-1} \circ a^{-1}) = f \circ (ab)^{-1} = (ab)f.$$

We denote the representation defined by this action of G on $F(M)$ by r^M. So we may write

$$(af)(x) = [r^M(a)f](x) = f(a^{-1}x).$$

The representation r^M is not irreducible, unless M consists of a single element, because the constant functions on M always form a one-dimensional invariant subspace, on which G acts by the trivial (identity) representation.

As an illustration, we construct representations of D_4 by letting the group act on the space of functions on the vertices of the square. As a convenient notation, we denote each function by a diagram showing its value on each vertex. Thus, for example,

$$\begin{matrix} 4 & 3 \\ - & \cdot \\ + & - \\ 1 & 2 \end{matrix}$$

is the function with $f(1) = 1$, $f(2) = -1$, $f(3) = 0$, $f(4) = -1$. Then the action of a group element on a function may be determined simply by letting the element act on the diagram which represents the function. For example, if a is a 90° counterclockwise rotation, then the rotated diagram

$$\begin{matrix} 4 & 3 \\ \cdot & - \\ - & + \\ 1 & 2 \end{matrix}$$

correctly describes af. The diagram shows that $af(2) = +1$; and we check that $af(2) = f(a^{-1}2) = f(1) = +1$.

The space of functions on the vertices of the square is clearly four dimensional. One invariant subspace is spanned by

$$\begin{matrix} + & + \\ + & + \end{matrix}$$

the constant function; this gives rise to the trivial representation. Another invariant subspace is spanned by

$$\begin{matrix} - & + \\ + & - \end{matrix}$$

Clearly this gives rise to a representation in which the identity, the 180° rotation in the plane of the page, and rotations about the diagonals are represented by $+1$, while the 90° and 270° rotations and the rotations about lines parallel to the sides of the square are represented by -1. Finally, there is a two-dimensional invariant subspace, spanned by

$$f_1 = \begin{matrix} + & + \\ - & - \end{matrix} \quad \text{and} \quad f_2 = \begin{matrix} + & - \\ + & - \end{matrix}$$

The 90° rotation carries f_1 into f_2, f_2 into $-f_1$, and so is represented by $\begin{pmatrix} 0 & -1 \\ 1 & 0 \end{pmatrix}$.

Similarly, the reflection about the principal diagonal carries f_1 into f_2, f_2 into f_1, and is represented by $\begin{pmatrix} 0 & 1 \\ 1 & 0 \end{pmatrix}$. Thus, we obtain the two-dimensional representation of D_4. Thus the four-dimensional space of functions in the vertices of the square breaks up into two one-dimensional and one two-dimensional representations.

There are two one-dimensional representations of D_4 which are *not* obtained by this procedure. One is the representation in which $+1$ is assigned to the four elements in the C_4 subgroup, -1 to the other four elements. The last is a representation in which $+1$ is assigned to the elements of the D_2 subgroup: the identity, and the $180°$ rotations about the x, y and z axes, while -1 is assigned to the $90°$ and $270°$ rotations and the $180°$ rotations about the diagonals. In the next section we shall see that when we let a group act on functions on the group itself, the resulting representation contains *all* irreducible representations in its decomposition.

Suppose that M decomposes into orbits under the action of G:

$$M = M_1 \cup \cdots \cup M_k.$$

Then we have a corresponding decomposition

$$r^M = r^{M_1} \oplus \cdots \oplus r^{M_k}.$$

Indeed, we can identify r^{M_i} with the subrepresentation of r^M given by the action of G on functions which vanish outside of M_i. Any function on M can be written uniquely as $f = f_1 + \cdots + f_k$, where each f_j vanishes outside of M_i. Let us compute the character of the representation r^M. For this purpose we introduce a convenient basis into the space $\mathbb{C}^M$: Let δ_x be the function on M defined by

$$\delta_x(y) = \begin{array}{ll} 1 & \text{if} \quad y = x \\ 0 & \text{if} \quad y \neq x \end{array}.$$

Then

$$(a\delta_x)(y) = \delta_x(a^{-1}y) = \begin{array}{ll} 1 & \text{if} \quad a^{-1}y = x, \text{ i.e. } y = ax \\ 0 & \text{if} \quad a^{-1}y \neq x, \text{ i.e. } y \neq ax \end{array}$$

so that

$$a\delta_x = \delta_{ax}.$$

The functions δ_x are clearly independent and they span $\mathscr{F}(M)$ since any function, f, can be written as $f = \sum_{x \in M} f(x)\delta_x$. For this basis, the diagonal elements of $r^M(a)$ will be one or zero according as $ax = x$ or $ax \neq x$. Thus

$$\chi^{r^M}(a) = \sum_{ax = x} 1 = \#(\text{fixed points of } a)$$

$$= \#FP(a). \tag{5.1}$$

This formula is the prototype of all character formulas. In fact, all the character formulas that we shall present will, in a sense, be generalizations of (5.1) in one form or another. We shall return to this point later.

Suppose that G acts on the two finite sets M and N. We wish to study the space

$\mathrm{Hom}\,(\mathscr{F}(M),\mathscr{F}(N))$ and the action of G on it. Notice that

$$\dim \mathrm{Hom}\,(\mathscr{F}(M),\mathscr{F}(N)) = \dim \mathscr{F}(M) \times \dim \mathscr{F}(N)$$
$$= (\#M)\cdot(\#N)$$
$$= \#(N \times M)$$
$$= \dim \mathscr{F}(N \times M).$$

We claim that there is a natural isomorphism between $\mathscr{F}(N \times M)$ and $\mathrm{Hom}\,(\mathscr{F}(M),\mathscr{F}(N))$. Indeed, given any function K on $N \times M$ define the operator $T_K:\mathscr{F}(M) \to \mathscr{F}(N)$ by

$$(T_K f)(y) = \sum_{x \in M} K(y,x) f(x).$$

The map sending K into T_K is one-to-one; indeed, for any $u \in M$

$$(T_K \delta_u)(y) = K(y,u)$$

so if $T_K = 0$, then $K(y,u) = 0$ for all y and u, i.e. $K = 0$. Since $\mathscr{F}(N \times M)$ and $\mathrm{Hom}\,(\mathscr{F}(M),\mathscr{F}(N))$ have the same dimension, we conclude that the map sending K to T_K is an isomorphism of vector spaces.

The group G acts on both $\mathrm{Hom}\,(\mathscr{F}(M),\mathscr{F}(N))$ and on $\mathscr{F}(N \times M)$. We claim that the isomorphism just described is a G morphism, i.e. that the representations of G on these two spaces are equivalent. Indeed, letting r^{Hom} denote the representation of G on $\mathrm{Hom}\,(\mathscr{F}(M),\mathscr{F}(N))$, we have

$$r^{\mathrm{Hom}}(a)T_K = r^N(a)T_K r^M(a)^{-1}.$$

Now

$$r^M(a)^{-1}f(x) = f(ax)$$

so

$$(T_K r^M(a)^{-1}f)(y) = \sum K(y,x)f(ax)$$

and

$$(r^{\mathrm{Hom}}(a)T_K f)(y) = \sum K(a^{-1}y,x)f(ax)$$
$$= \sum K(a^{-1}y,a^{-1}x)f(x).$$

But

$$(r^{N \times M}(a)K)(y,x) = K(a^{-1}y,a^{-1}x)$$

So

$$r^{\mathrm{Hom}}(a)T_K = T_{r^{N \times M}(a)K}$$

which was to be proved.

Now $\mathrm{Hom}_G(\mathscr{F}(M),\mathscr{F}(N))$ is the space of elements in $\mathrm{Hom}\,(\mathscr{F}(M),\mathscr{F}(N))$ which satisfy

$$r^{\mathrm{Hom}}(a)T = T$$

for all $a \in G$. If $T = T_K$, the preceding equation shows that

$$K(a^{-1}y,a^{-1}x) = K(y,x)$$

for all $a \in G$. In other words, the function K must be constant on orbits of G on $N \times M$. Thus

$$\dim \operatorname{Hom}_G(\mathscr{F}(M), \mathscr{F}(N)) = \# \text{ of } G \text{ orbits on } N \times M. \qquad (5.2)$$

In particular, on taking $M = N$ we see that

$$\dim \operatorname{Hom}_G(\mathscr{F}(M), \mathscr{F}(M)) = \# \text{ of } G \text{ orbits on } M \times M$$
$$= p_1^2 + \cdots + p_k^2 \qquad (5.3)$$

where p_i is the number of times that the ith irreducible representation of G occurs in $\mathscr{F}(M)$.

Example

Consider the group S_n acting on the n-element set $M = \{1, \ldots, n\}$. On $M \times M$ there are two orbits

$$\{(x, y) | x \neq y\} \quad \text{and} \quad \{(x, x)\}.$$

Indeed, if $x \neq y$ and $z \neq w$ we can find a permutation σ such that $\sigma(x) = z$ and $\sigma(y) = w$ Thus, the set $\{(x, y) | x \neq y\}$ is a single orbit in $M \times M$. Similarly the set $\{(x, x)\}$ is a single orbit. Thus,

$$\dim \operatorname{Hom}_G(\mathscr{F}(M), \mathscr{F}(M)) = 2 = p_1^2 + \cdots + p_k^2$$

so $k = 2$ and $p_1 = p_2 = 1$. Thus, $\mathscr{F}(M)$ is the direct sum of two irreducible representations. We already know one of them – the trivial one-dimensional representation, corresponding to the constant functions. The other must then be $n - 1$ dimensional. Thus

$$\mathscr{F}(M) = \quad V_1 \quad + \quad V_2$$
$$\uparrow \qquad\quad \uparrow$$
$$\text{one} \qquad n - 1$$
$$\text{dimensional} \quad \text{dimensional}$$

2.6   The regular representation

We apply the results of the preceding section to the special case where $M = G$ and G acts on itself by left multiplication. The corresponding representation, r^G, of G on $\mathscr{F}(G)$ is called the *regular* representation. It is defined by $[r^G(a)f](b) = f(a^{-1}b)$. We have

$$\#G = \dim \mathscr{F}(G) = \sum p_i n_i$$

where p_i is the number of times that the ith irreducible representation occurs in $\mathscr{F}(G)$, while n_i is the dimension of the ith irreducible representation. Also

$$\dim \operatorname{Hom}_G(\mathscr{F}(G), \mathscr{F}(G)) = \sum p_i^2$$

$$= \# \text{ of } G \text{ orbits on } G \times G.$$

We compute the number of orbits as follows: we can always act on (a, b) by a^{-1} to get $(e, a^{-1}b)$. Thus, each orbit of G in $G \times G$ contains a point of the form (e, c). But this is the only element of this form in its orbit, since multiplying by d sends (e, c) into (d, dc). Thus each orbit contains a unique representative of the form (e, c), and hence the number of orbits is equal to $\#G$. Thus

$$\#G = \sum p_i^2.$$

Since $\sum p_i^2 = \sum p_i n_i$ we are led to guess that $p_i = n_i$, i.e. that each irreducible representation, W, occurs in $\mathscr{F}(G)$ with a multiplicity equal to its dimension, i.e. that

$$\dim \operatorname{Hom}_G(W, \mathscr{F}(G)) = \dim W. \tag{6.1}$$

We shall prove this fact by constructing an isomorphism between W^*, the dual space of W, and $\operatorname{Hom}_G(W, \mathscr{F}(G))$. To each $l \in W^*$ and to each $\mathbf{w} \in W$ we assign the function $f_{\mathbf{w}}^l$ on G defined by

$$f_{\mathbf{w}}^l(a) = \langle r(a^{-1})\mathbf{w}, l \rangle.$$

Here $r(a^{-1})\mathbf{w} \in W$ and $l \in W^*$, and $\langle \mathbf{v}, l \rangle$ denotes the value of the linear function $l \in W^*$ at the element $\mathbf{v}$ of V. For fixed l the map sending $\mathbf{w}$ into $f_{\mathbf{w}}^l$ is a map from W to $\mathscr{F}(G)$. Thus each $l \in V^*$ defines an element of $\operatorname{Hom}(W, \mathscr{F}(G))$. We must show that this element lies in $\operatorname{Hom}_G(W, \mathscr{F}(G))$, i.e. that

$$f_{r(b)\mathbf{w}}^l = r^G(b) f_{\mathbf{w}}^l$$

or that

$$f_{r(b)\mathbf{w}}^l(a) = f_{\mathbf{w}}^l(b^{-1}a) \quad \text{for all } a, b \in G.$$

But

$$\begin{aligned} f_{r(b)\mathbf{w}}^l(a) &= \langle r(a)^{-1}r(b)\mathbf{w}, l \rangle \\ &= \langle r(a^{-1}b)\mathbf{w}, l \rangle \\ &= \langle r(b^{-1}a)^{-1}\mathbf{w}, l \rangle \\ &= f_{\mathbf{w}}^l(b^{-1}a) \end{aligned}$$

as required.

Furthermore, $f_{\mathbf{w}}^l(e) = \langle \mathbf{w}, l \rangle$ cannot be zero for all $\mathbf{w}$ unless $l = 0$. Thus the map of W^* into $\operatorname{Hom}_G(W, \mathscr{F}(G))$ that we have defined is injective. It follows that

$$p_i = \dim \operatorname{Hom}_G(W, \mathscr{F}(G)) \geq \dim W^* = n_i.$$

But it now follows from the two equations

$$\#G = \sum p_i n_i = \sum p_i^2$$

that we must have $p_i = n_i$ so (6.1) holds. Thus

$$\#G = \sum n_i^2. \tag{6.2}$$

Equations (6.1) and (6.2) have the following useful corollary. Suppose that we have found inequivalent irreducible representations $(r_1, W_1) \cdots (r_k, W_k)$ of G, with $\dim W_i = n_i$, such that $\sum_i^d n_i^2 = \#G$. Then it follows from (6.2) that there can be no other irreducible representation (i.e. ones not equivalent to the ones we already have). In other words, we have found, up to equivalence, all the irreducible representations of G. This justifies the conclusions that we drew in the examples of Section 1.

We can extract some further useful information from the proof of (6.1), but for this we must introduce some new notions. Let G and H be groups. Their direct product $G \times H$ consists of all pairs (a, b) with the multiplication law

$$(a, b)(c, d) = (ac, bd).$$

Suppose that (r, U) is a representation of G and (s, V) is a representation of H. We can form the tensor product, $U \otimes V$, of the two vector spaces U and V[†]. Recall from the theory of tensor products that if $A \in \text{Hom}(U, U)$ and $B \in \text{Hom}(V, V)$, then there is a unique transformation $A \otimes B$ on $U \otimes V$ such that

$$(A \otimes B)(\mathbf{u} \otimes \mathbf{v}) = A\mathbf{u} \otimes B\mathbf{v}.$$

Also

$$\text{tr}(A \otimes B) = (\text{tr } A) \cdot (\text{tr } B).$$

Furthermore, if $C \in \text{Hom}(U, U)$ and $D \in \text{Hom}(V, V)$, then

$$(A \otimes B)(C \otimes D) = AC \otimes BD.$$

This shows that we get a representation $r \otimes s$ of $G \times H$ on $U \otimes V$ by setting

$$(r \otimes s)(a, b) = r(a) \otimes s(b),$$

and that

$$\chi^{r \otimes s}(a, b) = \chi^r(a)\chi^s(b).$$

If $(\ ,\ )_{G \times H}$ denotes the scalar product on $G \times H$, and $\|\ \ \|_{G \times H}$ the corresponding norm, then

$$\|\chi^{r \otimes s}\|_{G \times H}^2 = (\chi^{r \otimes s}, \chi^{r \otimes s})_{G \times H}$$

$$= \frac{1}{\#(G \times H)} \sum_{\substack{a \in G \\ b \in H}} \chi^{r \otimes s}(a, b)\overline{\chi^{r \otimes s}(a, b)}$$

$$= \frac{1}{\#G} \cdot \frac{1}{\#H}\left(\sum_{a \in G} \chi^r(a)\overline{\chi^r(a)}\right)\left(\sum_{b \in H} \chi^s(b)\overline{\chi^s(b)}\right)$$

$$= \|\chi^r\|_G^2 \|\chi^s\|_H^2.$$

In particular, if $\|\chi^r\|_G^2 = \|\chi^s\|_H^2 = 1$, then $\|\chi^{r \otimes s}\|_{G \times H}^2 = 1$. Thus,

> If r is an irreducible representation of G and s is an irreducible representation of H, then $r \otimes s$ is an irreducible representation of $G \times H$.

Let r be a representation of G on W. We can construct a representation $\hat{r}$ of G on the dual space W^* by defining

$$\hat{r}(a)l = r(a)^{*-1}l.$$

[†] See Appendix B for a summary and presentation of the basic facts about tensor products.

This is a representation because

$$r(ab)^{*-1} = (r(a)r(b))^{*-1}$$
$$= (r(b)^*r(a)^*)^{-1}$$
$$= r(a)^{*-1}r(b)^{*-1}$$
$$= \hat{r}(a)\hat{r}(b).$$

We thus get a representation of $G \times G$ on $W \otimes W^*$. If the representation of G on W is irreducible, then so is the representation of $G \times G$ on $W \otimes W^*$.

Now let us go back to the proof of (6.1). We have seen how to attach a function $f_{\mathbf{w}}^l$ on G to each pair $(\mathbf{w}, l)$ with $\mathbf{w} \in W$ and $l \in W^*$. Since $f_{\mathbf{w}}^l$ depends linearly on $\mathbf{w}$ for l fixed, and linearly on l for $\mathbf{w}$ fixed, we have thus defined a map

$$W \otimes W^* \to \mathscr{F}(G)$$
$$w \otimes l \rightsquigarrow f_{\mathbf{w}}^l.$$

Now the group $G \times G$ acts on G by right and left multiplication:

$$(a, b)c = acb^{-1}$$

and hence we get a corresponding representation $\hat{r}^G$ on $\mathscr{F}(G)$

$$[\hat{r}^G(a, b)f](c) = f(a^{-1}cb).$$

Notice that

$$f_{r(a)\mathbf{w}}^{\hat{r}(b)l}(c) = \langle r(c)^{-1}r(a)\mathbf{w}, r(b)^{*-1}l \rangle$$
$$= \langle r(b)^{-1}r(c)^{-1}r(a)\mathbf{w}, l \rangle$$
$$= \langle r(a^{-1}cb)^{-1}\mathbf{w}, l \rangle$$
$$= f_{\mathbf{w}}^l(a^{-1}cb).$$

In other words, the map from $W \otimes W^*$ to $\mathscr{F}(G)$ is a morphism for the action of $G \times G$.

Now decompose $\mathscr{F}(G)$ into irreducibles under the action of $G \times G$: For each irreducible representation W_i of G, we know that $W_i \otimes W_i^*$ occurs as an irreducible component under $G \times G$ on $\mathscr{F}(G)$. Under G, the space $W_i \otimes W_i^*$ decomposes into a direct sum of n_i copies of W_i. In particular, no $W_i \otimes W_i^*$ has any component in common with $W_j \otimes W_j^*$ for $i \neq j$. A fortiori, $W_i \otimes W_i^*$ and $W_j \otimes W_j^*$ are inequivalent as irreducible representations of $G \times G$. Thus $W_1 \otimes W_1^* \oplus \cdots \oplus W_k \otimes W_k^*$ occurs as a summand of $\mathscr{F}(G)$, where $W_1, \ldots, W_k$ are all the irreducible representations of G. But the dimension of this summand is $\sum n_i^2 = \dim \mathscr{F}(G)$. Thus

$$\mathscr{F}(G) = W_1 \otimes W_1^* \oplus \cdots \oplus W_k \otimes W_k^*$$

gives the decomposition of $\mathscr{F}(G)$ into irreducibles of $G \times G$, where each summand occurs once.

Thus

$$\dim \mathrm{Hom}_{G \times G}(\mathscr{F}(G), \mathscr{F}(G)) = \underbrace{1^2 + \cdots + 1^2}_{k \text{ times}} = k.$$

We know that this dimension must equal the number of orbits of $G \times G$ acting on

$G \times G$ by the rule

$$(a, b)(c, d) = (acb^{-1}, adb^{-1}).$$

As before, we can always find an element of the form (e, d) on any orbit. But now $(a, b)(e, d) = (ab^{-1}, adb^{-1})$ will have the same form if $b = a$. Thus (e, d) and (e, ada^{-1}) lie on the same orbit, and hence

> the number of orbits of $G \times G$ on $G \times G$ is equal to the number of conjugacy classes of G.

Thus $\dim \operatorname{Hom}_{G \times G}(\mathscr{F}(G), \mathscr{F}(G)) = k = \#$ of conjugacy classes.

We have proved that

> the number of distinct irreducible representations is equal to the   (6.3)
> number of conjugacy classes.

Let C denote the space of functions which are constant on conjugacy classes, and let $\chi_1, \ldots, \chi_k$ be the distinct irreducible characters. We already know that the functions $\chi_i \in C$ are mutually orthogonal and have length one. Since $k = \#$ of conjugacy classes $= \dim C$, they form an orthonormal basis of C. Any $f \in C$ can be expanded in terms of the basis $\chi_1, \ldots, \chi_p$:

$$f = (f, \chi_1)\chi_1 + \cdots + (f, \chi_p)\chi_p.$$

Let us apply this formula to the function f_j, which equals one on the jth conjugacy class and vanishes on all the others. Then $(f, \chi_i) = (\#C_j/\#G)\overline{\chi_i(j)}$, where $\#C_j$ is the number of elements in the jth conjugacy class, C_j, and $\chi_i(j)$ is the value of χ_i on any element of this class. Substituting into the above formula and evaluating at a point in the jth conjugacy class, C_j, we get

$$1 = (\#C_j/\#G)(\chi_1(j)\overline{\chi_1(j)} + \cdots + \chi_p(j)\overline{\chi_p(j)}).   \tag{6.4}$$

Evaluating at a different conjugacy class gives

$$0 = \chi_1(k)\overline{\chi_1(j)} + \cdots + \chi_p(k)\overline{\chi_p(j)}   \quad \text{if } j \neq k.   \tag{6.5}$$

There is one further useful piece of information that we can extract from the above analysis. We have proved that for any irreducible W, the map of $W \otimes W^* \to \mathscr{F}(G)$ sending $u \otimes l \to f_u^l \in \mathscr{F}(G)$ is injective. This means that the transpose of this map is surjective. Let us write down what this transpose is. Let us use the trace on $W \otimes W^* = \operatorname{Hom}(W, W)$ to identify any $A \in W \otimes W^*$ with a linear function on $\operatorname{Hom}(W, W)$. So we think of A as the linear function which assigns to every B the number $\operatorname{tr} AB$. For each function f on $\mathscr{F}(G)$, and for a representation, ρ, of G on W, let us define $\hat{\rho}(f) \in \operatorname{Hom}(W, W)$ by

$$\hat{\rho}(f)u = \sum_{a \in G} f(a)\rho(a)u$$

or, more symbolically,

$$\hat{\rho}(f) = \sum_{a \in G} f(a)\rho(a).$$

Then

$$\operatorname{tr}\hat{\rho}(f)(u\otimes l)=\sum f(a)\operatorname{tr}\rho(a)(u\otimes l)$$
$$=\sum f(a)\langle au,l\rangle$$
$$=\sum f(a)f_u^l(a^{-1}).$$

We may identify $\mathscr{F}(G)$ with $\mathscr{F}(G)^*$ if we consider $h\in\mathscr{F}(G)$ as a linear function on $\mathscr{F}(G)$ sending f into $\sum f(a)h(a^{-1})$. Then the above equation says

$\hat{\rho}:\mathscr{F}(G)\to W\otimes W^*$ is the transpose of the map $W\otimes W^*\to\mathscr{F}(G)$

which we know to be injective. But this means that $\hat{\rho}$ is surjective. But the image of $\hat{\rho}$ consists of linear combinations of the $\rho(a)$. We have thus proved the important

if (ρ,W) is an irreducible representation, then the $\rho(a),a\in G$ span all of   (6.6)
Hom (W,W).

In fact, we have proved more: let $W_1,\ldots,W_k$ be any family of inequivalent irreducible representations. Then the map of $W_1\otimes W_1^*\oplus\cdots\oplus W_k\otimes W_k^*\to\mathscr{F}(G)$ is injective, so the transpose is surjective. This means that

given $T_1\in\operatorname{Hom}(W_1,W_1),T_2\in\operatorname{Hom}(W_2,W_2),\ldots,T_k\in\operatorname{Hom}(W_k,W_k)$,   (6.7)
there is an $f\in\mathscr{F}(G)$ such that

$$\hat{\rho}_1(f)=T_1,\ldots,\hat{\rho}_k(f)=T_k.$$

2.7   Character tables

Let $\chi_1,\ldots,\chi_p$ be the distinct irreducible characters of the group G, and let $C_1,\ldots,C_p$ denote the distinct conjugacy classes. We denote by $\chi_i(j)$ the (constant) value of the character χ_i on any element of the conjugacy class C_j. Then

$$(\chi_i,\chi_k)=(1/\#G)\sum_{a\in G}\chi_i(a)\overline{\chi_k(a)}=(1/\#G)\sum_{j=1}^{p}(\#C_j)\chi_i(j)\overline{\chi_k(j)}.$$

We can thus write the orthogonality relations (4.6) and (4.7) as

$$(\#C_1)\chi_i(1)\overline{\chi_k(1)}+\cdots+(\#C_p)\chi_i(p)\overline{\chi_k(p)}=\begin{cases}\#G & \text{if}\quad i=k\\ 0 & \text{if}\quad i\neq k\end{cases}.$$ (7.1)

We can write the orthogonality relations (6.4) and (6.5) as

$$(\#C_j)[\chi_1(j)\overline{\chi_1(l)}+\cdots+\chi_p(j)\overline{\chi_p(l)}]=\begin{cases}\#G & \text{if}\quad j=l\\ 0 & j\neq l\end{cases}.$$ (7.2)

Equations (7.1) and (7.2) can be summarized in the form of a table. We label the columns by the conjugacy classes, indicating, alongside C_j, the number of its elements $\#C_j$. We label the rows by the characters χ_i, and place the value $\chi_i(j)$ in the i,j position. Then (7.1)

Table 6.

$6S_3$	$1C_1$	$3C_2$	$2C_3$
χ_1	1	1	1
χ_2	2	0	-1
χ_3	1	-1	1

says that the 'scalar product' of two distinct rows is zero, and of a row with itself is $\#G$, provided that we weight the jth column by $\#C_j$. Similarly, (7.2) says the same thing about the scalar product of the columns, again weighting the columns by $\#C_j$. The table so obtained is called the character table of the group. In a sense, it contains all the information about the representations of the group. The first conjugacy class, C_1, is usually taken to be the one-element conjugacy class $[e]$. Thus, the elements of the first column consist of the values $\chi_i(e) = n_i$, the degree of the ith irreducible representation. The character χ_1 is usually taken to be the trivial representation, so that the entries of the first row are all ones.

As a first illustration, let us work out the character table for the symmetric group S_4. In the case, $\#S_3 = 6$. There are three conjugacy classes: $C_1 = e$ with $\#C_1 = 1$; C_2, consisting of all two cycles (reflections), with $\#C_2 = 3$; and C_3, consisting of all three cycles (rotations through 120° or 240°), with $\#C_3 = 2$. We let χ_1 be the trivial representation. For our second representation, we take the representation of S_3 on the plane, coming from the action of S_3 as symmetries of an equilateral triangle. We have seen that this representation is irreducible. The degree of this representation is two. The elements of C_2 act as reflections about a line, and so have trace zero. Thus $\chi_2(2) = 0$. The elements of C_3 act as rotations through angle $\pm 2\pi/3$, and hence have trace $2\cos(2\pi/3) = -1$. Thus the entries of the second row are $2, 0, -1$. Since $1^2 + 2^2 = 5$, we know from (6.2) that the one remaining is one dimensional and is, of course, the sign representation. Thus $\chi_3(1) = 1, \chi_3(2) = -1$ and $\chi_3(3) = 1$. The character table is as in Table 6.

Equation (7.1) for $i = k$ says

$$1^2 + 3\cdot 1^2 + 2\cdot 1^2 = 6$$

$$2^2 + 3\cdot 0^2 + 2(-1)^2 = 6$$

and

$$1^2 + 3\cdot(-1)^2 + 2\cdot 1^2 = 6.$$

For $i = 1$ and $k = 3$, equation (7.1) says

$$1\cdot 1 + 3\cdot 1\cdot(-1) + 2\cdot 1\cdot 1 = 0.$$

Equation (7.2) for $j = l = 1$ reduces to (6.2). This is true for any character table. For $j = 2, l = 3$ it says

$$3[1\cdot 1 + 0\cdot(-1) + (-1)\cdot 1] = 0.$$

Table 7.

nC_n	$1[e]$	$1[a]$	$1[a^2]$	$\cdots$	$1[a^{n-1}]$
χ_1	1	1	1	$\cdots$	1
χ_2	1	ε	ε^2	$\cdots$	ε^{n-1}
χ_3	1	ε^2	ε^4	$\cdots$	$\varepsilon^{2(n-1)}$
.	.	.	.	$\cdots$	.
.	.	.	.	$\cdots$	.
.	.	.	.	$\cdots$	.
χ_n	1	ε^{n-1}	$\varepsilon^{2(n-1)}$	$\cdots$	$\varepsilon^{(n-1)^2}$

For $j = l = 2$, and $j = l = 3$ (7.2) says, respectively,

$$3[1^2 + 0^2 + (-1)^2] = 6$$

and

$$2[1^2 + (-1)^2 + 1^2] = 6.$$

Notice that all the entries of the character table for S_3 happen to be integers. This is not true for the general finite group, but happens to be true for all the symmetric groups. We will discuss the representations of the general symmetric group, S_n, in the next section and in Appendix C.

The purpose of this section is to derive the character table for many interesting groups. We begin with some general remarks. Suppose that the group G is Abelian. Then each element makes up its own conjugacy class, so that there are $\#G$ conjugacy classes in all. Then (6.2) and (6.3) imply that all the $n_i = 1$, so that all irreducible representations of G are one dimensional. Conversely, suppose that all the $n_i = 1$. Then (6.2) can hold only if $k = \#G$. This means that there are $\#G$ distinct conjugacy classes. Thus, each conjugacy class contains exactly one element, i.e. G is Abelian. We have thus proved

> G is Abelian if and only if all its irreducible representations are one $\qquad$ (7.3)
> dimensional.

We can now write down the character table for the cyclic group C_n. Let a be a generator for C_n, so that the conjugacy classes are the various $[a^{j-1}]$, $j = 1, 2, \ldots, n$. Let ε be a primitive nth root of unity. Then the characters χ_i, determined by $\chi_i(a) = \varepsilon^{i-1}$, $i = 1, 2, \ldots, n$, are all distinct, and thus give all the characters. The character table of C_n is thus given by Table 7.

Before proceeding further, we record the following useful fact:

> Suppose that the group G has a commutative subgroup H. Then $\qquad$ (7.4)
> any irreducible representation of G has degree at most $\#G/\#H$.

Proof Let r be an irreducible representation of G on the vector space V. Then r_H, the restriction of r to H, gives a representation of the Abelian group H on V. Let W

Table 8.

$12T$	$[e]$	$4[r_3]$	$4[r_3^2]$	$3[r^2]$
χ_1	1	1	1	1
χ_2	1	ε	ε^2	1
χ_3	1	ε^2	ε	1
χ_4	3	0	0	-1

$\varepsilon = \exp 2\pi i/3$

be an irreducible subspace of V under H, so that W is one dimensional by (7.3). The space spanned by all the $r(a)W a \in G$, is clearly invariant and hence must be all of V. Let G_W be the subgroup of G which stabilizes W, i.e. $G_W = \{a \in G : r(a)W \subset W\}$. Then the number of distinct subspaces, among the $r(a)W$, is given by $\#(G/G_W)$, and therefore the maximal number of linearly independent such subspaces is at most this amount. Since $H \subset G_W$, we conclude (7.4).

Let us now compute the character table of the group T. This group is of order 12, and has a three-dimensional representation as the symmetries of the tetrahedron. The trace of any rotation through angle ϕ in $\mathbb{R}^3$ (or $\mathbb{C}^3$) is $1 + 2\cos\phi$. Thus, for this three-dimensional representation we have

$$\chi(e) = 3, \quad \chi(R_{120°}) = \chi(R_{240°}) = 0 \quad \text{and} \quad \chi(R_{180°}) = -1.$$

Thus

$$(\#G)\|\chi\|^2 = 9 + 3 \cdot 1 = 12$$

since there are three rotations through 180° and four each through 120° and 240°. We see that this three-dimensional representation is irreducible. Since the sum of squares of the degrees of all irreducible representations is 12, and $3^2 = 9$, there must also be three one-dimensional representations. These can be found as follows: let H be the subgroup of T consisting of the identity and the rotations through 180°. Then H is a normal subgroup and hence any representation of the quotient group, T/H, lifts to a representation of T. But T/H is just the cycle group C_3, which has three one-dimensional representations. Thus the character table of T is given by Table 8.

For the group O, observe again that the subgroup H consisting of the identity and the 180° rotations is a normal subgroup. This time, the quotient group is S_3, whose character table we have already computed. The representations of the quotient group S_3 give rise to two one-dimensional representations and one two-dimensional representation. Since $24 - 6 = 18$, and there are five conjugacy classes altogether, we conclude that there are also two three-dimensional representations. One of these is the representation of O as the group of symmetries of the cube in three-space. The other is the representation of O as the group T_d on three-space. The character table of O is thus given as in Table 9. Notice that all the entries in the table are integers. This is because $O \sim S_4$.

Table 9.

$24O$	$[e]$	$6[r_4]$	$3[r_4^2]$	$8[r_3]$	$6[r_2]$
χ_1	1	1	1	1	1
χ_2	2	0	2	-1	0
χ_3	1	-1	1	1	-1
χ_4	3	1	-1	0	-1
χ_5	3	-1	-1	0	1

Table 10.

$4D_2$	$[e]$	$[r_x]$	$[r_y]$	$[r_z]$
χ_1	1	1	1	1
χ_2	1	1	-1	-1
χ_3	1	-1	1	-1
χ_4	1	-1	-1	1

Table 11.

$8D_4$	$[e]$	$2[r_4]$	$[r_4^2]$	$2[r_2+]$	$2[r_2\times]$
χ_1	1	1	1	1	1
χ_2	1	1	1	-1	-1
χ_3	1	-1	1	-1	1
χ_4	1	-1	1	1	-1
χ_5	2	0	-2	0	0

To complete our discussion of character tables for the crystallographic rotation groups, we need only consider the dihedral groups D_2, D_4 and D_6.

D_2 is Abelian: it has four elements, four conjugacy classes, and four one-dimensional representations. In each representation, other than the trivial representation, one of the 180° rotations is represented by $+1$, the other two by -1. The character table is therefore as in Table 10.

We have worked out the irreducible representations of D_4 in Section 2.5; there are four one-dimensional representations and one two-dimensional representation. The characters of the two-dimensional representation may be found from the traces of the 2×2 matrices which describe rotation and reflection symmetries of the square: 0 for the 90° and 270° rotations and for all the reflections, 2 for the identity, -2 for the 180° rotation. The complete character table is therefore as in Table 11.

Table 12.

$12D_6$	$[e]$	$2[r_6]$	$2[r_6^2]$	$[r_6^3]$	$3[r_2]$(solid line)	$3[r_2]$(broken line)
χ_1	1	1	1	1	1	1
χ_2	1	1	1	1	-1	-1
χ_3	1	-1	1	-1	1	-1
χ_4	1	-1	1	-1	-1	1
χ_5	2	1	-1	-2	0	0
χ_6	2	-1	-1	2	0	0

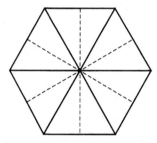

Fig. 2.1

Note that representation χ_2 has $+1$ for elements of the normal subgroup C_4, while χ_3 and χ_4 has $+1$ for elements of the two non-conjugate normal subgroups D_2.

For D_6 there are four one-dimensional representations and *two* two-dimensional representations (note that $4 \cdot 1^2 + 2 \cdot 2^2 = 12$). As with D_4, we may obtain the one-dimensional representations other than the trivial one by assigning $+1$ to each of the three normal subgroups of index 2: C_6 (rotations in plane of page), D_3 ($0°, 120°$, and $240°$ rotations in plane of page and $180°$ rotations about solid lines in Fig. 2.1), D_3 ($0°, 120°$, and $240°$ rotations in plane of page and $180°$ rotations about broken lines in Fig. 2.1). One two-dimensional representation with character χ_5 is obviously obtained by writing down the 2×2 rotation and reflection matrices. The character of this representation is 0 for reflection, $2\cos\phi$ for rotations. The other two-dimensional representation is difficult to discover by inspection (we shall construct it in the next chapter) but its character χ_6 is easily written down by orthogonality. The complete character table is shown in Table 12.

We have now computed the character tables for all the groups in the first column of Table 4. The groups in the third column are isomorphic as abstract groups to those in the first column, so we have their character tables as well. The groups in the second column are all direct products of the groups in the first column with Z_2. We get their character tables as a result of the following:

Table 13.

$12D_{3d}$	$[e]$	$2[r_3]$	$3[r_2]$	$[-I]$	$2[s_6]$	$3[\tau_d]$
χ_1	1	1	1	1	1	1
χ_2	2	-1	0	2	-1	0
χ_3	1	1	-1	1	1	-1
χ_4	1	1	1	-1	-1	-1
χ_5	2	-1	0	-2	1	0
χ_6	1	1	-1	-1	-1	1

Proposition 7.1

Let r^1 and r^2 be irreducible representations of the groups G_1 and G_2 on the vector spaces V_1 and V_2. Then the representation $r^1 \otimes r^2$ of $G_1 \times G_2$ on $V_1 \otimes V_2$ given by

$$(r^1 \otimes r^2)_{(g_1, g_2)}(v_1 \otimes v_2) = r^1_{g_1} v_1 \otimes r^2_{g_2} v_2,$$

is irreducible and has character $\chi^1(g_1)\chi^2(g_2)$, where χ^i is the character of r^i on G_i. Furthermore, every irreducible representation of $G_1 \times G_2$ is such a tensor product.

We have already proved the first two assertions. The last follows from the fact that the sums of the squares of the degrees of these representations equals $\#(G_1 \times G_2)$ and hence these are all the irreducibles.

Thus the character tables for the groups in the second column of Table 4 are obtained by 'doubling' the character table for the corresponding group in the first column. For instance, the character table for D_{3d} is given by Table 13, where the upper left-hand corner is the character table for D_3.

The character tables are useful in the explicit computation of the reduction of a reducible representation into its irreducible components. If ϕ is the character of the representation then, letting $\phi(k)$ denote the value of ϕ on any element in the kth conjugacy class, we have

$$(\phi, \chi_i) = (1/\#G)(\#C_1\phi(1)\overline{\chi_i(1)} + \cdots + \#C_p\phi(p)\overline{\chi_i(p)}),$$

and the values of the $\chi_i(j)$ can be read from the table. We shall make use of this technique in the study of molecular vibrations in the next chapter.

As an application, we determine the reduction of the representation of D_6 obtained by letting it act on the space of functions on the vectors of a hexagon. The character ϕ of this six-dimensional representation is easily determined from (5.1): it is six for the identity, which has six fixed points, two for the three $180°$ rotations about axes through opposite vertices, zero for all other elements. Using the character table for

D_6 we find:

$$(\phi, \chi_1) = \tfrac{1}{12}(6 \cdot 1 + 3 \cdot 2 \cdot 1) = 1$$
$$(\phi, \chi_2) = \tfrac{1}{12}(6 \cdot 1 + 3 \cdot 2(-1)) = 0$$
$$(\phi, \chi_3) = \tfrac{1}{12}(6 \cdot 1 + 3 \cdot 2 \cdot 1) = 1$$
$$(\phi, \chi_4) = \tfrac{1}{12}(6 \cdot 1 + 3 \cdot 2(-1)) = 0$$
$$(\phi, \chi_5) = \tfrac{1}{12}(6 \cdot 2) = 1$$
$$(\phi, \chi_6) = \tfrac{1}{12}(6 \cdot 2) = 1.$$

So each of the two-dimensional representations occurs once, the trivial representation occurs once, and the representation with character χ_3 occurs once. The invariant subspace for the representation with character χ_3 is clearly spanned by the function represented by the diagram

$$+ \qquad -$$
$$- \qquad\qquad +,$$
$$+ \qquad -$$

which is carried into itself by 120° and 240° rotations and by 180° rotations about axes through opposite vertices, but into its negative by the other six group elements.

2.8 The representations of the symmetric group

In this section we show how the methods of Section 2.4 and 2.5 allow us to determine, at least in principle, all the irreducible representations of the symmetric group S_n. We insert this material here since it depends only on the theory developed in this chapter. We shall need these results for physical and mathematical applications in Chapter 5, and the reader might prefer to postpone reading this section until then.

The symmetric group S_n is the group of all permutations, that is of all one-to-one transformations of the set $\{1, \ldots, n\}$ of n elements. There are various convenient notations for writing an element s of S_n. One is to write it out as

$$\begin{pmatrix} 1 & 2 & 3 & \cdots & n \\ s(1) & s(2) & s(3) & \cdots & s(n) \end{pmatrix}.$$

Thus

$$\begin{pmatrix} 12345 \\ 53124 \end{pmatrix}$$

is the element of S_5 which sends 1 to 5 and 2 to 3, etc. Another convenient notation is the 'cycle notation'. We start with any number, write its image to the right, and keep going until a cycle is completed. We then pick some number not in the first cycle, write its image to its right, and continue until a second cycle is completed. This procedure is repeated until the entire permutation has been described. Thus the above element of S_5 would be written as

$$(15423)$$

since 1 goes to 5, 5 goes to 4, etc. Similarly the element

$$\begin{pmatrix} 12345 \\ 31254 \end{pmatrix}$$

would be written in terms of two cycles as

$$(132)(45).$$

It is the product of (45) with (132).

Let $s = (132)(45)$ and let t be some other element of S_5. Then

$$(tst^{-1})(t(1)) = t(s(1)) = t(3),$$
$$(tst^{-1})(t(3)) = t(s(3)) = t(2), \text{etc.}$$

The general rule is that $tst^{-1}(t(i)) = t(s(i))$, so we may write tst^{-1} in terms of cycles as

$$tst^{-1} = (t(1)t(3)t(2))(t(4)t(5)).$$

The same argument shows that, in general, for a permutations $s \in S_n$ which is written in cycle form, the element tst^{-1} is obtained by replacing each integer i in the cycle form of s by $t(i)$. Conversely, if s_1 and s_2 have the same cycle structure, so that there is a permutation t relating the entry i in s_1 to $t(i)$ in s_2, then $s_2 = ts_1 t^{-1}$. Thus, two elements of S_n are conjugate if and only if they have the same cycle structure. Thus, for example, the five conjugacy classes of S_4 are

$$\{e\}$$
$$\{(ab)\} = \{(12),(13),(14),(23),(24),(34)\}$$
$$\{(ab)(cd)\} = \{(12)(34),(13)(24),(14)(23)\}$$
$$\{(abc)\} = \{(123),(124),(132),(134),(142),(143),(234),(243)\}$$
$$\{(abcd)\} = \{(1234),(1243),(1324),(1342),(1423),(1432)\}.$$

Notice that any *cyclic* permutation of the entries of a cycle does not change the permutation: (123) and (231) are the same permutation, as follows from the definition of cycle. Also, interchanging two disjoint cycles gives the same permutation. (12)(34) and (34)(12) are the same. A conjugacy class of S_n is thus determined by $[v_1, \ldots, v_n]$, where v_1 is the number of one-cycles, v_2 is the number of two-cycles, etc. Of course, the v's are constrained by

$$v_1 + 2v_2 + \cdots + nv_n = n.$$

The number of elements in a conjugacy class is given by $\#S_n/\#H$, where H is the isotropy subgroup of some element s in the conjugacy class, i.e. $H = \{t \mid tst^{-1} = s\}$. Suppose s has the cycle structure $[v_1, \ldots, v_n]$. Then t cannot interchange entries coming from cycles of different length. Within the set of cycles of a fixed length, t can act as a cyclic permutation within each cycle and can permute cycles as a whole. Thus, considering the cycles of different length independently, we see that

$$\#H = 1^{v_1}v_1!2^{v_2}v_2!3^{v_3}v_3!\cdots n^{v_n}v_n!$$

where. for example, 3^{v_3} is present because there are three cyclic permutations within each of the v_3 three-cycles and $v_3!$ is present because there are $v_3!$ permutations of these three-cycles among themselves. Thus,

$$\begin{array}{l}\text{the number of elements}\\ \text{in the conjugacy class}\\ \text{given by } [v_1,\ldots,v_n]\end{array} = \frac{n!}{1^{v_1}v_1!2^{v_2}v_2!\cdots n^{v_n}v_n!}.$$

So, for S_4

$$\#\{e\} = \#[4,0,0,0] = \frac{4!}{4!} = 1$$

$$\#\{(a,b)\} = \#[2,1,0,0] = \frac{4!}{2\cdot 2!} = 6$$

$$\#\{(a,b)(c,d)\} = \#[0,2,0,0] = \frac{4!}{2^2\cdot 2!} = 3$$

$$\#\{(abc)\} = \#[1,0,1,0] = \frac{4!}{3} = 8$$

$$\#\{(abcd)\} = \#[0,0,0,1] = \frac{4!}{4} = 6.$$

We can also set

$$\begin{aligned}\lambda_1 &= v_1 + v_2 + \cdots + v_n\\ \lambda_2 &= v_2 + v_3 + \cdots + v_n\\ &\ \,\vdots\\ \lambda_n &= v_n\end{aligned}$$

Thus $\lambda_1 \geqslant \lambda_2 \cdots \geqslant \lambda_n$, and it follows from

$$v_1 + 2v_2 + \cdots + nv_n = n$$

that

$$\lambda_1 + \cdots + \lambda_n = n.$$

For example, the permutation $(1)(23)(45)(678)\in S_8$ has $\lambda_1 = 4$, $\lambda_2 = 3$, $\lambda_3 = 1$. The set $\lambda = (\lambda_1,\ldots,\lambda_n)$ is called a *partition* of n. It is conveniently represented by a *Young diagram*.

We draw the diagram as an array of boxes with λ_1 boxes in the first row, λ_2 boxes in the second row, etc. For example, if $n = 7$ then $\lambda = (3,2,1,1)$ is drawn as and similarly $(5,2) = (5,2,0,0)$ (we usually drop the zeros) is

Given $\lambda = (\lambda_1, \ldots, \lambda_n)$ with $\lambda_i \geqslant \lambda_{i+1}$ and $\lambda_1 + \cdots + \lambda_n = n$, we recover v_i by setting

$$v_i = \lambda_i - \lambda_{i+1}.$$

For example, the first diagram corresponds to $v_1 = 1$, $v_2 = 1$, $v_3 = 0$, $v_4 = 1$; the second to $v_1 = 3$, $v_2 = 2$. Clearly $v_1 + 2v_2 + \cdots + nv_n = n$. Thus the number of conjugacy classes of S_n, which is the same as the number of inequivalent irreducible representations of S_n, is the same as the number of Young diagrams. Our task is to attach a distinct irreducible representation on each Young diagram.

From now on we consider a fixed n. We put a partial order on the diagrams by saying that $\lambda \geqslant \mu$ if, for all i, the total number of boxes in the first i rows of λ is no less than the total number of boxes in the first i rows of μ; i.e. if

$$\lambda_1 \geqslant \mu_1$$
$$\lambda_1 + \lambda_2 \geqslant \mu_1 + \mu_2$$
$$\lambda_1 + \lambda_2 + \lambda_3 \geqslant \mu_1 + \mu_2 + \mu_3, \text{ etc.}$$

For example, the partial ordering (down is decreasing) for S_6 is given by:

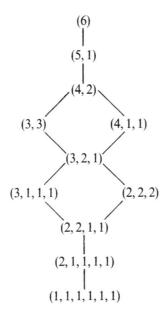

By a Young *tabloid* corresponding to the diagram $\lambda = (\lambda_1, \ldots, \lambda_n)$ we mean a decomposition of the set $\{1, \ldots, n\}$ into a union of disjoint sets where the first set contains λ_1 elements, the second set contains λ_2 elements, etc. Thus

$$\{3,5,2\}\{1,7\}\{4\}\{6\} \quad \text{or} \quad \left\{\begin{array}{ll}\boxed{3}\boxed{5}\boxed{2}\\ \boxed{1}\boxed{7}\\ \boxed{4}\\ \boxed{6}\end{array}\right\}$$

is a Young tabloid corresponding to the Young diagram $(3, 2, 1, 1)$. The individual subsets are unordered so

$$\{2,3,5\}\{7,1\}\{4\}\{6\} \quad \text{or} \quad \left\{\begin{array}{ll}\boxed{2}\boxed{3}\boxed{5}\\ \boxed{7}\boxed{1}\\ \boxed{4}\\ \boxed{6}\end{array}\right\}$$

is the same tabloid. However, the order of the subsets is important,

$$\{3,5,2\}\{1,7\}\{6\}\{4\} \quad \text{or} \quad \left\{\begin{array}{ll}\boxed{3}\boxed{5}\boxed{2}\\ \boxed{1}\boxed{7}\\ \boxed{6}\\ \boxed{4}\end{array}\right\}$$

is a different tabloid. We can think of a tabloid as a way of putting the number $\{1, 2, \ldots, n\}$ into the boxes of a Young diagram, where the order of numbers within each row does not matter.

We let M_λ denote the set of all tabloids corresponding to a Young diagram λ. If $\{t\}$ is a fixed tabloid corresponding to λ, the isotropy group of $\{t\}$ is clearly isomorphic to $S_{\lambda_1} \times S_{\lambda_2} \times \cdots \times S_{\lambda_p}$, the subgroup which permutes elements within each row of the diagram. Since S_n acts transitively on M_λ, we see that

$$\#M_\lambda = \frac{n!}{\lambda_1! \cdots \lambda_p!}.$$

Since S_n acts on the set M_λ, we get a representation of S_n on $\mathscr{F}(M_\lambda)$. For example, $M_{(n)}$ contains only one element,

$$\{1, \ldots, n\} \quad \text{or} \quad \{\boxed{1}\boxed{2}\boxed{3}\boxed{\cdots}\boxed{n}\}.$$

All permutations carry this tabloid into itself, so the representation of S_n on $\mathscr{F}(M_{(n)})$ is the trivial representation. An element of the set $M_{(n-1,1)}$ is of the form $\{1, \ldots, \hat{k}, \ldots, n\} \cup \{k\}$, where the symbol $\hat{k}$ means that k is *missing*. So $M_{(n-1,1)}$ can be identified with the set $\{1, \ldots, n\}$, where $k \in \{1, \ldots, n\}$ corresponds to the missing $\{k\}$. For example, if $n = 3$ there are three tabloids, $\{23\}, \{1\}; \{13\}, \{2\};$ and $\{12\}, \{3\}$, which may be identified with 1, 2, and 3, respectively. We have seen that S_n, acting on $M_{(n-1,1)} \times M_{(n-1,1)}$, has two orbits, so that

$$\mathscr{F}(M_{(n-1,1)}) = \mathbb{C} \oplus F_{(n-1,1)},$$

where $F_{n-1,1}$ is an irreducible space of dimension $n-1$. Notice that the first component, the constant functions, is just $\mathscr{F}(M_{(n)})$.

The set $M_{(n-2,2)}$ $(n>3)$ can be identified with the space of all two-element subsets of $\{1,\ldots,n\}$, where we look at the entries in the second subset, so that $\{1,\ldots,\hat{k},\ldots,\hat{l},\ldots,n\}$, $\{k,l\}$ is identified with $\{k,l\}$. For example, if $n=5$, $M_{(n-2,2)}$ has ten elements. The element $\{3,4,5\}$, $\{1,2\}$ is associated with $\{1,2\}$, the element $\{2,4,5\}$, $\{1,3\}$ with $\{1,3\}$, and so on. A pair of two-element subsets may have either zero, one or two elements in common. Thus, S_n has three orbits when acting on $M_{(n-2,2)} \times M_{(n-2,2)}$, and so $\mathscr{F}(M_{(n-2,2)})$ breaks up into three irreducible components. We claim that two of these components are $\mathscr{F}(M_{(n)})$ and $\mathscr{F}(M_{(n-1,1)})$. Indeed, we now describe a map from $\mathscr{F}(M_{(n-1,1)})$ to $\mathscr{F}(M_{(n-2,2)})$ which commutes with the action of S_n and is injective: we must find a map, T, which goes from functions, f, on $\{1,\ldots,n\}$ to functions on two-element subsets. Take T to be given by

$$(Tf)(\{a,b\}) = f(a) + f(b).$$

It is clear that T commutes with the action of S_n. Also $T(\text{constant}) = \text{constant}$ and $T\delta_a$ is not a constant (and, in particular, not zero). Thus T is not zero when restricted to each of the irreducible components of $\mathscr{F}(M_{(n-1,1)})$ and hence is injective. Thus

$$\mathscr{F}(M_{(n-2,2)}) = \mathbb{C} + T(F_{(n-1,1)}) + F_{(n-2,2)}$$

$$1 \qquad n-1 \qquad \frac{n(n-3)}{2}.$$

The dimension of $F_{(n-2,2)}$ is obtained by subtracting:

$$\dim F_{(n-2,2)} = \frac{n!}{(n-2)!2!} - n = \frac{n(n-3)}{2}.$$

We wish to prove the following: to each λ there corresponds a unique 'new' irreducible subrepresentation F_λ of $\mathscr{F}(M_\lambda)$. The space $\mathscr{F}(M_\lambda)$ decomposes into a direct sum of irreducible subrepresentations isomorphic to certain of the F_μ with $\mu \geqslant \lambda$ (and these may occur with multiplicity) together with the one unique new subrepresentation F_λ. Thus each Young diagram determines an irreducible representation of S_n.

By a Young *tableau* corresponding to λ we mean an assignment of the numbers $\{1,\ldots,n\}$ to each of the boxes of λ, one number to each box. In a tableau, the order in each row matters. Thus

$$\begin{array}{|c|c|c|}
\hline
3 & 5 & 2 \\
\hline
\end{array}$$
$$\begin{array}{|c|c|}
\hline
1 & 7 \\
\hline
\end{array}$$
$$\begin{array}{|c|}
\hline
4 \\
\hline
\end{array}$$
$$\begin{array}{|c|}
\hline
6 \\
\hline
\end{array}$$

is a $(3,2,1,1)$ tableau. Each tableau gives rise to a tabloid, by letting the entries in the first row belong to the first set, the entries of the second row correspond to the second set, etc. Two different tableaux, which differ by a permutation of the entries of their rows, give rise to the same tableau. If t is a tableau, the corresponding tabloid will be denoted by $\{t\}$. Thus if t is the above tableau, then $\{t\} = \{3,5,2\}\{1,7\}\{4\}\{6\}$.

We now describe the F_λ. Let t be a tableau. Let C_t denote the subgroup of S_n consisting of those π which permute the numbers in the various columns of t among themselves. Thus, if

$$t = \begin{array}{|c|c|c|} \hline 3 & 5 & 2 \\ \hline 1 & 7 \\ \cline{1-2} 4 \\ \cline{1-1} \end{array}$$
$$\begin{array}{|c|} \hline 6 \\ \hline \end{array}$$

then

$$C_t = S_{\{3,1,4,6\}} \times S_{\{5,7\}}$$

where $S_{\{3,1,4,6\}}$ are the permutations of $\{3,1,4,6\}$, etc. Let

$$e_t = \sum_{\pi \in C_t} \operatorname{sgn}(\pi)\delta_{\pi\{t\}}.$$

Notice that the e_t depends on t and not just the tabloid $\{t\}$

As an example we construct the e_t for the case of S_3, with the Young diagram

.

There are three tabloids in this case,

$$\{2,3\}\{1\} \quad \text{(associated with 1)}$$
$$\{1,3\}\{2\} \quad \text{(associated with 2)}$$

and

$$\{1,2\}\{3\} \quad \text{(associated with 3)}$$

and the space $\mathscr{F}(M_{(2,1)})$ is spanned by $\delta_1, \delta_2, \delta_3$. A typical tableau is

$$t = \begin{array}{|c|c|} \hline 1 & 2 \\ \hline 3 \\ \cline{1-1} \end{array} .$$

The subgroup C_t consists of the identity, whose sign is $+1$, and the permutation (13), whose sign is -1. We must determine what tabloid results from the action of each element of C_t on the tableau t. Clearly the identity gives the tabloid $\{1,2\}, \{3\}$, while (13) to converts t to

$$\begin{array}{|c|c|} \hline 3 & 2 \\ \hline 1 \\ \cline{1-1} \end{array} ,$$

which corresponds to the tabloid $\{2,3\}, \{1\}$. Hence

$$e_t = \delta_3 - \delta_1.$$

If we start with a different tableau, say

$$t' = \begin{array}{|c|c|} \hline 3 & 1 \\ \hline 2 \\ \cline{1-1} \end{array} ,$$

then $C_{t'}$ consists of the identity and (32), and $e_{t'} = \delta_2 - \delta_3$. Since the space spanned by δ_1, δ_2 and δ_3 has a one-dimensional irreducible subspace spanned by the constant function $\delta_1 + \delta_2 + \delta_3$, the elements e_t and $e_{t'}$ which we have constructed span the complementary two-dimensional subspace F_λ. There are four other tableaux for which we could construct e_t; they would just lead to linear combinations of what we have already found.

Since $\sigma\delta_{\{t\}} = \delta_{\sigma\{t\}}$, we have

$$\sigma e_t = \sum_{\pi \in C_t} \mathrm{sgn}\,(\pi)\delta_{\sigma\pi\{t\}}.$$

We can write this as

$$\sigma e_t = \sum_{\pi \in C_t} \mathrm{sign}\,(\sigma\pi\sigma^{-1})\delta_{\sigma\pi\sigma^{-1}\sigma\{t\}}.$$

But $C_{\sigma t} = \sigma C_t \sigma^{-1}$ so we can write the last sum as

$$\sum_{\rho \in C_{\sigma t}} \mathrm{sgn}\,(\rho)\delta_{\rho\{\sigma t\}} = e_{\sigma t}.$$

We conclude that

$$\sigma e_t = e_{\sigma t}.$$

Thus the space spanned by the e_t is invariant under S_n. We define this space to be F_λ. Thus

$$F_\lambda = \{\text{linear span of all the } e_t\}.$$

Consider, for example, the action of $\sigma = (23)$ on each of the two elements e_t and $e_{t'}$ which we constructed earlier. For

$$t = \begin{array}{|c|c|} \hline 1 & 2 \\ \hline 3 \\ \cline{1-1} \end{array}, \quad e_t = \delta_3 - \delta_1,$$

while for

$$t' = \begin{array}{|c|c|} \hline 3 & 1 \\ \hline 2 \\ \cline{1-1} \end{array}, \quad e_{t'} = \delta_2 - \delta_3.$$

Now $\sigma e_t = e_{\sigma t}$, where $\sigma t = \begin{array}{|c|c|} \hline 1 & 3 \\ \hline 2 \\ \cline{1-1} \end{array}$. We find $e_{\sigma t} = \delta_2 - \delta_1 = e_{t'} + e_t$.

Also, $\sigma e_{t'} = e_{\sigma t'}$ where $\sigma t' = \begin{array}{|c|c|} \hline 2 & 1 \\ \hline 3 \\ \cline{1-1} \end{array}$, so we have $e_{\sigma t'} = \delta_3 - \delta_2 = -e_{t'}$.

Thus, relative to the basis $\{e_t, e_{t'}\}$, (23) is represented by the matrix $\begin{pmatrix} 1 & 0 \\ 1 & -1 \end{pmatrix}$.

Continuing in this manner, we can construct the entire two-dimensional representation of S_3.

As another example, if $\lambda = (1, 1, \ldots, 1)$, then $C_t = S_n$ and, up to sign, there is only one e_t, and it is

$$e_t = \sum_{\pi \in S_n} \text{sgn}(\pi)\delta_{\pi t}$$

where we may take

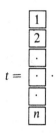

$$t = \begin{array}{|c|} \hline 1 \\ \hline 2 \\ \hline \cdot \\ \hline \cdot \\ \hline \cdot \\ \hline n \\ \hline \end{array} \ .$$

The representation in this case is the one-dimensional sign representation. Also in this case $S_{\{t\}} = \{e\}$, so we may identify $M = S_n/S_{\{t\}}$ with S_n and thus $\mathscr{F}(M)$ with $\mathscr{F}(S_n)$. We know that the regular representation contains any irreducible representation with multiplicity equal to its dimension and hence contains $F_{(1,\dots,1)}$ once. Also $(1,\dots,1)$ is the last diagram on our list. This supports the contention of the theorem.

We now prove the following results about the Young tableaux and the representations of S_n.

(1) Let λ and μ be diagrams and let t be a λ tableau and s be a μ tableau. Suppose that for every i, the numbers from the ith row of s belong to different columns of t. Then $\lambda \geqslant \mu$.

Proof The numbers in the first row of s all lie in different columns of t. Hence, λ has at least μ_1 columns (i.e. $\lambda_1 \geqslant \mu_1$). Suppose $\lambda_1 = \mu_1$. The numbers in the second row of s all lie in different columns of t. We see that $\lambda_1 + \lambda_2 \geqslant \mu_1 + \mu_2$, etc.   QED.

(2) With the same notations as in (1), suppose that

$$\sum_{\pi \in C_t} \text{sgn}(\pi)\delta_{\pi\{s\}} \neq 0.$$

Then $\lambda \geqslant \mu$, and if $\lambda = \mu$ then

$$\sum_{\pi \in C_t} \text{sgn}(\pi)\delta_{\pi\{s\}} = \pm e_t = \sum_{\pi \in C_t} \text{sgn}(\sigma\pi)\delta_{\pi\{t\}},$$

where $\sigma \in C_t$ and $\{s\} = \sigma\{t\}$.

Proof Let

$$A_t = \sum_{\pi \in C_t} \text{sgn}(\pi)\pi$$

as an operator, so that, for example,

$$A_t \delta_{\{t\}} = \sum_{\pi \in C_t} \text{sgn}(\pi)\delta_{\pi\{t\}} = e_t.$$

Thus we can rewrite (2) as saying that

$$A_t \delta_{\{s\}} = 0 \quad \text{if} \quad \mu \not\leqslant \lambda$$

and if $\mu = \lambda$ then

$$A_t \delta_{\{s\}} = \begin{cases} 0 & \text{if } \{s\} \neq \sigma\{t\} \text{ for some } \sigma \in C_t \\ \text{sgn}(\sigma)e_t & \text{if } \{s\} = \sigma\{t\} \text{ for } \sigma \in C_t. \end{cases}$$

Suppose that

$$A_t \delta_{\{s\}} \neq 0.$$

Under this hypothesis, we claim that two numbers x and y which lie in the same row of s cannot lie in the same column of t. For to say that x and y lie in the same row of s implies that $(xy)\{s\} = \{s\}$. If x and y lie in the same column of t, then $(xy) \in C_t$ and, since $\text{sgn}(xy) = -1$, we could have

$$A_t = \sum_{\pi \in C_t} \text{sgn}(\pi)\pi = \sum_{\pi \in C_t} \text{sgn}(\pi \cdot (xy))\pi \cdot (xy) = -\sum \text{sgn}(\pi)\pi(xy) = -A_t(xy).$$

Thus

$$A_t \delta_{\{s\}} = -A_t(xy)\delta_{\{s\}} = -A_t \delta_{(xy)\{s\}} = -A_t \delta_{\{s\}}$$

contradicting the assumption that $A_t \delta_{\{s\}} \neq 0$.

This implies that $\lambda \geqslant \mu$. If $\lambda = \mu$, all numbers in the first row of s occur in different columns of t. So we can find some $\pi \in C_t$ such that πt has the same first row as s. All elements of the second row of s occur in different columns of πt and below the first row. So we can find a $\pi' \in C_{\pi t} = C_t$, leaving the numbers in the first row of s fixed with $\pi'\pi t$ having the same first two rows, as s, etc. This shows that $\{s\} = \{\sigma t\}$ for some $\sigma \in C_t$. But then $A_t \delta_{\{s\}} = \text{sgn}(\sigma)e_t$. Thus we have proved (2). We conclude that for any $\{s\}$ whatsoever in M_λ, we have

$$A_t \delta_{\{s\}} = \begin{cases} e_t & \text{if} \quad \{s\} = \sigma\{t\} \quad \text{sgn}(\sigma) = 1 \\ 0 & \text{if} \quad \{s\} \neq \sigma\{t\} \\ -e_t & \text{if} \quad \{s\} = \sigma\{t\} \quad \text{sgn}(\sigma) = -1. \end{cases}$$

Now every $f \in \mathscr{F}(M_\lambda)$ is a linear combination of the $\delta_{\{s\}}$ as $\{s\}$ ranges over the λ tabloids. Hence (3).

(3) For any $f \in \mathscr{F}(M_\lambda)$,

$$A_t f = c_f e_t$$

where c_f is a scalar, i.e. $A_t f$ is a multiple of e_t for any f.

Let us put a scalar product (,) on $\mathscr{F}(M_\lambda)$ by taking the $\delta_{\{t\}}$ as an orthonormal basis. This is clearly S_n invariant. Now for any $u, v \in \mathscr{F}(M_\lambda)$,

$$(A_t u, v) = \sum_{\pi \in C_t} (\text{sgn}(\pi)\pi u, v)$$

$$= \sum_{\pi \in C_t} (u, \text{sgn}(\pi^{-1})\pi^{-1}v), \text{ since } \text{sgn}(\pi) = \text{sgn}(\pi^{-1})$$

$$= \sum_{\pi \in C_t} (u, \mathrm{sgn}(\pi)\pi v)$$

$$= (u, A_t v).$$

(4) Let U be an invariant subspace of $\mathscr{F}(M_\lambda)$. Then either $U \supset F_\lambda$ or $U \subset F_\lambda^\perp$. In particular, F_λ is irreducible.

Proof Let $u \in U$ and let t be a λ tableau. Then $A_t u$ is a multiple of e_t. If for some t and u this multiple is not zero, then $A_t u \in U$ and $A_t u = c_u e_t$, and since F_λ is generated by the σe_t as $\sigma \in S_n$, we see that $F_\lambda \subset U$. If these multiples are zero for all t and u, then $0 = (A_t u, \delta_{\{t\}}) = (u, A_t \delta_{\{t\}}) = (u, e_t)$ for all u and t, so $U \subset F_\lambda^\perp$.

(5) Let $T: \mathscr{F}(M_\lambda) \to \mathscr{F}(M_\mu)$ be any element of $\mathrm{Hom}_{S_n}(\mathscr{F}(M_\lambda), \mathscr{F}(M_\mu))$. Suppose that $F_\lambda \not\subset \ker T$. Then $\lambda \geqslant \mu$. If $\lambda = \mu$, then the restriction of T to F_λ is a scalar multiple of the identity.

Proof By (4), $\ker T \subset F_\lambda^\perp$. Let t be any λ tableau. Then $0 \neq T e_t = T A_t \delta_{\{t\}} = A_t T \delta_{\{t\}}$. But $T \delta_{\{t\}} \in \mathscr{F}(M_\mu)$ is some combination of $\delta_{\{s\}}$ for μ tabloids $\{s\}$ and $A_t \delta_{\{s\}} = 0$ unless $\lambda \geqslant \mu$. The second part follows from Schur's lemma and (2), since $A_t F_\lambda(M_\lambda) \subset F_\lambda$.

(6) $\mathrm{Hom}_{S_n}(F_\lambda, F_\mu) = 0$ unless $\lambda = \mu$. In particular, since the number of diagrams = the number of partitions = the number of conjugacy classes of S_n, the F_λ are exactly all the irreducible representations of S_n.

Proof Any $T \in \mathrm{Hom}_{S_n}(F_\lambda, F_\mu)$ can be extended to an element of $\mathrm{Hom}_{S_n}(\mathscr{F}(M_\lambda), \mathscr{F}(M_\mu))$ by setting it equal to zero on $F_\lambda^\perp$. By (5) this shows that if $T \neq 0$, then $\lambda \geqslant \mu$. Since F_λ and F_μ are irreducible, by Schur's lemma, if $T \neq 0$ then T is invertible and working with T^{-1} shows that $\mu \geqslant \lambda$, hence $\lambda = \mu$.

If we now decompose $\mathscr{F}(M_\mu)$ into irreducibles, we see that the only possible irreducible components, by (5), are the F with $\lambda \geqslant \mu$. This completes the proof of our assertions.

As an example of the procedure just described, we construct all the irreducible representations of S_4.

The first Young diagram is

$$\lambda = (4) = \square\square\square\square.$$

In this case, there is only one tabloid, $\{1, 2, 3, 4\}$, and we denote the unit function on this tabloid by δ. To construct a basis, we choose any tableau,

$$t = \boxed{1}\,\boxed{2}\,\boxed{3}\,\boxed{4}.$$

The subgroup C_t consists only of the identity, and so

$$e_t = \delta.$$

Since $\sigma\{t\} = \{t\}$ for any $\sigma \in S_4$, we have $\sigma e_t = e_{\sigma t} = e_t$: we have constructed the trivial one-dimensional identity representation.

The next Young diagram in order is $(3, 1)$ or

There are four tabloids, $\{2,3,4\}\{1\}$, $\{1,3,4\}\{2\}$, $\{1,2,4\}\{3\}$, and $\{1,2,3\}\{4\}$. We denote the basis elements of the space of functions on these tabloids by $\delta_1, \delta_2, \delta_3, \delta_4$. To construct a basis for the new irreducible representation of S_4 corresponding to this diagram, we choose three tableaux:

$$t_1 = \boxed{2}\ \boxed{3}\ \boxed{4}\ ; \quad e_{t_1} = \delta_1 - \delta_2$$
$$\boxed{1}$$

$$t_2 = \boxed{3}\ \boxed{2}\ \boxed{4}\ ; \quad e_{t_2} = \delta_1 - \delta_3$$
$$\boxed{1}$$

$$t_3 = \boxed{4}\ \boxed{2}\ \boxed{3}\ ; \quad e_{t_3} = \delta_1 - \delta_4 .$$
$$\boxed{1}$$

Clearly $\{e_{t_1}, e_{t_2}, e_{t_3}\}$ forms a basis for a three-dimensional invariant subspace of $\mathscr{F}(M_{(3,1)})$. The other invariant subspace is one dimensional, spanned by $\delta_1 + \delta_2 + \delta_3 + \delta_4$; on it we obtain the identity representation which we have already found.

The next Young diagram is $(2, 2)$ or

Now there are six tabloids, which we label by the numbers in the second row. Thus, for example, δ_{12} is the function which is one on $\{3,4\}\{1,2\}$ but zero on other tabloids such as $\{2,4\}\{1,3\}$. To construct a new two-dimensional irreducible representation on an invariant subspace $F_{(2,2)}$ of the six-dimensional space spanned by $\delta_{12}, \delta_{13}, \delta_{14}, \delta_{23}, \delta_{24}, \delta_{34}$, we choose two tableaux:

$$t_1 = \boxed{\begin{array}{cc} 3 & 4 \\ 1 & 2 \end{array}}: \quad e_{t_1} = \delta_{12} - \delta_{23} - \delta_{14} + \delta_{34}$$

$$t_2 = \boxed{\begin{array}{cc} 4 & 3 \\ 1 & 2 \end{array}}: \quad e_{t_2} = \delta_{12} - \delta_{24} - \delta_{13} + \delta_{34} .$$

Other tableaux would lead to linear combinations of these two basis elements: for example

$$t_3 = \boxed{\begin{array}{cc} 2 & 4 \\ 1 & 3 \end{array}}: \quad e_{t_3} = \delta_{13} - \delta_{23} - \delta_{14} + \delta_{24} = e_{t_1} - e_{t_2}$$

In the six-dimensional space $\mathscr{F}(M_{(2,2)})$ there are two other invariant subspaces, corresponding to the one-dimensional and three-dimensional representations that we have already found.

The next Young diagram is $(2, 1, 1)$ or

Corresponding to this diagram are 12 tabloids, which we may label by the next-to-lowest and lowest rows. For example, $\delta_{1,2}$ is the function which is one on the tabloid $\{3, 4\}\{1\}\{2\}$, zero on all other tabloids such as $\{3, 4\}\{2\}\{1\}$ or $\{1, 4\}\{2\}\{3\}$.

Again, to construct a new irreducible representation, this time a three-dimensional one, we choose three tableaux:

$$t_1 = \begin{array}{|c|c|} \hline 3 & 4 \\ \hline \end{array}\!\!\begin{array}{|c|}\hline 1 \\ \hline 2 \\ \hline \end{array} \quad : \quad e_{t_1} = \delta_{1,2} - \delta_{2,1} + \delta_{3,1} - \delta_{1,3} + \delta_{2,3} - \delta_{3,2}$$

$$t_2 = \begin{array}{|c|c|} \hline 4 & 3 \\ \hline \end{array}\!\!\begin{array}{|c|}\hline 1 \\ \hline 2 \\ \hline \end{array} \quad : \quad e_{t_2} = \delta_{1,2} - \delta_{2,1} + \delta_{4,1} - \delta_{1,4} + \delta_{2,4} - \delta_{4,2}$$

$$t_3 = \begin{array}{|c|c|} \hline 4 & 2 \\ \hline \end{array}\!\!\begin{array}{|c|}\hline 1 \\ \hline 3 \\ \hline \end{array} \quad : \quad e_{t_3} = \delta_{1,3} - \delta_{3,1} + \delta_{4,1} - \delta_{1,4} + \delta_{3,4} - \delta_{4,3}.$$

The elements e_{t_1}, e_{t_2} and e_{t_3} form a basis for a three-dimensional representation which is *not* equivalent to the one which arose from the diagram

An easy demonstration of this fact is to calculate the character of $\sigma = (34)$. In this representation

$$\sigma e_{t_1} = e_{\sigma t_1} = e_{t_2}$$
$$\sigma e_{t_2} = e_{\sigma t_2} = e_{t_1}$$
$$\sigma e_{t_3} = e_{\sigma t_3} = -e_{t_3}$$

so the character of σ is -1. In the representation from

we had

$$\sigma e_{t_1} = e_{\sigma t_1} = e_{t_1}$$
$$\sigma e_{t_2} = e_{\sigma t_2} = e_{t_3}$$
$$\sigma e_{t_3} = e_{\sigma t_3} = e_{t_2}$$

so the character of σ was $+1$.

Of course, the other representations which we constructed from earlier Young diagrams also occur on the 12-dimensional space which we are now considering: the one- and two-dimensional ones once, and the three-dimensional one twice.

The final Young diagram is

$$(1, 1, 1, 1) \quad \text{or} \quad \square .$$

Corresponding to this diagram are 24 tabloids, and the representation on the space $\mathscr{F}_{(1,1,1,1)}$ is the regular representation. The new representation is the one-dimensional sign representation; each of the other representations also occurs, with a multiplicity equal to its dimension.

This example shows that it is useful to know the dimension of the new irreducible representation of S_n associated with a given Young diagram, so that we will know when we have found enough independent elements e_t to span the representation space. It would also be useful to be able to calculate characters without having to choose a basis explicitly. We now state without proof two useful computational recipes which achieve these goals. Proofs will be given in Appendix C.

The hook formula for dimensions

At any position in a Young diagram, we define its *hook length* to be the sum of the number of positions to its right plus the number of positions below it plus one. For example, we have placed the hook length in the corresponding position below:

Then,

$$\dim F^\lambda = \frac{n!}{\prod (\text{all hook lengths in } \lambda)}.$$

For example, we may compute the dimensions of the representations of S_4 as follows (the number in each box is the hook length):

$$\dim F_{(4)} = \frac{4!}{4!} = 1$$

$$\dim F_{(3,1)} = \frac{4!}{4 \cdot 2 \cdot 1 \cdot 1} = 3$$

$$\dim F_{(2,2)} = \frac{4!}{3.2.2} = 2$$

4	1
2	
1	

$$\dim F_{(2,1,1)} = \frac{4!}{4\cdot2\cdot1\cdot1} = 3$$

4
3
2
1

$$\dim F_{(1,1,1,1)} = \frac{4!}{4!} = 1.$$

For S_8 the hook lengths of the entries of $(4,3,1)$ and $\dim F_{(4,3,1)}$ are

6	4	3	1
4	2	1	
1			

$$\dim F_{(4,3,1)} = \frac{8!}{6\cdot4\cdot4\cdot3\cdot2} = 70.$$

The Murnaghan–Nakayama rule

This is a rule for calculating a single entry in the character table of S_n; that is, for evaluating χ^λ on the conjugacy class having cycle structure corresponding to the partition $\mu = (\mu_1, \mu_2, \ldots)$, where $\mu_1 \geqslant \mu_2 \geqslant \cdots \geqslant \mu_p$ and $\mu_1 + \cdots + \mu_p = n$.

A *skew hook* is a connected part of the rim of a diagram which can be removed so that the remaining boxes form a (smaller) diagram. Thus

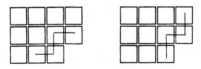

show the only two skew 4-hooks in $(4,4,3)$. If a skew hook starts on the ith row and ends on the jth row, then $i - j = l$ is called the leg length of the skew hook. Thus, the leg length of the skew hooks are 1 and 2, respectively.

To find the value of χ^λ on $\mu = (\mu_1, \ldots, \mu_p)$, proceed as follows: draw all possible ways of removing a skew μ_1 hook from λ so as to obtain a diagram λ'. Then

$$\chi^\lambda(\mu) = \sum (-1)^l \chi^{\lambda'}(\bar{\mu})$$

where

$$\bar{\mu} = (\mu_2, \mu_3, \ldots, \mu_p),$$

and l is the leg length of the skew hook. For example, if $\lambda = (5, 4, 4)$ and $\mu = (5, 4, 3, 1)$, then

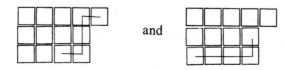

and

are the only skew 5-hooks so

$$\chi^{(5,4,4)}(5,4,3,1) = \chi^{3,3,2}(4,3,1) - \chi^{5,3}(4,3,1).$$

Applying the rule again we see that

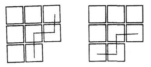

are the only skew 4-hooks in $(3,3,2)$ so

$$\chi^{3,3,2}(4,3,1) = \chi^{2,1,1}(3,1) - \chi^{3,1}(3,1)$$

and

is the only skew 4-hook in $(5,3)$ so

$$\chi^{5,3}(4,3,1) = \chi^{2,2}(3,1).$$

Now there is no skew 3-hook in $(2,1,1)$ (since

does not yield an admissible diagram as it has 0 boxes in the first two rows) and similarly there is no skew 3-hook in $(3,1)$. Therefore

$$\chi^{2,1,1}(3,1) = 0 = \chi^{3,1}(3,1).$$

As

is the only skew 3-hook in $(2,2)$ we see that

$$\chi^{2,2}(3,1) = -\chi^{1}(1)$$

and

$$\chi^{1}(1) = 1.$$

So finally

$$\chi^{(5,4,4)}(5,4,3,1) = -1$$

by repeated application of the Murnaghan–Nakayama rule.

Since there is no skew 8-hook in $(5,4,4)$, we conclude that $\chi^{(5,4,4)}$ vanishes on any conjugate class containing an 8-cycle.

Table 14. $n = 3$

	Conjugacy class		
	1	3	2
	(1, 1, 1)	(2, 1)	(3)
Partition			
(3)	1	1	1
(2, 1)	2	0	− 1
(1, 1, 1)	1	− 1	1

Table 15. $n = 4$

	Conjugacy class				
	1	6	8	3	6
	(1, 1, 1, 1)	(2, 1, 1)	(3, 1)	(2, 2)	(4)
Partition					
(4)	1	1	1	1	1
(3, 1)	3	1	0	− 1	− 1
(2, 2)	2	0	− 1	2	0
(2, 1, 1)	3	− 1	0	− 1	1
(1, 1, 1, 1)	1	− 1	1	1	− 1

Table 16. $n = 5$

	Conjugacy class						
	1	10	20	15	30	20	24
	(1, 1, 1, 1, 1)	(2, 1, 1, 1)	(3, 1, 1)	(2, 2, 1)	(4, 1)	(3, 2)	(5)
Partition							
(5)	1	1	1	1	1	1	1
(4, 1)	4	2	1	0	0	− 1	− 1
(3, 2)	5	1	− 1	1	− 1	1	0
(3, 1, 1)	6	0	0	− 2	0	0	1
(2, 2, 1)	5	− 1	− 1	1	1	− 1	0
(2, 1, 1, 1)	4	− 2	1	0	0	1	− 1
(1, 1, 1, 1, 1)	1	− 1	1	1	− 1	− 1	1

We list the character tables for S_n for low n in Tables 14–17. The reader can check any of the entries against the Murnaghan–Nakayama rule. Each irreducible representation F_λ is described by a partition λ. Each conjugacy class is described by its cycle structure. The number of elements is given above the cycle structure. The dimension of each F_λ is the character of the identity.

Table 17. $n = 6$

	Conjugacy class										
	1	15	40	45	90	120	144	15	90	40	120
	(1^6)	$(2,1^4)$	$(3,1^3)$	$(2,2,1,1)$	$(4,1,1)$	$(3,2,1)$	$(5,1)$	$(2,2,2)$	$(4,2)$	$(3,3)$	(6)
Partition											
(6)	1	1	1	1	1	1	1	1	1	1	1
$(5,1)$	5	3	2	1	1	0	0	-1	-1	-1	-1
$(4,2)$	9	3	0	1	-1	0	-1	3	1	0	0
$(4,1,1)$	10	2	1	-2	0	-1	0	-2	0	1	1
$(3,3)$	5	1	-1	1	-1	1	0	-3	-1	2	0
$(3,2,1)$	16	0	-2	0	0	0	1	0	0	-2	0
$(2,2,2)$	5	-1	-1	1	1	-1	0	3	-1	2	0
$(3,1,1,1)$	10	-2	1	-2	0	1	0	2	0	1	-1
$(2,2,1,1)$	9	-3	0	1	1	0	-1	-3	1	0	0
$(2,1^4)$	5	-3	2	1	-1	0	0	1	-1	-1	1
(1^6)	1	-1	1	1	-1	-1	1	-1	1	1	-1

3

MOLECULAR VIBRATIONS
AND HOMOGENEOUS VECTOR BUNDLES

As a physical problem to guide us through the mathematics of this chapter, consider the following situation. Suppose we have a system of point masses held together by forces which keep the system fairly rigid. We may think of this system as a model of a molecule. (We say 'model' because, for the moment, we want to treat the problem as one in classical mechanics. Of course, we are interested in actual molecules, and hence should really use quantum mechanics. We shall discuss the quantum version later in this chapter.)

3.1 Small oscillations and group theory

We assume that we know the number of point masses but not the shape of the system at equilibrium. The system is in a 'black box' which we can shake at various frequencies and so determine the resonant frequencies of the system, which are the same as its frequencies of free oscillation about equilibrium. We want to show how to use this information, together with group theory, to determine the shape of the 'molecule' at equilibrium.

Let $\mathbf{q}$ be the vector which describes the deviation of the system from equilibrium by specifying the displacement of each 'atom'. Thus, if there are N 'atoms', $\mathbf{q}$ will be a vector in a $3N$-dimensional space, since each atom can move independently in three dimensions. (It is sometimes convenient to take out overall translational and rotational motion, i.e. remove six degrees of freedom so $\mathbf{q}$ varies in a $(3N - 6)$-dimensional space.) From the theory of small oscillations we know that the behaviour of the system near equilibrium is described by a second-order differential equation of the form

$$\frac{\mathrm{d}^2\mathbf{q}}{\mathrm{d}t^2} + F\mathbf{q} = 0$$

where F is a self-adjoint operator. The eigenvalues of F determine the vibrational frequencies, and the eigenvectors determine the associated 'normal mode configuration'.

As an example, suppose that the 'molecule' consists of three identical 'atoms', each of mass m arranged in a line and joined by identical springs of force constant k (see

Fig. 3.1

Fig. 3.1). For simplicity, we assume that the atoms can be displaced only in one dimension, so that the vector

$$\mathbf{q} = \begin{pmatrix} q_1 \\ q_2 \\ q_3 \end{pmatrix}$$

completely describes the deviation of the molecule from equilibrium. Then Newton's law says that

$$m\frac{d^2 q_1}{dt^2} = k(q_2 - q_1)$$

$$m\frac{d^2 q_2}{dt^2} = k(q_3 - q_2) - k(q_2 - q_1)$$

$$m\frac{d^2 q_3}{dt^2} = -k(q_3 - q_2)$$

so that

$$\frac{d^2\mathbf{q}}{dt^2} + F\mathbf{q} = 0$$

where

$$F = \frac{k}{m} \begin{pmatrix} 1 & -1 & 0 \\ -1 & 2 & -1 \\ 0 & -1 & 1 \end{pmatrix}.$$

The symmetry of the matrix F indicates that it represents a self-adjoint operator. The eigenvalues and eigenvectors of F correspond to the 'normal modes' of vibration of the system as in Table 18.

If the 'molecule' possesses a symmetry group G, then G acts on the space of displacements from equilibrium. (An exact description of this action will be presented in the next section.) As the group G acts as symmetries of the forces holding the 'molecule' together, the operator F must commute with the action of G. In the preceding example, G includes the interchange of molecules 1 and 3, and the matrix

$$\begin{pmatrix} 0 & 0 & 1 \\ 0 & 1 & 0 \\ 1 & 0 & 0 \end{pmatrix},$$

which interchanges the displacements q_1 and q_3, commutes with the matrix F. We are thus in the following situation. We are given a representation r of a group G on a vector space V and we are given a self-adjoint operator F on V which commutes with the

Table 18.

Eigenvalue	Eigenvector	Mode
0	$\begin{pmatrix} 1 \\ 1 \\ 1 \end{pmatrix}$	uniform translation
$\dfrac{k}{m}$	$\begin{pmatrix} 1 \\ 0 \\ -1 \end{pmatrix}$	symmetric vibration
$\dfrac{3k}{m}$	$\begin{pmatrix} -1 \\ 2 \\ -1 \end{pmatrix}$	anti-symmetric vibration

action of G. Let us first look at the extreme case where the action of G on V is irreducible. Then by Schur's lemma F must be a scalar multiple of the identity, i.e. F has just one eigenvalue. On the other hand, if we pick an operator F at 'random' on V, not requiring that it commute with G, then we would expect that F has n distinct eigenvalues, where $n = \dim V$. As usual in science, we argue backwards: if we observe only one eigenfrequency, where there 'should be' n, we conclude that there must be some reason; most likely that there is a group G which acts irreducibly on the space and commutes with F, forcing all the eigenvalues of F to coincide. More generally, let us drop the requirement that V be irreducible. Let $\mathbf{v}$ be an eigenvector of F with eigenvalue λ, so that

$$F\mathbf{v} = \lambda\mathbf{v}.$$

Then

$$Fr(a)\mathbf{v} = r(a)F\mathbf{v} = \lambda r(a)\mathbf{v}, \quad \text{for any } a \in G.$$

In other words, the space of all eigenvectors of F corresponding to the same eigenvalue λ is invariant under G. Since F is self-adjoint, we can decompose V into a direct sum of eigenspaces corresponding to the various distinct eigenvalues of F;

$$V = Z_1 \oplus \cdots \oplus Z_s.$$

Each of the eigenspaces Z_j is invariant under G. We can further decompose each Z_j into irreducibles under G. Then

$$V = V_1 \oplus V_2 \oplus \cdots \oplus V_p$$

is a decomposition of V into irreducibles, where each V_i lies in some Z_j and hence F has a constant eigenvalue on V_i. Thus F has at most p distinct eigenvalues, where p is the number of irreducible components in the decomposition. But the number of irreducible components is independent of the particular choice of decomposition, and has nothing to do with F. It depends only on the representation.

Suppose, for the sake of illustration, that V has two possible groups acting on it, G_1 and G_2. Under the representation of G_1, V decomposes into

$$V = V_1 + V_2 + V_3 + V_4 + V_5 + V_6,$$

six irreducible components, while under the representation of G_2

$$V = U_1 + U_2 + U_3 + U_4,$$

four irreducible components. Suppose that by observing the resonant frequencies, we see six distinct eigenvalues. Then we know that G_2 cannot be the correct group. If we observe just four frequencies, this is evidence (although not conclusive) that G_2 is the correct group. Thus molecular spectroscopy is an application of Schur's lemma!

The above description of a system in a 'box' which we shake to determine the resonant frequencies does not apply directly to actual molecules, since the laws of classical mechanics must be replaced by quantum mechanics. Instead of shaking the molecule, we shine some electromagnetic radiation on it and observe the frequencies at which the radiation is absorbed (or the displacement in the frequencies). We will explain the modification needed for the quantum theory (in particular the 'selection' rules which prevent some frequencies from being observed) in Section 3.6.

3.2 Molecular displacements and vector bundles

Consider the space of motions of (our molecular model of) carbon tetrachloride. At equilibrium the carbon atom lies at the center, and the four chlorine atoms at the vertices of a regular tetrahedron. We shall label the central carbon atom by C and the four chlorine atoms by $1, 2, 3, 4$ (Fig. 3.2). In a small displacement from equilibrium, each of the atoms moves in its own three-dimensional vector space which describes the displacement of that atom from its equilibrium position. We label these spaces as

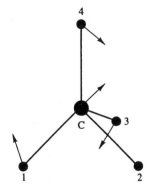

Fig. 3.2

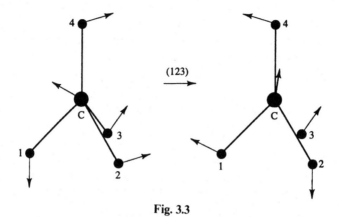

Fig. 3.3

E_C, E_1, E_2, E_3 and E_4. A displacement of the molecule as a whole moves each of the atoms, and so is a function f such that

$$f(C)\in E_C \quad \text{and} \quad f(i)\in E_i, \quad i=1,2,3,4,$$

which tells how each atom has been displaced from its equilibrium.

Now we study how the group S_4 acts on the set of displacements. Consider, for example, the action of the element $(1,2,3)\in S_4$. On the molecule itself, at equilibrium, $(1,2,3)$ leaves C fixed, rotates the chlorine atoms 1, 2 and 3 and leaves 4 fixed (Fig. 3.3). It carries a small displacement of 1 (a vector in E_1) into a rotated displacement of 2 (a vector in E_2). Thus

$$
\begin{aligned}
&E_1 \rightarrow E_2\\
&E_2 \rightarrow E_3\\
(1,2,3): \quad &E_3 \rightarrow E_1\\
&E_4 \rightarrow E_4\\
&E_C \rightarrow E_C.
\end{aligned}
$$

Although the carbon atom is carried into itself, the transformation determined by $(1,2,3)$ on E_C is not the identity transformation, but a rotation. We can now begin to see how the group S_4, or the particular element $(1,2,3)$, acts on displacements of the molecule as a whole: the *new*, rotated, displacement at atom 2 depends on the old, original, displacement at atom 1. There are two things happening: the atoms are interchanged and the displacements themselves are rotated. In order to describe what is going on, it is simpler to pass to a more abstract setting, the notion of a vector bundle over a finite set.

Let M be a finite set. A *vector bundle* over M consists of a collection of vector spaces E_x, one vector space for each point x of M. We let $E = \bigcup_{x\in M} E_x$ (union, not direct sum) and sometimes talk of E as being the vector bundle. E is *not* a vector space. We define the map $\pi: E \rightarrow M$ by setting $\pi(\mathbf{v}) = x$ if $\mathbf{v}\in E_x$. That is, any $\mathbf{v}\in E$ belongs to one of the vector spaces E_x, and we let $\pi(\mathbf{v})$ be this particular x. The space $E_x = \pi^{-1}(x)$ is sometimes called the *fiber* over x, Fig. 3.4(a).

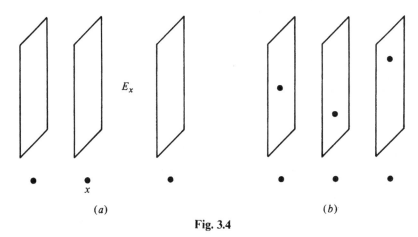

Fig. 3.4

Let E be a vector bundle over M. A *section* of E is a function f, which assigns a vector $f(x) \in E_x$ to each $x \in M$. Thus, f is a map from M to E with the property that

$$\pi \circ f = \text{identity}.$$

In the example of Section 3.1, we introduced a vector

$$\mathbf{q} = \begin{pmatrix} q_1 \\ q_2 \\ q_3 \end{pmatrix}$$

to describe the complete displacement of the three-atom molecule. In keeping with the previous discussion, we should think of $\mathbf{q}$ as the *function* which assigns the displacement q_1 in space E_1 to atom 1, the displacement q_2 in the space E_2 to atom 2, and the displacement q_3 in the space E_3 to atom 3. When we draw a 'picture' of $\mathbf{q}$, e.g.

$$\begin{array}{ccc} \bullet\!\!\rightarrow & \leftarrow\!\!\bullet & \bullet\!\!\rightarrow \\ 1 & 2 & 3 \end{array}$$

we are representing the *function* $\mathbf{q}$ by its value on the set on which it acts.

More generally, let f_1 and f_2 be sections. Since $f_1(x)$ and $f_2(x)$ lie in the same vector space, E_x, it makes sense to add them. We can thus define the sum of two sections by

$$(f_1 + f_2)(x) = f_1(x) + f_2(x).$$

Similarly, if c is any complex number and f is any section, we define cf by

$$(cf)(x) = cf(x).$$

In this way, *the space of all sections becomes a vector space*. We denote this vector space by $\Gamma(E)$. So $\Gamma(E)$ has the same dimension as the direct sum of the spaces E_x, but it is not to be thought of as the same space. The distinction will become crucial when we consider vector bundles over sets which are not discrete.

Suppose that the group G acts on M and that E is a vector bundle over M. We say that G acts as a group of vector bundle morphisms on E, or that E is a *homogeneous vector bundle* under G, if

(1) G acts on E;
(2) the map $\pi: E \to M$ is a G morphism, i.e. $a\pi(v) = \pi(av)$ for all $a \in G$ and $v \in E$.

Condition (2) is the same as saying that $a: E_x \to E_{ax}$. We require also that

(3) the map $a: E_x \to E_{ax}$ is linear for each $a \in G$ and $x \in M$.

Thus, each a permutes the various vector spaces according to its action on M, and is a linear map from one vector space to another. Convince yourself that this action of a corresponds to the physical operation of performing a symmetry operation on a molecule whose atoms are displaced from equilibrium.

We can now let the group G act on $\Gamma(E)$ by setting

$$r(a)f(x) = a[f(a^{-1}x)]. \tag{2.1}$$

Notice that $f(a^{-1}x) \in E_{a^{-1}x}$ so that the right-hand side of (2.1) is an element of E_x, and thus $r(a)f$ is indeed again a section of E. It is clear that the map $f \rightsquigarrow r(a)f$ is a linear transformation, and so we obtain a linear representation of G on $\Gamma(E)$. In the case that M is the set of atoms of a molecule and E_x denotes the displacement of the atom x from its equilibrium position, $\Gamma(E)$ is the space of displacements of the molecule as a whole, and the action of G on $\Gamma(E)$ is our desired action of the symmetry group on the space of displacements.

As an example, let M be the atoms of a square planar 'molecule', and let $f(x)$, an element of $\Gamma(E)$, be the function which assigns to each atom the displacement shown in Fig. 3.5(a). If a is counter-clockwise rotation through 90°, then $r(a)f(x) = a[f(a^{-1}x)]$ tells us, for example, that $r(a)f(1) = af(4)$ and $r(a)f(2) = af(1)$. We obtain the displacement assigned to atom 1 by $r(a)f$ by performing a 90° counter-clockwise rotation on the displacement assigned to atom 4 by f, and so on. The net effect of this procedure is simply that Fig. 3.5(b) representing $r(a)f$, is obtained by rotating Fig. 3.5(a) representing f, counter-clockwise through 90°.

As a second example, consider the case of the 'trivial line bundle' where $E = M \times \mathbb{C}$. Over each point we have a copy of $\mathbb{C}$: the points of E_x are pairs (x, z), where z is a complex number, and $a \in G$ sends (x, z) into (ax, z). In other words, the group does not

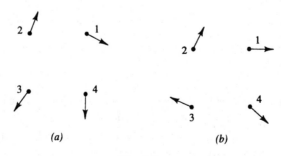

(a)   (b)

Fig. 3.5

act on the z component at all. In this case, a section f is a function of the form $f(x) = (x, g(x))$, where g is an ordinary complex-valued function on M. We may simply identify f with g, so that (2.1) becomes $r(a)f(x) = f(a^{-1}x)$. This is the representation of G on $F(M)$ that we studied in detail in Chapter 2. Our immediate task is to generalize many of the theorems we proved about the representation of G on $F(M)$ to the more general case of the representation of G on $\Gamma(E)$.

We shall denote the representation of G on $\Gamma(E)$ by r_E and its character by χ_E. In order to evaluate χ_E we introduce a convenient basis of $\Gamma(E)$, generalizing the δ function basis we used for $F(M)$. For each vector space, E_x, we introduce a basis $\mathbf{v}_{x1}, \ldots, \mathbf{v}_{xk_x}$, where $k_x = \dim E_x$. For any $\mathbf{v} \in E_x$ we define the section $f_{\mathbf{v}}$ by

$$f_{\mathbf{v}}(y) = \begin{matrix} \mathbf{v} & \text{if } y = x \\ 0 & \text{if } y \neq x \end{matrix}.$$

It now follows, as for the case of $F(M)$, that

$$r_E(a)f_{\mathbf{v}} = f_{a\mathbf{v}}. \tag{2.2}$$

For example, if M is a triangular molecule and a is a 120° counter-clockwise rotation, typical diagrams representing $f_{\mathbf{v}}$ and $r(a)f_{\mathbf{v}}$ are as in Fig. 3.6.

As x ranges over all of M and the $\mathbf{v}_{xi}$ over a basis of each E_x, the sections $f_{\mathbf{v}_{xi}}$ form a basis of $\Gamma(E)$: they simply assign independent displacements to each 'atom' in turn. If we use this basis, it is clear from (2.2) that the only non-zero entries on the diagonal of the matrix representing $r_E(a)$ can come from those $\mathbf{v}_{xi}$ for which $ax = x$. For each x with $ax = x$, the element a maps E_x into itself, and

$$af_{\mathbf{v}_{xi}} = f_{a\mathbf{v}_{xi}}.$$

Now if (A_{xij}) is the matrix of the map $a: E_x \to E_x$ with respect to the basis $\mathbf{v}_{xi}$, then

$$a\mathbf{v}_{xj} = \sum_i A_{xij}\mathbf{v}_{xi}$$

and hence

$$af_{\mathbf{v}_{xj}} = \sum_i A_{xij}f_{\mathbf{v}_{xi}}.$$

Thus, for each fixed x the sum of the diagonal elements of the linear transformation $f \rightsquigarrow af$ coming from the $\mathbf{v}_{xi}$ is $\sum_i A_{xii}$, which is just the trace of the linear transformation $a: E_x \to E_x$. Summing over all the fixed points, x, of M, we obtain the celebrated

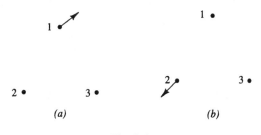

(a)   *(b)*

Fig. 3.6

Table 19. Character table of T_d

$24T_d$	$[e]$	$8[r_3]$	$3[r_2]$	$6[s_4]$	$6[\tau_d]$
χ_1	1	1	1	1	1
χ_2	2	-1	2	0	0
χ_3	1	1	1	-1	-1
χ_4	3	0	-1	1	-1
χ_5	3	0	-1	-1	1

Frobenius fixed point character formula:

$$\chi_E(a) = \sum_{ax=x} [\text{tr}(a:E_x \to E_x)]. \tag{2.3}$$

For the case of the trivial line bundle, where $\Gamma(E) = \mathscr{F}(M)$, the map $a:E_x \to E_x$ is the identity, so the trace at each fixed point is one. Equation (2.3) then reduces to (5.1), Chapter 2.

For the case of molecular vibrations, all the summands on the right of (2.3) are the same, and (2.3) reduces to Wigner's formula:

$$\chi_E(a) = [\# \text{ of fixed points of } a] \cdot \chi^r(a). \tag{2.4}$$

By combining Wigner's formula with the use of character tables, as described in Section 7, Chapter 2, we can find the decomposition of the space of displacements of any molecule into irreducibles.

Let us illustrate this by considering the space of possible displacements of our molecule of CCl_4, which is assumed to have the symmetry of a tetrahedron, so that its symmetry group is T_d. The chlorine molecules are assumed to lie at the vertices of the tetrahedron, and the carbon atom at the center. All elements of T_d keep the carbon atom fixed and act on the chlorine atoms as T_d acts on the vertices of its tetrahedron. The character table of T_d is presented in Table 19. The character χ_5 corresponds to the action of T_d on $\mathbb{R}^3$ as the symmetries of the tetrahedron, and so this is the χ^r that we want to use in the right-hand side of Wigner's formula (2.4). The number of fixed points is 5 for the identity $[e]$, 2 for any $120°$ rotation $[r_3]$, 1 for any $180°$ rotation $[r_2]$, which corresponds to a permutation like (12) (34) of the tetrahedron but holds the central carbon atom fixed, 1 for the elements of class $[s_4]$, which correspond to permutations like (1234) which hold only the carbon atom fixed, and 3 for any reflection $[\tau_d]$, which interchanges two chlorine atoms while holding the other two fixed, and also holds the carbon atom fixed. Using (2.4), we find that the character χ_E is represented by the row

$$15 \quad 0 \quad -1 \quad -1 \quad 3.$$

(The last entry, for example, arises because a reflection through a plane containing an edge leaves two vertices of the tetrahedron fixed. Together with the central atom, this gives three fixed points. Multiplying by the value of χ_5 gives 3.) We now find how many times each irreducible representation occurs in the decomposition of the representation of $\Gamma(E)$.

$$(\chi_E, \chi_1) = (1/24)(1 \cdot 15 + 0 + 3 \cdot (-1) + 6 \cdot (-1) + 6 \cdot 3) = 1$$
$$(\chi_E, \chi_2) = (1/24)(2 \cdot 15 + 0 + 3 \cdot 2 \cdot (-1) + 0 + 0) = 1$$
$$(\chi_E, \chi_3) = (1/24)(15 - 3 + 6 - 18) = 0$$
$$(\chi_E, \chi_4) = (1/24)(45 + 3 - 6 - 18) = 1$$
$$(\chi_E, \chi_5) = (1/24)(45 + 3 + 6 + 18) = 3$$

so that

$$\chi_E = \chi_1 + \chi_2 + \chi_4 + 3\chi_5,$$

which can be checked by adding up the columns with the appropriate multiplicity.

It is informative to visualize the subspaces corresponding to this decomposition as actual subspaces of the space of displacements of the molecule. Since χ_1, χ_2 and χ_4 each occur with multiplicity one, each corresponds to a unique subspace. For χ_5, there is no unique way to pick out any three-dimensional subspace with character χ_5, although, as we know, the nine-dimensional space consisting of the direct sum of any three independent χ_5-type three-dimensional subspaces *is* invariantly defined. The representation with character χ_1 is the trivial representation. If we do not move the carbon atom at all, and move the chlorine atoms out along the axes joining them to the carbon atom, moving each chlorine atom the same amount, we get a configuration that is clearly invariant under T_d, and there is a one-dimensional space of such possible motions, cf. Fig. 3.7.

The space of displacements in which the carbon atom is fixed and the chlorine atoms move along the axes is an invariant four-dimensional space which contains the above space as a one-dimensional invariant subspace. It therefore contains as an invariant complement, a three-dimensional subspace.

The space of displacements in which the carbon atom is fixed and the chlorine atoms are displaced perpendicularly to the axes forms an eight-dimensional invariant subspace, which must therefore split into a two, a three and a three. One of these three-

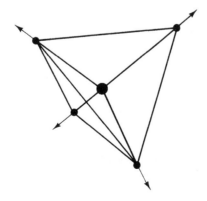

Fig. 3.7

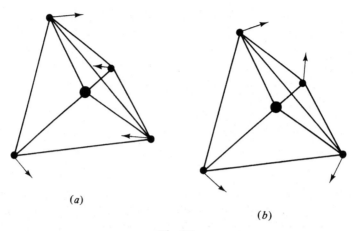

(a)

(b)

Fig. 3.8

dimensional subspaces can be identified as being given by an infinitesimal rotation of the molecule as a whole. It is an instructive exercise for the reader to verify that the group T_d acting on the space of infinitesimal rotations has character χ_4. (The space of infinitesimal rotations can be identified with the space of antisymmetric 3×3 matrices. These can be identified with vectors in three-space only if a choice of orientation is made. Reversing the orientation changes the identification by a factor of -1. That is why the character χ_4 differs from χ_5 by a factor of -1 precisely on the orientation reversing elements.) Thus the character of this eight-dimensional representation is $\chi_2 + \chi_4 + \chi_5$.

We draw two linearly independent displacements in the two-dimensional invariant subspace in Figs. 3.8(a) and (b).

The rigid translation of the molecule as a whole, where we draw the same vector in three-space at each molecule, clearly gives a three-dimensional invariant subspace with character χ_5. Actually, for physical reasons, we will want to separate the translations and the rotations of the molecule from the other displacements. This leaves us with a nine-dimensional space of vibrations, with character

$$\chi_1 + \chi_2 + 2\chi_5$$

There is no preferred way of picking the two three-dimensional vibrating subspaces.

3.3   Induced representations

Let $E \to M$ be a homogeneous vector bundle for the group G and let N be a subset of M. We can form $\pi^{-1}N = \bigcup_{x \in N} E_x$, which is now a vector bundle over N, which we denote by E_N. If N is mapped onto itself by all elements of G, i.e. is stable under G, then E_N is clearly a homogeneous vector bundle for G. We can identify $\Gamma(E_N)$ with a subspace of $\Gamma(E)$, namely the subspace consisting of all sections which vanish outside of N, i.e those f for which $f(y) = 0$ for $y \notin N$. If $M = \cup M_i$ is a disjoint union, then we have the direct sum decomposition

$$\Gamma(E) = \Gamma(E_{M_1}) \oplus \cdots \oplus \Gamma(E_{M_k}).$$

If the M_i are invariant, in particular if they are orbits, we obtain the Mackey decomposition formula:

$$r_E = r_{E_{M_1}} \oplus \cdots \oplus r_{E_{M_k}} \tag{3.1}_r$$

and the corresponding character decomposition

$$\chi_E = \chi_{E_{M_1}} + \cdots + \chi_{E_{M_k}}. \tag{3.1}_\chi$$

So we might as well look at the case where G acts transitively on M. Let m be a point of M and $H = G_m$, the isotropy group of m. Each element of H maps E_m into E_m and acts linearly. In other words, we get a representation of H on E_m. Let us denote this representation by s. Let n be some other point of M. Since G acts transitively, we can write

$$n = am$$

for some $a \in G$, and if $n = bm$, then $b = ah$ for some $h \in H$. Any vector in E_n is the image of some vector in E_m under the action of a. In other words, we can write any vector in E_n as

$$\mathbf{w} = a\mathbf{u} \quad \mathbf{u} \in E_m.$$

If also

$$\mathbf{w} = b\mathbf{v}$$

then $b\mathbf{v} = a\mathbf{u}$ or $a^{-1}b\mathbf{v} = \mathbf{u}$ or

$$\mathbf{u} = s(h)\mathbf{v}.$$

This shows that the representation s determines the vector bundle E. We now show how to construct a homogeneous vector bundle from a representation of a subgroup.

So suppose we start with G and a subgroup H. We can reconstruct M as the coset space G/H: a point of $M = G/H$ is a coset aH. Suppose that s is a representation of the subgroup H on a vector space V. On the space $G \times V$ we introduce the equivalence relation

$$(gh, \mathbf{v}) \sim (g, s(h)\mathbf{v})$$

and let E denote the set of all equivalence classes. We denote the equivalence class of $(g, \mathbf{v}) \in G \times V$ by $[(g, \mathbf{v})]$. We write

$$E = G \underset{H}{\times} V.$$

The map sending $(g, \mathbf{v}) \rightsquigarrow gH \in M$ is clearly constant on equivalence classes, and hence defines a map from $E \to M$. Suppose that $x = gH$. Let E_x consist of all equivalence classes of elements $(g, \mathbf{v})$ as $\mathbf{v}$ ranges over V. We have an identification of the set E_x with the vector space V, which depends on the choice of g:

$$\phi_g \colon V \to E_x, \quad \phi_g(\mathbf{v}) = [(g, \mathbf{v})].$$

If we change our choice of g by replacing g by gh, we get

$$\phi_{gh}(\mathbf{v}) = [(gh, \mathbf{v})] = [(g, s(h)\mathbf{v})] = \phi_g(s(h)\mathbf{v})$$

so that

$$\phi_{gh} = \phi_g \circ s(h). \tag{3.2}$$

This equation shows that we may use the map ϕ_g to define a vector space structure on E_x which is independent of the particular choice of g. If $\mathbf{e}_i = \phi_g(\mathbf{v}_i)$, we define

$$\mathbf{e}_1 + \mathbf{e}_2 = \phi_g(\mathbf{v}_1 + \mathbf{v}_2).$$

In view of (3.2) and the fact that $s(h)$ acts linearly on V, the value of this sum does not depend on the choice of g, and similarly for the definition of multiplication of a vector by a scalar.

We have thus made E into a vector bundle over M. We define the action of the group G on this vector bundle by multiplication on the left:

$$a[(g, \mathbf{v})] = [(ag, \mathbf{v})].$$

It is obvious that this definition is independent of the choice of representative. It defines a linear map from $E_x \to E_{ax}$ by the very definition of the linear structure on these fibers. Thus E is a homogeneous vector bundle for G, and we obtain a representation of G on $\Gamma(E)$. This representation was obtained from the subgroup H and the representation s of H. It is called the representation of G *induced* from the representation s, and we will denote it by $(s \uparrow G)$.

Before proceeding with the general theory of induced representations, let us illustrate the concept with a specific example. Let $G = D_k$ be the dihedral group of order $2k$, and let $H = C_k$ be our choice of subgroup. Thus C_k is of index two in D_k, so $M = G/H = D_k/C_k$ consists of two points. If we choose some $g \notin H$, then the two points can be denoted by

$$x = H \quad \text{and} \quad y = gH.$$

The action of G on M is given by

$$hx = x \qquad hy = y \qquad h \in H$$
$$(gh)x = y \quad (gh)y = x \quad h \in H, \quad g \notin H.$$

We can write $H = \{e, a, a^2, ..., a^{k-1}\}$, where a is the rotation through angle $2\pi/k$. Let us choose a one-dimensional representation, s, of H, so that

$$s(a) = \varepsilon,$$

where ε is a kth root of unity, $\varepsilon^k = 1$, and so

$$s(a^j) = \varepsilon^j.$$

Thus E consists of two complex lines, one sitting over each point of M, as shown in Fig. 3.9. Choosing a $\mathbf{v} \in \mathbb{C}$ and a $g \notin H$ then picks out a basis, $\mathbf{u} = [(e, \mathbf{v})]$ of E_x and $\mathbf{w} = [(g, \mathbf{v})]$ of E_y. We thus get a basis, $f_{\mathbf{u}}, f_{\mathbf{w}}$ of $\Gamma(E)$, where

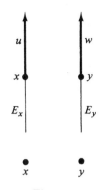

Fig. 3.9

$$f_{\mathbf{u}}(x) = \mathbf{u}, \quad f_{\mathbf{u}}(y) = 0$$

and

$$f_{\mathbf{w}}(x) = 0, \quad f_{\mathbf{w}}(y) = \mathbf{w}.$$

To compute further, we observe that the multiplication in G gives

$$ga^i g^{-1} = a^{-i} \quad g \notin H.$$

Now

$$a^i \mathbf{u} = a^i[(e, \mathbf{v})] = [(a^i, \mathbf{v})] = [(e, s(a^i)\mathbf{v})] = [(e, \varepsilon^i \mathbf{v})]$$
$$= \varepsilon^i \mathbf{u}$$

so

$$a^i f_{\mathbf{u}} = \varepsilon^i f_{\mathbf{u}}$$

while

$$a^i \mathbf{w} = a^i[(g, \mathbf{v})] = [(a^i g, \mathbf{v})] = [(g a^{-i}, \mathbf{v})]$$
$$= [(g, s(a^{-i})\mathbf{v})] = [(g, \varepsilon^{-i}\mathbf{v})]$$
$$= \varepsilon^{-i}\mathbf{w}.$$

so

$$a^i f_{\mathbf{w}} = \varepsilon^{-i} f_{\mathbf{w}}.$$

Notice that a^i fixes both points x and y. Thus the matrix of a^i on $\Gamma(E)$ with respect to the basis $f_{\mathbf{u}}$, $f_{\mathbf{w}}$ is

$$\begin{pmatrix} \varepsilon^i & 0 \\ 0 & \varepsilon^{-i} \end{pmatrix}.$$

For the element ga^i, we have $(ga^i)x = y$ and $(ga^i)y = x$ and

$$ga^i \mathbf{u} = ga^i[(e, \mathbf{v})] = [(ga^i, \mathbf{v})] = [(g, s(a^i)\mathbf{v})] = [(g, \varepsilon^i \mathbf{v})]$$
$$= \varepsilon^i \mathbf{w}$$

and

$$ga^i \mathbf{w} = ga^i[(g, \mathbf{v})] = [(ga^i g, \mathbf{v})] = [(a^{-i}, \mathbf{v})] = [(e, s(a^{-i})\mathbf{v})]$$
$$= [(e, \varepsilon^{-i}\mathbf{v})] = \varepsilon^{-i}\mathbf{u}$$

and the matrix of ga^i is

$$\begin{pmatrix} 0 & \varepsilon^{-i} \\ \varepsilon^i & 0 \end{pmatrix}.$$

Since every element of G is either of the form a^i or ga^i, we have completely determined the matrices of the induced representation $s \uparrow G$:

$$a^i \rightsquigarrow \begin{pmatrix} \varepsilon^i & 0 \\ 0 & \varepsilon^{-i} \end{pmatrix} \quad ga^i \rightsquigarrow \begin{pmatrix} 0 & \varepsilon^{-i} \\ \varepsilon^i & 0 \end{pmatrix}.$$

The character of this representation is clearly given by

$$\chi(a^i) = \varepsilon^i + \varepsilon^{-i} \quad \text{and} \quad \chi(ga^i) = 0,$$

in agreement with (2.3). Notice that this representation is irreducible, unless $\varepsilon = \pm 1$, for if $\varepsilon \neq \pm 1$, the only subspaces invariant under a are the lines spanned by f_u or f_w and these are interchanged by g. For $\varepsilon = \pm 1$, $s \uparrow G$ is reducible.

We return to the general theory of induced representations. Let s be a representation of the subgroup H of the group G, and let σ be its character. We shall denote the character of the induced representation, $s \uparrow G$, by $\sigma \uparrow G$. The formula for $\sigma \uparrow G$ is given by (2.3). It is sometimes convenient to rewrite (2.3) in a slightly different form. The sum in (2.3) is over fixed points. To say that $x = gH$ is fixed under $a \in G$ means that $g^{-1}ag \in H$. Of course the element g is only determined by x up to right multiplication by an arbitrary element of H. The action of a on E_x can be described as follows: if $\mathbf{u} \in E_x$ is given by $\mathbf{u} = \phi_g \mathbf{v}$ for $\mathbf{v} \in V$, then

$$a\mathbf{u} = \phi_g(s_{g^{-1}ag})\phi_{g^{-1}}\mathbf{u}$$

so that

$$\text{tr}\,[a : E_x \to E_x] = \text{tr}\,s_{g^{-1}ag} = \sigma(g^{-1}ag).$$

The expression on the right makes sense because $g^{-1}ag \in H$, and does not depend on which g we pick with $x = gH$. For purposes of counting, it is sometimes easier to sum over *all* g with $g^{-1}ag \in H$, instead of summing over the fixed points, but then we will have counted each fixed point $\#H$ times. We divide by $\#H$ to compensate for the overcounting, and (2.3) becomes

$$\sigma \uparrow G(a) = (1/\#H) \sum_{\substack{g \in G \\ g^{-1}ag \in H}} \sigma(g^{-1}ag). \tag{3.3}$$

Let χ be a character of the group G, and let $\chi|_H$ denote the restriction of χ to H. Let us compute $(\sigma \uparrow G, \chi)_G$. We have

$$(\sigma \uparrow G, \chi)_G = \frac{1}{\#G} \sum (\sigma \uparrow G)(a)\overline{\chi(a)}$$

$$= \frac{1}{\#G} \frac{1}{\#H} \sum_{\substack{a \in G \\ h = b^{-1}ab \in H}} \sigma(b^{-1}ab)\overline{\chi(a)} \quad \text{by (3.3)}$$

$$= \frac{1}{\#G}\frac{1}{\#H} \sum_{\substack{h\in H \\ b\in G}} \sigma(h)\overline{\chi(bhb^{-1})}$$

$$= \frac{1}{\#H} \sum_{h\in H} \sigma(h)\overline{\chi(h)}$$

since $\chi(bhb^{-1}) = \chi(h)$

$$= (\sigma, \chi_H)_H.$$

Thus we have proved the *Frobenius reciprocity formula*:

$$(\sigma\uparrow G, \chi)_G = (\sigma, \chi|_H)_H. \tag{3.4}$$

If χ is the character of a representation of G on a vector space W, the left-hand side of (3.4) is just dim $\operatorname{Hom}_G(W, \Gamma(E))$. If σ is the character of a representation of H on a vector space F (so the fiber of E over H is F), the right-hand side of (3.4) is just dim $\operatorname{Hom}_H(W, F)$. Thus we can rewrite (3.4) as

$$\dim \operatorname{Hom}_G(W, \Gamma(E)) = \dim \operatorname{Hom}_H(W, F). \tag{3.5}$$

In fact, we can say more: that there is a natural identification of the two vector spaces $\operatorname{Hom}_G(W, \Gamma(E))$ and $\operatorname{Hom}_H(W, F)$. For this purpose it is convenient to have a different description of the space $\Gamma(E)$ and the action of G on it.

Let $x\in M = G/H$. Recall that for each $b\in G$ such that $x = bH$, we have identified E_x with the set of all (b, v), where $v\in F$. If $f\in\Gamma(E)$, then $f(x)\in E_x$, so we can write

$$f(x) = [(b, \hat{f}(b))],$$

where $\hat{f}(b)\in F$. If $c\in H$, then $f(x) = f(bH) = f(bcH)$, and

$$f(bcH) = [(bc, \hat{f}(bc))].$$

Thus

$$\hat{f}(bc) = s(c)^{-1}\hat{f}(b) \quad \text{for } c\in H. \tag{3.6}$$

Conversely, any function $\hat{f}:G\to F$ satisfying (3.6) defines a section of $\Gamma(E)$. Thus *we may identify $\Gamma(E)$ with the space of all functions from G to F satisfying (3.6)*. Let us denote this space by $\hat{\Gamma}$. We now compare the action of G on both spaces. On $\Gamma(E)$, the representation r_E is given by

$$(r_E(a)f)(x) = af(a^{-1}x)$$

while on the space of functions the representation $\hat{r}$ is given by

$$\hat{r}(a)\hat{f}(g) = \hat{f}(a^{-1}g).$$

If $x = bH$, then

$$f(x) = [(b. \hat{f}(b))]$$

and

$$(r_E(a)f)(x) = a[(a^{-1}b, \hat{f}(a^{-1}b))]$$
$$= [(b, \hat{f}(a^{-1}b))]$$

so the function corresponding to $r_E(a)f$ is $\hat{r}(a)\hat{f}$ as required.

We can thus think of the induced representation either way. The section point of view is geometrical and convenient for certain computations. The function point of view is useful for certain proofs.

The Frobenius reciprocity theorem
Let u be an element of F. Define $c_u \in \hat{\Gamma}$ by

$$c_u(a) = \begin{matrix} 0 & \text{if} & a \notin H \\ s(a^{-1})u & \text{if} & a \in H \end{matrix}.$$

The map sending u to c_u is a map from F to $\hat{\Gamma}$ and, in fact, $c \in \mathrm{Hom}_H(F, \hat{\Gamma})$ since

$$c_{s(h)u}(a) = \begin{matrix} 0 & \text{if} & a \notin H \\ s(a^{-1}h)u & \text{if} & a \in H \end{matrix} = c_u(h^{-1}a) = (\hat{r}(h)c_u)(a).$$

Now let (t, W) be any representation of G. Then, by restriction, we can think of W as a representation space of H. Let $S:F \to W$ be an element of $\mathrm{Hom}_H(F, W)$. Define

$$T_S:\hat{\Gamma} \to W$$

by

$$T_S\hat{f} = \frac{1}{\#H} \sum_{a \in G} t(a)S\hat{f}(a)$$

Then

$$t(b)T_S\hat{f} = \frac{1}{\#H} \sum_{a \in G} t(ba)S\hat{f}(a)$$

$$= \frac{1}{\#H} \sum_{c \in G} t(c)S\hat{f}(b^{-1}c) \qquad a = b^{-1}c$$

$$= T_S\hat{r}(b)\hat{f}$$

so

$$T_S \in \mathrm{Hom}_G(\hat{\Gamma}, W).$$

Also

$$T_S \cdot c_u = \frac{1}{\#H} \sum_{a \in H} t(a)Ss^{-1}(a)u$$

$$= \frac{1}{\#H} \sum_{a \in H} Ss(a)s^{-1}(a)u = Su.$$

Given $S \in \mathrm{Hom}_H(F, W)$, we thus get a $T_S \in \mathrm{Hom}_G(\hat{\Gamma}, W)$ with

$$T_S \circ c = S.$$

This gives an injection of $\mathrm{Hom}_H(F, W)$ into $\mathrm{Hom}_G(\hat{\Gamma}, W)$ which is an isomorphism, since we already know that the dimensions are the same.

$$\Gamma(E) \sim \hat{\Gamma} \xrightarrow{\;T_S\;} W$$

This is the Frobenius reciprocity theorem:

$$\operatorname{Hom}_H(F, W) \sim \operatorname{Hom}_G(\Gamma(E), W). \tag{3.7}$$

The Frobenius reciprocity theorem says that, in a sense, the operations of restriction and induction are reciprocal to one another. (They are adjoint, in the language of category theory.) We can formulate (3.7) as, say, that the induced representation can be viewed as the solution of a 'universal problem': given an H morphism, S, from F to the G representation space, W, there is a unique G morphism, T_S, from $\Gamma(E)$ to W such that $S = T_S \circ c$, where c is the map of $F \to \hat{\Gamma}(E)$ defined above.

For many purposes it is convenient to be able to replace the space $\Gamma(E)$ by a space of functions from M to a fixed vector space V, i.e. to replace the vector bundle E by a 'trivial bundle' $M \times V$. Let us describe one situation where this is always possible. Suppose, as before, that G acts transitively on M, that H is the isotropy group of some point $m \in M$ and that E is induced from some representation (ρ, V) of H. Thus, a typical point of E is an equivalence class

$$[(a, v)]$$

with

$$(ah^{-1}, \rho(h)v) \sim (a, v)$$

$a \in G$, $h \in H$, $v \in V$. Now suppose that

there is a representation, τ, of G on V such that $\tau(h) = \rho(h)$ for all $h \in H$.

Then let us define the map $\phi : E \to M \times V$ by

$$\phi[(a, v)] = (am, \tau(a)v).$$

This is well defined since

$$ah^{-1}m = am$$

and

$$\begin{aligned} \tau(ah^{-1})\rho(h)v &= \tau(a)\tau(h^{-1})\rho(h)v \\ &= \tau(a)\rho(h^{-1})\rho(h)v \\ &= \tau(a)v \end{aligned}$$

It is clear that ϕ is an isomorphism of vector bundles. Let us define the action of G on $M \times V$ by

$$b(m, w) = (bm, \tau(b)w).$$

Then

$$\phi(b[(a, v)]) = \phi([(ba, v)])$$

$$= (bam, \tau(ba)v)$$
$$= b(am, \tau(a)v).$$

Thus

$$b\phi = \phi b,$$

i.e. ϕ is a G morphism.

At the level of sections, our identification sends the function $\hat{f}$ in $\hat{\Gamma}$ into a V-valued function ψf on M by

$$(\psi \hat{f})(m) = \tau(a)\hat{f}(a) \quad \text{if} \quad m = aH.$$

Then

$$(\psi \hat{r}(b)\hat{f})(m) = \tau(a)\hat{f}(b^{-1}a) = \tau(b)\tau(b^{-1}a)\hat{f}(b^{-1}a) = \tau(b)(\psi \hat{f})(b^{-1}m).$$

If we think of $(\psi \hat{f})$ as defining a section of the bundle $M \times V$, this last expression on the right is exactly the action of b on $\psi \hat{f}$, evaluated at the point m. Thus ψ defines a G morphism between $\Gamma(E) \sim \hat{\Gamma}$ and $\Gamma(M \times V)$.

To summarize:

> suppose that the representation (ρ, V) of H extends to a represent-
> ation, τ, of G on V. Then there is a G isomorphism ψ between $\qquad$ (3.8)
>
> $$\Gamma(E) \quad \text{and} \quad \mathscr{F}(M, V)$$

where $\mathscr{F}(M, V)$ denotes the space of functions from M to V. The morphism is defined as follows: if an $s \in \Gamma$ is given by

$$s(m) = [(a, u(a))], \quad \text{where } m = am_0$$

then

$$(\psi s)(m) = \tau(a)u(a). \qquad (3.9)$$

Here the group G acts on $\mathscr{F}(M, V)$ by

$$(bh)(m) = \tau(b)h(b^{-1}m), \quad h \in \mathscr{F}(M, V)$$

3.4 Principal bundles

Let us go back and examine once more the construction of the vector bundle E from the representation of H on V. In the construction, we used the action of H on G by right multiplication, in order to define the action of H on $G \times V$ which gave our equivalence relation. But we did not use the full fact that G was a group until we wanted to define the action of G on E. It is best to separate these two phenomena, and discuss the construction of E in a more general setting. For this we postulate the necessary properties of the right action that we need. Assume that we are given a set P together with a right action of H on P, i.e. a map

$$P \times H \to P, \quad (p, h) \to ph$$

with

$$p(h_1 h_2) = (ph_1)h_2.$$

We further assume that the action is *free* in the sense that

$$ph_1 = ph_2 \text{ for some } p \in P \text{ implies } h_1 = h_2.$$

The set P is called a principal H bundle. It is a bundle over $M = P/H$. In other words, let M be the set of orbits of the right action of H on P. We let π denote the projection of $P \to M$ which assigns to each point its orbit. Given an $x \in M$ and two points p_1 and $p_2 \in \pi^{-1}(x)$, there is a unique $h \in H$ with $p_1 = p_2 h$. Conversely, given $p_1 \in \pi^{-1}(x)$ and $h \in H$, the element $p_2 = p_1 h$ again belongs to $\pi^{-1}(x)$. Thus each $\pi^{-1}(x)$ looks like a copy of H but with no preferred identity element. It is much like the relation between an affine space and a vector space. We say that *P is a principal bundle over M, with structure group H*. Now suppose that we are given some representation, σ, of H on a vector space, V. As before, we can define the (left) action of H on $P \times V$ by

$$h(p, v) = (ph^{-1}, \sigma(h)v).$$

As before, the quotient space

$$E = (P \times V)/H$$

is a vector bundle over M. It is called the vector bundle *associated* to the representation σ, or, more succinctly, an associated bundle.

We will now show that there is a natural identification between sections of E and certain kinds of functions from P to V. Let s be a section of E. Define the function f_s: $P \to V$ by

$$s(\pi(p)) = [(p, f_s(p))].$$

In other words, what we are saying is as follows. For each $x \in M$ the element $s(x)$ lies in E_x and can be represented as the equivalence class $s(x) = [(p, w)]$ for some w in V. The choice of w depends on which p we pick in $\pi^{-1}(x)$, but having picked a p the w is determined, and we define the function $f_s(p)$ by setting $f_s(p) = w$. If we chose a different element of $\pi^{-1}(x)$, say the element ph for some h in H, then we would have to replace w by $\sigma^{-1}(h)w$. In other words, the function f_s satisfies the identity

$$f_s(ph) = \sigma(h^{-1})f_s(p). \tag{4.1}$$

Conversely, suppose f is a function satisfying this identity. Then define the section s of E by setting

$$s(x) = [(p, f(p))] \text{ for any } p \text{ with } \pi(p) = x.$$

The identity satisfied by f guarantees that this is well defined.

In particular, taking the case $P = G$, we see that the space $\Gamma(E)$ can be identified with the subspace of the space of V valued functions, namely those functions satisfying (4.1), on G. This was the construction we used in the preceding section.

Before going on to the next topic, we mention some definitions connected to principal bundles. Although we will not make any immediate use of these definitions, they play a key role in the present-day theory of elementary particles. Let $P \to M$ be a principal bundle with structure group H. A map $\phi: P \to P$ is called an *automorphism* of P if it is one-to-one and surjective and commutes with the action of H, i.e.

$$\phi(ph) = \phi(p)h \quad \text{for all} \quad h \in H \quad \text{and} \quad p \in P.$$

In other words, an automorphism of P is an H morphism which is bijective. If ϕ is such an automorphism, then $\pi(\phi(p))$ clearly depends only on $\pi(p)$. Thus ϕ determines a one-to-one transformation of M onto itself, which we denote by $\bar{\phi}$. The group of all automorphisms of P will be denoted by Aut P. We have shown that there is a homomorphism from Aut P to $S(M)$, the group of all one-to-one transformations of M, the homomorphism sends ϕ into $\bar{\phi}$. The kernel of this homomorphism is the group of those ϕ which induce the identity transformation on M, i.e. those $\phi \in$ Aut P which satisfy

$$\pi \circ \phi = \pi.$$

This subgroup of Aut P is called the *gauge group* of P and is denoted by Gau P.

It is interesting to observe that the elements of Gau P can be identified with sections of an associated bundle. Let H act on itself by conjugation, and consider the corresponding associated bundle, $H(P)$. Since conjugation preserves group multiplication, each fiber of $H(P)$ is a group – it makes sense to multiply two elements of $H(P)$ which lie over the same point of M. For each x, the group sitting over x is isomorphic to H, but there is no preferred isomorphism. The space of sections $\Gamma(H(P))$ becomes a group under pointwise multiplication (just as the space of sections of a vector bundle became a vector space). If we think of a section as a function from P to H, the product of two sections is given as

$$(f_1 f_2)(p) = f_1(p) f_2(p).$$

We claim that each section $f \in \Gamma(H(P))$ defines an element $\phi = \phi_f$ of Gau P by the formula

$$\phi(p) = p f(p).$$

This certainly defines ϕ as a transformation of P with $\pi \circ \phi = \pi$. To check that ϕ commutes with the action of H, observe that

$$\phi(ph) = (ph) f(ph)$$
$$= (ph) \sigma(h^{-1}) f(p)$$
$$= (ph) h^{-1} f(p) h = p f(p) h = \phi(p) h.$$

It is clear that $\phi_{f_1 f_2} = \phi_{f_1} \circ \phi_{f_2}$ and that $\phi_{f^{-1}} = (\phi_f)^{-1}$ so that we have a homomorphism from the group $\Gamma(H(P))$ to Gau P. Conversely, given $\phi \in$ Gau P and $p \in P$, we know that $\phi(p)$ projects onto the same point as p, and hence we can write $\phi(p) = ph$ for some $h \in H$. This h depends on p, so we write it as $f(p)$. The preceding equations can now be read backwards to conclude that $f(ph) = h^{-1} f(p) h$, i.e. that $f \in \Gamma(H(P))$. Thus the mapping between $\Gamma(H(P))$ and Gau P is an isomorphism.

3.5   Tensor products

Let (ρ_1, V_1) and (ρ_2, V_2) be representations of the same group G. We can construct a representation $\rho_1 \otimes \rho_2$ on the space $V_1 \otimes V_2$ by defining*

$$(\rho_1 \otimes \rho_2)(a) = \rho_1(a) \otimes \rho_2(a).$$

It is easy to check that this is indeed a representation of G, which is called the tensor product of the two original representations. Let χ_1 and χ_2 be the characters of ρ_1 and ρ_2. Since

$$\text{tr}\,(A \otimes B) = \text{tr}\,A \cdot \text{tr}\,B$$

for any linear transformations A and B on V_1 and V_2, we conclude that the character of $\rho_1 \otimes \rho_2$ is simply the product function $\chi_1 \chi_2$. In many applications, it is necessary to decompose $V_1 \otimes V_2$ into irreducibles under G. Using the character tables, we have the tools at our disposal to solve this problem. For example, suppose we consider the group $S_4 = T_d$, whose character table is given in Table 9. Suppose we take $V_1 = V_2 = \mathbb{R}^3$ with the representation given by the standard action of T_d, whose character is χ_5, corresponding to the last row of the table. The character of the representation on $V_1 \otimes V_2 = \mathbb{R}^3 \otimes \mathbb{R}^3$ is then χ_5^2, which take on the values $(9, 0, 1, 1, 1)$ (in the order of conjugacy classes listed in Table 9). A direct computation then shows that

$$(\chi_5^2, \chi_1) = (1/24)(1 \cdot 9 \cdot 1 + 8 \cdot 0 \cdot 1 + 3 \cdot 1 \cdot 1 + 6 \cdot 1 \cdot 1 + 6 \cdot 1 \cdot 1) = 1$$

and similarly,

$$(\chi_5^2, \chi_2) = 1$$
$$(\chi_5^2, \chi_3) = 0$$
$$(\chi_5^2, \chi_4) = 1$$

and

$$(\chi_5^2, \chi_5) = 1.$$

So

$$\chi_5^2 = \chi_1 + \chi_2 + \chi_4 + \chi_5,$$

and this yields the corresponding decomposition of the space $\mathbb{R}^3 \otimes \mathbb{R}^3$.

Suppose that we are given a bilinear map $f : V_1 \times V_2 \to W$ which commutes with the action of G in that

$$f(\rho_1(a)v_1, \rho_2(a)v_2) = \rho(a)f(v_1, v_2)$$

where ρ is a given representation of G on W. Any such bilinear map corresponds to a

---

* The relation with the definition of tensor product given in Section 2.6 is the following. There we defined the 'external' tensor product of two representations of two groups. Applying that definition here would give us a representation of $G \times G$. Restricting to the 'diagonal' subgroup $\{(a, a)\}$ which is isomorphic to G gives the tensor product here.

linear map $F: V_1 \otimes V_2 \to W$ with

$$f(v_1, v_2) = F(v_1 \otimes v_2).$$

The map F is a G morphism. Suppose that f, and hence F, is surjective. This means that any *irreducible component of W must occur as an irreducible component of $V_1 \otimes V_2$* and its multiplicity in W is at most equal to its multiplicity in $V_1 \otimes V_2$. We shall have occasion to apply this remark in the next section where we discuss some quantum mechanical applications of group theory. Roughly speaking, the situation that we will encounter there will be where V_1 is some G invariant subspace of $\operatorname{Hom}(V, V)$ with $V = V_2$. The bilinear map will be the map which sends (A, v) into Av and W is the linear span of such Av.

3.6 Representative operators and quantum mechanical selection rules

In this section we present, in somewhat sketchy form, one of the principal tools used in the application of group theory to quantum mechanics. We begin with a description of the structure of quantum mechanics. A quantum mechanical system consists of a separable complex Hilbert space, $\mathscr{H}$. A *pure state* of the system consists of a one-dimensional subspace. Usually a unit vector, ϕ, is chosen in this subspace. We let P_ϕ denote the projection onto the subspace spanned by ϕ. A *physical observable* is represented by a self-adjoint operator, A, on the Hilbert space, $\mathscr{H}$. Each pure state assigns a probability distribution to the observed values of any observable. The expected value of the observable A in the state corresponding to ϕ is

$$\langle \phi, A\phi \rangle = \operatorname{tr} AP_\phi$$

(provided that $\phi \in \operatorname{dom}(A)$ so that the expression makes sense. From now on, we will not make explicit the hypotheses concerning domains, in order not to clutter up the discussion). Here $\langle \ , \ \rangle$ denotes the scalar product on $\mathscr{H}$. The variance of the probability distribution of the observed values of A in the state ϕ is given by

$$\langle \phi, A^2\phi \rangle - \langle \phi, A\phi \rangle^2.$$

In particular, the observable A will take on a definite value in the pure state, ϕ, if and only if ϕ is an eigenvector of A, i.e. $A\phi = \lambda\phi$, in which case λ is the observed value of A in the state, ϕ.

As a typical sort of example of the kind of Hilbert space that might arise, consider the space $L^2(\mathbb{R}^3)$ of all square integrable functions on Euclidean three space, with respect to the usual measure. A pure state is then a square integrable function of norm one. If B is any subset of $\mathbb{R}^3$, let 1_B be the indicator function of B, i.e. $1_B(x) = 1$ for $x \in B$ and $1_B(x) = 0$ for $x \notin B$. Let A_B denote the self-adjoint operator consisting of multiplication by 1_B. Then

$$\langle A_B\phi, \phi \rangle = \int_{\mathbb{R}^3} 1_B(x)|\phi(x)|^2 \, dx = \int_B |\phi(x)|^2 \, dx$$

is interpreted as the probability of being in the set B when in the state ϕ. Notice that if G is any group of Euclidean motions, then G acts as a group of unitary operators on $L^2(\mathbb{R}^3)$, where, as usual, $(a\phi)(x) = \phi(a^{-1}x)$.

More generally, let $\mathcal{H}$ be the Hilbert space of a quantum system. We say that G acts as a group of automorphisms if G acts on the pure states, and, for any two unit vectors, ϕ and ψ, we have

$$|\langle g\phi, g\psi \rangle|^2 = |\langle \phi, \psi \rangle|^2.$$

The meaning of this equation is the following. Each state ϕ defines an observable, P_ϕ. The self-adjoint operator, P_ϕ, can be thought of as the observable asserting that the system is in the state ϕ. Then $\langle P_\phi\psi, \psi \rangle = |\langle \phi, \psi \rangle|^2 = \operatorname{tr} P_\phi P_\psi$ is thought of as the probability of observing the system in the state ϕ when it is known to be in the state ψ. The equation $|\langle g\phi, g\psi \rangle|^2 = |\langle \phi, \psi \rangle|^2$ asserts that these probabilities are unchanged. It is a theorem due to Wigner (see Appendix D) that any such transformation of the states actually arises from either a unitary or an anti-unitary transformation of $\mathcal{H}$. That is, there exists either a unitary operator, U_g, of $\mathcal{H}$ such that $g\phi = U_g\phi$, or there exists an anti-unitary operator, V_g such that $g\phi = V_g\phi$. (To say that V_g is anti-unitary means that V_g is an anti-linear map, i.e. that

$$V_g(\alpha\phi + \beta\psi) = \bar{\alpha}V_g\phi + \bar{\beta}V_g\psi$$

for any complex numbers α and β and any vectors ϕ and ψ, that V_g is invertible, and that

$$\langle V_g\phi, V_g\psi \rangle = \langle \psi, \phi \rangle = \overline{\langle \phi, \psi \rangle}.$$

The operator U_g (or V_g) is not completely determined, since the vector ϕ itself is not quite determined by the state it defines. It turns out that U_g (or V_g) is determined up to scalar factor (the scalar being of absolute value one). The product of two anti-unitary transformations is a unitary transformation, and hence the subgroup G_U consisting of those U which correspond to unitary transformations is of index two in G. Let us restrict attention to this subgroup. The unitary operator U_g which is associated to $g \in G$ is only determined up to multiplication by a complex number of absolute value one. So all we can say, in general, is that, if we take any two elements g_1 and g_2 in G,

$$U_{g_1g_2} = c(g_1, g_2)U_{g_1}U_{g_2}$$

where $c(g_1, g_2)$ is some complex number of absolute value one depending on g_1 and g_2. If we could choose the U_g (by multiplying by an appropriate factor of absolute value one depending on g) so that $c(g_1, g_2) = 1$ for all g_1 and g_2, then we would get a unitary representation of G on $\mathcal{H}$. For many groups it is possible to prove that one can always make such a choice, and in many situations one is actually given a unitary representation. (In general the map $g \rightsquigarrow U_g$ is called a projective representation of G. It is always possible to enlarge the group G to a group G' so that a projective representation of G corresponds to an ordinary representation of G'. We do not want to go into these points here, although both anti-unitary operators (which arise in the study of time reversal) and projective representations (which arise in the study of the Galilean group) have important physical applications.)

With the above remarks in mind, we shall assume, for the rest of the discussion, that we are given a unitary representation of the finite group, G, on the Hilbert space $\mathscr{H}$.

Let ξ be any vector in $\mathscr{H}$. Then the set of all linear combinations of the vectors $a\xi$, as a ranges over G, clearly forms a finite-dimensional subspace of $\mathscr{H}$, which is invariant under the action of the group, G. Thus every vector lies in some finite-dimensional invariant subspace of $\mathscr{H}$. Of course, each such subspace can be decomposed into an orthogonal direct sum of irreducible finite-dimensional subspaces. Let W_i and W_j be irreducible (finite-dimensional) subspaces corresponding to two inequivalent representations of G. Let P denote the map from W_i to W_j given by orthogonal projection in $\mathscr{H}$: if $\xi \in W_i$, then $\xi' = P\xi \in W_j$ is the unique vector in W_j such that

$$\langle \xi, \eta \rangle = \langle \xi', \eta \rangle \quad \text{for all} \quad \eta \in W_j.$$

It follows that $gP = Pg$ for all $g \in G$ and hence, from Schur's lemma, that $P = 0$. In other words, *irreducible subspaces corresponding to inequivalent representations are orthogonal*.

Let F_i be the subset consisting of those η which belong to some irreducible subspace, W_j, with representation, r_j, inequivalent to the ith irreducible representation, r_i. Then $\mathscr{H}_i = (F_i)^\perp$ is a closed subspace, and every vector $\xi \in \mathscr{H}_i$ belongs to an irreducible representation isomorphic to r_i. We clearly have

$$\mathscr{H} = \mathscr{H}_1 \oplus \cdots \oplus \mathscr{H}_p$$

giving a decomposition of $\mathscr{H}$.

Let $L(\mathscr{H})$ denote the space of linear operators on $\mathscr{H}$. (In order to avoid complications about domains, the reader may prefer to replace $L(\mathscr{H})$ by the space of bounded linear operators or by some space or operators with a common dense domain.) The group G acts on $L = L(\mathscr{H})$ by sending $T \in L$ into gTg^{-1}. This gives a representation of G on L.

Let r be some representation of G on some auxiliary vector space, V (usually finite dimensional). By a *representation operator* belonging to r, or, more simply, an r-operator, we shall mean some element, R, of $\mathrm{Hom}_G(V, L)$. Thus $R: V \to L$ is a linear map such that

$$R(gv) = gR(v)g^{-1} \quad \text{for} \quad g \in G \quad \text{and} \quad v \in V.$$

(In the physics literature, a representation operator is frequently called a *tensor operator*.) The value, $R(v)$, of the r-operator at some $v \in V$ we shall call a *representative operator*. Thus a representative operator is an actual operator on $\mathscr{H}$, i.e. an element of L, whereas the r-operator, R, is not an operator but a map from V to the space of operators.

Let W_1 and W_2 be irreducible subspaces of $\mathscr{H}$, and let $R: V \to L$ be a representation operator. The set of all vectors of the form

$$R(v)\eta, \quad v \in V, \quad \eta \in W_2$$

clearly forms an invariant subspace of $\mathscr{H}$ equivalent to the image of $V \times W_2$ under the bilinear map of $V \times W_2 \to \mathscr{H}$ sending (v, η) to $R(v)\eta$. We know that W_1 will be orthogonal to this subspace, unless $V \otimes W_2$ contains a subrepresentation equivalent to

W_1. We thus obtain the fundamental quantum mechanical selection rule

> if W_1 and W_2 are irreducible subspaces and $R: V \to L$ is a representation operator, then
>
> $$\langle \eta_1, R(v)\eta_2 \rangle = 0 \quad \text{for all} \quad \eta_i \in W_i, v \in V \tag{6.1}$$
>
> unless $V \otimes W_2$ contains a subrepresentation equivalent to W_1.

The simplest kinds of representative operators correspond to the trivial representation of G. These are just the operators which commute with the action of G. They are called 'scalar' operators. Suppose that H is a self-adjoint scalar operator, and let

$$U(t) = \exp(-itH/\hbar)$$

(here $\hbar = h/2\pi$, where h is Planck's constant) be the associated one parameter group of unitary operators. It then follows that $U(t)$ is again a scalar operator, i.e. commutes with the action of G. In quantum mechanics, one is frequently given a self-adjoint operator, H, called the Hamiltonian, whose associated one parameter group describes the time evolution of the system. To say that the Hamiltonian is a scalar operator means that the time evolution of the system commutes with the action of G.

Suppose that H is a scalar operator and that ξ is an eigenvector of H with eigenvalue λ:

$$H\xi = \lambda\xi.$$

Then

$$Ha\xi = aH\xi = a\lambda\xi = \lambda(a\xi)$$

so that the space of eigenvectors with eigenvalue λ is invariant under the action of G. Thus, eigenvalues of scalar operators will tend to be degenerate, with multiplicity equal (at least) to the dimensions of the various irreducible representations of G.

One of the most important ways of applying (6.1) is in conjunction with time-dependent perturbation theory which we now describe. Suppose that the Hamiltonian of the system has the form

$$H = H_0 + H_1,$$

where $H_1 = H_1(t)$ should be thought of as representing the effect of some external disturbance. For example, we might want to consider the behavior of a vibrating molecule under the influence of some electromagnetic radiation. Then H_0 would be the Hamiltonian describing the time evolution of the unperturbed vibrations of the molecule, while H_1 describes the effect of the radiation.

We suppose that we know the time evolution of the unperturbed Hamiltonian, i.e. that we know the one parameter group

$$U(t) = \exp(-itH_0/\hbar),$$

and wish to find the one parameter family of unitary transformations corresponding to H. That is, we are looking for a one parameter family, $V(t)$, of unitary transformations which satisfy the differential equation

$$i\hbar V'(t) = HV(t).$$

Let us proceed formally to try to solve this equation by the method of variation of constants. That is, we set

$$V(t) = U(t)B(t)$$

so that

$$V'(t) = U'(t)B(t) + U(t)B'(t)$$

and hence

$$i\hbar V'(t) = H_0 V(t) + i\hbar U(t)B'(t).$$

Thus B must satisfy the differential equation

$$i\hbar B'(t) = (U^{-1} H_1 U)B(t).$$

Suppose that H_0 has a complete set of eigenstates ϕ_n with eigenvalues E_n, so that

$$H_0 \phi_n = E_n \phi_n$$

and therefore

$$U(t)\phi_n = \exp(-\mathrm{i}t E_n/\hbar)\phi_n.$$

For any $\xi \in \mathcal{H}$, we can write $B(t)\xi = \sum_n c_n(t)\phi_n$ and the above differential equation becomes

$$i\hbar \sum c'_n(t)\phi_n = (U^{-1} H_1 U)(\sum c_n(t)\phi_n).$$

Take the scalar product of both sides with ϕ_m. On the left we get $i\hbar c'_m(t)$, when the ϕ_n are chosen to the orthonormal. On the right we get $\sum \langle U^{-1} H_1 U \phi_n, \phi_m \rangle = \sum \langle H_1 U(t)\phi_n, U(t)\phi_m \rangle = \sum \exp[\mathrm{i}(E_m - E_n)t/\hbar]\langle H_1 \phi_n, \phi_m \rangle$. So we obtain the differential equations

$$i\hbar c'_m(t) = \sum \exp[\mathrm{i}(E_m - E_n)t/\hbar]\langle H_1 \phi_n, \phi_m \rangle c_n(t). \tag{6.2}$$

Let us assume that we start with the state ϕ_k at time 0, so that we have the initial conditions $c_k(0) = 1$, $c_j(0) = 0$ for $j \neq k$. We are assuming that H_1 is 'small' so that a first approximation to the solution can be obtained by regarding the c's on the right as taking on their constant initial values. We write $c_m(t) = \delta_{km} + c_m^{(1)}(t) + \cdots$, where the approximation $c_m^{(1)}(t)$ is given by

$$i\hbar c_m^{(1)}(t) = \int_0^t \exp[\mathrm{i}(E_m - E_k)t/\hbar]\langle H_1 \phi_m, \phi_k \rangle. \tag{6.3}$$

There are now several types of $H_1(t)$'s that arise in applications of (6.3). In the case of radiation, for example, it is reasonable to assume that H_1 is a periodic function of the time as a sum or integral of operators of the form:

$$H_1(t) = A\mathrm{e}^{-\mathrm{i}\omega t} + B\mathrm{e}^{\mathrm{i}\omega t}, \quad \omega = 2\pi\nu.$$

The fact that H_1 is to be Hermitian requires $B = A^*$. Let us set

$$\omega_{mk} = (E_m - E_k)/\hbar.$$

Then the integral in (6.3) becomes

$$i\hbar c_m^{(1)}(t) = \langle A\phi_k, \phi_m \rangle \frac{\exp\left[i(\omega_{mk} - \omega)t\right] - 1}{\omega_{mk} - \omega} + \langle A^*\phi_k, \phi_m \rangle \frac{\exp\left[i(\omega_{mk} + \omega)t\right] - 1}{\omega_{mk} + \omega} \quad (6.4)$$

for those values of ω and t for which the left-hand side is sufficiently small, so that it still may be considered a good approximation to the solution of (6.2). This will be the case if ω is not too close to $\pm \omega_{mk}$ or if t is not too large. Suppose, to fix the ideas, that we consider the situation where $E_m > E_k$. Then $|c_m(t)|^2$ can be regarded as the 'transition probability at time t' for unperturbed state, ϕ_k, whose energy was E_k, to 'absorb the quantum of energy $h\nu$' and pass into the state ϕ_m whose energy is E_m in the unperturbed system. According to (6.4), for moderate values of t, this transition probability will be non-negligible only for ω close to ω_{mk}. Plugging into (6.4) and ignoring the second term, we obtain the approximation

$$\hbar^2 |c_m^{(1)}(t)|^2 = t^2 |\langle A\phi_m, \phi_n \rangle|^2 \left| \frac{\sin^2 \zeta(\omega, t)}{\zeta^2(\omega, t)} \right|, \quad \text{where } \zeta = (\omega_{mk} - \omega)t/2. \quad (6.5)$$

This is the formula for a fixed frequency, ω. If we are given a whole spectrum of perturbations of the form $\int A(\omega)e^{i\omega t}\,d\omega + \text{(adjoint)}$, we must integrate (6.5) with respect to ω. Again, the major contribution comes from the values of ω near ω_{mk}, so that if A is approximately constant near ω_{mk}, we can pull it out of the integral sign to obtain (with A now denoting $A(\omega_{mk})$)

$$t^2 |\langle A\phi_m, \phi_k \rangle|^2 \int \frac{\sin^2 \zeta}{\zeta^2}\,d\omega.$$

By a change of variables, the integral becomes

$$4t^{-1} \int (\sin^2 x / x^2)\,dx = 4\pi t^{-1}$$

so that

$$|c_m(t)|^2 = 4\pi t |\langle A\phi_m, \phi_k \rangle|^2 / \hbar^2. \quad (6.6)$$

Thus, up to constant factors, and to the extent that the approximations are valid, the expression $|\langle A\phi_m, \phi_k \rangle|^2$ gives the transition probabilities per unit time.

That transition probabilities are non-negligible only for ω close to ω_{mk} can be formulated as saying that the system described by H_0 can only 'absorb' or 'emit' energy to the vibrating system in discrete units corresponding to differences in the energy levels on H_0. It took about a century of investigation of spectra to realize that the observed spectral lines of atoms or molecules could best be described as differences between various values. The Planck–Einstein relation between energy and frequency, $E = h\nu$, led to the understanding that one was really dealing with differences of energy. Not *all* differences $E_k - E_m$ make their appearance in the spectrum, so this raised the question of why certain differences do occur and others do not; i.e. what are the 'selection rules'? In terms of (6.4) or (6.5), why should $\langle Aq_k, q_n \rangle$ vanish? The answer of course is group theoretical:

It will usually happen in our applications that A is a representative operator for some group, so that we can apply (6.1) to conclude that certain of the $\langle A\phi_m, \phi_k \rangle$ must vanish. We speak of the corresponding transition as being *forbidden* by symmetry considerations. Frequently the argument is used in reverse: we notice that certain transitions do not occur, and attempt to explain their absence on the grounds of symmetry. In other words, we use the observed transitions to deduce the structure of the underlying symmetry group.

Another application of (6.5) is to the situation where $\omega = 0$, but where one averages over the final states, instead of over ω. This is the situation that arises in a crude approximation to the analysis of collision processes. Here we assume that some perturbing 'particle' passes near our system. We approximate the situation by assuming that there is a constant perturbation applied for a short period of time. Thus $\omega = 0$, and (6.5) becomes

$$|c_m(t)|^2 = 4|\langle A\phi_m, \phi_k \rangle|^2 \frac{\sin^2 \omega_{mk}t/2}{\omega_{mk}^2}.$$

We now assume that there is a whole family of states whose energies are all close to one another and close to the energy of the state ϕ_k, and for which the values of $\langle A\phi_m, \phi_k \rangle$ are approximately constant. Then the transition probability from ϕ_k to any one of these states is obtained by summing the above formula over m to obtain

$$4|\langle A\phi_m, \phi_k \rangle|^2 \sum_m \frac{\sin^2 \omega_{mk}t/2}{\omega_{mk}^2}.$$

We assume that the energy levels are sufficiently closely spaced to justify approximating the sum by an integral. We assume that the number of states whose energy is in an interval about E_m is given by the integral of the density $\rho(E_m)$, and, since $dE_m = \hbar\, d\omega_{mk}$, the above sum is replaced by the integral

$$\hbar \int \frac{\sin^2 \omega_{mk}t/2}{\omega_{mk}^2} \rho(E_m)\, d\omega_{mk}.$$

For moderately large values of t (but not too large to render the approximations invalid), the major contribution of the integral will come from the neighborhood of $\omega_{mk} = 0$ and we obtain the formula

$$\text{rate of transition} = (2\pi/\hbar)|\langle A\psi, \phi_k \rangle|^2 \rho(E_k) \tag{6.7}$$

where ψ is one state among a family of states having energies close to E_k, and for which the $\langle A\psi, \phi_k \rangle$ is approximately constant, and ρ gives the density of states per energy level for this family. Equation (6.7) is known as Fermi's 'golden rule no. 2' and plays a key role in the study of collision processes. Again we see that (6.1) implies that certain transitions are forbidden if A is a representative operator for a suitable group.

In any specific application, we must determine the representation associated to the operator A and the representation associated to the unperturbed eigenstates ϕ_j. For example, we shall show in the next section that the semiclassical theory of radiation suggests that the operator A to be taken for induced emission or absorption of radiation is a 'vector operator', i.e. A transforms like an element of $\mathbb{R}^3$ under a given action of G on $\mathbb{R}^3$. (The operator A is associated with the electric dipole moment of the system, and the dipole moment is a vector in $\mathbb{R}^3$.)

In a given situation, we must also determine the eigenvalues, $E_0, E_1, \ldots$ of H_0. In principal, any allowed transition from E_n to E_k can occur. But the system must be in state E_m and it must be given enough energy (or be able to give up enough energy to the oscillating perturbation) to get to E_k. In practice, of course, we observe many copies of the same system, not one; we shine light on a sample of a given substance, not on a single molecule. At certain temperatures, most of the copies may be in a given energy state. For instance, at low temperatures most of the molecules might be expected to be in the 'ground state', E_0. Thus we would be primarily interested in the $E_0 \to E_k$ transitions.

In the case of a (harmonic) oscillator, the eigenvalues of H_0 were worked out at the very beginning of quantum mechanics. For a one-dimensional oscillator whose classical frequency is ω_0, the quantum energy levels are

$$\tfrac{1}{2}\hbar, (\omega_0 + \tfrac{1}{2})\hbar, \quad (2\omega_0 + \tfrac{1}{2})\hbar, \ldots.$$

More generally, if we are given a finite-dimensional space V for the classical oscillator, then the Hilbert space for the quantum system can be written as the (Hilbert space) direct sum

$\mathbb{C}$	$\oplus$	V	$\oplus$	$S^2(V)$	$\oplus$	$S^3(V)$	$\oplus \cdots$
ground state		fundamental or one-phonon states		first harmonic or two-phonon states		three-phonon states	

where $S^k(V)$ denotes the kth symmetric power (the space of symmetric k-tensors) of V. Suppose that

$$V = V_1 \oplus \cdots \oplus V_k$$

where the V_i are eigenspaces of the classical oscillator with eigenvalues $\omega_1, \ldots, \omega_k$. Then

$$S^2(V) = S^2(V_1) \oplus \cdots \oplus S^2(V_k) \oplus V_1 \otimes V_2 + \cdots$$

and the eigenvalues of the quantum oscillator are $2\omega_1, 2\omega_2, \ldots 2\omega_k, \omega_1 + \omega_2$, etc. Similarly the eigenvalues of the quantum oscillator on $S^n(V)$ are $\omega_{i_1} + \cdots + \omega_{i_n}$.

For the case of a vibrating molecule at laboratory temperatures, the most important transitions are those from the ground state (of lowest energy) to a 'fundamental state', where the change in energy is (proportional to) an eigenfrequency of the classical vibrating system. Now the ground state is invariant under the group, i.e. transforms according to the trivial representation, while each of the fundamental states transforms according to the representation associated to the given eigenvalue of the vibrating system. So, if ϕ_0 denotes the ground state, then $A\phi_0$ transforms according to $\mathbb{R}^3$, while a fundamental state ϕ_f transforms according to an appropriate irreducible representation W_f occurring in the decomposition of the classical vibrating system. The fundamental transition $\phi_0 \to \phi_f$ will be *allowed* if W_f occurs in the decomposition of $\mathbb{R}^3$ (or rather, its complexification, $\mathbb{C}^3$) under the group G, and will be *forbidden* otherwise.

We can thus determine which of the irreducible components of a vibrating molecule makes its appearance as a fundamental transition.

Thus, for example, in a molecule like methane, whose symmetry group is T_d, we have computed the decomposition of the character of the nine-dimensional space of vibrations as

$$\chi_1 + \chi_2 + 2\chi_5.$$

Here χ_5 is the character of the $\mathbb{R}^3$ actions. So, although there will generically be four frequencies, only two of them will be observed as absorption lines corresponding to transitions from the ground state to a fundamental state.

The transitions we have described so far have to do with absorption. Experimentally, a red-hot body emits infra-red radiation which is collimated and passed through a sample of the substance to be studied. It is then dispersed by a prism or grating and the amount of radiation in each small portion of the spectrum is measured. This is compared to the amount emitted. In this way, one detects the absorption spectrum. There is another procedure discovered by Raman in 1928 which, fortunately, has different selection rules. In Raman spectroscopy, an intense monochromatic beam of light, of (any) frequency v, is brought to bear on the sample and observations are made at right angles to the beam. Most of the scattered light has the same frequency, v, as the incident beam. But there will be a small amount of light at frequency $v - v_R$ (and less at $v + v_R$), where v_R is known as the Raman line. The displacement v_R is independent of the incident frequency and is associated with a 'tensor' operator, i.e. a representative operator, transforming according to the representation on the space of symmetric tensors $S^2(\mathbb{R}^3)$. Again, at room temperatures, the only important Raman transitions will be from the ground state to the fundamental state. So we must compare the fundamental state with $S^2(\mathbb{R}^3)$ to check which Raman lines will be observed. For example, we have seen that the representation of T_d on $\mathbb{R}^3 \otimes \mathbb{R}^3$ has character

$$\chi_5^2 = \chi_1 + \chi_2 + \chi_4 + \chi_5,$$

and we know that χ_4 is the character of T_d on the three-dimensional space of anti-symmetric tensors. So the representation on $S^2(\mathbb{R}^3)$ has character

$$\chi_1 + \chi_2 + \chi_5.$$

So all ground state to fundamental transitions can occur in Raman spectroscopy.

The combination of infra-red and Raman spectroscopy can be very effective. For example, suppose that we have a molecule of the form AB_4 and we consider three possible shapes (Fig. 3.10). We can evaluate the number of fundamental eigenvalues and which are infra-red or Raman admissible. We let the reader check that we obtain Table 20 (we have already worked out the result for T_d).

In principle, we should be able to distinguish experimentally between these (or other) alternatives. (In practice, the experimental results may not be clear cut.)

In the next section we shall sketch the proof that the infra-red selection rules are associated to vector operators. We refer the reader to specialized treatises for the proof that the Raman selection rules are associated to $S^2(\mathbb{R}^3)$ second order perturbation theory.

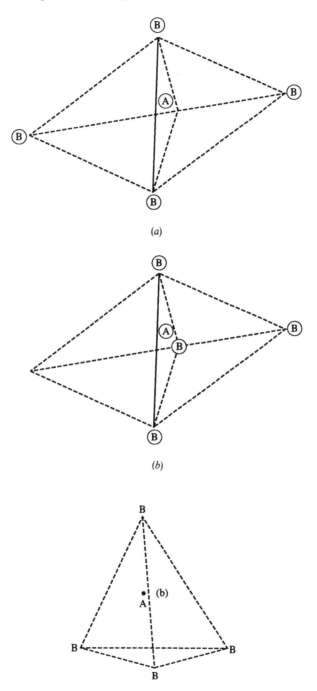

(a)

(b)

(c)

Fig. 3.10

Table 20.

Group	Number of eigenvalues	Infra-red	Raman
C_{2v}	9	8	9
C_{3v}	6	6	6
T_d	4	2	4

Let us apply the results of the results we have derived in this chapter to describe the infra-red and Raman spectra of the buckyball. We know that the space of vibrational states has dimension $174 = 180 - 6$. We wish to prove the following.

(1) That there are at most 46 distinct vibrational modes. In other words, the space of classical vibrational states decomposes into 46 irreducible representations.

(2) Of these (at most) four are visible in the infra-red. In other words, exactly four of the 46 irreducibles are equivalent to the representation of the group I_h on the (complexification of) ordinary three-dimensional space, $\mathbb{R}^3$.

(3) (At most) ten lines are visible in the Raman spectrum. More precisely, the complexification of the space $S^2(\mathbb{R}^3)$, which is six-dimensional, decomposes into a direct sum of the trivial representation (given by the scalar matrices) and a five-dimensional representation (corresponding to the traceless tensors). This five-dimensional representation is irreducible and occurs eight times in the space of vibrational states, whereas the trivial representation occurs twice.

The above facts concerning the spectrum of the buckyball have all been verified experimentally. The two different kinds of Raman spectra, corresponding to the trivial and to the five-dimensional representation, can be distinguished through the use of polarized light.

To begin, we must list all the irreducibles of I_h. Since I_h is the direct product of I with $\mathbb{Z}_2$, it is enough to find the irreducibles of I and then label each with a $+$ or $-$ sign according to the trivial or sign representation of the $\mathbb{Z}_2$ component.

Now $I \sim A_5$, which is a subgroup of S_5. The elements of A_5 consist of the identity, 20 three-cycles, 15 elements of the form $(ab)(cd)$, and 24 five-cycles. Any two three-cycles are conjugate by an even permutation, since the odd permutation, (de), commutes with (abc). So the 20 three-cycles form a single conjugacy class in A_5. Similarly, the odd permutation (ab) commutes with $(ab)(cd)$ and hence the 15 elements of the form $(ab)(cd)$ form a single conjugacy class. On the other hand, a five-cycle such as $\sigma = (12345)$ does not commute with any odd element, since it cannot carry the elements of any two-cycle or any four-cycle into itself. Now $\sigma^2 = (13524)$ is conjugate to σ in S_5 by the odd element (2354) and hence by no even element. Thus the 24 five-cycles split up in A_5 into two conjugacy classes, those conjugate to σ and those conjugate to σ^2. Thus there are five conjugacy classes in all, and so five inequivalent irreducible representations.

We now go back to Table 16 of Chapter 2 and see what happens when we restrict the irreducible representations of S_5 to A_5. Direct computation shows that the first three lines restrict to irreducibles of A_5. The fourth line, the six-dimensional representation of S_5, does not remain irreducible; the character satisfies $(\chi, \chi)_{A_5} = 2$. So it splits into two irreducibles of dimensions, say, d_1 and d_2. Since

$$1 + 16 + 25 + d_1^2 + d_2^2 = 60,$$

we must have

$$d_1 = d_2 = 3.$$

One of these three-dimensional representations we already know – the complexification of the representation identifying A_5 with I acting as rotational symmetries of the icosahedron. Since all elements are rotations, having trace $1 + 2\cos\theta$, we see that, for this representation,

$$\chi(abc) = 0$$

and

$$\chi((ab)(cd)) = -1.$$

If σ corresponds to rotation through angle $2\pi/5$, then

$$\chi(\sigma) = 1 + 2\cos 2\pi/5, \quad \chi(\sigma^2) = 1 + 2\cos 4\pi/5.$$

A direct check shows that this is an irreducible character, and interchanging the roles of σ and σ^2 in this last equation gives the character of the other three-dimensional representation, as can be checked by adding the two together to get the character of the reducible six-dimensional representation coming from S_5.

We may label the irreducibles of I as $\mathbf{1}, \mathbf{4}, \mathbf{5}, \mathbf{3}$, and $\mathbf{3}'$. Thus the regular representation of I decomposes into

$$1 + 4 + 5 + 3 + 3 = 16$$

irreducible representations.

Let $E \to B$ denote the vector bundle describing the displacements of the carbon atoms in the buckyball from their equilibrium positions. It is a homogeneous vector bundle with respect to the groups I and I_{h}. The group I has a trivial isotropy group at any point of B. Hence the space $\Gamma(E)$ transforms as the direct sum of three copies of the regular representation of I. It thus decomposes into 48 irreducibles. We remove two copies of $\mathbf{3}$ when we subtract off overall translations and overall rotations. Hence there are (at most) 46 distinct vibrational modes.

Let X be a vertex of the buckyball and let H be the isotropy group of X in I. So H consists of two elements, the identity and the reflection, r_X, in the plane passing through X and bisecting the buckyball. The representation of I_{h} given by (the complexification of) its action on three-dimensional space is labeled as $\mathbf{3}^-$, since the inversion operator (the generator of the $\mathbb{Z}_2$ component) acts as $-Id$. The vector bundle E is induced from the restriction of this representation to H. The group H has only two irreducible representations, both one-dimensional: the trivial representation, $\mathbf{1}_+$, and the 'sign' representation, $\mathbf{1}_-$, which assigns the value -1 to r_X. Thus the three-dimensional representation $\mathbf{3}^- \downarrow$ of H must decompose into a sum of these one-dimensional representations. To see what this decomposition is, choose coordinates so that r_X is reflection in the y, z plane. Then the matrix representing r_X is

$$\begin{pmatrix} -1 & 0 & 0 \\ 0 & 1 & 0 \\ 0 & 0 & 1 \end{pmatrix},$$

which shows that we have the block decomposition

$$\mathbf{3}^- \downarrow = \mathbf{1}_- \oplus \mathbf{1}_+ \oplus \mathbf{1}_+. \tag{6.8}$$

But now Frobenius reciprocity says that $\mathbf{3}^-$ occurs once in the representation induced from $\mathbf{1}_-$ and twice in the representation induced from $\mathbf{1}_+$. Hence it follows from (6.8) that dim $\mathrm{Hom}(\mathbf{3}^-, \Gamma(E)) = 5$. Subtracting one copy of $\mathbf{3}^-$ corresponding to overall translations shows that there are exactly four vibrational lines in the infra-red. This proves statement (2) above.

We can use Frobenius reciprocity together with Table 16 to compute the multiplicities of all the irreducibles in the induced representations. Indeed, the reflection, r_X, is the product of the inversion $-Id$ with a rotation ρ_X through $180°$ lying in I:

$$r_X = -\rho_X.$$

Let $K = \{e, \rho_X\}$ be the corresponding two element subgroup of I, and $\mathbf{1}_{K\pm}$ its one-dimensional representations. Then for any $+$ representation of I_h we have

$$\dim \mathrm{Hom}_H(\mathbf{1}_\pm, \mathbf{k}^+ \downarrow) = \dim \mathrm{Hom}_K(\mathbf{1}_{K\pm}, \mathbf{k}^+ \downarrow),$$

whereas for the $-$ representations the signs are reversed:

$$\dim \mathrm{Hom}_H(\mathbf{1}_\pm, \mathbf{k}^- \downarrow) = \dim \mathrm{Hom}_K(\mathbf{1}_{K\mp}, \mathbf{k} \downarrow).$$

So, for example,

$$\dim \mathrm{Hom}_{I_h}(\mathbf{5}^+, \mathbf{1}_+ \uparrow) = \dim \mathrm{Hom}_H(\mathbf{1}_+, \mathbf{5}^+ \downarrow) = \dim \mathrm{Hom}_K(\mathbf{1}_{K+}, \mathbf{k}^+ \downarrow) = \tfrac{1}{2}(5+1) = 3.$$

Proceeding in this way we compute the decompositions

$$\mathbf{1}_+ \uparrow = \mathbf{1}^+ \oplus 2 \times \mathbf{3}^- \oplus \mathbf{3}^+ \oplus 2 \times \mathbf{3'}^- \oplus \mathbf{3'}^+ \oplus 2 \times \mathbf{4}^+ \oplus 2 \times \mathbf{4}^- \oplus 3 \times \mathbf{5}^+ \oplus 2 \times \mathbf{5}^- \tag{6.9}$$

and

$$\mathbf{1}_- \uparrow = \mathbf{1}^- \oplus 2 \times \mathbf{3}^+ \oplus \mathbf{3}^- \oplus 2 \times \mathbf{3'}^+ \oplus \mathbf{3'}^- \oplus 2 \times \mathbf{4}^+ \oplus 2 \times \mathbf{4}^- \oplus 3 \times \mathbf{5}^- \oplus 2 \times \mathbf{5}^+. \tag{6.10}$$

We can use (6.9) and (6.10) to determine the number of lines in the Raman spectrum. In the Raman experiment the operator, A, transforms like an element of $S^2(\mathbb{R}^3)^+$. Using the characters of I, a direct computation shows that

$$\mathbf{3} \otimes \mathbf{3} = \mathbf{3} \oplus \mathbf{1} \oplus \mathbf{5}.$$

The first summand on the left is just the space of anti-symmetric tensors. So we see that the complexification of $S^2(\mathbb{R}^3)$ transforms, under I_h, like $\mathbf{1}^+ \oplus \mathbf{5}^+$. (The $\mathbf{5}^+$ corresponds to traceless tensors and the $\mathbf{1}^+$ corresponds to multiples of the identity. The reason for the $+$ is that the parity operator, P, has no effect on tensors of even degree.) Let us examine these two components separately. Notice that $\mathbf{1}^+$ does not occur at all on the right-hand side of (6.10), and occurs once on the right-hand side of (6.9). Hence we conclude from (6.8) and Frobenius reciprocity that

$$\dim \mathrm{Hom}_{I_h}(\mathbf{1}^+, Vib) = 2. \tag{6.11}$$

In other words there should be two Raman lines corresponding to the $\mathbf{1}^+$ representation of I_h. The $\mathbf{5}^+$ occurs twice on the right-hand side of (6.10), and three times on the right-hand side of (6.9). Hence we conclude from (6.8) and Frobenius reciprocity that

$$\dim \mathrm{Hom}_{I_h}(\mathbf{5}^+, Vib) = 2 + 3 + 3 = 8. \tag{6.12}$$

So there should be eight lines corresponding to $\mathbf{5}^+$. All ten lines have been observed. In fact that $\mathbf{1}^+$ lines can be distinguished, experimentally, from the $\mathbf{5}^+$ lines through the use of polarized light.

We should point out that decompositions such as (6.9) and (6.10) together with Schur's lemma offer a powerful tool for the location of the eigenvalues of invariant operators and not merely for counting the number of distinct eigenvalues, which is the use we have made so far. For example, the 60-dimensional space, $\mathbf{1}^+$, plays a central role in the electronic structure of the buckyball, and, hopefully, in understanding the high temperature superconductivity of doped buckyball crystals. So it will be important to analyze invariant operators, R, on this space, that is matrices, R, which lie in $\mathrm{Hom}_{I_h}(\mathbf{1}^+, \mathbf{1}^+)$. Such a matrix has size 60×60. But Schur's lemma implies that there cannot be any interaction between inequivalent components. In other words, R must have a block decomposition corresponding to the right-hand side of (6.9), where the size of the blocks is given by the multiplicities of the various irreducible representations, i.e. by the various coefficients occurring on the right-hand side of (6.9). The largest such coefficient is 3. So, although R is a 60×60 matrix, in searching for the eigenvalues of R, the worst computation that we have to do is diagonalizing a 3×3 matrix, and this can be done analytically.

3.7   The semiclassical theory of radiation

In this section we regard the ambient radiation as giving a periodic perturbation of the quantum mechanical system describing the atom or molecule, as described in the preceding section. In particular, we will be concerned with describing the 'first order transition operator', A, which arises from first order perturbation theory. The main result of this section is that, for absorption or emission spectra, the operator A transforms as an element of $\mathbb{R}^3$ (i.e. is a 'vector operator'). The reader may want to accept this assertion on faith and immediately proceed to the next section.

According to Lorentz, the force $\mathbf{F}$ acting on a (spinless) particle of charge q in an electromagnetic field is given by

$$\mathbf{F} = q\left[\mathbf{E} + \frac{1}{c}(\mathbf{v} \times \mathbf{B})\right]$$

where $\mathbf{E}$ is the electric field, $\mathbf{B}$ is the magnetic field, and $\times$ denotes vector product. (Later on we shall discuss how this equation is to be modified for particles which possess an intrinsic angular momentum, i.e. a spin.) We can treat $\mathbf{E}$ as a differential form of degree one: $\mathbf{E} = E_x\,dx + E_y\,dy + E_z\,dz$ on (x, y, z, t) space and $\mathbf{B}$ as a differential form of degree two: $\mathbf{B} = B_x\,dy \wedge dz - B_y\,dx \wedge dz + B_z\,dx \wedge dy$. We can then write the electromagnetic field as a two-form

$$\Omega = \mathbf{E} \wedge c\,dt + \mathbf{B}$$
$$= c(E_x\,dx \wedge dt + E_y\,dy \wedge dt + E_z\,dz \wedge dt)$$
$$+ B_z\,dy \wedge dz - B_y\,dx \wedge dz + b_z\,dx \wedge dy.$$

Then the Lorentz force, $\mathbf{F}$, is the space component of

$$\frac{-q}{c}\,i(\mathbf{w})\Omega$$

where $\mathbf{w}$ is the four-vector

$$\mathbf{w} = \mathbf{v} + (\partial/\partial t)$$
$$= (v_x, v_y, v_z, 1).$$

The Maxwell equations for the electromagnetic field without sources are

$$d\Omega = 0, \quad d \maltese \Omega = 0,$$

where $\maltese$ denotes the star operator relative to the Lorentz metric in which

$$(dx, dx) = (dy, dy) = (dz, dz) = -1,$$
$$(dx, dy) = 0, \text{ etc. } (dx, dt) = 0, \text{ etc. } \text{ and } (c\,dt, c\,dt) = 1$$

so that, explicitly,

$$\maltese\,dx = -dy \wedge dz \wedge c\,dt$$
$$\maltese\,dy = dx \wedge dz \wedge c\,dt$$
$$\maltese\,dz = -dx \wedge dy \wedge c\,dt$$
$$\maltese\,c\,dt = dx \wedge dy \wedge dz$$

and

$$(dx \wedge dy) = -dz \wedge c\,dt, \text{ etc.}$$
$$(dx \wedge c\,dt) = dy \wedge dz, \text{ etc.}$$

The first of the Maxwell equations, $d\Omega = 0$, implies that we can write

$$\Omega = d\Xi$$

for some form

$$\Xi = A_x\,dx + A_y\,dy + A_z\,dz - c\phi\,dt,$$

where $\mathbf{A} = (A_x, A_y, A_z)$ is called the vector potential and ϕ is called the scalar potential. (Of course Ξ is only determined up to the addition of $d\psi$ for some function ψ — replacing Ξ by $\Xi + d\psi$ is known as a guage transformation.) Explicitly,

$$E_x = -(1/c)(\partial A_x/\partial t) - \partial\phi/\partial x,$$

etc. and

$$B_x = \partial A_z/\partial y - \partial A_y/\partial z,$$

etc. The equations of motion become (for a particle of mass m)

$$m\frac{d^2 x}{dt^2} = -q\frac{\partial\phi}{\partial x} - \frac{q}{c}\frac{\partial A_x}{\partial t} + \frac{q}{c}\left[\frac{dy}{dt}\left(\frac{\partial A_y}{\partial x} - \frac{\partial A_x}{\partial y}\right) + \frac{dz}{dt}\left(\frac{\partial A_z}{\partial x} - \frac{\partial A_x}{\partial z}\right)\right]$$

etc., which can be written as

$$\frac{dx}{dt} = \frac{1}{m}\left(p_x - \frac{q}{c}A_x\right)$$

$$\frac{dp_x}{dt} = \frac{q}{mc}\left(\frac{dx}{dt}\frac{\partial A_x}{\partial x} + \frac{dy}{dt}\frac{\partial A_y}{\partial x} + \frac{dz}{dt}\frac{\partial A_z}{\partial x}\right) - q\frac{\partial\phi}{\partial x}$$

etc., or

$$\frac{dx}{dt} = \frac{\partial H}{\partial p_x}, \quad \frac{dp_x}{dt} = -\frac{\partial H}{\partial x}$$

etc., where

$$H = \frac{1}{2m}\left[\left(p_x - \frac{q}{c}A_x\right)^2 + \left(p_y - \frac{q}{c}A_y\right)^2 + \left(p_z - \frac{q}{c}A_z\right)^2\right] + q\phi.$$

If we compare this with the free Hamiltonian

$$H_0 = \frac{1}{2m}[p_x^2 + p_y^2 + p_z^2],$$

we see that H may be obtained from H_0 by replacing the energy momentum vector (p_x, p_y, p_z, E) by $(p_x - (q/c)A_x, \ldots, E + q\phi)$ in the Hamiltonian. This scheme is the one we shall use in the quantum mechanical formulation of the problem as well. In quantum mechanics the momentum, p_x, becomes the operator $-i\hbar(\partial/\partial x)$, so that the expression $(p_x - (q/c)A_x)^2$ becomes the operator

$$\left[-i\hbar(\partial/\partial x) - \frac{q}{c}A_x\right]^2 = -\hbar^2\frac{\partial^2}{\partial x^2} + i\hbar\frac{q}{c}A_x\frac{\partial}{\partial x} + i\hbar\frac{q}{c}\frac{\partial A_x}{\partial x} + \frac{q^2}{c^2}A_x^2.$$

Now the second Maxwell equation in free space, $d\Omega = 0$, implies (as is easily seen) that we can, by an appropriate gauge transformation, arrange that $\phi = 0$, and $\partial A_x/\partial x + \partial A_y/\partial y + \partial A_z/\partial z = 0$. Thus, substituting this choice of gauge into the quantum mechanical expression for H, we obtain

$$\frac{1}{2m}\left[-\hbar^2\Delta + i\hbar\frac{q}{c}\left(A_x\frac{\partial}{\partial x} + A_y\frac{\partial}{\partial y} + A_z\frac{\partial}{\partial z}\right) + \frac{q^2}{c^2}\|A\|^2\right]$$

as the Hamiltonian of a particle of mass m and charge q in the presence of an electromagnetic field in free space. For the case of an electromagnetic field of weak to moderate strength, the last term in this expression is negligible in comparison with the first two. If we have a system of such charged particles, with an internal energy V in the presence of an electromagnetic field, the Hamiltonian can be written as $H = H_0 + H_1$, where now

$$H_0 = -\sum \frac{\hbar^2}{2m_j} \Delta_j + V$$

and

$$H_1 = -\sum \frac{q_j}{m_j c} A_j \cdot p_j$$

with $A_j \cdot p_j$ denoting the operator $-i\hbar(A_x \partial/\partial x + A_y \partial/\partial y + A_z \partial/\partial z)$ for the jth particle. Let us first assume that we are dealing with an electromagnetic field given by a plane wave,

$$\mathbf{A}(v, t) = \tfrac{1}{2} \| A \| u[\exp i(\omega t - k \cdot v) + \exp i(k \cdot v - \omega t)],$$

where $\| A \|$ is a constant, u is a (constant) vector giving the polarization of the radiation, and k is the wave vector determining the direction of propagation and the wavelength of the radiation. Now the wavelengths of visible and ultra-violet radiation are of the order of magnitude 10^{-5} cm at the least, while a rough estimate of molecular dimensions gives a diameter of about 10^{-8} cm. Thus $|k \cdot v| < 10^{-3}$ as v varies over a region of size 10^{-8}, i.e. in the expansion

$$e^{ik \cdot v} = 1 + ik \cdot v + \cdots$$

we can ignore all but the constant term. Thus, in the formula for H_1, we may assume that the A_j's take on the same constant value. In equation (6.6), we must evaluate a sum of terms of the form

$$\langle (u \cdot p_j)\phi_m, \phi_m \rangle = u \cdot \langle p_j \phi_m, \phi_n \rangle$$

where, on the right, p_j denotes the vector operator corresponding to the momentum of the jth particle. Now ϕ_m and ϕ_n are eigenvectors of H_0 with eigenvalues E_m and E_n, while we have the commutation relations

$$r_j H_0 - H_0 r_j = \frac{i\hbar}{m_j} p_j,$$

where r_j denotes the position operator for the jth particle. Therefore,

$$\left\langle \left(\sum \frac{q_j}{m_j c} p_j\right)\phi_m, \phi_n \right\rangle = (i\hbar c)^{-1} \langle (\sum q_j(r_j H_0 - H_0 r_j))\phi_m, \phi_n \rangle$$

$$= (i\hbar c)^{-1}(E_m - E_n)\langle (\sum q_j r_j)\phi_m, \phi_n \rangle.$$

Now the expression $\beta = \sum q_j r_j$ is the total dipole moment of the system, and is clearly a vector in $\mathbb{R}^3$, classically, and a vector operator quantum mechanically. Substituting into (6.6) of the preceding section gives, for the transition probability per unit time,

$$\frac{\pi^2 v_{mn}^2}{c^2 \hbar^2} \| A \|^2 |u \cdot \langle \beta \phi_m, \phi_n \rangle|^2, \tag{7.1}$$

where, as usual, $E_m - E_n = hv_{mn}$. Averaging over possible directions of polarization, u (and arbitrary initial phases), gives, for the induced transition probability per unit time, under the assumption of homogeneous radiation,

$$\frac{\pi^2 v_{mn}^2}{c^2 \hbar^2} \| A \|^2 \| \langle \beta \phi_m, \phi_n \rangle \|^2 \tag{7.2}$$

where $\langle \beta \phi_m, \phi_n \rangle \in \mathbb{C}^3$ is the expectation value of the vector operator β between the states ϕ_m and ϕ_n and $\| \langle \beta \phi_m, \phi_n \rangle \|$ is the length of this expectation. Formula (7.1) shows that the transition probabilities for induced emission or absorption is given by a vector operator, as indicated at the beginning of this section. We can give an intuitive interpretation of (7.2) as follows: it follows from the equation $E = -(1/c)\partial A/\partial t - d\phi$, with $\phi = 0$ and $A = \| A^0 \| u \cos 2\pi v_{mn} t$, that

$$E = \frac{2\pi v_{mn}}{c} \| A^0 \| u \sin 2\pi v_{mn} t,$$

for radiation of frequency v_{mn}. We can then rewrite (7.1) as

$$\frac{1}{2\hbar^2} \overline{|\langle (E \cdot \beta) \phi_m, \phi_n \rangle|^2}$$

where $\overline{}$ denotes the time average, since the average value of $\sin^2 2\pi v_{mn} t$ is $\frac{1}{2}$. Now, classically, $E \cdot \beta$ is the energy of the dipole β in the electric field, E. Thus, we can say that within the approximations that we have adopted it is the 'electric dipole energy' which is responsible for the transitions from one stationary state to another with corresponding absorption or emission of light.

Let us return to the formulas for induced emission and absorption, (7.1) and (7.2). The time average of the density of radiation associated with a plane wave is, according to classical electromagnetic theory, given by $(1/8\pi) \| E^0 \|^2$, where $E = E^0 \sin 2\pi vt$. Taking the time average of (7.2) then gives us a formula relating the transition probabilities, P_{mn}, from the state ϕ_m to the state ϕ_n and the density, $\rho(v_{mn})$, of radiation at frequency v_{mn}. Namely,

$$P_{mn} = B_{mn}\rho(v_{mn}) \quad \text{where} \quad B_{mn} = \frac{2\pi}{\hbar^2} \| \langle \beta \phi_m, \phi_n \rangle \|^2.$$

The coefficients, $B_{mn} = B_{nm}$, are known as the Einstein transition probability coefficients. They were introduced by Einstein in 1917 as part of an argument relating induced absorption and emission with spontaneous emission, using thermodynamic considerations and the Planck radiation law. We sketch his argument. According to the Boltzmann distribution law, the probability for a system to be in a state of energy E_m is proportional to $\exp - E_m/kT$, where k is Boltzmann's constant and T is the absolute temperature. If N_m denotes the number of systems in the state with energy E_m, then we expect that

$$N_m/N_n = \exp - (E_m - E_n)/kT = \exp - hv_{mn}/kT.$$

Now a system in an excited state can be expected to emit radiation even in the absence of an ambient electromagnetic field. Suppose that we let Q_{mn} denote the probability of spontaneous transition from the state ϕ_m to the state ϕ_n, where we assume that $E_m > E_n$. Then the expected number of systems making the transition from the state ϕ_m to the state ϕ_n as a result of both spontaneous and induced emission will be given by

$$N_m\{Q_{mn} + B_{mn}\rho(v_{mn})\},$$

while the expected number of systems making the reverse transition as a result of absorption of radiation is

$$N_n B_{nm}\rho(v_{mn}) = N_n B_{mn}\rho(v_{mn}).$$

At equilibrium, the number of expected transitions in both directions should be equal, yielding the equation

$$\exp - hv_{mn}/kT = \frac{B_{mn}\rho(v_{mn})}{Q_{mn} + B_{mn}\rho(v_{mn})},$$

or

$$Q_{mn}/B_{mn} = [\exp(hv_{mn}/kT) - 1]\rho(v_{mn}).$$

In the above expression, we can substitute the Planck law of radiation,

$$\rho(v) = (8\pi hv^3/c^3)[\exp(hv/kT) - 1]^{-1}$$

to obtain

$$Q_{mn} = (8\pi hv^3/c^3)B_{mn} = (16\pi^2 v_{mn}^3/c^3 h)\|\langle \beta\phi_m, \phi_n \rangle\|^2. \tag{7.3}$$

Thus, to within the accuracy of the approximations that we have been using, the probability for spontaneous emission depends on only the matrix element of the dipole moment between the stationary states. It is not difficult to check that a rough estimate of the next terms in the expansion of the electromagnetic field shows that they can be expected to be of much smaller magnitude than $(\varepsilon a)^2$, where ε is the electronic charge and a is of the size of a typical radius. If we use this expression for our estimate of the order of magnitude of the dipole transitions, we see that the next order terms are considerably smaller. Of course, some of the matrix elements will vanish on account of selection rules. However, the contribution of the 'dipole transitions' will predominate.

We will not present a detailed discussion of the theory of Raman scattering. Roughly speaking, the transitions are due to the interaction with the ambient electromagnetic field of the dipole moments induced by the field. This depends on the so-called polarizability tensor of the molecule, which measures its response to the electromagnetic field. This is a symmetric tensor, which is why the representation space associated with the Raman effect is $S^2(\mathbb{R}^3)$.

The actual analysis of real spectra is, of course, much more complicated than the simple scheme described above and requires much experience and intuition. As some allowed fundamental lines may in fact be relatively weak, there will be confusion with overtone transitions. The practice of spectroscopy thus becomes a profession which cannot be adequately described within the confines of a book like this.

3.8   Semidirect products and their representations

Let G be a group and N a subgroup. Recall that N is called a *normal* subgroup if $aNa^{-1} = N$ for all $a \in G$. Then we can define the quotient group G/N. We multiply aN and bN by the rule

$$aN \cdot bN = (ab)N.$$

This is well defined since

$$anbn' = abb^{-1}nbn'$$

$$= (ab)n''n'$$

with $n'' = b^{-1}nb \in N$.

For example, consider the group of Euclidean motions in $\mathbb{R}^3$. It consists of all 4×4 matrices of the form

$$\begin{pmatrix} A & v \\ 0 & 1 \end{pmatrix} \qquad \begin{matrix} A \in SO(3) \\ v \in \mathbb{R}^3. \end{matrix}$$

Such a matrix sends the point x in $\mathbb{R}^3$ into $Ax + v$. This can be written as a matrix multiplication:

$$\begin{pmatrix} A & v \\ 0 & 1 \end{pmatrix}\begin{pmatrix} x \\ 1 \end{pmatrix} = \begin{pmatrix} Ax + v \\ 1 \end{pmatrix}.$$

The inverse of

$$\begin{pmatrix} A & v \\ 0 & 1 \end{pmatrix} \quad \text{is} \quad \begin{pmatrix} A^{-1} & -A^{-1}v \\ 0 & 1 \end{pmatrix}$$

as can be checked by matrix multiplication. Let N denote the subgroup consisting of all translations. So N consists of all matrices of the form

$$\begin{pmatrix} I & w \\ 0 & 1 \end{pmatrix}$$

where I is the (3×3) identity matrix. Then

$$\begin{pmatrix} A & v \\ 0 & 1 \end{pmatrix}\begin{pmatrix} I & w \\ 0 & 1 \end{pmatrix}\begin{pmatrix} A^{-1} & -A^{-1}v \\ 0 & 1 \end{pmatrix} = \begin{pmatrix} I & Aw \\ 0 & 1 \end{pmatrix}. \tag{8.1}$$

So N is a normal subgroup. Notice that there are two further special properties that hold for this example:

(i) N is Abelian.

(ii) There is a subgroup $H\left(\text{consisting of all } \begin{pmatrix} A & 0 \\ 0 & 1 \end{pmatrix}\right)$ such that every element of G can be written uniquely as hn with $h \in H$ and $n \in N$.

Suppose that G is any group with a normal Abelian subgroup, N, and some other subgroup H such that every element of G can be written in a unique way as hn, where $h \in H$ and $n \in N$. If G is finite, this implies that $\#G = \#H \cdot \#N$ and $H \cap N = \{e\}$. Conversely, if these last two conditions hold, then every element of G can be written uniquely as hn; indeed, we cannot have $hn = e$ unless $h = e$ and $n = e$, since $hn = e$ implies $h = n^{-1} \in H \cap N$. Therefore, $hn = h'n'$ implies $h = h'$ and $n = n'$ and so there are $\#H \cdot \#N$ distinct elements hn. Since $\#H \cdot \#N = \#G$, this implies that every element of G can be uniquely written in the desired form.

Examples

$$G = D_n, N = C_n, H = \{e, r\}, \text{ where } r \notin C_n.$$
$$G = S_4, N = \{e, a, b, c\}, \text{ where } a = (12)(34),$$
$$b = (13)(24), c = (14)(23) \text{ and}$$
$$H \sim S_3 = \{e, (12), (13), (23), (123), (132)\}.$$

We say that G is the *semidirect product* of H and N, and write $G = H \circledS N$.

The group H acts on N by conjugation:

$$h \text{ sends } n \text{ into } hnh^{-1} \in N.$$

For the case of the Euclidean group, (8.1) gives an explicit formula for this action. We can generalize the Euclidean example as follows: let H be any group and let us be given a representation τ of H on a vector space N. We let G consist of all pairs (h, n) and define multiplication by

$$(h_1, n_1)(h_2, n_2) = (h_1 h_2, n_1 + \tau(h_1)n_2). \tag{8.2}$$

(Of course, in writing (8.2) we are really writing the multiplication law for 'matrices'

$$\begin{pmatrix} h & n \\ 0 & 1 \end{pmatrix}$$

with h_1 acting on n_2 via τ.) Then

$$(h, n)^{-1} = (h^{-1}, -\tau(h^{-1})n)$$

and

$$(h, m)(I, n)(h, m)^{-1} = (I, \tau(h)n). \tag{8.3}$$

We can then identify N with the subgroup of G consisting of all (I, n), $n \in N$. It becomes a normal subgroup. We can identify H with the subgroup of G consisting of all $(h, 0)$, $h \in H$. Then it is immediate that G is the semidirect product of H and N. In this case, taking $m = 0$ in (8.3) shows that the conjugation action of H on N is just the representation τ.

A very important case for us is where $H = Sl(2, \mathbb{R})$, $N = \mathbb{R}^{1,3}$ and τ is the homomorphism from $Sl(2, \mathbb{R})$ to the Lorentz group defined at the very beginning of this book. The resulting semidirect product is then the *Poincaré group*.

Let us return to the finite case. Since N is Abelian, all of its irreducible representations are one dimensional. Let N^* denote the space of all one-dimensional characters on N. Now N^* is a set of functions on N. Since H acts on N, it also acts on the space of all functions, $\mathscr{F}(N)$ by the rule

$$hf(n) = f(n')$$

where n' is the image of n under h^{-1},
i.e.

$$hf(n) = f(h^{-1}nh).$$

In particular, if $\chi \in N^*$ is a character, so is $h\chi$, where

$$h\chi(n) = \chi(h^{-1}nh). \tag{8.4}$$

We will now show how to find all the irreducible representations of a semidirect product.

Suppose that G is the semidirect product of H and N. Let (ρ, V) be an irreducible representation of G. Decompose V under N:

$$V = V_{\chi_1} + \cdots + V_{\chi_k}$$

where

$$V_{\chi_j} = m_j W_j$$

and (σ_j, W_j) is the irreducible (one-dimensional) representation of N with character χ_j. Thus V_{χ_j} consists of those vectors $v \in V$ which satisfy

$$\rho(n)v = \chi_j(n)v.$$

For any $a \in G$ and $v \in V_{\chi_j}$

$$\begin{aligned} \rho(n)\rho(a)v &= \rho(a)\rho(a)^{-1}\rho(n)\rho(a)v \\ &= \rho(a)\rho(a^{-1}na)v \\ &= \rho(a)\chi_j(a^{-1}na)v \\ &= (a\chi_j)(n)\rho(a)v \end{aligned}$$

so

$$\rho(a)V_{\chi_j} = V_{a\chi_j}.$$

Thus, if $V_{\chi_j} \neq \{0\}$ and if N_j^* denotes the orbit of G acting on N^* containing χ_j then the subspace

$$V_{\chi_{l_1}} + \cdots + V_{\chi_{l_p}} \qquad \chi_{l_i} \in N_j^*$$

is an invariant subspace of V. Since V is irreducible, we conclude that

$$V = V_{\chi_{l_1}} + \cdots + V_{\chi_{l_p}} \qquad \chi_{l_i} \in N_j^*.$$

In other words $m_i = 0$ unless $\chi_i \in N_j^*$. If we think of V_{χ_i} as a vector space situated over

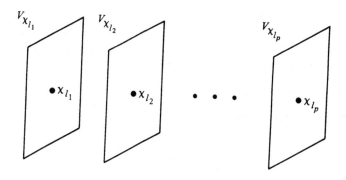

Fig. 3.11

$\chi_i \in N_j^*$, we have constructed a vector bundle E over N_j^* on which G acts (see Fig. 3.11). So we have

$$E$$
$$\downarrow$$
$$N_j^*$$

and (ρ, V) is equivalent to the action of G on $\Gamma(E)$.

Let G_j be the isotropy group of $\chi_j \in N_j^*$. Then G_j is the semidirect product of L_j and N, where $L_j \subset H$ is the isotropy group of H acting on N^* at χ_j. G_j is represented on V_{χ_j} and each element $n \in N$ acts on V_{χ_j} by $\chi_j(n)I$. In order for G acting on $\Gamma(E)$ to be irreducible, the group G_j must act irreducibly on V_{χ_j}. This means that L_j must act irreducibly on V_{χ_j}. We have thus shown that every irreducible representation of G is constructed as follows:

(1) Decompose N^* into orbits under G:

$$N^* = N_1^* \cup \cdots \cup N_r^*$$

and pick a $\chi_j \in N_j^*$ for each orbit.

(2) Let $L_j \subset H$ be the subgroup fixing χ_j and let $G_j = L_j \; \circledS \; N \subset G$. Choose an irreducible representation of L_j, say (ρ_j, V_j) and consider it as a representation of G_j by

$$\rho_j(hn)v = \chi_j(n)\rho(h)v.$$

(3) Construct the corresponding vector bundle E and let G act on $\Gamma(E)$.

(4) The space V_j can be identified with the δ-sections concentrated at χ_j and can be characterized as those sections s which satisfy $\rho(n)s = \chi_j(n)s$; in other words

$$V_j = (\Gamma(E))_{\chi_j}.$$

If we start with an irreducible representation (ρ_j, V_j) of L_j and construct $\Gamma(E)$, then $\Gamma(E)$ must be irreducible. In fact, if $\Gamma(E) = \Gamma(E)' + \Gamma(E)''$, then we must have

$$\Gamma(E)_{\chi_j} = \Gamma(E)'_{\chi_j} + \Gamma(E)''_{\chi_j}$$

as a decomposition into invariant subspaces under L_j. But $\Gamma(E)_{\chi_j} = V_j$ is irreducible.

Table 21.

	e	R_1	R_2	R_3
χ_0	1	1	1	1
χ_1	1	i	-1	$-i$
χ_2	1	-1	1	-1
χ_3	1	$-i$	-1	i

Hence $\Gamma(E)''_{\chi_j} = \{0\}$ and hence $\Gamma(E)''_\chi = 0$ for all $\chi \in N_j^*$ so $\Gamma(E)'' = 0$. Thus $\Gamma(E)$ *is irreducible.*

Example The irreducible representations of D_4

Step 1. $D_4 = H \circledS N$, where $H = \{e, \alpha_1\}$ and $N = \{e, R_1, R_2, R_3\}$.

Step 2 Form N^*, the set of characters of N ($\cong C_4$) (see Table 21).

Step 3 Let H act on N^* by $h\chi_i(n) = \chi_i(h^{-1}nh)$ (see Table 22). The action breaks up N^* into three orbits as in Fig. 3.12.

Step 4 Pick an orbit and a point in the orbit. Find L_i, the isotropy subgroup, which fixed the point. For the single-element orbits $L_i = H$. For the double-element orbit $L_i = \{e\}$.

Step 5 Consider the two-element orbit and choose χ_1 to be identified with the coset N. Then χ_3 is identified with the coset $\alpha_1 N$. Construct a vector bundle over the two points by taking as basis elements $e_1 = [(e, v_0)]$ and $e_2 = [(\alpha_1, v_0)]$.

Step 6 Calculate representation matrices by letting $G = D_4$ act on basis elements. Use the rule $[(b, v_0)] \sim [(b, \sigma(l)v_0)]$, where $\sigma(l) = \chi_1(n)\rho(h)$ and $l = hn$. Since $H = \{e\}$, $\rho(h) \equiv 1$. So

$$R_1[(e, v_0)] = [(R_1, v_0)] = [(eR_1, v_0)] = [(e, \sigma(R_1)v_0)]$$
$$= [(e, \chi_1(R_1)v_0)] = [(e, iv_0)] = i[(e, v_0)] = ie_1$$

Fig. 3.12

Table 22.

	χ_0	χ_1	χ_2	χ_3
e	χ_0	χ_1	χ_2	χ_3
α_1	χ_0	χ_3	χ_2	χ_1

and

$$R_1[(\alpha_1, v_0)] = [(R_1\alpha_1, v_0)] = [(\alpha_1 R_3, v_0)] = [(\alpha_1, \sigma(R_3)v_0)]$$
$$= [(\alpha_1, \chi_1(R_3)v_0)] = [(\alpha_1, -iv_0)] = -ie_2.$$

Thus R_1 is represented by $\begin{pmatrix} i & 0 \\ 0 & -i \end{pmatrix}$.

For α_1 we have

$$\alpha_1[(e, v_0)] = [(\alpha_1, v_0)] = e_2$$

and

$$\alpha_1[(\alpha_1, v_0)] = [(e, v_0)] = e_1.$$

So α_1 is represented by $\begin{pmatrix} 0 & 1 \\ 1 & 0 \end{pmatrix}$.

Since $R_1\alpha_1 = \beta_1$ we can, with these two matrices, recover all of the matrix representations of the group. So this orbit gave the two-dimensional representation of D_4.

Step 6'   We apply Step 6 to each of the two remaining orbits. Suppose we consider the orbit with χ_2. Now $L_2 = H$ is not trivial and there are two representations of the little group as shown in Table 23.

The choice of using either ρ_1 or ρ_2 in $\sigma(l) = \chi_2(n)\rho_i(h)$ will give *two* distinct one-dimensional representations.

There is only one basis element now, $[(e, v_0)]$.

$$R_1[(e, v_0)] = [(eR_1, v_0)] = [(e, \chi_2(R_1)\rho_i(e)v_0)]$$
$$= -[(e, v_0)]$$

for both representations, since $\rho_i(e) = 1$ for $i = 1, 2$. However,

$$\alpha_1[(e, v_0)] = [(e\alpha_1, v_0)] = [(e, \chi_2(e)\rho_i(\alpha_1)v_0)]$$
$$= [(e, v_0)]$$

since $\rho_1(\alpha_1) = 1$ but $\rho_2(\alpha_1) = -1$. Hence our two representations are as shown in Table 24.

Again, these will generate the whole group, so we are done.

Table 23.

	e	α_1
ρ_1	1	1
ρ_2	1	-1

Table 24.

	R_1	α_1
3	-1	1
4	-1	-1

It remains to do the third orbit, which consists of χ_0. The little group is H, and R_1 is obviously represented by $+1$ since $\chi_0 \equiv 1$. For α, we have again

$$\alpha_1[(e, v_0)] = [(e, \rho_i(\alpha_1)v_0)] = \pm [(e, v_0)].$$

So the other two one-dimensional representations are as in Table 25.

This is summarized in Table 26, which of course checks with the known character table for D_4.

Let M be an orbit of H in N^*. Let $H_0 \in H$ be the isotropy group of $\chi_0 \in M$. Let (ρ, V) be a representation of H_0. For future use, let us write once again the action of various elements of G on $\Gamma(E)$, where E is the associated vector bundle: if $s \in \Gamma(E)$ and $n \in N$, then

$$(ns)(\chi) = \chi(n)s(\chi). \tag{8.5}$$

To evaluate the action of $h \in H$ on $\Gamma(E)$ in a convenient way, we first make some preliminary remarks. We know that the most general element of E can be written as an equivalence class $[(a, v)]$, where $a \in H$ and $v \in V$, cf. Section 3.3. Let us choose some 'cross section'

$$r: M \to H \text{ such that } r(\chi)\chi_0 = \chi.$$

We can write any section s of E as

$$s(\chi) = [(r(\chi), \psi(\chi))]$$

where $\psi: M \to V$. Then if $a \in H$, we have

$$\begin{aligned}(as)(\chi) &= as(a^{-1}\chi) \\ &= [(ar(a^{-1}\chi), \psi(a^{-1}\chi))] \\ &= [(r(\chi)r(\chi)^{-1}ar(a^{-1}\chi), \psi(a^{-1}\chi))] \\ &= [(r(\chi), \rho(r(\chi)^{-1}ar(a^{-1}\chi))\psi(a^{-1}\chi))].\end{aligned}$$

Table 25.

	R_1	α_1
1	1	1
2	1	-1

Table 26.

	e	R_1, R_3	R_2	$\alpha_1, \alpha_2 = R_2\alpha_1$	$\beta_1 = R_1\alpha_1, \beta_2 = R_3\alpha_1$
1	1	1	1	1	1
2	1	1	1	-1	-1
3	1	-1	1	1	-1
4	1	-1	1	-1	1
5	$\begin{pmatrix} 1 & 0 \\ 0 & 1 \end{pmatrix}$	$\begin{pmatrix} i & 0 \\ 0 & -i \end{pmatrix}, \begin{pmatrix} -i & 0 \\ 0 & i \end{pmatrix}$	$\begin{pmatrix} -1 & 0 \\ 0 & -1 \end{pmatrix}$	$\begin{pmatrix} 0 & 1 \\ 1 & 0 \end{pmatrix}, \begin{pmatrix} 0 & -1 \\ -1 & 0 \end{pmatrix}$	$\begin{pmatrix} 0 & i \\ -i & 0 \end{pmatrix}, \begin{pmatrix} 0 & -i \\ i & 0 \end{pmatrix}$

Thus, *as* corresponds to the function ψ_a, where

$$\psi_a(\chi) = \rho(r^{-1}(\chi)ar(a^{-1}\chi))\psi(a^{-1}\chi). \tag{8.6}$$

This rather formidable expression simplifies for elements of H_0 if r has nice properties. Suppose that

$$r(a\chi) = ar(\chi)a^{-1}, \quad a \in H_0 \tag{8.7}$$

i.e. that the section r is equivariant for the conjugation action of a on H. Then

$$r^{-1}(\chi)ar(a^{-1}\chi) = r^{-1}(\chi)aa^{-1}r(\chi)a = a$$

and (8.6) simplifies to

$$\psi_a(\chi) = \rho(a)\psi(a^{-1}\chi), \quad a \in H_0 \tag{8.8}$$

We recall from the end of Section 3.3 that if the representation, ρ, of H_0 extends to a representation τ of H, then we can identify $\Gamma(M)$ with $\mathscr{F}(M, V)$ and (the analogue of) (8.8) is true for all $a \in H$:

$$a\psi(\chi) = \tau(a)\psi(a^{-1}\chi). \tag{8.9}$$

3.9 Wigner's classification of the irreducible representation of the Poincaré group

Let us now see how to formulate the results of the preceding section for the case of groups like the Euclidean group or the Poincaré group. We first must decide how to define N^* and then describe the action of H on N^*. For the finite case, a character χ *must* take on values of absolute value one. For the infinite case, we must add this as an additional condition in the definition. We also require that χ be a continuous function. Thus N^* consists of all continuous functions $\chi: N \to \mathbb{C}$ such that $|\chi| \equiv 1$ and

$$\chi(n_1 n_2) = \chi(n_1)\chi(n_2). \tag{9.1}$$

Now

$$n_1 = \begin{pmatrix} I & w_1 \\ 0 & 1 \end{pmatrix} \quad n_2 = \begin{pmatrix} I & w_2 \\ 0 & 1 \end{pmatrix}$$

$$n_1 n_2 = \begin{pmatrix} I & w_1 + w_2 \\ 0 & 1 \end{pmatrix}, \quad w_L \in V,$$

where V is a finite-dimensional vector space on which H acts. So writing

$$\chi(w) \text{ instead of } \chi\begin{pmatrix} I & w \\ 0 & 1 \end{pmatrix}$$

(9.1) becomes

$$\chi(w_1 + w_2) = \chi(w_1)\chi(w_2). \tag{9.2}$$

This implies that χ must be of the form χ_l where, for $l \in V^*$,

$$\chi_l(w) = e^{il(w)}.$$

It follows from (8.3) (with $m = 0$) that under the conjugation action of H on N, the element A^{-1} sends

$$n = \begin{pmatrix} I & w \\ 0 & 1 \end{pmatrix} \text{ into } \begin{pmatrix} I & \tau(A^{-1})w \\ 0 & 1 \end{pmatrix}.$$

So the action of H on N^* is given by

$$A\chi_l = \chi_{\tau(A)^{*-1}l}. \tag{9.3}$$

Suppose that the $\tau(A)$ preserve some non-degenerate scalar product on N. Then we can use this scalar product to identify N with N^*. Thus $l(w)$ becomes $l \cdot w$

$$\chi_l(w) = e^{il \cdot w}$$

and $\tau(A)^{*-1} = \tau(A)$ so (9.3) simplifies to

$$A\chi_l = \chi_{\tau(A)l}.$$

Up to now in this book we have only discussed representations of finite groups, not groups such as the Poincaré group. We will give some of the basic definitions and theorems in the next chapter. Suffice it to say that we will require all our representations to be unitary and to be continuous (in the appropriate sense). For groups such as the Poincaré group or the Euclidean group, it turns out that the typical irreducible representation will be infinite dimensional. Nevertheless, it is true that the method of the preceding section applies to give all irreducible representations of a semidirect product. This was proved by Wigner for the case of the Poincaré group and generalized to the case of a general semidirect product by Mackey.

Of course, in order to describe the representations involved, we must replace the finite and discrete by the continuous. We give some indication of how to do this in the next chapter. For the moment, let us follow the prescription of the preceding section if we want to label the representations. We must

(i) describe the orbits of $Sl(2, \mathbb{C})$ acting on $\mathbb{R}^{1,3}$;

(ii) pick a point on each orbit and find its isotropy group;

(iii) list all the representations of the groups in (ii).

Recall from Chapter 1 that an $A \in Sl(2, \mathbb{C})$ acts on

$$P = \begin{pmatrix} p_0 + p_3 & p_1 + ip_2 \\ p_1 - ip_2 & p_0 - p_3 \end{pmatrix}$$

by

$$\tau(A)P = APA^*,$$

and that

$$m^2 = \det P = p_0^2 - p_1^2 - p_2^2 - p_3^2$$

is preserved. Thus each orbit must be contained in a level set of m^2. For $m^2 > 0$, the hyperboloid $\det P = m^2$ has two sheets corresponding to $p_0 > 0$ and $p_0 < 0$ (see Fig. 3.13). For $m^2 = 0$, we have the forward 'light cone' with $p_0 > 0$, the backward light cone with $p_0 < 0$ and the isolated point orbit consisting of the origin alone. For $m^2 < 0$, the hyperboloids $\det P = m^2$ are connected and are hence orbits. We therefore get the table of orbits, Table 27, where we have indicated a representative point on each orbit.

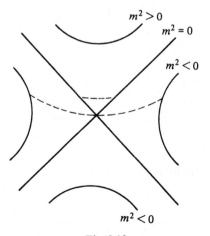

Fig. 3.13

Table 27.

Orbit	Representative point
$m^2 > 0, p_0 > 0$	$\begin{pmatrix} m & 0 \\ 0 & m \end{pmatrix}$
$m^2 > 0, p_0 < 0$	$\begin{pmatrix} -m & 0 \\ 0 & -m \end{pmatrix}$
$m^2 = 0, p_0 > 0$	$\begin{pmatrix} 2 & 0 \\ 0 & 0 \end{pmatrix}$
$m^2 = 0, p_0 < 0$	$\begin{pmatrix} -2 & 0 \\ 0 & 0 \end{pmatrix}$
$\{0\}$	$\begin{pmatrix} 0 & 0 \\ 0 & 0 \end{pmatrix}$
$m^2 < 0$	$\begin{pmatrix} 0 & |m|i \\ -|m|i & 0 \end{pmatrix}$

For each of these types, let us compute the isotropy group of the representative point.

(i)
$$A\begin{pmatrix} m & 0 \\ 0 & m \end{pmatrix}A^* = AA^*\begin{pmatrix} m & 0 \\ 0 & m \end{pmatrix}$$

so the condition that

$$A\begin{pmatrix} m & 0 \\ 0 & m \end{pmatrix}A^* = \begin{pmatrix} m & 0 \\ 0 & m \end{pmatrix}$$

is the same as $AA^* = I$ so the isotropy group for the first two cases is $SU(2)$. Let us now turn to case (ii).

(ii)
$$\begin{pmatrix} a & b \\ c & d \end{pmatrix}\begin{pmatrix} 2 & 0 \\ 0 & 0 \end{pmatrix}\begin{pmatrix} \bar{a} & \bar{c} \\ \bar{b} & \bar{d} \end{pmatrix} = \begin{pmatrix} 2|a|^2 & 2a\bar{c} \\ 2c\bar{a} & 2|c|^2 \end{pmatrix}$$

so

$$A\begin{pmatrix} 2 & 0 \\ 0 & 0 \end{pmatrix}A^* = \begin{pmatrix} 2 & 0 \\ 0 & 0 \end{pmatrix}$$

holds if and only if

$$|a|^2 = 1 \text{ and } c = 0.$$

We are thus dealing with the group of all matrices of the form

$$\begin{pmatrix} e^{i\theta} & b \\ 0 & e^{-i\theta} \end{pmatrix}.$$

The product of two such matrices is given by

$$\begin{pmatrix} e^{i\theta} & b \\ 0 & e^{-i\theta} \end{pmatrix}\begin{pmatrix} e^{i\theta'} & b' \\ 0 & e^{-i\theta'} \end{pmatrix} = \begin{pmatrix} e^{i(\theta+\theta')} & e^{i\theta}b' + be^{-i\theta'} \\ 0 & e^{-i(\theta+\theta')} \end{pmatrix}$$

and we have the conjugation

$$\begin{pmatrix} e^{i\theta} & 0 \\ 0 & e^{-i\theta} \end{pmatrix}\begin{pmatrix} 1 & b \\ 0 & 1 \end{pmatrix}\begin{pmatrix} e^{-i\theta} & 0 \\ 0 & e^{i\theta} \end{pmatrix} = \begin{pmatrix} 1 & be^{2i\theta} \\ 0 & 1 \end{pmatrix}.$$

If we write $b = x + iy$, we see that the isotropy group in question is just the semidirect product of the circle $SO(2)$ and the plane, with θ acting as rotation through angle 2θ. Thus the group in question is the double cover (on account of the factor 2) of the group of Euclidean motions. Let us denote this group by $\tilde{E}(2)$. The same computation gives $\tilde{E}(2)$ as the isotropy group of the point $\begin{pmatrix} -2 & 0 \\ 0 & 0 \end{pmatrix}$. The isotropy group of $\begin{pmatrix} 0 & 0 \\ 0 & 0 \end{pmatrix}$ is clearly the whole of $Sl(2, \mathbb{C})$. So the remaining case (iii) is $\begin{pmatrix} 0 & |m|i \\ -|m|i & 0 \end{pmatrix}$.

(iii) The set of A satisfying

$$A\begin{pmatrix} 0 & |m|i \\ -|m|i & 0 \end{pmatrix}A^* = \begin{pmatrix} 0 & |m|i \\ -|m|i & 0 \end{pmatrix}$$

is clearly the same as the set of A satisfying

$$A\begin{pmatrix} 0 & 1 \\ -1 & 0 \end{pmatrix}A^* = \begin{pmatrix} 0 & 1 \\ -1 & 0 \end{pmatrix}.$$

Let $A^t = \begin{pmatrix} a & c \\ b & d \end{pmatrix}$ denote the transpose of $A = \begin{pmatrix} a & b \\ c & d \end{pmatrix}$. Then

$$A\begin{pmatrix} 0 & 1 \\ -1 & 0 \end{pmatrix}A^t = \begin{pmatrix} 0 & ad-bc \\ -(ad-bc) & 0 \end{pmatrix} = \begin{pmatrix} 0 & 1 \\ -1 & 0 \end{pmatrix}.$$

Thus we must have

$$A\begin{pmatrix} 0 & 1 \\ -1 & 0 \end{pmatrix}A^* = A\begin{pmatrix} 0 & 1 \\ -1 & 0 \end{pmatrix}A^t$$

which can happen if and only if

$$A^* = A^t,$$

i.e. all the entries of A are real. Thus the isotropy group in this case is $Sl(2.\mathbb{R})$.

In the physics literature, the isotropy group of a representative point on an orbit is called the 'little group' (of the point or of the orbit). We can thus complete Table 27 to obtain Table 28.

For the next step, we must classify the irreducible representations of the little groups. We will compute all the irreducible representations of $SU(2)$ in the next chapter. We will find that they are parametrized by a number s which can take on

Table 28.:

Orbit	Representative point	Little group
$m^2 > 0, p_0 > 0$	$\begin{pmatrix} m & 0 \\ 0 & m \end{pmatrix}$	$SU(2)$
$m^2 > 0, p_0 < 0$	$\begin{pmatrix} -m & 0 \\ 0 & -m \end{pmatrix}$	$SU(2)$
$m^2 = 0, p_0 > 0$	$\begin{pmatrix} 2 & 0 \\ 0 & 0 \end{pmatrix}$	$\tilde{E}(2)$
$m^2 = 0, p_0 < 0$	$\begin{pmatrix} -2 & 0 \\ 0 & 0 \end{pmatrix}$	$\tilde{E}(2)$
$\{0\}$	$\begin{pmatrix} 0 & 0 \\ 0 & 0 \end{pmatrix}$	$Sl(2, \mathbb{C})$
$m^2 < 0$	$\begin{pmatrix} 0 & \lvert m \rvert i \\ -\lvert m \rvert i & 0 \end{pmatrix}$	$Sl(2, \mathbb{R})$

non-negative half integer values:

$$s = 0, \tfrac{1}{2}, 1, \tfrac{3}{2}, \dots.$$

Thus, corresponding to our first family of orbits, the representations are parametrized by a continuous parameter $m^2 > 0$ and the discrete parameter s. In physical terms, m is the 'rest mass'

$$m^2 = p_0^2 - p_1^2 - p_2^2 - p_3^2$$

and s can be identified with the 'intrinsic spin' of a point particle. We shall discuss its meaning and implication in the next chapter.

The group $E(2)$ is itself a semidirect product, so we can compute its representations by the same method. The orbits of $SO(2)$ acting on $\mathbb{R}^2$ are circles and the one-point orbit consisting of the origin. It turns out that the representations corresponding to the circles do not arise in physics. The little group of $\{0\}$ is $SO(2)$. Its irreducible unitary representations are all one dimensional and are given by

$$\theta \to e^{in\theta} \quad n = 0, \pm 1, \pm 2, \dots.$$

Since θ corresponds to rotation by 2θ it is usual to write $\theta = \psi/2$, $0 \leq \psi \leq 4\pi$ and $n = 2s$ so the irreducibles corresponding to the orbit $\{0\}$ are labeled by

$$s = 0, \pm\tfrac{1}{2}, \pm 1, \dots.$$

For $m^2 < 0$, we must list all the irreducible unitary representations of $Sl(2, \mathbb{R})$.

Outside of the trivial representation these are all infinite dimensional and their description lies outside the scope of this book. Fortunately, no particles are known to exist with $m^2 < 0$. (They would be 'tachyons' travelling faster than the speed of light, although some recent theories demand their existence.)

Thus, the physically relevant representations of $Sl(2, \mathbb{C}) \circledS \cdot \mathbb{R}^{1,3}$ with $p_0 > 0$ are parametrized by

$$m^2 > 0 \quad s = 0, \tfrac{1}{2}, 1, \tfrac{3}{2}, \ldots$$

and

$$m^2 = 0 \quad s = 0, \pm \tfrac{1}{2}, \pm 1, \pm \tfrac{3}{2}, \ldots$$

It is time to pause and consider the significance of this result, which was proved by Wigner some 50 years ago.

Speculation as to the ultimate constituents of matter occurs in the earliest scientific thought. The concept of an element as a 'substance from which all other bodies are made or derived' was held at the very beginning of occidental philosophy. Thales of Miletus (about 640–547 Before the Common Era) regarded 'water' as the basis of all things. (His ideas may have been influenced by still earlier doctrines having their origin in ancient Egypt.) His followers accepted his idea of a primordial substance as the basis of all bodies, but they endeavored to determine some other general element such as 'fire' or 'spirit' or 'love' or 'hatred'. Around 500 BCE Heraclitus suggested that the fundamental elements were earth, air, fire and water, and that none could be obtained from one another by physical means. A little later Democritus (460–370 BCE) claimed that matter was composed of many different kinds of minute hard particles or 'atoms', and similar ideas were put forward by Epicurus (341–276 BCE). Aristotle (384–322 BCE) in his *Physics*, adopted the point of view of Heraclitus. Furthermore, he attributed the motions of matter to the tendency of the constituent elements to return to their 'natural' place.

In any attempt to classify the elementary constituents of matter we must come to grips with two problems: what is the meaning of the word 'elementary' and when should two objects be regarded as the 'same'? The second of these questions immediately brings into consideration the theory of groups, and, in particular, those groups which are intimately related to our conception of space. For example, we may wish to regard a particle as unchanged if we pick it up and place it down at some other location in a horizontal plane. Thus, the location of a particle in a plane is not an 'intrinsic' property of a particle; the group of symmetries of the Euclidean plane enters as a 'group of symmetries' of the classification problem. It would not make sense to talk of the 'natural' horizontal location of a particle. In a geocentric conception of the universe, the 'natural' vertical location of a particle (or, more precisely, its 'natural' distance from the center of the earth) makes perfectly good sense. Hence, in Aristotle's theory, one could conceive that each element had its own (vertical) 'natural' place. With the overthrow of the geocentric conception, and with the advent of Newtonian physics, we tend to believe, in some form or other, in the isotropy of space and can no longer single out a vertical or radial direction as being preferred. Similarly, our conceptions about space time will change the kinds of properties that can be allowed for the description of particles. We regard two freely moving bodies which differ only in their kinetic energy as being the 'same'. This reflects the fact that we do not believe in the existence of an 'absolute rest frame'. By a geometrically admissible transformation of space time (either in Galilean relativity or in Einstein's special relativity) we can arrange that a freely moving massive body be at rest. It is thus clear that in any theory of classification of elementary particles there must be some group, G, in the background of our thoughts, and that this group should contain the group associated to our geometric conceptions of space and time. Of course, the group might be larger, reflecting symmetries that go beyond geometry. Needless to say, our conceptions about the geometry of space and time are strongly influenced by our theory of the nature of matter.

The first problem, that of defining the meaning of the word 'elementary' is much more difficult to discuss at this level of generality, without specifying a more detailed choice of theory. The way in which objects are 'put together' will tend to vary from one theory to another. Nevertheless, a brief glance at some examples might prove instructive even at this stage. If we believe in the undulatory theory of light, then, in an intuitive sense that can be made quite precise, the most 'elementary' kind of wave is a plane wave. (Here the elementary waves are 'put together' by Fourier analysis.) Now a plane wave cannot be localized in space. If we try by physical means (say a screen with an aperture) to localize (a physical approximation to) a plane wave, we find that the localized wave is much more complicated than the plane wave we started with, due to diffraction effects. We learn two lessons: our 'particles' may tend to be somewhat ideal, and we may have to give up on the notion of localizability in space. A similar situation obtains in currently popular theories which are based on 'relativistic quantum mechanics'. In these theories it requires a great deal of energy to confine a particle in a small region of space, and when these energies become sufficiently great, there is spontaneous production of new particles. Thus we may have to be more Aristotelian than Democritean in our outlook.

All of the recent theories of elementary particles have been shaped by the paper by Wigner, containing the classification of the irreducible representations of $Sl(2, \mathbb{R}) \circledS \mathbb{R}^{1,3}$ described above. It is difficult to overestimate the importance of this paper, which will certainly stand as one of the great intellectual achievements of our century. It has not only provided a framework for the physical search for elementary particles, but has also had a profound influence on the development of modern mathematics, in particular the theory of group representations. From our point of view, we can summarize Wigner's main points as follows. The logic of physics is quantum mechanics. Hence, a symmetry group of the system manifests itself as a unitary (or possibly anti-unitary) representation. Ignoring the anti-unitaries (for example, by considering connected groups), one posits that

(i) an elementary particle 'is' an irreducible unitary representation of the group, G of physics, where these representations are required to satisfy certain physically reasonable restrictions, and where

(ii) the group G of physics is the group $Sl(2, \mathbb{C}) \circledS \mathbb{R}^{1,3}$, the semidirect product of $Sl(2, \mathbb{C})$ and the translation group in Minkowski space, i.e. the group G is the double (universal) covering of the Poincaré group, P, where P denotes the group of those transformations of special relativity which can be continuously deformed to the identity.

Wigner computes all the irreducible representations of G and finds, as we have seen, that
(iii) the representations satisfying the physical conditions can be parametrized by two parameters m and s, where m is a non-negative real number and where s is constrained by

$$s = 0, \tfrac{1}{2}, 1, \tfrac{3}{2}, \text{ etc. if } m \text{ is positive}$$

and

$$s = 0, \pm\tfrac{1}{2}, \pm 1, \pm\tfrac{3}{2}, \text{ etc. if } m = 0.$$

Here the parameter m can be identified with the rest mass of the particle, and the parameter s can be identified with its spin (or spin and helicity if $m = 0$). The spin is a notion related to the 'intrinsic angular momentum' of the particle.

Wigner also considers the minor modification in this scheme that has to be made if we replace the connected group of symmetries of special relativity by disconnected groups. (The full groups have four components, corresponding to the possibilities of space inversion, time inversion, and their product. The current belief is that the correct group has two components corresponding to simultaneous space and time inversion, but that this transformation must be accompanied by reversal of all electric charges.)

In the past 50 years or so since Wigner's work various modifications have been introduced. First of all, even at the time of publication of Wigner's article, the scheme was not completely satisfactory in so far as, although it explained the observed spins, it failed to explain why only certain mass values were observed and not others. Also, it did not deal with electric charge, a clearly conserved property of observed particles on the same footing as mass and spin. In the intervening years various other invariant 'quantum numbers' of particles have been registered, such as isotopic spin, strangeness, etc. We will discuss these in Chapters 4 and 5. The main thrust of recent theories (or at least most of them) has been to modify (ii) by enlarging the group G. The most successful of such theories, involving quarks and electroweak unification, indicates that the group $SU(3)$ should be contained, in some way, in the group G. Recent speculation concerning the so-called 'supersymmetries' suggests the notion of a group might have to be slightly enlarged. However, point (i) in the dogma remains unchanged. This is because in quantum mechanics the 'state' of a system is a unit vector in Hilbert space and any symmetry manifests itself as a unitary (or anti-unitary) transformation of this Hilbert space. To the extent that we accept quantum mechanics as providing the underlying foundational theory of physics, we cannot make major changes in (i). We shall enlarge on these points in the next two chapters.

3.10 Parity

One issue that is convenient to discuss here is that of parity: do the basic laws of physics distinguish between right and left handedness? Put another way, let $\mathcal{P}$ denote the operator of spatial inversion:

$$\mathscr{P} = \begin{pmatrix} x_0 \\ x_1 \\ x_2 \\ x_3 \end{pmatrix} \rightarrow \begin{pmatrix} x_0 \\ -x_1 \\ -x_2 \\ -x_3 \end{pmatrix}.$$

In terms of 2×2 matrices, the operation $\mathscr{P}$ has a very simple expression. For any matrix

$$M = \begin{pmatrix} c & d \\ f & g \end{pmatrix}$$

let M^a denote its 'cofactor matrix'

$$M^a = \begin{pmatrix} g & -d \\ -f & c \end{pmatrix}$$

so

$$MM^a = (\det M) \begin{pmatrix} 1 & 0 \\ 0 & 1 \end{pmatrix}$$

and

$$(M_1 M_2)^a = M_2^a M_1^a$$

for any two matrices M_1 and M_2. Now if

$$X = \begin{pmatrix} x_0 - x_3 & x_1 + ix_2 \\ x_1 - ix_2 & x_0 + x_3 \end{pmatrix}$$

then

$$X^a = \begin{pmatrix} x_0 + x_3 & -x_1 - ix_2 \\ -x_1 + ix_2 & x_0 - x_3 \end{pmatrix}.$$

So we see that when we represent points of Minkowski space as self-adjoint 2×2 matrices, we have

$$\mathscr{P}(X) = X^a.$$

Notice that $A^a = A^{-1}$ if $\det A = 1$, so

$$\mathscr{P}(AXA^*) = (A^*)^a X^a A^a$$
$$= A^{*-1} X^a A^{-1}$$

for $A \in Sl(2, \mathbb{C})$.

Notice that if the group H in our semidirect $H \circledS N$ product is enlarged from $H = Sl(2, \mathbb{C})$ so as to include $\mathscr{P}$, then we see from the preceding equation that

$$\mathscr{P}A = A^{*-1}\mathscr{P}$$

or

$$(\mathscr{P}A\mathscr{P}^{-1})X = A^{*-1}X \tag{10.1}$$

for $A \in Sl(2, \mathbb{C})$ and X self-adjoint.

Similarly, are the laws of physics invariant under the time reversal transformation $\mathscr{T}$ given by

$$\begin{pmatrix} x_0 \\ x_1 \\ x_2 \\ x_3 \end{pmatrix} \rightsquigarrow \begin{pmatrix} -x_0 \\ x_1 \\ x_2 \\ x_3 \end{pmatrix}?$$

If the laws of nature are invariant under both parity and time reversal, then they are also invariant under their product $\mathscr{P}\mathscr{T}$. This would mean that we would be replacing the connected component of the Lorentz group by all four components. So we are taking up the question raised in the preceding section as to the number of connected components of the Lorentz group that occur as symmetries of the laws of physics.

If we include more components in the Lorentz group, this means that the group H in our semidirect product, $H \circledS N$, must be enlarged from $Sl(2, \mathbb{C})$ to a group whose action on space time includes various components of the Lorentz group. We would then be interested in the representations of $H \circledS N$. So our first question should be – what are the possibilities for H? This is a rather subtle question. We shall return to it at the end of this section. For the moment we will investigate some consequences of an H whose action on space time includes $\mathscr{P}$ and explain how this led to the remarkable discovery in 1956 that parity is *not* a symmetry of nature.

Since at the moment we are interested only in $\mathscr{P}$, we may restrict to a subgroup if necessary and assume that the action of H on Minkowski space has exactly two components. That is, if we let π denote the homomorphism

$$\pi: H \to Lor$$

of H into the Lorentz group, we are assuming that $\pi(H)$ has two components, the non-identity component containing $\mathscr{P}$. We also assume that π is two-to-one, so the identity component of H is $Sl(2, \mathbb{C})$. Let $\mathscr{Q}$ be an element of H such that

$$\pi(\mathscr{Q}) = \mathscr{P}.$$

It follows from (10.1) that

$$\pi(\mathscr{Q}A\mathscr{Q}^{-1}) = \pi(A^{*-1})$$

so either

$$\mathscr{Q}A\mathscr{Q}^{-1} = A^{*-1}$$

or

$$\mathscr{Q}A\mathscr{Q}^{-1} = -A^{*-1}$$

for all $A \in Sl(2, \mathbb{C})$. Since $Sl(2, \mathbb{C})$ is connected and conjugation in any group leaves the identity fixed, we conclude that

$$\mathscr{Q}A\mathscr{Q}^{-1} = A^{*-1}, \text{ for all } A \in Sl(2, \mathbb{C}). \tag{10.2}$$

Since $\pi(\mathscr{Q}^2) = \mathscr{P}^2 = \mathrm{id}$ we conclude that either

$$\mathscr{Q}^2 = I \tag{10.3a}$$

or

$$\mathscr{Q}^2 = -I. \tag{10.3b}$$

Both of these possibilities exist for very natural choices of the group H, as we shall see at the end of this section, so we have to take both of them into account.

Let us now look at the irreducible representations of $H \circledS N$ by the orbit method for semidirect products. Each of the orbits that we computed in the preceding section for the group $Sl(2, \mathbb{C})$ is mapped into itself by $\mathscr{P}$ as can be checked directly. Hence the orbits for H are the same as for $Sl(2, \mathbb{C})$; what changes is the isotropy group. (This would not be the case if we included all four components, since time reversal interchanges the two sheets of the double sheeted hyperboloid, for example.)

So let us look at the orbits. The cases $m^2 > 0$ and $m^2 = 0$ behave differently. For $m^2 > 0$ the transformation $\mathscr{P}$ preserves the representative point

$$ml = \begin{pmatrix} m & 0 \\ 0 & m \end{pmatrix}.$$

Furthermore, $A^{*-1} = A$ for $A \in SU(2)$. Thus the isotropy group contains $\mathscr{Q}$ and $SU(2)$ and $\mathscr{Q}$ commutes with all elements of $SU(2)$. There are thus two possibilities. In *case (a)* the isotropy group is the direct product

$$\mathbb{Z}_2 \times SU(2) \text{ generated by } \mathscr{Q} \text{ and } SU(2) \text{ with } \mathscr{Q}A = A\mathscr{Q}, A \in SU(2) \mathscr{Q}^2 = I. \quad (10.4a)$$

In *case (b)* it is not the direct product but the

$$\text{group generated by } \mathscr{Q} \text{ and } SU(2) \text{ with } \mathscr{Q}A = A\mathscr{Q}, A \in SU(2) \mathscr{Q}^2 = -I. \quad (10.4b)$$

We must now investigate the irreducible representations of these isotropy groups. Since $\mathscr{Q}$ lies in the center of the isotropy group in both cases, it must be represented by a scalar matrix. In case (a) this scalar value must be ± 1 and either possibility can occur. In other words, the labels of the irreducible representations are now $(s, \pm)$. The symbol $\pm$ is called the *intrinsic parity* of the representation. Thus $(1, -)$ labels a representation of spin one and negative intrinsic parity. Let us now examine case (b). If s is an integer, then $-I$ is represented by the identity matrix and it follows from (10.4b) that the scalar representing $\mathscr{Q}$ is ± 1. On the other hand, if s is not integral, then $-I$ is represented by $-\text{id}$ and hence $\mathscr{Q}$ is represented by $\pm i$. So in case (b) the representations are also labeled by $(s, \pm)$ but $\pm$ means ± 1 for integer s and $\pm i$ for half-integer s. If the group of nature includes the parity transformation, then every positive-mass particle would have its own intrinsic parity, and total parity would be conserved in all interactions. But the meaning of this total parity conservation would be different for fermions, according to whether case (a) or (b) holds. Since $(\pm 1)^3 = \pm 1$, whereas $(\pm i)^3 = -(\pm i)$, cases (a) and (b) are really different. Assuming that there is a fermion with no other conserved quantum numbers, then a decay of one particle into three would be permitted in case (a) but not in case (b).

Let us now turn to the mass-zero orbits. The transformation $\mathscr{P}$ does not preserve the point $\begin{pmatrix} 2 & 0 \\ 0 & 0 \end{pmatrix}$; indeed

$$\mathscr{P} \begin{pmatrix} 2 & 0 \\ 0 & 0 \end{pmatrix} = \begin{pmatrix} 0 & 0 \\ 0 & 2 \end{pmatrix}.$$

Let $U = \begin{pmatrix} 0 & -1 \\ 1 & 0 \end{pmatrix}$. Then the operator $\pi(U)\mathscr{P}$ does preserve $\begin{pmatrix} 2 & 0 \\ 0 & 0 \end{pmatrix}$ since

$$\begin{pmatrix} 0 & -1 \\ 1 & 0 \end{pmatrix}\begin{pmatrix} 0 & 0 \\ 0 & 2 \end{pmatrix}\begin{pmatrix} 0 & 1 \\ -1 & 0 \end{pmatrix} = \begin{pmatrix} 2 & 0 \\ 0 & 0 \end{pmatrix}.$$

A direct computation shows that $\pi(U)\mathscr{P}$ acts as reflection about the line through $\mathbf{e}_1$ in the $\mathbf{e}_1$, $\mathbf{e}_2$ plane. So the isotropy group is now generated by $\tilde{E}(2)$ and the element $U(\mathscr{Q})$. Let us define $\mathscr{R} = U\mathscr{Q}$. A direct check using (10.2) shows that conjugation by $U(\mathscr{Q})$ sends every element of $\tilde{E}(2)$ into its complex conjugate. Furthermore, U commutes with $\mathscr{Q}$ since U is unitary. Once again we must consider two cases. In case (a)

$$\mathscr{R}^2 = U^2\mathscr{Q}^2 = -I \tag{10.5a}$$

so the isotropy group is a twisted form of double cover of the group of Euclidean motions. In case (b) we have

$$\mathscr{R}^2 = U^2\mathscr{Q}^2 = I. \tag{10.5b}$$

Notice that the roles of cases (a) and (b) are reversed here. Case (b) is now the direct product $\mathbb{Z}_2$ with $\tilde{E}(2)$. In both cases the isotropy groups are semidirect products of the group of translations in the plane with a group acting on the plane linearly. As in the preceding section, we will, for 'physical grounds', only be interested in those representations of the 'little group' which correspond to zero orbit in the plane. The 'little little group', i.e. the subgroup of the isotropy group which fixes the origin, is either (case b)

$$D_\infty = \mathbb{Z}_2 \circledS U(1)$$

generated by $\mathscr{R}$ and $e^{i\theta} \in U(1)$ with $\mathscr{R}e^{i\theta}\mathscr{R}^{-1} = e^{-i\theta}$ and $\mathscr{R}^2 = 1$ (10.6b)

or (case a)

generated by $\mathscr{R}$ and $e^{i\theta} \in U(1)$ with $\mathscr{R}e^{i\theta}\mathscr{R}^{-1} = e^{-i\theta}$ and $\mathscr{R}^2 = -1$. (10.6a)

The group D_∞ is itself a semidirect product so we can use the Wigner–Mackey technique a third time. In fact D_∞ is a continuous analogue of the dihedral groups; it is the group of symmetries of a two sided disk. The characters of $U(1)$ are the integers (which we label as $2s$, where $s = 0, \frac{1}{2}, -\frac{1}{2}, 1, -1$, etc.) and the element $\mathscr{R}$ sends s into $-s$. So, just as in the case of the odd dihedral groups, there is a family of two-dimensional representations parametrized by $j = \frac{1}{2}, 1, \frac{3}{2}$, etc. And there are two one-dimensional representations corresponding to $s = 0$. The two-dimensional representations split into the sum of the one-dimensional $s = j$ and $s = -j$ representations. The story for case (a) is much the same, except that there will be four one-dimensional representations instead of two. In any event we see that if the group of nature does *not* contain the parity transformation then the $m = 0, s \neq 0$ particles come in 'right handed' ($s > 0$) and 'left handed' ($s < 0$) versions. If parity is a symmetry then it interchanges this handedness and we must combine the $\pm j$ representations to get an irreducible of the full group. Let us summarize (in a form common to both cases (a) and (b)) the effects of the presence or absence of parity as a symmetry of nature in Table 29.

Table 29.

	For $m^2 > 0$	For $m^2 = 0$
If the laws of nature are invariant under $\mathscr{P}$	There is a property called 'intrinsic parity' conserved under interactions	Right handedness and left handedness makes no sense: the $s \neq 0$ irreducible representations of the little group are two dimensional
If the laws of nature are not invariant under $\mathscr{P}$	There is no intrinsic parity	The $m = 0$, $s \neq 0$ particles come in right- and left-handed versions. The irreducible representations of the little group are all one dimensional

Until 1956 it was a commonly held dogma that the laws of nature should be invariant under $\mathscr{P}$. Thus, values of intrinsic parity were assigned to various particles, and corresponding selection rules were observed for many interactions. But it was then observed that while *some* interactions preserved parity, others did not. Whether we want to include $\mathscr{P}$ or not depends on the type of interaction.

In 1956, Madame Wu (after the suggestion of Lee and Yang) performed a crucial experiment proving that radioactive decay does *not* conserve parity. One such experiment is the following. Some cobalt 60 is placed inside a coil of wire. The Co^{60} nucleus is unstable – one of its neutrons decays into a proton emitting an electron and an anti-neutrino; this is known as β-decay. When an electric current is passed through the coil, more electrons are emitted in one direction perpendicular to the area element of the coil than in the other (Fig. 3.14). Reversing the direction of the current changes the preferred direction. Now an area element such as described by current can be related to a direction (i.e. a vector) only through a choice of orientation of space. (In more mathematical language, an area element is an element of $\Lambda^2(\mathbb{R}^3)$ while a vector is an element of $\mathbb{R}^3$. We have seen that these spaces are equivalent under the action of $SO(3)$ but *not* of $O(3)$.) Thus parity is not conserved in β-decay.

This experiment could, in principle, have been performed any time after the discovery of β-decay. In fact, it was not done because everyone took for granted on

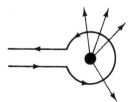

Fig. 3.14

some *a priori* grounds that parity was conserved. The conservation of parity was called into question by a bold suggestion of Lee and Yang to resolve the so-called 'θ-τ puzzle' as follows. In the early 1950s, new particles were discovered which decay, with a lifetime of about 10^{-10} s, into pions. Some of these particles decayed into two pions and were called θ-particles. Others decayed into three pions and were called τ-particles. Now, previous work on the pions indicated that they were particles of spin 0 and negative intrinsic parity. Hence, two pions (the tensor product of two pions) would have positive intrinsic parity, while three would have negative intrinsic parity again. Thus, assuming conservation of parity, the θ-particle would have positive intrinsic parity while the τ-particle would have negative intrinsic parity. However, in all other respects, the θ and τ seemed exactly alike, in particular they had exactly the same mass. The bold suggestion of Lee and Yang was that perhaps they were the same particle, but that parity is not conserved in this type of decay. In fact, it is now generally believed that parity is not conserved in the 'weak interactions'.

Although parity is not conserved in β-decay, there *is* a discrete symmetry that is conserved. Classical electromagnetism is invariant under a change of sign of all electric charges. The fields reverse their signs, but there is a compensating change of sign in the Lorenz law so the forces are unchanged. Suppose there is an operation C which maps each type of particle into its anti-particle and which has the effect of changing the sign of all 'internal quantum numbers' such as electric charge. (The notion of an anti-particle was one of the more startling predictions of Dirac's theory of the electron in the late 1920s.) Then symmetry is restored in the Wu experiment if we apply CP instead of P.

Experiments done in 1964 by Fitch and Cronin seem to indicate that CP is not conserved. I do not fully understand the issues involved in the correct interpretation of this experiment, which clearly shows that CP and CPT are not *both* conserved. It follows from the locality axioms, that quantum field theory implies that CPT is a symmetry of nature, and hence that CP is violated in the Fitch–Cronin experiment. But other, group theoretical, hypotheses might favor CP. I have my own views on the subject, which I will not expand on in this book.

Perhaps we should pause here to introduce some terminology. There are four fundamental types of forces or 'interactions'.

(i) The 'nuclear force' or *strong interaction* is the force that binds the atomic nucleus together. It is obviously much stronger than the electromagnetic force because it overcomes the mutual electrical repulsion of the protons. In fact, it is about 100 times as strong as the electromagnetic force. It is also of short range, acting only over distances of order 10^{-13} cm. This accounts for the instability of larger nuclei, since the nuclear force acts only on nearest neighbors while the electrical repulsion acts between all pairs of protons. The time scale of a strong interaction is, roughly, the time it takes to traverse 10^{-13} cm with the speed of light, i.e. about 10^{-23} s.

(ii) Electromagnetic forces.

(iii) The *weak interaction* responsible for radioactive β-decay. They are 10^{-12} times as strong as the electromagnetic interaction. They are also short range and their typical time scale is about 10^{-10} s.

(iv) Gravitational forces. These are long range but only 10^{-35} times as strong as the electromagnetic interactions. Due to the cancellation of positive and negative charges, and the absence of such cancellation in gravity, it is the cumulative effect of gravity that we experience in everyday life. But, on the scale of elementary particles, it can be ignored.

Not all particles participate in all forces. Particles which participate in the strong interactions are called *hadrons*. Those hadrons which have half integral spin, $s = \frac{1}{2}, \frac{3}{2}, \ldots$ are called *baryons*. Those with integer spin $s = 0, 1, 2, \ldots$ are called *mesons*. Thus the nucleons p and n are baryons while the pions π^+, π^0, π^- are mesons. The hadrons also engage in the electromagnetic and weak interactions. Particles of spin $\frac{1}{2}$ which have weak but not strong interactions are called *leptons*. Examples are the electron and the neutrino. The photon is a mass 0, $s = \pm 1$ particle.

Let us now return to the discussion of some of the issues involved in considering the discrete symmetry groups of space time. Wigner's theorem, as we explained in Section 3.6, asserts that every symmetry of a quantum system can be realized by a unitary or an anti-unitary operator, and that this operator is determined only up to multiplication by a scalar phase factor. The product of two anti-unitaries is a unitary. So as long as we considered connected groups, where every element is the square of some other, no anti-unitaries can arise. In fact, Wigner chose to represent the time reversal symmetry by an anti-unitary transformation, and this choice has been accepted in the physics literature for the past 60 years. The mathematical reasons for this choice are quite persuasive: $\mathscr{T}$ carries the 'forward mass shell', the orbit $m^2 > 0$, $p_0 > 0$, onto the 'backward mass shell' $m^2 > 0$, $p_0 < 0$, and vice versa. Hence, if $\mathscr{T}$ were represented by a unitary operator, we would have to include representations corresponding to both mass shells if we want to get an irreducible representation of the enlarged group. Although this is not an insuperable difficulty, it can be avoided if we represent $\mathscr{T}$ by an anti-unitary transformation. This is because the representation of spin s corresponding to the backward mass shell is anti-unitarily equivalent to the forward one.

A second problem relates to the issue of ordinary versus projective transformations. Even if we only had to deal with unitary transformations, Wigner's theorem only guarantees that we have a projective representation of the group. That is, we can assign to each group element a an operator $\rho(a)$ which is only determined up to phase, and

$$\rho(ab) = c(a, b)\rho(a)\rho(b).$$

The function c is called a *multiplier* of the representation. It depends, to some extent, on how we choose ρ. If we replace $\rho(a)$ by $f(a)\rho(a)$, where f is some complex-valued function on the group with $|f| = 1$, then clearly c will be replaced by d, where

$$d(a, b) = c(a, b)f(a)f(b)/f(ab).$$

Now it is a theorem that for the group $Sl(2, \mathbb{C}) \circledS \mathbb{R}^{1,3}$ we can choose the operators in any projective representation so that the c is identically one, i.e. every projective representation is given by an ordinary representation. Hence our classification of the ordinary irreducible unitary representations of $Sl(2, \mathbb{C}) \circledS \mathbb{R}^{1,3}$ is, in fact, a classification of its projective irreducible representations as well. In fact, in Wigner's paper on the classification of the representations of the Poincaré group he classifies the irreducible projective representations of $Lor \circledS \mathbb{R}^{1,3}$ and allows for disconnected components.

For projective representations, however, a knowledge of the irreducible components (as projective representations) does not determine the representation. The reason for this is quite easy to see by example. Since all one-dimensional unitary transformations are just multiplications by a phase factor, and since we are allowed, in a projective representation, to change any operator by multiplication by a phase factor, it follows that all one-dimensional projective representations of a group G are the same. But now suppose that we are given a projective representation on a space which is a direct sum of a number (say even two) of one-dimensional invariant subspaces. We are now allowed only one overall phase factor for each group element, not a separate phase factor for each one-dimensional subspace. Hence two such representations on spaces of the same dimension need not be equivalent. In terms of our physical problem, a knowledge of the irreducible projective representations of the Poincaré group is insufficient once we want to consider interactions between particles, which will involve a direct sum of the irreducible representations.

The theory of projective representations (for finite groups and complex linear representations) was completely worked out by Schur in two astounding papers in 1904 and 1907, both in *J. für Math*. Each projective representation of a group G comes equipped with its own multiplier, and two projectively equivalent representations have multipliers differing in (multiplication by) a 'trivial multiplier'. The equivalence classes of multipliers, mod 'trivial multipliers', is what we call today $H^2(G, \mathbb{C}^*)$, the second cohomology of G with values in the non-zero complex numbers. Schur called this group $M(G)$ (the group of Multiplikatoren); we shall stick to Schur's notation to avoid confusion with topological cohomology groups. Thus, for example, a computation shows that

$$M(\mathbb{Z}_2 \times \mathbb{Z}_2) = \mathbb{Z}_2.$$

This computation will be important for us because we will want to think of $\mathbb{Z}_2 \times \mathbb{Z}_2$ as the group consisting of $I, \mathscr{P}, \mathscr{T}$ and $\mathscr{P}\mathscr{T}$. Schur shows that for any group G we can find a central extension, that is a group G^* and a homomorphism of $G^* \to G$ with kernel A lying in the center of G; we write this as

$$1 \to A \to G^* \to G \to 1,$$

such that any projective representation of G is projectively equivalent to one that can be lifted to an ordinary representation of G^* with the property that *the elements of A are represented by scalar operators*. Conversely any representation of G^* which satisfies the italicized condition clearly descends to give a projective representation of G. If G^* is as small as possible (essentially this means that A has the same size as $M(G)$) then Schur calls G^* a 'representation group' of G. The reason is that the study of projective representations of G can be reduced to the study of certain ordinary representations of G^*, namely those for which the italicized condition holds. (It holds automatically for irreducible representations by Schur's lemma.) But the representation group is not uniquely determined. This is again a homological question completely treated by Schur. For the case of $\mathbb{Z}_2 \times \mathbb{Z}_2$ we have the two possibilities for the group

G^* which must be of order eight:

$$1 \to \mathbb{Z}_2 \to G_2 \to \mathbb{Z}_2 \times \mathbb{Z}_2 \to 1$$

and

$$1 \to \mathbb{Z}_2 \to D_4 \to \mathbb{Z}_2 \times \mathbb{Z}_2 \to 1$$

where G_2 is the quaternionic group $G_2 = \{\pm 1, \pm \mathbf{i}, \pm \mathbf{j}, \pm \mathbf{k}\}$ of unit quaternions and D_4 is the dihedral group of order eight. In a sense, the group G_2 is nicer, because $M(G_2) = \{0\}$ whereas $M(D_4) = \mathbb{Z}_2$.

Now, projective representations are very unpleasant objects. You cannot take direct sums of two projective representations unless the multipliers agree; when you take tensor products you have to multiply the multipliers, and so on. There is a lot of messy bookkeeping involved. The natural thing to do, and one that is done automatically and instinctively, is to pass from the group G to the group G^*, and to *forget about the above italicized condition.* That is, we use G^* instead of G as the group of physics. We then want to consider all honest unitary representations of G^*. For the case of no discrete symmetries, this involved passing from the proper Lorentz group to $Sl(2, \mathbb{C})$ – there was no ambiguity in the choice of G^*. However, for the full Lorentz group there are several choices (in fact eight in all). Let us describe two very natural choices arising from the study of Clifford algebras. (For details about Clifford algebras see the treatment in Greub, *Multilinear algebra*, or in Bamberg and Sternberg, vol. II.)

Let $V = \mathbb{R}^{p,q}$ be real Cartesian $p + q$ space with a non-degenerate quadratic form of signature p pluses and q minuses. We denote its orthogonal group by $O(p, q)$. The corresponding Clifford algebra, denoted by $C(p, q)$, is the associative algebra generated by the elements of V subject to the relations

$$uv + vu = 2(u, v)\mathbf{1}$$

where $(,)$ denotes the scalar product associated to the bilinear form and $\mathbf{1}$ is the unit element in the algebra. The group $Pin(p, q)$ is the group generated multiplicatively in $C(p, q)$ by 'unit' vectors in V, that is by vectors e satisfying $(e, e) = \pm 1$. There is a two-to-one homomorphism, $\phi : Pin(p, q) \to O(p, q)$ determined by

$$\phi(e)v = eve^{-1} \text{ for all 'unit' vectors } e \text{ and all } v \in V.$$

Notice that if we take $v = e$ we get $\phi(e)e = eee^{-1} = e$, whereas if we take v orthogonal to e then the Clifford identities say that $ev = -ve$, so that $\phi(e)v = -vee^{-1} = -v$. Thus $\phi(e)$ is the negative of the reflection through the hyperplane perpendicular to e. In particular, in ordinary space time, if we take e to be a time-like vector, then $\phi(e)$ is just the parity operator $\mathscr{P}$. But this shows that the double covers

$$Pin(1, 3) \to O(1, 3) \quad \text{and} \quad Pin(3, 1) \to O(3, 1)$$

are different. In the first case $e^2 = \mathbf{1}$, whereas in the second case $e^2 = -\mathbf{1}$. Put another way, in each of these groups there will be two elements $\pm \mathscr{P}$ covering the parity transformation, $\mathscr{P}$. But in $Pin(1, 3)$ $\mathscr{P}^2 = id$, whereas in $Pin(3, 1)$ $\mathscr{P}$ is of order four, with $\mathscr{P}^2 = -I \in Sl(2, \mathbb{C})$. These are precisely the alternatives (a) and (b) discussed at the beginning of this section.

If we look at the inverse images of the four-element group $\{I, \mathscr{P}, \mathscr{T}, \mathscr{P}\mathscr{T}\}$ in the *Pin* groups, for *Pin*(3, 1) we get G_2 and for *Pin*(1, 3) we get D_2. We should point out (see the references cited above) that

$C(1, 3) \sim H(2)$, the algebra of 2×2 matrices over the quaternions

and

$C(3, 1) \sim R(4)$, the algebra of 4×4 matrices over the reals.

Thus, for example, the smallest dimension for a real representation of $C(1, 3)$ is eight, and the spin representation for $C(3, 1)$ has four real dimensions (the Majorana spinors).

The existence of the two alternatives (a) and (b) mentioned at the beginning of this section has an interesting history. The fact that there were two types of spinors seems to have been known to Racah in the 1930s. It was rediscovered by Fermi and Yang in the 1940s; they raised the question as to whether this could have an observable physical consequence. This, in turn, led to a celebrated paper by Wick, Wigner and Wightman in which the notion of a 'superselection rule' was introduced. This declared that this question might (or should) not have relevant physical consequences. My own feeling is that in the light of recent work on supersymmetry this question should be reexamined.

As we indicated above, from a mathematical point of view the groups *Pin*(1, 3) and *Pin*(3, 1) seem most natural as candidates for the cover of the four-component Lorentz group. However, for the sake of completeness we list all eight possibilities in the following.

The group generated by $\mathscr{P}$ and $\mathscr{T}$ is isomorphic to $\mathbb{Z}_2 \times \mathbb{Z}_2$. The overall possibilities for its inverse image in the double cover must be extensions of $\mathbb{Z}_2$ by $\mathbb{Z}_2 \times \mathbb{Z}_2$, so $1 \to G \to \mathbb{Z}_2 \times \mathbb{Z}_2 \to 1$. Each of $\mathscr{P}$, $\mathscr{T}$, $\mathscr{P}\mathscr{T}$ have two covering elements. Call them (temporarily) $\pm p, \pm t, \pm pt$. Each of these is of order either two or four, so there are eight possible choices for these orders (two for each of p, t, and pt). Elementary arguments show that each of these possibilities occurs for exactly one central extension. The possibilities are:

- the direct product $\mathbb{Z}_2 \times \mathbb{Z}_2 \times \mathbb{Z}_2$ (all elements of order two);

- the group D_4, where the kernel $\mathbb{Z}_2$ is identified as rotation through $180°$. In D_4 four elements have square *id* (the four reflections) and two (rotations through $\pm 90°$) have square -1 (the non-trivial element in the central $\mathbb{Z}_2$). So the $90°$ rotation element must cover one of the three elements, $\mathscr{P}$, $\mathscr{T}$, or $\mathscr{P}\mathscr{T}$. So there are three choices in all, corresponding to which of the $\pm p, \pm t, \pm pt$ is of order four;

- the group $\mathbb{Z}_2 \times \mathbb{Z}_4$, where the kernel $\mathbb{Z}_2$ is taken to be the $\mathbb{Z}_2$ subgroup of $\mathbb{Z}_4$. In this group there are four elements of order four and four of order two. So two of the $\pm p, \pm t, \pm pt$ are of order four and one of order two. Again three possibilities;

- the quaternionic group, $\mathscr{Q} = \{\pm 1, \pm \mathbf{i}, \pm \mathbf{j}, \pm \mathbf{k}\}$, with $\mathbf{i}^2 = \mathbf{j}^2 = \mathbf{k}^2 = -1, \mathbf{ij} = \mathbf{k} = -\mathbf{ji}$, etc. All the $\pm p, \pm t, \pm pt$ are of order four.

So there are eight possibilities in all. It turns out that all these can occur for the Lorentz group exactly once. Thus there are eight double covers for the Poincaré group.

Summary of the relation between the *Pin* groups and the discrete symmetries

Facts common to Pin(3,1) and Pin(1,3) e_0 covers the parity operator, $\mathscr{P}$; $\mu = e_0 e_1 e_2 e_3$ covers $-id = \mathscr{P}\mathscr{T}$ and $\mu^2 = -1$; $e_1 e_2 e_3$ covers $\mathscr{T}$.

Differences In $Pin(1,3)$ $e_0^2 = 1$, $(e_1 e_2 e_3)^2 = 1$. The group is D_4 with μ being rotation through $90°$, e_0 a bisector reflection, and $e_1 e_2 e_3$ a diagonal reflection.
  In $Pin(3,1)$ $e_0^2 = -1$, $(e_1 e_2 e_3)^2 = -1$. The group is G_2, the quaternionic group.

3.11 The Mackey theorems on induced representations, with applications to the symmetric group

In this section we return to the general study of the representation $s\uparrow G$ induced up to G from a representation s of a subgroup H on a vector space V. Thus, from the representation s we construct the vector bundle $E \to M$, where $M = G/H$, and $s\uparrow G$ is the corresponding action of elements of G on $\Gamma(E)$. All of this is described in Section 3.3. We let σ denote the character of s and $\sigma\uparrow G$ the character of $s\uparrow G$. We recall the Frobenius fixed point character formula, equation (3.3)

$$(\sigma\uparrow G)(a) = (1/\#H) \sum_{\substack{g\in G \\ g^{-1}ag\in H}} \sigma(g^{-1}ag). \tag{11.1}$$

We now want to consider the following situation. Let K be some other subgroup of G. We can consider the restriction of the representation $s\uparrow G$ to K. Let us denote it by $(s\uparrow G)\downarrow K$. The subgroup K need not act transitively on M.

If we let M_i denote the orbits of M under K, and $E_i = \pi^{-1}(M_i)$, then we can apply the decomposition (3.1) to the representation $(s\uparrow G)\downarrow K$, where we are now considering E as a homogeneous vector bundle for the subgroup, K. To see what (3.1) actually says in this case, we describe the orbits M_i more explicitly, and describe the vector bundles E_i as being induced from certain representations s_i of subgroups K_i of K. Here are the details. Let us pick a point x_i in each orbit, M_i. We can write $x_i = g_i H$ for some $g_i \in H$. Then we can write

$$M_i = K g_i H, \tag{11.2}$$

i.e. the orbits, M_i, of M under K are precisely the distinct double cosets $K g_i H$. To identify E_i as an induced vector bundle, we consider the isotropy group, K_{x_i} of the point, x_i. For notational convenience, let us denote this subgroup by K_i, so that

$$K_i = (g_i H g_i^{-1}) \cap K. \tag{11.3}$$

Let us set

$$V_i = E_{x_i}.$$

Recall from the beginning of Section 3.3 that the element g_i defines an isomorphism φ_{g_i}: $V \to V_i = E_{x_i}$ given by $\varphi_{g_i}(v) = [(g_i, v)]$.

Then $k_i \in K_i$ acts on V_i according to the representation defined by

$$k_i \varphi_{g_i} v = \varphi_{g_i} s(g_i^{-1} k_i g_i) v. \tag{11.4}$$

The right-hand side makes sense, since the element $g_i^{-1} k_i g_i$ belongs to H. Let us identify V with V_i via φ_{g_i}, and let us define the representation, s^i, of K_i by

$$s^i(k_i) = s(g_i^{-1} k_i g_i). \tag{11.5}$$

It follows from the above considerations that the homogeneous vector bundle, E_i, is isomorphic to the bundle induced over $M_i = K/K_i$ by the representation, s^i of K_i. Then (3.1) becomes the Mackey subgroup decomposition formula

$$(s \uparrow G) \downarrow K \sim (s^1 \uparrow K) \oplus \cdots \oplus (s^n \uparrow K) \tag{11.6}$$

where there is one summand for each double coset, $K g_i H$, and where the representations s^i are given by (11.5) on the subgroups K_i of K given by (11.3) and the $s^i \uparrow K$ are the corresponding induced representations of K.

We wish to apply (11.6) to the following very important special case. Let H' and H'' be two subgroups of a group, G. Then their direct product, $H' \times H''$ is a subgroup of the direct product, $G \times G$ of G with itself. We can consider G as the 'diagonal' subgroup of $G \times G$ consisting of all pairs of the form (g, g). Thus, in (11.6) G will play the role of K, $G \times G$ will play the role of G and $H = H' \times H''$ will be the H. If we are given a representation, s', of H' on V' and a representation, s'', of H'' on V'', we obtain a representation of $H' \times H''$ on the space $\text{Hom}(V', V'')$ by setting

$$(h', h'')T = h'' T h'^{-1}.$$

Let s be this representation, so that

$$S(h)T = h'' T h'^{-1} \quad \text{if} \quad h = (h', h'') \in H = H' \times H''.$$

We wish to apply (11.6), but before doing so we make some identifications. We can write

$$M = (G \times G)/H = (G \times G)/H' \times H'' = (G/H') \times (G/H'')$$
$$= M' \times M''$$

where we have set $M' = G/H'$ and $M'' = G/H''$. Let E be the vector bundle induced by s over M, and E', E'' the vector bundles corresponding to s', s''. Let (x', x'') be a point of M. There is a natural identification of

$$E_{(x', x'')} \text{ with } \text{Hom}(E'_{x'}, E''_{x''}).$$

Indeed, if $x' = g'H''$ and $x'' = g''H''$, then the left-hand side consists of all equivalence classes

$$[((g', g''), T)] \qquad T \in \text{Hom}(V', V'')$$

while the right-hand side becomes identified with $\text{Hom}(V', V'')$ if we use $\varphi_{g'}$ to identify V' with $E'_{x'}$ and $\varphi_{g''}$ to identify V'' with $E''_{x''}$. It is easy to check that the identification is independent of the particular choice of g' and g''.

Let $F \in \Gamma(E)$ be a section of E over M. We can think of F as assigning an element, $F(x', x'') \in \text{Hom}(E_{x'}, E_{x''})$ to the point (x', x'') in view of the above identification. Furthermore, we can regard such an F as giving a 'kernel' which defines a map from $\Gamma(E') \to \Gamma(E'')$. Indeed for any $u \in \Gamma(E')$ define $Fu \in \Gamma(E'')$ by the formula

$$(Fu)(x'') = \sum_{x' \in M'} F(x', x'')u(x').$$

In the sum on the right, $u(x') \in E'_{x'}$ and $F(x', x'')$ maps $E'_{x'} \to E''_{x''}$, and so we have indeed defined a section of E''. Thus each F defines a map of $\Gamma(E') \to \Gamma(E'')$ and the map so obtained depends linearly on F. It is easy to see that we have defined an isomorphism

$$\Gamma(E) \sim \text{Hom}(\Gamma(E'), \Gamma(E'')).$$

The group G acts on both sides, and this is an equivalence of the corresponding representations. Thus we have identified the representation $(s \uparrow (G \times G)) \downarrow G$ with the representation of G on $\text{Hom}(\Gamma(E'), \Gamma(E''))$.

In the case that the representations s' and s'' are both the trivial one-dimensional representations of H' and H'', the space $\Gamma(E')$ and $\Gamma(E'')$ reduce to the ordinary function spaces $F(M')$ and $F(M'')$. The identification that we have just described then becomes the identification of $F(M' \times M'')$ with $\text{Hom}(F(M'), F(M''))$ that we used in Section 5, Chapter 2. We will soon see that the Mackey formula (11.6) will give us a generalization to vector bundles of (5.2) which was so important in the proof of the basic facts about characters in Chapter 2.

So much for the left-hand side of (11.6) in the case we are considering. To deal with the right-hand side we must examine the G, H double cosets in $G \times G$ and choose a convenient representative for each of them. Given any $(g_1, g_2) \in G \times G$, its double coset consists of all products of the form

$$(g, g)(g_1, g_2)(h', h'') = (gg_1 h', gg_2 h'')$$

where $h' \in H'$ and $h'' \in H''$. If we choose $g = (g_1 h')^{-1}$ we get all elements of the form $(e, H'g_3 H'')$, where $g_3 = g_1^{-1}g_2$. Thus the G, H double coset in $G \times G$ gives rise to an H', H'' double coset in G. It is easy to see that this relation is one-to-one, the G, H double coset, M_i in $G \times G$ goes over into the H', H'' double coset in G given by

$$(e \times G) \cap M_i.$$

We may choose our representative in the form (e, g_i) and then it is easy to see that the subgroup K_i in (11.3), which is now a subgroup of G, is given by

$$G_i = H' \cap g_i H'' g_i^{-1}. \tag{11.7}$$

The representations s^i on $\text{Hom}(V', V'')$ given by (11.5) is

$$s^i(k)T = (g_i^{-1} k g_i)Tk^{-1} \quad k \in G_i. \tag{11.8}$$

(Notice that $k \in H'$, and so acts on V' via s', and $g_i^{-1}kg_i \in H''$, and so acts on V'' via s'', so that (11.8) makes sense.)

Putting the various identifications together into (11.6) we obtain the following formula (due to Mackey):

> the representation of G on $\mathrm{Hom}(\Gamma(E'), \Gamma(E''))$ is equivalent to the   (11.9)
> direct sum of the induced representations:
>
> $$(s^1 \uparrow G) \oplus \cdots \oplus (s^n \uparrow G)$$
>
> where the sum is taken over all H', H'' double cosets in G, and where
> the s^i are the representations given by (11.8) on the subgroups, G_i,
> defined by (11.7)

Let us be given two representations, r^1 and r^2, of the group G on the vector spaces W_1 and W_2. We denoted by $\mathrm{Hom}_G(W_1, W_2)$ the subspace of $\mathrm{Hom}(W_1, W_2)$ consisting of those T which satisfy

$$r_g^2 T = T r_g^2 \quad \text{for all} \quad g \in G.$$

Thus $\mathrm{Hom}_G(W_1, W_2)$ consists of all invariant elements of $\mathrm{Hom}(W_1, W_2)$ under the action of G given by the representation $\mathrm{Hom}(r^1, r^2)$. We wish to use (11.9) to compute $\mathrm{Hom}_G(\Gamma(E'), \Gamma(E''))$. Now (11.9) identifies the representation space $\mathrm{Hom}(\Gamma(E'), \Gamma(E''))$ as a direct sum. The space of invariants in a direct sum is the direct sum of the invariant elements in each summand. So to apply (11.9) we need a little lemma telling us how to compute the invariant elements in an induced representation.

Let s be a representation of a subgroup H of the group G. Then there is a vector space identification

$$\mathrm{Inv}_G(s \uparrow G) = \mathrm{Inv}_H s, \tag{11.10}$$

where the left-hand side is the space of invariant elements in the induced representation, $\Gamma(E)$ under the action of the group, G, and the right-hand side denotes the invariant elements under H of the representation s.

Proof Let f be an element of $\Gamma(E)$. To say that f is invariant means that

$$af(a^{-1}x) = f(x)$$

or

$$f(ax) = af(x)$$

for all $a \in G$. Since G acts transitively on $M = G/H$, this shows that an invariant f is determined by its values at one point, say the point $x = H$. Taking $a \in H$ shows that $f(x)$ must be invariant. We thus have a map, $f \rightsquigarrow f(x)$ of $\mathrm{Inv}_G(s \uparrow G) \rightsquigarrow \mathrm{Inv}_H s$, and it is easy to see this is a bijection.

If we now plug (11.10) into (11.9), we get the Mackey formula for invariants:

$$\mathrm{Hom}_G(\Gamma(E'), \Gamma(E'')) \sim \mathrm{Hom}_{G_1}(V', V'') \oplus \cdots \oplus \mathrm{Hom}_{G_n}(V', V''), \tag{11.11}$$

where the subgroup G_i is given by (11.7), and where $k \in G_i$ acts on V' via s'_k and on V'' via $s''(g_i^{-1}kg_i)$.

There are many important applications of the Mackey formula, (11.11). For example, suppose that we take s' and s'' to be the trivial one-dimensional representations of H' and H''. Then each of the summands on the right-hand side of (11.11) is one dimensional. The number of summands is exactly the number of G orbits on $M' \times M''$. Thus (11.11) reduces to

$$\dim \operatorname{Hom}(F(M'), F(M'')) = \#(\text{of } G \text{ orbits on } M' \times M'')$$

which is just equation (5.2), Chapter 2. At another extreme, let us consider the case where $H' = H$ is a subgroup of G and where we take $H'' = G$. In this case there is only one double coset so that in (11.9) and (11.11) we may take $g_1 = e$ and $G_1 = H$. We are now given a representation s of H and construct the vector bundle $\Gamma(E)$ which is the $\Gamma(E')$ in (11.11), and we are also given a representation r of G on some vector space W, and the $\Gamma(E'')$ in (11.11) is just W itself, and the induced representation is just r. The left-hand side is $\operatorname{Hom}_G(\Gamma(E), W)$ and the right-hand side reduces to the single summand $\operatorname{Hom}_H(V, W)$. Thus (11.11) reduces to the Frobenius reciprocity theorem

$$\operatorname{Hom}_G(\Gamma(E), W) \sim \operatorname{Hom}_H(V, W).$$

For the remainder of this section we wish to given some applications of the Mackey formula (11.11) to the theory of the representations of the symmetric group. We shall take $G = S_n$. Let λ be a Young diagram, and let t be a Young tableau of type λ, and let $\{t\}$ be the corresponding tabloid, so that $\{t\}$ is a partition of $\{1, \dots, n\}$ into disjoint subsets. We shall take H' and H'' in (11.11) be of the type $G_{\{t\}}$ for various choices of λ and t. Thus the spaces M' and M'' will be the spaces M_λ of Section 2.8.

In order to apply (11.11) we must first describe the S_n orbits on $M_\lambda \times M_\mu$ for two Young diagrams λ and μ. Consider two tabloids, $\{t\} \in M_\lambda$ and $\{s\} \in M_\mu$. Then $\{t\}$ is a disjoint union of subsets $\{t\} = t_1 \cup \cdots \cup t_n$ of subsets of $\{1, \dots, n\}$ with $\# t_i = \lambda_i$, and similarly for s and μ. Then $t_i \cap s_j$ is a subset of $1, \dots, n$, and the number

$$m_{ij} = \#(t_i \cap s_j)$$

is clearly a function on $M_\lambda \times M_\mu$ which is invariant under the action of S_n, and hence is constant on S_n orbits. Conversely, we claim that the matrix of entries (m_{ij}) completely determines the orbit: since the values m_{ij} determine λ, as $\lambda_i = \sum_j m_{ij}$, and the group S_n acts transitively on M_λ, it is clearly enough to prove the following – suppose that

$$\#(t_i \cap s_j) = \#(t_i \cap s_j').$$

Then there exists a permutation which preserves $\{t\}$ and carries $\{s\}$ into $\{s'\}$. But the condition implies that for each fixed i the decompositions

$$t_i = \bigcup (t_i \cap s_j) \quad \text{and} \quad t_i = \bigcup (t_i \cap s_j')$$

are partitions of the set t_i into subsets of equal sizes. Hence we can find a permutation of the set t_i which carries $t_i \cap s_j$ into $t_i \cap s_j'$. Considering one such permutation for each i gives a $\tau \in S_n$ such that

$$\tau(t_i \cap s_j) = \tau(t_i \cap s_j').$$

Thus

$$\tau(t) = t \quad \text{and} \quad \tau(s) = s'$$

as required. Thus we have proved that

> the set of G orbits on $M_\lambda \times M_\mu$ is in one-to-one correspondence with the set of all $n \times n$ matrices with non-negative integer entries (m_{ij}) such that the row sum $\sum_j m_{ij} = \lambda_i$ and the column sum $\sum_i m_{ij} = \mu_j$. The orbit through the point $(\{t\}, \{s\})$ is characterized by the matrix (m_{ij}), where

$$m_{ij} = \#(t_i \cap s_j). \tag{11.12}$$

> The isotropy group of the point $(\{t\}, \{s\})$ is the group of those permutations which preserve all the intersections $t_i \cap s_j$.

We are going to apply (11.11) to two types of vector bundles, that is to representations induced from two types of representations of $S_{\{t\}}$ – the one-dimensional trivial and sign representation. For the trivial representation, the corresponding $\Gamma(E)$ is just the function space $F(M)$. We shall denote the $\Gamma(E)$ coming from the sign representation by $\tilde{F}(M)$. Then an immediate consequence of (11.12) is

$$\dim \operatorname{Hom}_{S_n}(F(M_\lambda), F(M_\mu)) = \#(\text{non-negative integer matrices } (m_{ij})$$
$$\text{with row sums } \lambda_i \text{ and column sums } \mu_j). \tag{11.13}$$

Suppose we want to compute $\operatorname{Hom}_{S_n}(F(M_\lambda), \tilde{F}(M_\mu))$. On the right-hand side of (11.11) the various groups are the subgroups which preserve the $t_i \cap s_j$. If some $t_i \cap s_j$ contains more than one element, then the corresponding isotropy group contains at least one odd permutation, and the trivial and sign permutations are therefore inequivalent. Thus the only non-zero contributions to the right-hand side of (11.11) can come from pairs (t, s) for which all $t_i \cap s_j$ are either empty or contain one element, so the corresponding m_{ij} must be zero or one. Thus

$$\dim \operatorname{Hom}_{S_n}(F(M_\lambda), \tilde{F}(M_\mu)) = \#(\text{zero one matrices } (m_{ij}) \text{ with row sums}$$
$$\lambda_i \text{ and column sums } \mu_j). \tag{11.14}$$

For a Young diagram λ, we shall denote by $\hat{\lambda}$ the Young diagram obtained from λ by interchanging the rows and columns. Thus if

$$\lambda = \begin{array}{l}\square\square\square\square\\\square\end{array}$$

then

$$\hat{\lambda} = \begin{array}{l}\square\square\\\square\\\square\end{array}$$

Now to say that the column sums of (m_{ij}) equal the rows of $\hat{\lambda}$ is the same as saying that the column sums of (m_{ij}) equal the columns of λ. If all the entries m_{ij} are either zero or one, the only way that this can happen is for the ones to occur at the positions of the diagram, λ, and zeros elsewhere in the matrix. Thus there is only one such matrix and

$$\dim \mathrm{Hom}_{S_n}(F(M_\lambda), \tilde{F}(M_{\tilde{\lambda}})) = 1. \qquad (11.15)$$

Thus $F(M_\lambda)$ and $\tilde{F}(M_{\tilde{\lambda}})$ have exactly one irreducible representation in common. Now in Section 2.8 we saw that there was a distinguished irreducible representation, F^λ, that occurs with multiplicity one in $F(M_\lambda)$. Let t be a tableau for the diagram λ, and C_t the corresponding group of column permutations. Then the space $\tilde{F}(M_{\tilde{\lambda}})$ can be considered as being induced from the sign representation of C_t, that is C_t is the isotropy group of a point of $M_{\tilde{\lambda}}$. But the element e_t of F^λ determines a one-dimensional subspace of F^λ which transforms under C_t as the sign representation. Hence $\dim \mathrm{Hom}_{C_t}(F^\lambda, \mathrm{sign}) \geqslant 1$. But, by the Frobenius reciprocity theorem

$$\mathrm{Hom}_{C_t}(F^\lambda, \mathrm{sign}) = \mathrm{Hom}_{S_n}(F^\lambda, \tilde{F}(M_{\tilde{\lambda}})).$$

Thus F^λ occurs with multiplicity at least one in $\tilde{F}(M_{\tilde{\lambda}})$. Hence it occurs with multiplicity exactly one there, and can be identified with the unique irreducible representation that $F(M_\lambda)$ and $\tilde{F}(M_{\tilde{\lambda}})$ have in common. In short we have proved the following.

> The representation of S_n on $F(M_\lambda)$ and $\tilde{F}(M_{\tilde{\lambda}})$ have exactly one (11.16) irreducible representation in common, and that representation is (equivalent to) the irreducible representation, F^λ, that we have associated to the diagram λ in Section 2.

Before proceeding let us state a general fact about induced representations which is interesting in its own right. Let (r, V) be a representation of a group G and let (s, W) be a representation of a subgroup, H, of G. We can restrict the representation r to H so as to obtain the representation $r{\downarrow}H$ of H. We can then form the tensor product $(r{\downarrow}H)\otimes s$ of H and then form the induced representation $[(r{\downarrow}H)\otimes s]{\uparrow}G$. We claim that

$$r\otimes(s{\uparrow}G) \sim [(r{\downarrow}H)\otimes s]{\uparrow}G \qquad (11.17)$$

as representations of G. To prove this, it is convenient to use the description of the space of the induced representations as a space of functions on G. Thus the underlying space of the right-hand side of (11.17) consists of all functions F from G to $V \otimes W$ which satisfy

$$F(ah) = (r(h)^{-1} \otimes s(h)^{-1})F(a) \qquad h \in H$$

while the left-hand side has as its underlying space linear combinations of expressions of the form $v \otimes f$, where f is a function from G to W satisfying $f(ah) = s(h)^{-1}f(a)$. We can define a map φ from the left-hand side to the right-hand side by

$$\varphi(v \otimes f)(a) = r(a)^{-1}v \otimes f(a).$$

It is easy to check that this does define a linear map which is a G morphism and is in fact an isomorphism, proving (11.17). For the case of the trivial representation of H this is the isomorphism introduced at the end of Section 3.3. Let us apply (11.17) taking $r = A$ to be the sign representation of $S_n = G$ and H to be the subgroup S_t with s either the sign or the trivial representation. It then follows from (11.17) that

$$A \otimes F(M_\lambda) = \tilde{F}(M_\lambda)$$

and

$$A \otimes \tilde{F}(M_\lambda) = F(M_\lambda).$$

It therefore follows from (11.16) that

$$A \otimes F^\lambda = F^{\hat{\lambda}}. \tag{11.18}$$

In particular, the characters of the representations F^λ and $F^{\hat{\lambda}}$ agree on even elements and differ by a sign on odd elements.

3.12   Exchange forces and induced representations

One of the main computational problems in approximate methods of quantum mechanics is finding the eigenvalues of a linear operator T on a finite-dimensional vector space W. In this section we describe a method for calculating the eigenvalues of T under the assumptions that $W = \Gamma(E)$, where $E \to M$ is a homogeneous vector bundle for a group G, induced from a representation of a subgroup H. Here G acts transitively on M, and H is the isotropy group of a point of M, and it is assumed that T commutes with the action of G.

The simplest non-trivial case is when $G = S_2$, $H = \{e\}$, so that M is the two-element set and the representation of H is trivial. In this case $\Gamma(E)$ is two dimensional and the representation of G sends (12) into

$$\begin{pmatrix} 0 & 1 \\ 1 & 0 \end{pmatrix}.$$

The condition that

$$T = \begin{pmatrix} x & y \\ z & w \end{pmatrix}$$

commutes with the action of G then requires $x = w$ and $z = y$ so

$$T = \begin{pmatrix} x & y \\ y & x \end{pmatrix}.$$

We can write

$$x = t(e) \quad \text{and} \quad y = t((12)).$$

In case T is a matrix of 'forces' the element $y = t((12))$ is called the 'exchange force'. The eigenvectors of T are $\begin{pmatrix} 1 \\ 1 \end{pmatrix}$ and $\begin{pmatrix} 1 \\ -1 \end{pmatrix}$ and these, of course, span the subspaces transforming according to the character

$$\chi_1 = \chi_{\square\square} \quad \text{and} \quad \chi_2 = \chi_{\substack{\square \\ \square}}.$$

Finally, the eigenvalues of T are

$$x + y = \chi_1(e)t(e) + \chi_1((12))t((12))$$

and

$$x - y = \chi_2(e)t(e) + \chi_2((12))t((12)).$$

This section is devoted to deriving a generalization of these facts to the general case of induced representations. The formulation and proofs presented here are due to Mackey.

We begin with some generalities which could have been placed in the middle of Chapter 2. Let (ρ, N) be a representation of a (finite) group G whose irreducible characters are $\chi_1, \ldots, \chi_s$. Let

$$W = W_1 \oplus \cdots \oplus W_s$$

be the canonical decomposition of W where each W_i is a direct sum of m_i copies of the ith irreducible representation of G. Let P_i denote projection onto W_i.

Suppose that $T \in \mathrm{Hom}_G(W, W)$. Then T carries each W_i into itself so

$$TP_i = P_i T P_i$$

and we wish to compute the eigenvalues of TP_i. If these eigenvalues were $\lambda_1, \ldots, \lambda_{m_i}$ then

$$\mathrm{tr}\, TP_i = d_i(\lambda_1 + \cdots + \lambda_{m_i}),$$

where

$$d_i = \chi_i(e) = \text{dim of ith irreducible} \tag{12.1}$$

and similarly

$$\mathrm{tr}\, T^2 P_i = d_i(\lambda_1^2 + \cdots + \lambda_{m_i}^2)$$

etc. So knowing all the $\mathrm{tr}\, T^k P_i$ will determine the sums of the powers of the λ's. From these sums of powers we can compute the λ's themselves. (We get the elementary symmetric functions which give the coefficients of the polynomial satisfied by the λ's and hence by solving a polynomial equation we get the λ's. For example, knowing $\lambda_1 + \lambda_2$ and $\lambda_1^2 + \lambda_2^2$ gives $\lambda_1 + \lambda_2$ and $\lambda_1\lambda_2 = \frac{1}{2}[(\lambda_1 + \lambda_2)^2 - \lambda_1^2 - \lambda_2^2]$ from which we can determine λ_1 and λ_2.)

Thus we would like a convenient group theoretical way of computing

$$\frac{1}{d_i} \mathrm{tr}\, TP_i,$$

without necessarily knowing the P_i explicitly. By Schur's lemma,

$$\mathrm{tr}\, T\rho(a)P_i = c\chi_i(a)$$

since the diagonal component of T lying in any irreducible is a constant times the identity matrix. Taking $a = e$ gives

$$\mathrm{tr}\, TP_i = cd_i$$

so

$$\mathrm{tr}\, T\rho(a)P_i = \left(\frac{1}{d_i}\mathrm{tr}\, TP_i\right)\chi_i(a)$$

and summing over i gives

$$\mathrm{tr}\, T\rho = \sum_i \left(\frac{1}{d_i}\mathrm{tr}\, TP_i\right)\chi_i.$$

Thus the numbers we are looking for are just the 'Fourier coefficients' of the function $\text{tr}\, T\rho$. We can find them by the usual recipe:

$$\frac{1}{d_i}\text{tr}\, TP_i = (\text{tr}\, T\rho, \chi_i)_G. \tag{12.2}$$

So our problem becomes one of calculating the function $\text{tr}\, T\rho$.

We shall do this for the case of an induced representation. So $W = \Gamma(E)$, where E is the vector bundle over $M = G/H$ induced from a representation (σ, V) of H. The most general linear operator on $\Gamma(E)$ is given by a 'kernel':

$$(Ts)(x) = \sum_y T(x, y)s(y), \quad \text{where}\ \ T(x, y)\in\text{Hom}\,(E_y, E_x).$$

If $x = aH$ and $y = bH$, we can write the most general element of E_r as $[(b, v)]$ and

$$T(x, y)[(b, v)] = [(a, t(a, b)v)],$$

where $t(a, b)\in\text{Hom}\,(V, V)$. We can replace a by ah_1^{-1} and b by bh_2^{-1} (where h_1 and h_2 are in H) and

$$T(x, y)[(bh_2^{-1}, \sigma(h_2)v)] = [(ah_1^{-1}, t(ah_1^{-1}, bh_2^{-1})\sigma(h_2)v)]$$
$$= [(a, \sigma(h_1)^{-1}t(ah_1^{-1}, bh_2^{-1})\sigma(h_2)v)]$$

so we must have the consistency condition

$$t(ah_1^{-1}, bh_2^{-1}) = \sigma(h_1)t(a, b)\sigma(h_2)^{-1}. \tag{12.3}$$

Conversely, given any $t: G \times G \to \text{Hom}\,(V, V)$ we get an element of $\text{Hom}\,(\Gamma(E), \Gamma(E))$. In terms of the identification of $\Gamma(E)$ with the space of functions from G to V satisfying

$$f(ah) = \sigma(h)^{-1}f(a)$$

we can write the operator T as

$$(Tf)(a) = \frac{1}{\#H}\sum t(a, b)f(b).$$

It is clear that $T\in\text{Hom}_G\,(\Gamma(E), \Gamma(E))$ if and only if

$$t(ca, cb) = t(a, b) \quad \text{for all}\ \ c\in G.$$

We can thus write

$$t(a, b) = t(e, a^{-1}b). \tag{12.4}$$

Setting

$$t(b) = t(e, b) \tag{12.5}$$

we see that (12.3) becomes

$$t(h_1 bh_2) = \sigma(h_1)t(b)\sigma(h_2). \tag{12.6}$$

Conversely, any function t satisfying (12.6) gives an invariant T satisfying (12.3) by (12.4) and (12.5).

Now for $T \in \mathrm{Hom}_G(\Gamma(E), \Gamma(E))$ and $s \in \Gamma(E)$ we have

$$[T\rho(a)s](x) = \sum T(x, y)as(a^{-1}y)$$
$$= \sum T(x, ay)as(y)$$

where $a: E_y \to E_{ay}$ so $T(x, ay)a: E_y \to E_x$. Thus

$$\mathrm{tr}\, T\rho(a) = \sum_x \mathrm{tr}\, T(x, ax)a. \tag{12.7}$$

If $x = bH$ we have

$$a: [(b, v)] \to [(ab, v)]$$
$$T(x, ax): [(ab, v)] \to [(b, t(b, ab)v)]$$

so

$$\mathrm{tr}\, T(x, ax)a = \mathrm{tr}\, t(b, ab)$$
$$= \mathrm{tr}\, t(e, b^{-1}ab)$$
$$= \mathrm{tr}\, t(b^{-1}ab).$$

Summing over all b overcounts by a factor $\#H$ so

$$\mathrm{tr}\, T\rho(a) = \frac{1}{\#H} \mathrm{tr}\, t(b^{-1}ab). \tag{12.8}$$

Substituting into (12.2) gives

$$\mathrm{tr}\, TP_i = (\mathrm{tr}\, T\rho, \chi_i)_G$$

$$= \frac{1}{\#G} \sum_a \mathrm{tr}\, T\rho(a)\overline{\chi_i(a)}$$

$$= \frac{1}{\#G} \frac{1}{\#H} \sum_{a \in G} \sum_{b \in G} \overline{\chi_i(a)}\, \mathrm{tr}\, t(b^{-1}ab)$$

$$= \frac{1}{\#G} \frac{1}{\#H} \sum_{a \in G} \sum_{b \in G} \overline{\chi_i(bab^{-1})}\, \mathrm{tr}\, t(a)$$

$$= \frac{1}{\#H} \sum_{a \in G} \overline{\chi_i(a)}\, \mathrm{tr}\, t(a).$$

This is our final result: any $T \in \mathrm{Hom}(\Gamma(E), \Gamma(E))$ comes from a function $t: G \to \mathrm{Hom}(V, V)$ satisfying (12.6) and we have the formula

$$\mathrm{tr}\, TP_i = \frac{1}{\#H} \sum_{a \in G} \overline{\chi_i(a)}\, \mathrm{tr}\, t(a). \tag{12.9}$$

This clearly generalizes the formulas for the case $G = S_2$ derived at the beginning of the section.

4

COMPACT GROUPS
AND LIE GROUPS

The theorems that we proved in Chapter 2 all had to do with finite groups. Many of them, however, carry over with little change to groups such as $SO(3)$, $SU(N)$, etc. The technique of averaging over the group becomes a process of integration with respect to an 'invariant volume element' or, more succinctly, 'invariant integration', a concept that we must explain. It will turn out that the 'total volume' of such groups is finite, which makes the averaging process work. We will also be interested in groups such as $Sl(d, \mathbb{C})$, the Poincaré groups, the Euclidean group, etc., for which the 'total volume' is not finite, and so no averaging process exists.

In order to formulate the results in some generality, we will need to use some terminology and theorems with which some readers may be unfamiliar. Our advice to those readers is simply to stick to the examples we use and ignore the general formulation.

For instance, we want to define a 'topological group' as a group G, which is also a 'topological space' such that the map $G \times G \to G$ is continuous, and the map $a \to a^{-1}$ of $G \to G$ is continuous. Now the reader may or may not know what the words 'topological space' and the attendant phrase 'continuous map' mean. For the reader who does not, we do not propose to make a large detour to explain the fundamental definitions and theorems of point set topology. All we will say is that 'topological space' is a concept that axiomatically describes a space with a notion of 'closeness' between its points. For example, the group $O(3)$ (and, in fact, almost all the groups we shall study) is explicitly given as a group of matrices. Two matrices are said to be close if all the corresponding matrix entries are close. If A is close to A' and B is close to B', then AB is close to $A'B'$, and A^{-1} is close to A'^{-1}. This is what is meant by saying that $O(3)$ (or $SU(N)$ or $Sl(n, \mathbb{C})$, etc.) are all topological groups.

A more restrictive notion than topological group is the notion of a Lie group. This depends on the concept of 'differentiable manifold'. A differentiable manifold is a space on which it is possible to introduce coordinates near each point in a consistent manner. For example, we saw how to introduce local coordinates for the groups $SO(3)$ and $SU(2)$ in Chapter 1. For differentiable manifolds, the concept of a differentiable map makes sense. A group G is called a Lie group if it is a differentiable manifold and the multiplication map $G \times G \to G$ is differentiable, as is the inverse map $G \to G$, sending a to a^{-1}.

If $V = \mathbb{R}^n$ is a real vector space, we have the group, $Gl(n, \mathbb{R})$, of all invertible real

$n \times n$ matrices. This is an open set in the space of all $n \times n$ matrices, which gives it its structure as a differentiable manifold of dimension n^2. Multiplication and inverse are clearly differentiable maps. For complex $V = \mathbb{C}^n$, we have the group $Gl(n, \mathbb{C})$ which is an open subset of $\mathbb{C}^{n^2}$. We may identify $\mathbb{C}^{n^2}$ with $\mathbb{R}^{2n^2}$ by taking the real and imaginary part of each coordinate. Again, $Gl(n, \mathbb{C})$ is a Lie group, of dimension $2n^2$.

It is a theorem in differential calculus (the implicit function theorem) that if

$$F : X \to W$$

is a map between vector spaces X and W such that

$$\mathrm{d}F_x \text{ is surjective when } F(x) = 0,$$

then $F^{-1}(0)$ is a submanifold of X. Here $\mathrm{d}F_x$ denotes the differential of F at x defined by

$$\mathrm{d}F_x(y) = \lim_{t=0} \frac{1}{t} \{ F(x + ty) - F(x) \}.$$

If we examine many of the groups we have been considering, we can see that they were in fact defined by the vanishing of some F. For example, $Sl(n, \mathbb{R})$ was defined by the vanishing of

$$F(A) = \det A - 1,$$

where F maps the vector space of all matrices into $\mathbb{R}$.

A group that is given as a (closed) submanifold of one of the groups $Gl(V)$ is called a *linear Lie group*. Thus, for example, the groups $Sl(n, \mathbb{R})$, $O(n)$, $O(1, 3)$ were defined as submanifolds of $Gl(n, \mathbb{R})$ and hence are linear Lie groups, while $Sl(n, \mathbb{C})$, $U(n)$, and $SU(n)$ were defined as (real) submanifolds of $Gl(n, \mathbb{C})$ and hence are also linear Lie groups.

4.1 Haar measure

Our first item of business is to introduce the notion of invariant integral. A continuous function f on G is said to have compact support if it vanishes outside a compact set. (This means, for the kinds of matrix groups that we are studying, that f vanishes whenever any of the matrix entries of the group exceeds some bound. For a group such as $O(n)$, where all the matrix entries are bounded (indeed ≤ 1), this is no restriction at all on f.) We let $\mathscr{F}_0(G)$ denote the space of continuous functions of compact support. We can think of integration as a linear function on $\mathscr{F}_0(G)$ which satisfies the appropriate axioms (for example, is non-negative on non-negative functions, etc.). Thus it is a rule which assigns to each $f \in \mathscr{F}_0(G)$ a number, $\int_G f(a)\,\mathrm{d}a$.

We say that the integration is invariant if $r(b)f$ and f yield the same values, i.e.

$$\int_G (r(b)f)(a)\,\mathrm{d}a = \int_G f(a)\,\mathrm{d}a$$

for all $f \in \mathscr{F}_0(G)$ and all $b \in G$. Since $r(b)f(a) = f(b^{-1}a)$, we can also write this condition as

$$\int_G f(b^{-1}a)\,\mathrm{d}a = \int_G f(a)\,\mathrm{d}a.$$

It is a theorem that every compact group has a unique invariant integral (up to scalar multiple). It has the property that the total volume of G, i.e. $\int 1\,\mathrm{d}a$ is finite. It then becomes completely determined once we normalize the total volume of the group to be unity:

$$\int_G 1\,\mathrm{d}a = 1.$$

In this section we explain how one actually computes this integral in given cases.

We recall some facts from the calculus on manifolds. (Again, the uninitiated reader can skip to the end of this discussion where the computational method is exhibited.) Let M and N be differentiable manifolds and $\phi\colon M \to N$ a differentiable map. If ω is a differential form on N, then $\phi^*\omega$, its 'pull back', is a differential form on M. The map ϕ^* preserves all algebraic operations, in particular,

$$\phi^*(\omega_1 \wedge \omega_2) = \phi^*\omega_1 \wedge \phi^*\omega_2,$$

where $\wedge$ denotes exterior multiplications. If ϕ is a diffeomorphism and Ω is an n-form on N where $n = \dim N$, then

$$\int_M \phi^*\Omega = \int_N \Omega.$$

Here we are assuming that M and N are oriented, ϕ is orientation preserving and Ω is of compact support so that the integrals make sense. If Ω does not have compact support, we can multiply it by a function f of compact support and write

$$\int_M \phi^* f\phi^*\Omega = \int_N f\Omega.$$

We will be interested in the case where $M = N = G$ and $\phi = L_b$ is left multiplication by b^{-1}. Then $\phi^* f = r(b)f$ and the above equation then becomes

$$\int_G (r(b)f)L_b^*\Omega = \int_G f\Omega.$$

If we could find an n-form Ω such that

$$L_b^*\Omega = \Omega$$

for all b in G, then $\int f\Omega$ would be our invariant integral. We shall replace the problem of finding an invariant n-form by the harder-looking problem of finding n invariant one-forms. That is, we will look for n one-forms $\omega_1, \ldots, \omega_n$ on G such that

$$L_b^*\omega_i = \omega_i$$

for all $b \in G$ and $i = 1, \ldots, n$. Then

$$\Omega = \omega_1 \wedge \cdots \wedge \omega_n$$

is an invariant n-form, and if it does not vanish, it gives us our invariant integral.

Let us illustrate the method first for the multiplicative group $G = \mathbb{R}^+$. A point of the group is given by a positive real number x, and the most general linear differential form is given as $k(x)\,dx$. Then $L_b x = b^{-1}x$ and

$$L_b^*(k(x)\,dx) = k(b^{-1}x)\,db^{-1}x = k(b^{-1}x)b^{-1}\,dx.$$

We wish to choose k so that $k(b^{-1}x)b^{-1} = k(x)$ and the solution is clear. Take

$$k(x) = x^{-1}.$$

Thus

$$\omega = \Omega = x^{-1}\,dx$$

is the left invariant form, and the Haar integral is given by

$$\int_G f = \int_{\mathbb{R}^+} f(x)x^{-1}\,dx.$$

The method generalizes as follows. Suppose we are given a representation of our group on $\mathbb{R}^k$ or $\mathbb{C}^k$. Thus we are given a matrix valued function, A, on the group G. Then

$$dA$$

is a $k \times k$ matrix of differential forms on G: if

$$A = (A_{ij}) \quad \text{then} \quad dA = (dA_{ij})$$

by definition. Also

$$A^{-1}\,dA$$

is a $k \times k$ matrix of linear differential forms. I claim that each of the entries of $A^{-1}\,dA$ is a left invariant differential form! Indeed,

$$L_b^*(A) = B^{-1}A,$$

where B is the (constant) matrix representing b. Then

$$L_b^*\,dA = d(B^{-1}A) = B^{-1}\,dA$$

since B is a fixed constant matrix and

$$L_b^* A^{-1} = (B^{-1}A)^{-1} = A^{-1}B$$

so

$$L_b^*(A^{-1}\,dA) = A^{-1}BB^{-1}\,dA = A^{-1}\,dA.$$

If A is a complex matrix, then each of the real and imaginary parts of the matrix entries will be invariant.

Let us illustrate the method for the group $SU(2)$. We can write each element of $SU(2)$

as a complex 2×2 matrix

$$A = \begin{pmatrix} \bar{\alpha} & -\beta \\ \bar{\beta} & \alpha \end{pmatrix}, \quad \text{where } |\alpha|^2 + |\beta|^2 = 1.$$

Then

$$dA = \begin{pmatrix} d\bar{\alpha} & -d\beta \\ d\bar{\beta} & d\alpha \end{pmatrix}$$

and

$$A^{-1} = A^* = \begin{pmatrix} \alpha & \beta \\ -\bar{\beta} & \bar{\alpha} \end{pmatrix}$$

so

$$A^{-1}\,dA = \begin{pmatrix} \alpha & \beta \\ -\bar{\beta} & \bar{\alpha} \end{pmatrix}\begin{pmatrix} d\bar{\alpha} & -d\beta \\ d\bar{\beta} & d\alpha \end{pmatrix} = \begin{pmatrix} \alpha\,d\bar{\alpha} + \beta\,d\bar{\beta} & -\alpha\,d\beta + \beta\,d\alpha \\ -\bar{\beta}\,d\bar{\alpha} + \bar{\alpha}\,d\bar{\beta} & +\bar{\beta}\,d\beta + \bar{\alpha}\,d\alpha \end{pmatrix}.$$

(Notice that the matrix $A^{-1}\,dA$ is skew-adjoint. This is no accident, as we shall see.) Each of the real and imaginary parts of the entries is left invariant. But let us multiply these entries directly:

$$(\bar{\alpha}\,d\alpha + \bar{\beta}\,d\beta) \wedge (-\alpha\,d\beta + \beta\,d\alpha) \wedge (-\bar{\beta}\,d\bar{\alpha} + \bar{\alpha}\,d\bar{\beta})$$
$$= (-\alpha\bar{\alpha} - \bar{\beta}\beta)\,d\alpha \wedge d\beta \wedge (-\bar{\beta}\,d\bar{\alpha} + \bar{\alpha}\,d\bar{\beta})$$
$$- d\alpha \wedge d\beta \wedge (-\bar{\beta}\,d\bar{\alpha} + \bar{\alpha}\,d\bar{\beta})$$

since $|\alpha|^2 + |\beta|^2 = 1$. We can simplify this expression by differentiating the equation

$$\alpha\bar{\alpha} + \beta\bar{\beta} = 1$$

to obtain

$$\alpha\,d\bar{\alpha} + \bar{\alpha}\,d\alpha + \beta\,d\bar{\beta} + \bar{\beta}\,d\beta = 0$$

or, for $\beta \neq 0$,

$$d\bar{\beta} = -\frac{1}{\beta}(\alpha\,d\bar{\alpha} + \bar{\alpha}\,d\alpha + \bar{\beta}\,d\beta).$$

The terms involving $d\alpha$ and $d\beta$ disappear when we multiply by $d\alpha \wedge d\beta$, so writing

$$-\bar{\beta}d\bar{\alpha} \text{ as } -\frac{1}{\beta}\beta\bar{\beta}\,d\bar{\alpha}$$

we get

$$\frac{1}{\beta}\,d\alpha \wedge d\beta \wedge d\bar{\alpha},$$

using $\alpha\bar{\alpha} + \beta\bar{\beta} = 1$ once again. So $(1/\beta)\,d\alpha \wedge d\beta \wedge d\bar{\alpha}$ is a left invariant form. Let us write this out in 'polar coordinates'

$$\alpha = w + iz$$
$$\beta = x + iy \qquad x^2 + y^2 + z^2 + w^2 = 1$$
$$w = \cos\theta$$
$$z = \sin\theta\cos\psi$$

$$x = \sin\theta\sin\psi\cos\phi$$
$$y = \sin\theta\sin\psi\sin\phi$$

so that

$$\alpha = \cos\theta + i\sin\theta\cos\psi$$
$$\beta = \sin\theta\sin\psi\,e^{i\phi}.$$

Here $0 \leqslant \theta \leqslant \pi$, $0 \leqslant \psi \leqslant \pi$, and $0 \leqslant \phi \leqslant 2\pi$. Now, if u and v are any real functions

$$d(u + iv) \wedge d(u - iv) = -2i\,du \wedge dv$$

so

$$d\alpha \wedge d\bar{\alpha} = -2i\,d(\cos\theta) \wedge d(\sin\theta\cos\psi)$$
$$= 2i\sin\theta\,d\theta \wedge \sin\theta\,d\cos\psi$$
$$= -2i\sin^2\theta\sin\psi\,d\theta \wedge d\psi.$$

Thus

$$d\alpha \wedge d\beta \wedge d\bar{\alpha} = -d\alpha \wedge d\bar{\alpha} \wedge d\beta$$
$$= 2i\sin^2\theta\sin\psi\,d\theta \wedge d\psi \wedge i\beta\,d\phi$$

since the other terms in $d\beta$ will disappear when multiplied by $d\theta \wedge d\psi$. Thus

$$\frac{1}{\beta}d\alpha \wedge d\beta \wedge d\bar{\alpha} = -2\sin^2\theta\sin\psi\,d\theta \wedge d\psi \wedge d\phi.$$

We can multiply this expression by any constant. We choose the constant so that the total volume is unity and so

$$\Omega = \frac{1}{2\pi^2}\sin^2\theta\sin\psi\,d\theta \wedge d\psi \wedge d\phi$$

gives the Haar measure.

If $(f_1, f_2)_G$ denotes the corresponding L_2 inner product for functions on $G = SU(2)$, then

$$(f_1, f_2)_G = \int_G f_1\bar{f}_2\Omega = \frac{1}{2\pi^2}\int f_1(\theta, \psi, \phi)\overline{f_2(\theta, \psi, \phi)}\sin^2\theta\sin\psi\,d\theta \wedge d\psi \wedge d\phi.$$

Recall that f is called a central function if $f(A)$ depends only on the conjugacy class of A. Now the conjugacy class of $A \in SU(2)$ determines and is determined by $\operatorname{tr} A = \alpha + \bar{\alpha} = 2\cos\theta$. Thus, a function f is central if and only if it depends only on $\cos\theta$. Thus, for central functions,

$$(f_1, f_2)_G = \frac{2}{\pi}\int_0^\pi f_1(\theta)\overline{f_2(\theta)}\sin^2\theta\,d\theta.$$

4.2 The Peter–Weyl theorem

Let G be a topological group. By a representation we mean a continuous homomorphism, ρ, of G into $\operatorname{Hom}(V, V)$, where V is a vector space. In case V is finite dimensional, this definition is unambiguous. In case V is an infinite topological vector space, the topology on $\mathrm{Hom}\,(V, V)$ must be specified. The weakest notion of continuity we would want to consider is the following:

for each $x \in V$ and each continuous linear function l on V, the function

$$f_x^l(a) = \langle \rho(a)^{-1}x, l \rangle \qquad (2.1)$$

is continuous.

The cases of most interest to us will be where V is a Hilbert space and ρ a unitary representation. In that case, standard theorems in Hilbert space theory guarantee that the various definitions of continuity coincide. We will not press this point here.

If $\mathscr{F}(G)$ denotes the space of continuous functions on G, we have, as before, the regular representation, r, of G on $\mathscr{F}(G)$:

$$(r(b)f)(a) = f(b^{-1}a).$$

The above definition of (weak) continuity implies that any $l \in V^*$ defines a G morphism

$$\begin{aligned}\phi_l &: V \to \mathscr{F}(G) \\ \phi_l(x) &= f_x^l,\end{aligned} \qquad (2.2)$$

as in the case of finite groups.

We say that V is topologically irreducible if the only closed invariant subspaces of V are V itself and $\{0\}$. Thus, if V is finite dimensional, it is topologically irreducible if and only if it is irreducible in the old fashioned sense.

Now $\ker \phi_l$ is clearly a closed invariant subspace of V. So if V is topologically irreducible, then ϕ_l must be injective. Thus $\phi_l(V) \subset \mathscr{F}(G)$ is equivalent to V.

If V is finite dimensional and irreducible, then $G \times G$ acts irreducibly on $V \otimes V^*$. (We proved this theorem using characters in Chapter 2. See Appendix E for a more general proof.) Hence

$$\begin{aligned}\phi &: V \otimes V^* \to \mathscr{F}(G) \\ \phi(x \otimes l) &= f_x^l\end{aligned} \qquad (2.3)$$

is injective. Thus each finite-dimensional irreducible representation of G occurs as a subrepresentation of the regular representation of G on $\mathscr{F}(G)$, with a multiplicity at least equal to its dimension.

Suppose we pick some function $f \in \phi_l(V)$. Then $r(a)f \in \phi_l(V)$ and the linear combinations of all the $r(a)f$ span an invariant subspace of $\phi_l(V)$. If V is finite dimensional, then this is some finite-dimensional subspace of $\phi_l(V)$. If V is irreducible (and $f \neq 0$), then this will be all of $\phi_l(V)$.

If we pick some $f \in \mathscr{F}(G)$, there is no reason to expect that set of all $r(a)f$ should lie in a finite-dimensional space. If they do, then f is called a *representative* function.

So far we have managed without the technique of averaging over the group. If we want to apply this technique, we must assume that the total volume of this group, i.e.

$$\int_G 1$$

is finite. For compact groups this is always the case. (We have verified this explicitly for $SU(2)$.) So for the rest of this section we assume that G is compact and that we have normalized the measure so that

$$\int_G 1 = 1.$$

The reader can check that all the results of Section 2.3 having to do with *finite-dimensional* representations carry over with no modification, provided that we replace

$$\frac{1}{\#G} \sum_{a \in G}$$

by integrals

$$\int_G \cdot$$

The one place where something new happens is where we consider the regular representation (or more generally a representation on $\mathscr{F}(M)$), since $\mathscr{F}(G)$ is no longer finite dimensional. We make $\mathscr{F}(G)$ into a pre-Hilbert space by defining the scalar product

$$(f_1, f_2)_G = \int_G f_1(a)\overline{f_2(a)}\, da$$

and complete to obtain a Hilbert space called $L^2(G)$. The regular representation then becomes a unitary representation of G on $L^2(G)$.

The basic theorem is the Peter–Weyl theorem which we will prove in Appendix E. Here are the key assertions of the theorem:

Peter–Weyl theorem

Let G be a compact group.

(1) The representative functions are dense in $L^2(G)$.

(2) The space $L^2(G)$ decomposes into a Hilbert space direct sum of irreducible representations of G, each of which is finite dimensional.

(3) Every irreducible representation of G is finite dimensional

(4) Each irreducible representation of G occurs in $L^2(G)$ with a multiplicity equal to its dimension.

(5) Any unitary representation of G on any Hilbert space decomposes into a Hilbert space direct sum of finite-dimensional irreducible representations.

As in the finite case, the characters of irreducible representations form a basis of the space of central functions, but in the L^2 sense:

the irreducible characters form an orthonormal basis of the Hilbert space of square integrable central functions.

(Again, the proof is given in Appendix E.)

Let us show how the fact that irreducible characters have unit length in $L^2(G)$ determines the possible form for characters of $SU(2)$. Indeed, let (ρ, V) be some irreducible unitary representation of $SU(2)$. Then

$$\rho \begin{pmatrix} e^{i\theta} & 0 \\ 0 & e^{-i\theta} \end{pmatrix}$$

is a unitary matrix, so V decomposes into a direct sum of eigenspaces. So we can find a basis $v_1, \ldots, v_n$ of V such that

$$\rho \begin{pmatrix} e^{i\theta} & 0 \\ 0 & e^{-i\theta} \end{pmatrix} v_j = \lambda_j(\theta) v_j.$$

By the representation property, we must have

$$\lambda_j(\theta_1 + \theta_2) = \lambda_j(\theta_1)\lambda_j(\theta)$$

so

$$\lambda_j = \exp i c_j \theta$$

for some integer c_j. If χ denotes the character of the representation, then

$$\chi(\theta) = \sum \exp i c_j \theta$$

Our problem is to determine the possibilities for the c_j. For example, since χ is a function of $\cos\theta$, we know that for each c_j that occurs, $-c_j$ must also occur. Now consider the condition

$$1 = (\chi, \chi) = \frac{2}{\pi} \int_0^\pi \chi(\theta) \bar\chi(\theta) \sin^2\theta \, d\theta$$

$$= \frac{1}{\pi} \int_0^{2\pi} \chi(\theta)\overline{\chi(\theta)} \sin^2\theta \, d\theta.$$

Write

$$\sin^2\theta = \tfrac{1}{2}(1 - \cos 2\theta) = \tfrac{1}{4}(2 - \exp 2i\theta - \exp(-2i\theta))$$

so

$$1 = (\chi, \chi) = \frac{1}{4} \cdot \frac{1}{\pi} \sum_{j=1}^{n} \sum_{k=1}^{n} \int_0^{2\pi} \exp i c_j \theta \cdot \exp(-i c_k \theta)(2 - \exp 2i\theta - \exp(-2i\theta)) \, d\theta.$$

Now $\int_0^{2\pi} \exp m i\theta \, d\theta = 0$ if $m \neq 0$ and $= 2\pi$ if $m = 0$. So in evaluating the above expression, we multiply out and only keep terms which multiply to one. We have

n terms where $j = k$ each giving $\dfrac{1}{4} \cdot \dfrac{1}{\pi} \cdot 2\pi \cdot 2$ yielding n;

$2a$ terms from $j \neq k$ where $c_j = c_k$ giving $2a$;

$2b$ terms where $j \neq k$ and $c_j = c_k + 2$ each yielding $\dfrac{1}{4} \cdot \dfrac{1}{\pi} \cdot 2\pi \cdot (-1)$

yielding $-b$.

So $(\chi|\chi) = n + 2a - b$. Now b can be at most $n - 1$, so $n - b + 2a \geqslant 1 + 2a$ and this can equal 1 only if $a = 0$ and $b = n - 1$. The fact that the c_js are symmetric about the origin implies that the c_j are given as

$$\{-n+1, -n+3, \ldots, n-3, n-1\}. \tag{2.4}$$

It is conventional to label these characters (and the representations which we shall see exist) by $s = (n-1)/2$,

$$\begin{aligned}
\chi(\theta) &= 1 & s &= 0 \\
\chi(\theta) &= \exp(-i\theta) + \exp i\theta & s &= \tfrac{1}{2} \\
\chi(\theta) &= \exp(-2i\theta) + 1 + \exp 2i\theta & s &= 1 \\
\chi(\theta) &= \exp(-3i\theta) + \exp(-i\theta) + \exp i\theta + \exp 3i\theta & s &= \tfrac{3}{2}
\end{aligned}$$

etc.

4.3   The irreducible representations of $SU(2)$

We still must show that each of the functions

$$\chi_0 = 1$$
$$\chi_{1/2} = \exp(-i\theta) + \exp i\theta$$
$$\chi_1 = \exp(-2i\theta) + 1 + \exp 2i\theta$$
$$\vdots$$
$$\chi_s = \exp(-2si\theta) + \exp(-2s+2)i\theta + \cdots + \exp 2si\theta \tag{3.1}$$
$$\vdots$$

are indeed characters of (irreducible) representations of $SU(2)$. We do so by explicitly exhibiting the representation. The group $SU(2)$ acts on $\mathbb{C}^2$. Hence, it has a representation on the space of all functions on $\mathbb{C}^2$. Let V_s denote the space of homogeneous polynomials of degree s. Thus, a basis for the space V_s are the monomial functions

$$z_1^{2s}, z_1^{2s-1}z_2, z_1^{2s-2}z_2^2, \ldots, z_2^{2s}. \tag{3.2}$$

If A is any invertible 2×2 matrix and if $f \in V_s$, then clearly $\rho(A)f = f(A^{-1}\cdot)$ is again a homogeneous polynomial of degree s. We thus get a representation of $SU(2)$ on V_s. The diagonal matrix

$$U_{-\theta} = \begin{pmatrix} e^{-i\theta} & 0 \\ 0 & e^{i\theta} \end{pmatrix}$$

clearly has each monomial as eigenvector:

$$\rho(U_{-\theta})z_1^{2s-k}z_2^k = [\exp i(2s - 2k)\theta]z_1^{2s-k}z_2^k$$

and hence has trace $\chi_s(\theta)$. Since $(\chi_s, \chi_s) = 1$, we conclude that the restriction of ρ to V_s is irreducible, and by the results of the preceding section, we know that these are *all* the irreducible representations of $SU(2)$.

Let us write out the matrix of this representation for the monomial basis z_1, z_2 for the simplest case, $s = \frac{1}{2}$: If

$$A = \begin{pmatrix} \bar{a} & -b \\ \bar{b} & a \end{pmatrix} \quad \text{then} \quad A^{-1} = \begin{pmatrix} a & b \\ -\bar{b} & \bar{a} \end{pmatrix}$$

so

$$A^{-1} \begin{pmatrix} z_1 \\ z_2 \end{pmatrix} = \begin{pmatrix} az_1 + bz_2 \\ -\bar{b}z_2 + \bar{a}z_2 \end{pmatrix}.$$

Then

$$\rho(A)z_1 = az_1 + bz_2$$
$$\rho(A)z_2 = -\bar{b}z_1 + \bar{a}z_2$$

so the matrix representing A is

$$\begin{pmatrix} a & -\bar{b} \\ b & \bar{a} \end{pmatrix} = \bar{A}.$$

It is conventional to take the matrix representing A in the $\frac{1}{2}$ representation to be A and not $\bar{A}$, that is to define

$$\rho_{\frac{1}{2}}(A) = \rho(\bar{A})|_{V_s}. \tag{3.3}$$

Notice that

$$\begin{pmatrix} 0 & -1 \\ 1 & 0 \end{pmatrix} \begin{pmatrix} \bar{a} & -b \\ \bar{b} & a \end{pmatrix} \begin{pmatrix} 0 & 1 \\ -1 & 0 \end{pmatrix} = \begin{pmatrix} a & -\bar{b} \\ b & \bar{a} \end{pmatrix};$$

in other words, A is conjugate to $\bar{A}$. Thus $\rho(\bar{A})|_{V_s}$ is equivalent to $\rho(A)|_{V_s}$ (as we know they must be since they have the same characters).

(This conjugacy between A and $\bar{A}$ is not true for $SU(n)$ when $n > 2$. The matrices

$$\begin{pmatrix} e^{i\theta} & 0 & 0 \\ 0 & e^{i\theta} & 0 \\ 0 & 0 & e^{-2i\theta} \end{pmatrix} \quad \text{and} \quad \begin{pmatrix} e^{-i\theta} & 0 & 0 \\ 0 & e^{-i\theta} & 0 \\ 0 & 0 & e^{2i\theta} \end{pmatrix}$$

have different eigenvalues and so cannot be conjugate. In the next chapter, we will show how to construct all the irreducible representations of $SU(n)$ and will find two inequivalent n-dimensional representations given by A and $\bar{A}$.)

The basis (3.2) of V_s is orthogonal. For many purposes, we would like to have an orthonormal basis. We claim that the functions

$$e^m(z) = e_s^m(z) = \frac{z_1^{s+m} z_2^{s-m}}{[(s+m)!(s-m)!]^{1/2}} \quad m = s, s-1, \ldots, -s \tag{3.4}$$

all have the same length relative to the unique (up to scale) $SU(2)$ invariant scalar product on V_s. Hence if we choose the arbitrary scale factor so that they have length one, they form an orthonormal basis of V_s.

To prove this, observe that

$$\sum_{m=-s}^{s} e^m(z)\overline{e^m(w)} = \sum \frac{z_1^{s+m} z_2^{s-m} \bar{w}_1^{s+m} \bar{w}_2^{s-m}}{(s+m)!(s-m)!}$$

$$= \frac{1}{(2s)!}(z_1\bar{w}_1 + z_2\bar{w}_2)^{2s} = \frac{1}{(2s)!}(z,w)^{2s}$$

where (z,w) denote the Hermitian scalar product in $\mathbb{C}^2$, $(z,w) = z_1\bar{w}_1 + z_2\bar{w}_2$. Now $A \in SU(2)$ preserves $(,)$ so

$$\sum [\rho_s(A)e^m](z)\overline{[\rho_s(A)e^m](w)} = \sum e^m(\bar{A}^{-1}z)\overline{e^m(\bar{A}^{-1}w)}$$

$$= \frac{1}{2s!}(\bar{A}^{-1}z, \bar{A}^{-1}w)^{2s} = \frac{1}{2s!}(z,w)^{2s} = \sum e^m(z)\overline{e^m(w)}.$$

If we work out the matrices $r(A)$ relative to the e^m, i.e.

$$\rho_s(A)e^m = \sum r_j^m(A)e^j,$$

the preceding equation says

$$\sum e^m(z)\overline{e^m(w)} = \sum r_j^m(A)e^j(z) \sum \overline{r_k^m(A)e^k(w)}$$

or

$$\sum_{j,k}\left(\sum_{m=-s}^{s} r_j^m(A)\overline{r_k^m(A)} - \delta_{jk} \right)e^j(z)\overline{e^k(w)} = 0.$$

Now the functions $e^j(z)\overline{e^k(w)}$ are clearly linearly independent of one another for different values of j and k, when considered as functions on $\mathbb{C}^2 \times \mathbb{C}^2$. Thus

$$\sum r_j^m(A)\overline{r_k^m(A)} = \delta_{jk}.$$

But this is precisely the condition that

$$r(A)(r(A))^* = I,$$

i.e. the matrix of $\rho_s(A)$ relative to the e basis is unitary. Thus we have an orthonormal basis.

For the first few values of s, we have

$$s = 0 \qquad 1$$

$$s = \tfrac{1}{2} \qquad z_1, \quad z_2$$

$$s = 1 \qquad \frac{z_1^2}{2^{\frac{1}{2}}}, \quad z_1 z_2, \quad \frac{z_2^2}{2^{\frac{1}{2}}}$$

$$s = \frac{3}{2} \qquad \frac{z_1^3}{6^{\frac{1}{2}}} \quad \frac{z_1^2 z_2}{2^{\frac{1}{2}}} \quad \frac{z_1 z_2^2}{2^{\frac{1}{2}}} \quad \frac{z_2^2}{6^{\frac{1}{2}}}$$

We can use our explicit formula for the characters χ_s of the representations (ρ_s, V_s) to obtain the decomposition of $V_{s_1} \otimes V_{s_2}$ into irreducibles, known as the Clebsch–

Gordan decomposition. For example,

$$\chi_0 \cdot \chi_s = 1 \cdot \chi_s = \chi_s$$

so

$$V_0 \otimes V_s = V_s.$$

Similarly,

$$\chi_{\frac{1}{2}} \cdot \chi_{\frac{1}{2}} = (\exp -i\theta + \exp i\theta)(\exp -i\theta + \exp i\theta)$$
$$= \exp -2i\theta + 1 + \exp 2i\theta + 1$$
$$= \chi_1 + \chi_0$$

so

$$V_{\frac{1}{2}} \otimes V_{\frac{1}{2}} = V_0 \oplus V_1.$$

We now consider the general problem of analyzing the tensor product $V_{s_1} \otimes V_{s_2}$. Let us set $\zeta = \exp 2i\theta$ so that

$$\chi_s(\theta) = \zeta^{-s} + \zeta^{-s+1} + \cdots + \zeta^s = \frac{\zeta^{s+1} - \zeta^{-s}}{\zeta - 1}$$

so that for $s_1 \leqslant s_2$ we have

$$\chi_{s_1}(\theta) \cdot \chi_{s_2}(\theta) = (\zeta^{-s_1} + \cdots + \zeta^{s_1}) \cdot \frac{\zeta^{s_2+1} - \zeta^{-s_2}}{\zeta - 1}$$

$$= \frac{\zeta^{s_1+s_2+1} - \zeta^{-s_1-s_2}}{\zeta - 1} + \cdots + \frac{\zeta^{s_2-s_1+1} - \zeta^{-(s_2-s_1)}}{\zeta - 1}$$

so

$$\chi_{s_1} \cdot \chi_{s_2} = \chi_{s_2-s_1} + \chi_{s_2-s_1+1} + \cdots + \chi_{s_2+s_2}. \tag{3.5}$$

Thus, for any values of s_1 and s_2, we have

$$V_{s_1} \otimes V_{s_2} = V_{|s_2-s_1|} + V_{|s_2-s_1|+1} + \cdots + V_{s_2+s_1}. \tag{3.6}$$

For example,

$$V_{\frac{1}{2}} \otimes V_{\frac{1}{2}} = V_0 + V_1$$
$$V_{\frac{1}{2}} \otimes V_1 = V_{\frac{1}{2}} + V_{\frac{3}{2}}$$
$$V_1 \otimes V_1 = V_0 + V_1 + V_2$$
$$V_1 \otimes V_2 = V_1 + V_2 + V_3$$
$$V_2 \otimes V_2 = V_0 + V_1 + V_2 + V_3 + V_4, \text{ etc.}$$

These formulas are the Clebsch–Gordan decomposition formulas for $SU(2)$.

4.4 The irreducible representations of SO(3) and spherical harmonics

We can now also determine all the irreducible representations of $SO(3)$. Recall from Chapter 1 that we have a surjective homomorphism of $SU(2)$ onto $SO(3)$ that we will now denote by τ:

$$SU(2) \overset{\tau}{\to} SO(3),$$

$$\tau(A) = \tau(B) \quad \text{if and only if } A = \pm B.$$

If (σ, W) is a representation of $SO(3)$, then $\sigma \circ \tau$ sending A into $\sigma(\tau(A))$ is a representation of $SU(2)$ on W. If σ is irreducible, then so is $\sigma \circ \tau$. Thus, if σ is an irreducible representation of $SO(3)$, then $\sigma \circ \tau$ must be (equivalent to) one of the ρ_s. So our question becomes: which of the ρ_s give rise to representations of $SO(3)$. Clearly, a necessary condition is that

$$\rho_s(-I) = \text{id}.$$

But this is also sufficient since we can then define

$$\sigma_s(\tau(A)) = \rho_s(A)$$

without ambiguity, to get an irreducible representation of $SO(3)$.
Now

$$\rho_s(-I)z_1^{2s-k}z_2^k = (-1)^{2s}z_1^{2s-k}z_2^k$$

so

$$\rho_s(-I) = \text{id} \quad \text{if and only if } s \text{ is an integer.}$$

Thus the representations $\sigma_s, s = 0, 1, 2, \ldots,$ given by

$$\sigma_s(R) = \rho_s(\tau^{-1}(R))$$

give all the irreducible representations of $SO(3)$
For historical reasons to be explained in Appendix F the various low values of s are given letter names according to

$$
\begin{array}{cccccc}
0 & 1 & 2 & 3 & 4 & 5 \\
\text{S} & \text{P} & \text{D} & \text{F} & \text{G} & \text{H.}
\end{array}
$$

We would like to have a more direct description of these representations in terms of ordinary three-dimensional Euclidean space. For this purpose, we give a method of identifying the representation (ρ_s, V_s) – a method for which we will have further use later on.

Lemma 4.1
Let (ρ, V) be any finite-dimensional representation of $SU(2)$ such that

$$\dim V = 2s + 1$$

and

$$\exp 2is\theta \quad \text{occurs as an eigenvalue of } \rho(U_\theta).$$

Then (ρ, V) is equivalent to (ρ_s, V_s).

Proof Decompose V into irreducibles:

$$V = V_{s_1} \oplus \cdots \oplus V_{s_p}.$$

No $s_p > s$ can occur since dim $V_{s_p} = 2s_p + 1 > 2s + 1$. If only $s_p < s$ occur, then all the eigenvalues of $\rho(U_\theta)$ would be of the form $\exp im\theta$ with $|m| < s$. Thus one of the s_p must equal s. But dim $V_s = 2s + 1 = $ dim V. So $V = V_s$, proving the lemma.

Let Δ denote the Laplace operator

$$\Delta = \frac{\partial^2}{\partial x^2} + \frac{\partial^2}{\partial y^2} + \frac{\partial^2}{\partial z^2}$$

acting on smooth functions on $\mathbb{R}^3$. Since Δ is defined in terms of the Euclidean metric, it commutes with all Euclidean motions: for any Euclidean motion, A, we have

$$\Delta \sigma(A) f = \sigma(A) \Delta f$$

where, as usual,

$$\sigma(A) f(v) = f(A^{-1} v).$$

In particular,

$$\Delta \sigma(R) f = \sigma(R) \Delta f \quad \text{for any } R \in SO(3).$$

In terms of polar coordinates, we can write

$$\Delta = \frac{\partial^2}{\partial r^2} + \frac{2}{r} \frac{\partial}{\partial r} + \frac{1}{r^2} \Delta_{S^2} \tag{4.1}$$

where

$$\Delta_{S^2} = \frac{1}{\sin\theta} \frac{\partial}{\partial\theta} \sin\theta \frac{\partial}{\partial\theta} + \frac{1}{\sin^2\theta} \frac{\partial}{\partial\phi^2} \tag{4.2}$$

is the 'spherical Laplacian'. Each of the three terms in the (4.2) decomposition of Δ, in particular the spherical Laplacian Δ_{S^2}, commutes with $\sigma(R)$ for all $R \in SO(3)$.

Let P^k denote the space of (complex-valued) polynomials on $\mathbb{R}^3$ which are homogeneous of degree k. Thus

$$\begin{aligned}
P^0 &= \{1\} & \dim P^0 &= 1 \\
P^1 &= \{x, y, z\} & \dim P^1 &= 3 \\
P^2 &= \{z^2, xz, yz, xy, x^2, y^2\} & \dim P^2 &= 6
\end{aligned}$$

In general, any homogeneous polynomial of degree k in $\mathbb{R}^3$ can be written as

$$p(x, y, z) = \sum_{i=0}^{k} p_i(x, y) z^{k-i}$$

where p_i is homogeneous of degree i in x and y. The space of p_i has dimension $i + 1$ so

$$\dim P^k = 1 + 2 + \cdots + k + 1 = \frac{(k+1)(k+2)}{2}. \tag{4.3}$$

The space P^k is clearly invariant under the $\sigma(R)$ for $R \in SO(3)$. It is not irreducible: let $H^k \subset P^k$ be defined by

$$H^k = \ker \Delta$$

where Δ is the Laplace operator

$$\Delta : P^k \to P^{k-2}.$$

The space H^k is known as the space of harmonic ploynomials of degree k. It is clearly invariant under $SO(3)$. We will show that it is irreducible, and, in fact, equivalent as a representation space of $SO(3)$ to (σ_k, V_k).

For this purpose, observe that the map $\Delta : P^k \to P^{k-2}$ is surjective as can easily be checked by looking at monomials and using induction. Thus

$$\dim H^k = \dim P^k - \dim P^{k-2} = 2k + 1. \tag{4.4}$$

On the other hand,

$$\frac{\partial^2}{\partial x^2} + \frac{\partial^2}{\partial y^2} = \left(\frac{\partial}{\partial x} - i\frac{\partial}{\partial y} \right) \left(\frac{\partial}{\partial x} + i\frac{\partial}{\partial y} \right)$$

and

$$\left(\frac{\partial}{\partial x} + i\frac{\partial}{\partial y} \right)(x + iy)^k = k(x + iy)^{k-1} \left(\frac{\partial}{\partial x} + i\frac{\partial}{\partial y} \right)(x + iy) = 0$$

so

$$(x + iy)^k \in H^k.$$

Also, we know that

$$\begin{pmatrix} e^{i\theta} & 0 \\ 0 & e^{-i\theta} \end{pmatrix}$$

maps into $R^z_{2\theta}$, rotation through angle 2θ about the z axis and hence multiplies $(x + iy)^k$ by $\exp 2ik\theta$. Thus $(x + iy)^k$ is an eigenvector with eigenvalue $\exp 2i\theta$. The conditions of our lemma are satisfied. Hence

$$H^k \text{ is equivalent to } V_k.$$

Thus

$$H^0 = P^0 = \text{constants}$$
$$H^1 = P^1 = \text{linear functions}$$
$$H^2 = \{ax^2 + by^2 + cz^2 + exy + fxz + gyz \,|\, a + b + c = 0\}.$$

Notice that

$$P^2 = H^2 \oplus \{r^2\}$$

where $r^2 = x^2 + y^2 + z^2$. We can prove inductively that

$$P^k = H^k \oplus r^2 P^{k-2}.$$

Indeed, let $Q: P^{k-2} \to P^k$ be the operation of multiplication by r^2. Then Q is clearly an injection and an $SO(3)$ morphism. Thus, by induction, the image of Q contains no component equivalent to H^k. Hence, decomposing P^k into irreducibles, we find that

$$P^k = H^k \oplus H^{k\perp}$$

where $H^{k\perp}$ contains all the irreducibles in P^k not equivalent to H^k. Since dim $H^{k\perp} =$ dim $P^{k-2} =$ dim im Q, we conclude that

$$P^k = H^k \oplus \mathrm{im}\, Q$$

so

$$P^k = H^k \oplus r^2 P^{k-2}.$$

Thus

$$P^k = H^k \oplus r^2 H^{k-2} \oplus r^4 H^{k-4} \oplus \cdots. \tag{4.5}$$

The functions in H^k are all homogeneous of degree k. They are thus all determined by their values on the unit sphere: in terms of polar coordinates, we can write an element of H^k as

$$f(r, \theta, \phi) = r^k Y(\theta, \phi) \tag{4.6}$$

where Y is a function on S^2. We let $\tilde{H}^k$ denote the space of all such Y's, i.e. the restriction of the elements of H^k to the unit sphere. The elements of H^k are called the *spherical harmonics* of degree k. Notice that

$$0 = \Delta f = \left(\frac{\partial^2}{\partial r^2} + \frac{2}{r} \frac{\partial}{\partial r} + \frac{1}{r^2} \Delta_{S^2} \right) r^k Y$$

$$= (k(k-1) + 2k + \Delta_{S^2}) r^{k-2} Y.$$

Thus

$$\Delta_{S^2} Y = -k(k+1) Y \quad Y \in \tilde{H}^k. \tag{4.7}$$

Thus each of the spaces $\tilde{H}^k \subset L^2(S^2)$ is an eigenspace for the operator Δ_{S^2} with eigenvalue $-k(k+1)$ and with multiplicity $2k + 1$. Since the operator Δ_{S^2} is symmetric, we conclude that the spaces $\tilde{H}^k$ and $\tilde{H}^l$ are orthogonal in $L^2(S^2)$. We claim that

$$L^2(S) = \overline{\oplus \tilde{H}^k} \tag{4.8}$$

(Hilbert space direct sum). Indeed, it is enough to show that the direct sum of the $\tilde{H}^k$ is dense in the space of continuous functions on S^2. But we can approximate any continuous function on S^2 by a polynomial in $\mathbb{R}^3$. We can write any polynomial as a sum of its homogeneous components, and we can write each homogeneous component as $f + r^2 f_1 + r^4 f_2 + \ldots$, where the $f_i \in H^{k-2i}$. But on the unit sphere, $r \equiv 1$. So the $\tilde{H}^k$ are dense in $L^2(S^2)$.

Notice that by a pure use of group theory we have computed the spectrum of the operator Δ_{S^2}! It has purely discrete spectrum with eigenvalues $-k(k+1), k = 0, 1, \ldots$. Any function f in $L^2(S^2)$ has a spherical harmonic expansion

$$f = a_0 Y^0 + a_1 Y^1 + \cdots \quad Y^i \in \tilde{H}^i.$$

It is conventional to choose an orthonormal basis of H^i consisting of eigenvectors of U_t. Thus

$$Y_0^0(\theta, \phi) = \frac{1}{(4\pi)^{\frac{1}{2}}} \quad \text{is a basis of } H^0$$

$$\left.\begin{array}{l}
Y_1^1(\theta, \phi) = -\left(\dfrac{3}{8\pi}\right)^{\frac{1}{2}} \sin\theta \exp i\phi = -\left(\dfrac{3}{8\pi}\right)^{\frac{1}{2}}\left(\dfrac{x+iy}{r}\right) \\[3ex]
Y_0^1(\theta, \phi) = \left(\dfrac{3}{4\pi}\right)^{\frac{1}{2}} \cos\theta \\[3ex]
Y_{-1}^1(\theta, \phi) = \left(\dfrac{3}{8\pi}\right)^{\frac{1}{2}} \sin\theta \exp -i\phi = \left(\dfrac{3}{8\pi}\right)^{\frac{1}{2}}\left(\dfrac{x-iy}{r}\right)
\end{array}\right\} \quad \text{is a basis of } H^1$$

etc. Then every $f \in L^2(S^2)$ has an expansion in 'spherical harmonics'

$$f = a_0^0 Y_0^0 + a_1^1 Y_1^1 + a_1^0 Y_0^1 + a_1^{-1} Y_{-1}^1 + \ldots .$$

Equation (4.8) says that $L^2(S^2)$ decomposes into a direct sum of irreducibles, with each irreducible of $SO(3)$ occurring exactly once. Now

$$S^2 = G/K \tag{4.9}$$

where $G = SO(3)$ and $K = SO(2) = $ rotations about the z axis, say. The representation of G on $L^2(S^2)$ can be thought of as an induced representation – induced from the trivial representation of K. We should therefore expect (4.8) from the reciprocity theorem: for this we must ask how many times the trivial representation of K occurs in each irreducible of G. But we know that for integer s, the eigenvalue 1, i.e. the trivial representation, occurs with multiplicity 1. Thus

$$\dim \mathrm{Hom}_G(V_s, L^2(S)) = \dim \mathrm{Hom}_K(V_s, \mathbb{C}) \tag{4.10}$$

where $\mathbb{C}$ denotes the trivial representation, and both sides equal 1. Notice we could take $G = SU(2)$ and

$$K = \left\{ \begin{pmatrix} e^{i\theta} & 0 \\ 0 & e^{-i\theta} \end{pmatrix} \right\}$$

is (4.8). Then again (4.10) would hold for all s, but both sides would be zero for non-integral s and one for integer s. We shall present a general formulation and proof of the Frobenius reciprocity theorem for compact groups in Appendix E.

4.5 The hydrogen atom

The best known use of (4.7) is the method of 'separation of variables' in elementary quantum mechanics, in the study of the central potential problem. One wants to analyse the operator

$$H = \frac{-\hbar^2}{2\mu}\Delta + V, \qquad \hbar = \frac{h}{2\pi} \tag{5.1}$$

where μ is the (reduced) mass and $V = V(r)$ is a spherically symmetric potential. Introducing polar coordinates means that we can identify $L^2(\mathbb{R}^3)$ with $L^2(\mathbb{R}^+)\hat{\otimes}L^2(S)$. The fact that H commutes with $SO(3)$ implies that H leaves invariant each of the subspaces $L^2(\mathbb{R}^+)\otimes\tilde{H}_l$. Furthermore, the contribution of the $\tilde{H}_l$ component is just a constant and so the eigenvalue problem for H is restricted to each of these spaces. Indeed, by (4.1)

$$\Delta = \frac{1}{r^2}\frac{\partial}{\partial r}\left(r^2\frac{\partial}{\partial r}\right) + \frac{1}{r^2}\Delta_{S^2}$$

and therefore by (4.7) H reduces to the operator $K_l\otimes\text{id}$ when restricted to $L^2(\mathbb{R}^+)\otimes\tilde{H}_l$ where K_l is the ordinary differential operator

$$K_l = \frac{-\hbar^2}{2\mu r^2}\left[\frac{\partial}{\partial r}\left(r^2\frac{\partial}{\partial r}\right) - l(l+1)\right] + V(r). \tag{5.2}$$

Let us denote the eigenvalues of K_l by $\lambda_{l1} \leqslant \lambda_{l2}\cdots \leqslant \lambda_{ln} \leqslant \cdots$. Then the discrete spectrum (the set of eigenvalues) of H is parametrized by the double index l, n. The multiplicity of each $\lambda_{l,n}$ will be $2l + 1$. Of course, if some of the $\lambda_{l,n}$'s coincide, there will be higher 'degeneracy'. This is precisely what happens for the case of the hydrogen atom where the potential is

$$V(r) = \frac{-e^2}{r}. \tag{5.3}$$

In that case it turns out that the eigenvalues of K_l are given by

$$\frac{-2\pi^2\mu e^4}{n^2 h^2} \quad n = l+1, l+2,\ldots \tag{5.4}$$

as we shall sketch below. Thus the eigenvalue (5.4) occurs for $l = 0, 1,\ldots, n-1$ and hence has multiplicity

$$1 + 3 + 5 + \cdots + 2n - 1 = n^2.$$

The energy levels are as in Fig. 4.1.

There is, in fact, a group theoretical explanation for these 'accidental degeneracies' involving the group $SO(4)$. We shall sketch the principal ideas in Section 4.8. Let us briefly recall here the computations leading to (5.4). Since these are standard, the reader may prefer to skip them and go directly to Section 4.6.

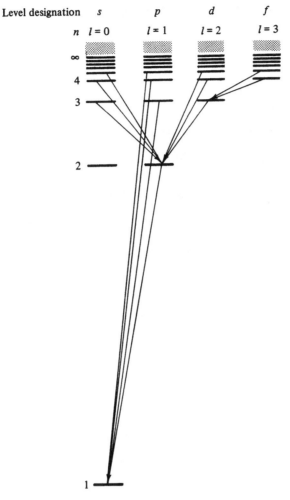

Fig. 4.1

Substituting (5.3) into (5.2) the equation

$$K_l R = ER$$

becomes

$$\frac{\mathrm{d}^2 R}{\mathrm{d}r^2} + \frac{2}{r}\frac{\mathrm{d}R}{\mathrm{d}r} + \left[\frac{8\pi^2\mu}{h^2}\left(E + \frac{e^2}{r}\right) - \frac{l(l+1)}{r^2} R\right] = 0. \tag{5.5}$$

Let us first consider the case where $E < 0$. Anticipating the final answer, let us introduce a new real parameter n by the equation

$$E = -\frac{2\pi^2\mu e^4}{n^2 h^2}$$

and a new independent variable x defined by

$$r = \frac{nh^2}{8\pi^2 \mu e^2} x.$$

Then (5.5) becomes

$$\frac{d^2R}{dx^2} + \frac{2}{x}\frac{dR}{dx} + \left(-\frac{1}{4} + \frac{n}{x} - \frac{l(l+1)}{x^2}\right)R = 0. \tag{5.6}$$

We wish to show that (5.6) implies that n is an integer $\geqslant l+1$.

If we look for a solution of the form

$$R = u(x)x^l \exp(-x/2)$$

we find that $u(x)$ must satisfy the differential equation

$$x\frac{d^2u}{dx^2} + (2l+2-x)\frac{du}{dx} + (n-l-1)u = 0. \tag{5.7}$$

If we put

$$\beta = 2l+1 \quad \text{and} \quad \alpha = n+l$$

we get the equation

$$x\frac{d^2u}{dx^2} + (\beta+1-x)\frac{du}{dx} + (\alpha-\beta)u = 0. \tag{5.8}$$

Now 0 is a regular singular point of this equation so we can look for our solution as a series

$$u = a_0 x^L + a_1 x^{L+1} + \cdots.$$

Substituting into (5.8) gives the indicial equation

$$L(L+\beta) = 0$$

so

$$L=0 \quad \text{or} \quad L=-\beta.$$

Now $\beta \geqslant 1$ so $x^{-\beta}$ is not even locally square integrable at $x=0$. So we must discard this solution and examine the case $L=0$, so

$$u = a_0 + a_1 x + \cdots.$$

Substituting into (5.8) gives the recursion formula

$$(\nu+\beta+1)(\nu+1)a_{\nu+1} = (\nu+\beta-\alpha)a_\nu$$

so that

$$\lim_{\nu \to \infty} \frac{a_{\nu+1}}{a_\nu} = \frac{1}{\nu}.$$

The series always converges, but, as $v \to \infty$, the limiting ratio of the coefficients is the same as that in the series expansion of e^x, so $(\exp -x/2)x^{\beta/2}u = R$ cannot be in $L^2(\mathbb{R}^+)$ unless

$$v + \beta - \alpha = 0$$

for some v in which case all the $a_\mu = 0$ for $\mu > v$. This shows that

$$\alpha - \beta \quad \text{must be a positive integer}$$

or that

$$n \geqslant l + 1 \text{ is an integer,}$$

as was to be proved.

By the uniqueness theorem of differential equations we know that (5.8) has exactly one polynomial solution (up to constant factor). They are known as the associated Laguerre polynomials. An easy closed form for them can be found as follows. Consider the differential equation

$$x\frac{dy}{dx} + (x - \alpha)y = 0. \tag{5.9}$$

One solution of this equation is

$$y = x^\alpha e^{-x}.$$

Define the Laguerre polynomials L_α by the equation

$$L_\alpha(x) = e^x \frac{d^\alpha}{dx^\alpha}(x^\alpha e^{-x}). \tag{5.10}$$

Differentiating (5.9) $\alpha + 1$ times gives

$$x\frac{d^2z}{dx^2} + (x + 1)\frac{dz}{dx} + (\alpha + 1)z = 0 \tag{5.11}$$

where

$$z = \frac{d^\alpha y}{dx^\alpha} = \frac{d^\alpha}{dx^\alpha}(x^\alpha e^{-x}) = e^{-x}L_\alpha(x).$$

Substituting this value of z in (5.11) gives us the differential equation for the Laguerre polynomials

$$x\frac{d^2 L_\alpha}{dx^2} + (1 - x)\frac{dL_\alpha}{dx} + \alpha L_\alpha = 0. \tag{5.12}$$

Now define the associated Laguerre polynomials L_α^β by

$$L_\alpha^\beta = \frac{d^\beta}{dx^\beta}L_\alpha.$$

Differentiating (5.12) β times with respect to x gives us exactly (5.8). So the L_α^β are exactly the polynomial solutions of (5.8) that we were looking for. Thus the solution of (5.7) is given by

$$R(r) = cx^l(\exp - x/2)L_{n+l}^{2l+1}(x); \quad x = \frac{8\pi^2\mu e^2}{nh^2}r$$

where the constant c is determined so that the corresponding wave in $\mathbb{R}^3$ have unit norm, i.e. so that

$$\int_0^\infty R(r)^2 r^2 \, dr = 1,$$

and can be determined in a straightforward manner. It is customary to write

$$a_0 = \frac{h^2}{4\pi^2\mu e^2};$$

(it is the radius of the 'first Bohr orbit', see Appendix F). Then, putting in all the numerical factors gives

$$R(r) = -\left[\left(\frac{2}{na_0}\right)^3 \frac{(n-l-1)!}{2n[(n+l)!]^3}\right]^{\frac{1}{2}}\left(\frac{2r}{na_0}\right)^l \exp\left(\frac{-r}{na_0}\right)L_{n+l}^{2l+1}\left(\frac{2r}{na_0}\right). \quad (5.13)$$

Let us now consider (5.5) with $E > 0$. In this case we make the substitutions

$$E = \frac{2\pi^2\mu e^4}{k^2h^2}$$

$$r = \frac{kh^2}{8\pi^2\mu e^2}x.$$

Equation (5.6) then becomes

$$\frac{d^2R}{dx^2} + \frac{2}{x}\frac{dR}{dx} + \left[\frac{1}{4} + \frac{k}{x} - \frac{l(l+1)}{x^2}\right]R = 0. \quad (5.14)$$

For very large values of x this is approximately

$$\frac{d^2R}{dx^2} + \frac{1}{4}R = 0$$

which has the solutions $c\exp \pm ix/2$. At the origin we expand

$$R = a_0x^L + a_1x^{L+1} + \cdots$$

and get the indicial equation

$$L(L-1) + 2L - l(l+1) = 0$$

so

$$L = l \quad \text{or} \quad L = -(l+1).$$

The solution with $L = L$ is therefore smooth at all x and behaves as a 'plane wave' at ∞. It represents a 'scattering state', corresponding to the ionization of the hydrogen

Table 30.

n	l	$R_{n,l}$
1	0	$2\left(\dfrac{Z}{a_0}\right)^{\frac{3}{2}}\exp(-\rho)$
2	0	$\dfrac{1}{2(2^{\frac{1}{2}})}\left(\dfrac{Z}{a_0}\right)^{\frac{3}{2}}(2-\rho)\exp(-\rho/2)$
2	1	$\dfrac{1}{2(6^{\frac{1}{2}})}\left(\dfrac{Z}{a_0}\right)^{\frac{3}{2}}\rho\exp(-\rho/2)$
3	0	$\dfrac{2}{81(3^{\frac{1}{2}})}\left(\dfrac{Z}{a_0}\right)^{\frac{3}{2}}(27-18\rho+2\rho^2)\exp(-\rho/3)$
3	1	$\dfrac{4}{81(6^{\frac{1}{2}})}\left(\dfrac{Z}{a_0}\right)^{\frac{3}{2}}(6\rho-\rho^2)\exp(-\rho/3)$
3	2	$\dfrac{4}{81(30^{\frac{1}{2}})}\left(\dfrac{Z}{a_0}\right)^{\frac{3}{2}}\rho^2\exp(-\rho/3)$

$\rho=(z/a_0)r; \quad n\geqslant l+1$

atom. Thus we get a continuous spectrum for all $E>0$ describing the scattering states, and negative discrete spectrum (5.4) describing the bound states.

The study of ionized atoms with only one electron such as He^+, Li^{2+} is essentially identical to the hydrogen atom. The potential has Ze^2 in place of e^2, and the formula for the reduced mass, μ, now involves the electron mass and the nuclear mass. Then energy levels are given by

$$E_Z=\frac{-2\pi^2\mu_Z Z^2 e^4}{n^2 h^2}=Z^2\frac{\mu_Z}{\mu_H}E_H; \quad \mu_Z=\frac{M_Z m}{M_Z+m}. \tag{5.15}$$

The radial wave functions are

$$\begin{aligned}
R_{n,l}(r) &= -\left[\frac{(n-l-1)!}{2n[(n+l)!]^3}\left(\frac{2Z}{na_0}\right)^3\right]^{\frac{1}{2}}\left(\frac{2Zr}{na_0}\right)^l\exp\left(\frac{-Zr}{na_0}\right)L_{n+1}^{2l+1}\left(\frac{2Zr}{na_0}\right)\\
&= -\left[\frac{4(n-l-1)!}{n^4[(n+l)!]^3}\right]^{\frac{1}{2}}\left(\frac{Z}{a_0}\right)^{\frac{3}{2}}\left(\frac{2\rho}{n}\right)^l\exp\left(-\frac{\rho}{n}\right)L_{n+l}^{2l+1}\left(\frac{2\rho}{n}\right)
\end{aligned}$$

where $\rho=(Z/a_0)r$. The first few of them are given in Table 30. For potentials which are not exactly $-k/r$, the degeneracy among the various l's is broken. Thus, if $V(r)$ is a small perturbation of $-k/r$, one expects that the energy levels might look something like Fig. 4.2.

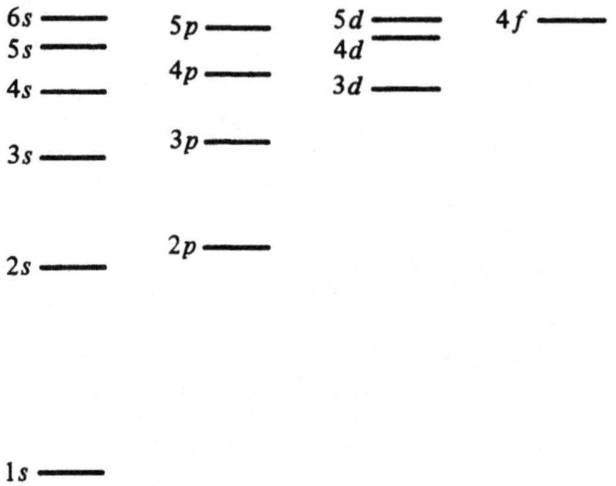

$$
\begin{array}{llll}
6s \text{ ——} & 5p \text{ ——} & 5d \text{ ——} & 4f \text{ ——} \\
5s \text{ ——} & & 4d \text{ ——} & \\
4s \text{ ——} & 4p \text{ ——} & 3d \text{ ——} & \\
\end{array}
$$

$3s$ —— $3p$ ——

$2s$ —— $2p$ ——

$1s$ ——

Fig. 4.2

If $V(r)$ is close to $-e/r$ we would expect these levels to be close to those of the hydrogen atom. Furthermore, we would expect some selection rules for the absorption spectrum. Indeed since the (dipole) operator for the absorption spectrum is a vector operator and

$$V_1 \otimes V_l = V_{l-1} \oplus V_l \oplus V_{l+1}$$

we expect only transitions with $\Delta l = \pm 1, 0$, i.e. between adjacent columns. Furthermore, $V(r)$ is invariant under inversion $x \to -x$ so we have selection rules for the group $Z_2 = \{I, -I\}$. Now the $\tilde{H}_l$ for l even come from even degree polynomials, hence correspond to the trivial representation of Z_2, while for l odd it corresponds to the non-trivial, odd representation. But V_1 is clearly odd, so the transition $\Delta l = 0$ is not possible and we must have

$$\Delta l = \pm 1.$$

We thus expect an absorption spectrum coming from transitions that look something like Fig. 4.3.

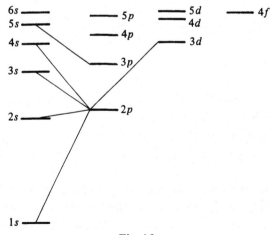

Fig. 4.3

The actual experimental energy levels have this general appearance, but show some remarkable deviation. For instance, we can consider the energy diagram for the 'outermost electron' in a sodium atom. (We will explain in the next section what this means.) The nuclear charge is $+11$ while there are 10 inner electrons. To a first approximation the outermost electron feels a charge $+1$, but this is not completely accurate as the inner electrons are distributed in space. Thus the potential is some perturbation of $-e/r$. The observed energy levels and transitions are given in Fig. 4.4.

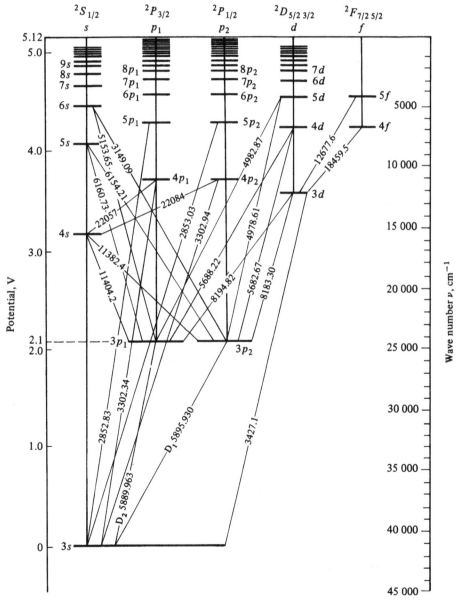

Fig. 4.4

Two things should be noticed. There are two sets of p levels instead of one. (In fact, the two $3s$–$3p$ transitions give the famous yellow D lines of sodium, which were so important in the last century, see Appendix F.) This is accounted for by the existence of the electron spin and its contribution to the energy. We shall not go into this point right now. Secondly, the energy levels start with the $3s$ state. The outermost electron cannot get into a state below $3s$. This is a consequence of the Pauli exclusion principle which we shall discuss in the next section.

4.6 The periodic table

The Schrodinger equation that we wrote down and solved in the preceding section was for the motion of the electron relative to the center of mass of the atom. Suppose that we were to try to do the same for a more complicated atom having several electrons.

The Hamiltonian is

$$H = -\frac{\hbar^2}{2m}\sum \Delta_i^2 - \sum \frac{Ze^2}{r_i} + \sum \frac{e^2}{r_{ij}} \tag{6.1}$$

where r_i is the distance of the ith electron to the center of mass which can be identified with the nucleus and r_{ij} the interelectron distances. The potential energy now involves the mutual repulsion of all the electrons. The problem of finding an exact solution to the Schrodinger equation, like the corresponding n-body problem in celestial mechanics, is completely intractable. One must resort to a suitable approximation. The approximation which is used most successfully is the so-called 'self-consistent field' method of Hartree. The idea is as follows: the true equation of motion of a single electron involves the repulsive forces of all the remaining electrons and these depend on their instantaneous positions. We try to replace these forces by a single electric field obtained by a suitable average over these instantaneous positions. We find this averaged field by a series of successive approximations. One begins by choosing as a first approximation an approximate wave function, ψ, for each of our n electrons each with some definite n, l value. Here ψ is chosen as some modification of the hydrogen like wave function. Each such ψ determines a charge density $e|\psi|^2$. Pick one electron. Compute the electrical field given by the charge distribution of all the other electrons and the nucleus. Average it with respect to $SO(3)$ so as to obtain a spherically symmetric field. This electric field then gives a potential V. Solve (5.1) for the single electron. (Of course this will also involve some approximation method in one variable.) Use this solution as a new improved wave function in computing the average field for the second electron. Continue this way to get a new improved wave function for the second electron etc. One goes through the electrons cyclically in this manner until successive iteration does not lead to any significant change in the wave functions.

One ends up with an approximate H_0 of the form

$$H_0 = -\frac{\hbar^2}{2m}\sum \Delta_i^2 + \sum V(r_i) \tag{6.2}$$

which is a sum of operators

$$\frac{\hbar^2}{2m}\Delta + V(r),\tag{6.3}$$

one for each electron.

By this method, or by cruder approximations to be described below, one obtains various wave functions ψ which are approximate eigenfunctions ψ for the single electron equation (6.3). Such wave functions are called *atomic orbitals*. Each atomic orbital will carry a label l describing to which representation of $SO(3)$ it belongs. So $l = 0, 1, 2, \ldots$ with the conventional notation s, p, d, f etc. For each l one then labels the eigenvalues in increasing order by $n = l + 1, l + 2$, etc. So the energy levels are labeled by the pair (n, l). The usual order of energies is

$$1s < 2s < 2p < 3s < 3p < 3d \sim 4s < 4p < 5s \sim 4d$$

so that one has the schematic energy diagram in Fig. 4.5.

The self-consistent field calculations of the atomic orbitals are quite difficult and frequently replaced by solutions to the central charge problem with Ze replaced by $(Z - S)e$ and where the 'shielding constant' S depends on the (n, l) label of the orbital and the number and type of other electron orbitals present. The rules for the choice of S, which are due to Slater, are empirical in nature and somewhat complicated.

Of course each energy level (n, l) occurs with multiplicity $2l + 1$. Individual orbitals within a given eigenspace can be further labeled by a basis of $\tilde{H}_l$, thus by an integer m with $-l \leqslant m \leqslant l$.

The electron is a spin $\frac{1}{2}$ particle. In the non-relativistic realm where all particles are essentially at rest (relative to the speed of light), all this means is that the ψ, instead of being a complex-valued function on $\mathbb{R}^3$, actually take values in a two-dimensional vector space (corresponding to the $\frac{1}{2}$ representation of $SU(2)$). For the present purposes, we can take this to be the *definition* of spin $\frac{1}{2}$. Thus we need an addition label, corresponding to a basis of this two-dimensional space. Usually $\frac{1}{2}$ and $-\frac{1}{2}$ or $\uparrow$ and $\downarrow$ are

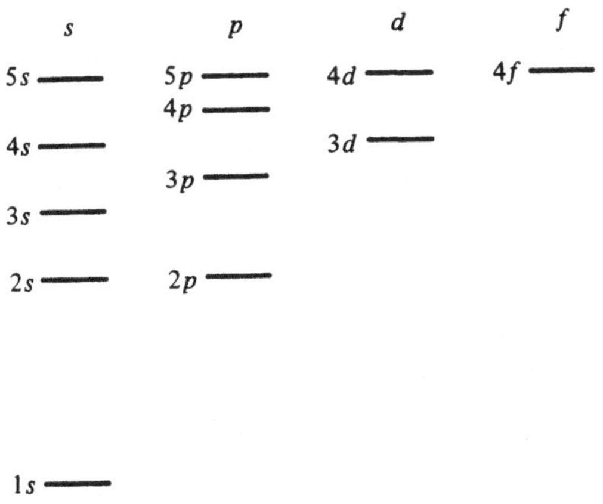

Fig. 4.5

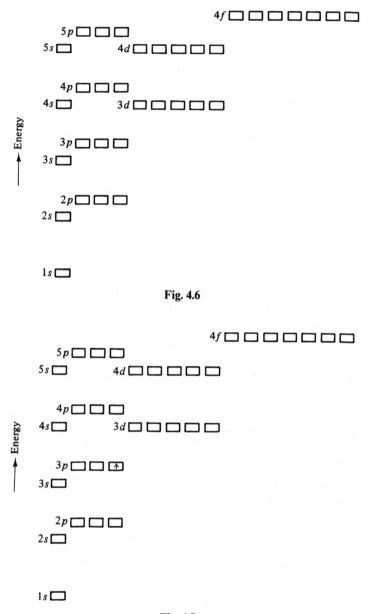

Fig. 4.6

Fig. 4.7

used. Thus, for example, we can let Fig. 4.6 denote the list of the (n, l) states up to $(5, p)$ or $(4, f)$ with the $2l + 1$ multiplicity exhibited and Fig. 4.7 denote the state ψ given by the third basis element of $\tilde{H}_l$ in the $3p$ level with the spin value $\uparrow$.

So far we have been considering the states of a single electron. The states the n-electron system as a whole will be given by elements in the (completed) tensor product space

$$H_1 \otimes \cdots \otimes H_n$$

where H_i is the space of all states of the *i*th electron. Of course, since the electrons are indistinguishable – all of these Hs are isomorphic – this is of the form

$$H \otimes H \otimes \cdots \otimes H$$

where H is the single electron state space.

Now the symmetric group S_n acts on the space $H \otimes \cdots \otimes H$ (*n* times) by interchanging the order in the tensor product. For example

$$(12)(\psi \otimes \varphi) = \varphi \otimes \psi$$

illustrates the action of S_2 on $H \otimes H$. A detailed study of this action will be given at the beginning of Chapter 5. We discuss there how the *n*-fold tensor product decomposes into irreducibles under S_n. The Schrodinger equation is clearly invariant under interchanges of the electrons, i.e. under the action of S_n. Therefore its eigenspaces are made up of (irreducible) invariant subspaces of S_n. There are two invariant subspaces of $H \otimes \cdots \otimes H$ which are particularly easy to describe. One is the space $S^n(H)$ of totally symmetric elements, the ones which transform according to the trivial representation of S_n, i.e. those which satisfy

$$\pi t = t \quad \text{for all} \quad \pi \in S_n.$$

For example, in $H \otimes H$ all linear combinations of elements of the form

$$\varphi \otimes \psi + \psi \otimes \varphi$$

give $S^2(H)$.

The other is the space of $\wedge^n(H)$ of totally anti-symmetric elements, the ones which transform according to the sign (sgn) representation of S_n, i.e. the elements which satisfy

$$\sigma t = (\text{sgn } \sigma)t.$$

For $n = 2$, $\wedge^2(H)$ will be spanned by elements of the form

$$\psi \otimes \varphi - \varphi \otimes \psi.$$

When $n > 2$ there will be other representations of S_n and, correspondingly, other types of subspaces of $H \otimes \cdots \otimes H$. We shall describe them all in Chapter 5.

We can now state one of the most remarkable facts in (relativistic) quantum mechanics, the 'spin statistics theorem'. It asserts that if we are given a system of *n* identical particles of spin *s*, the only allowable states in $H \otimes \cdots \otimes H$ are the totally symmetric states $S^n(H)$, if $s = 0, 1, 2, \ldots$ and the totally anti-symmetric states, $\wedge^n(H)$ if $s = \frac{1}{2}, \frac{3}{2}, \ldots$. The integer spin particles, which obey the symmetric 'statistics' are called *bosons* while the half integer spin particles obeying the anti-symmetric statistics are called *fermions*.

The electrons are spin $\frac{1}{2}$ particles and hence are fermions. This means, for example, that $\psi \otimes \psi$ is not an admissible state, or for that matter neither is $\psi \otimes \psi \otimes \varphi_3 \otimes \cdots \varphi_n$. A repeated ψ in a tensor product will give zero when anti-symmetrized. This is the famous *Pauli exclusion principle*: two electrons cannot occupy the same state.

Suppose that $\psi_1, \ldots, \psi_n$ are atomic orbitals. Then their anti-symmetric tensor

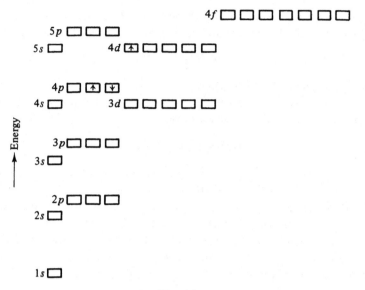

Fig. 4.8

product, $\psi_1 \wedge \cdots \wedge \psi_n$ is a eigenvector for the hamiltonian of the system of all electrons of the atom. The corresponding energy level is the sum of the levels of the ψ_i. We can let a typical such element be given by putting ↑ or ↓ in the appropriate boxes. Thus, Fig. 4.8 might be used to describe an electronic state of a three-electron atom (lithium). Its energy would be $2E(4p) + E(4d)$, where $E(n, l)$ denotes the energy of the orbitals (n, l). The Pauli exclusion principle says that each box can contain at most two arrows, one ↑ and one ↓. Thus a lowest energy state, the 'ground state' for lithium would be obtained by putting two arrows in the $1s$ box and one in the $2s$ box. This would correspond to the totally anti-symmetric state $\psi(1s, \uparrow) \wedge \psi(1, s, \downarrow) \wedge \psi(2, s, \uparrow)$. This state is labeled $(1s)^2(2s)$ and can be indicated diagrammatically as in Fig. 4.9. The ground state energy is then (approximately) $2E(1s) + E(2s)$, where $E(n, l)$ is the energy of the orbital. In this approximation the states of nitrogen shown in Fig. 4.10 would have the same energy levels. That is, the operator H_0 has a certain (20-fold) degeneracy. Explicitly, the space in question is

$$\wedge^2(V_0 \otimes V_{\frac{1}{2}}) \otimes \wedge^2(V_0 \otimes V_{\frac{1}{2}}) \otimes \wedge^3(V_1 \otimes V_{\frac{1}{2}}).$$

Here the $V_{\frac{1}{2}}$ is the spin space, the first V_0 is the one-dimensional space of $1s$, the second V_0 of $2s$ and the V_1 is the three-dimensional space of $2p$ states. Since $\wedge^2$ of a two-dimensional space is one dimensional (and here with the trivial representation) all the multiplicity comes from $\wedge^3(V_1 \otimes V_{\frac{1}{2}})$ which has dimension $(6 \cdot 5 \cdot 4)/(1 \cdot 2 \cdot 3) = 20$. (The reader can check that there are 20 ways of distributing three arrows among three boxes with the constraint that no two ↑ or ↓ lie in the same box.)

We shall describe how this degeneracy is removed – by using a more exact Hamiltonian – at the end of this section. It turns out that the chemical properties of nitrogen are, in the main, determined by the eigenvectors in this 20-dimensional space under the more refined Hamiltonians.

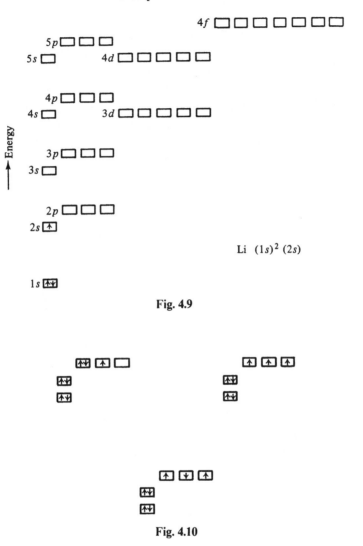

Li $(1s)^2 (2s)$

Fig. 4.9

Fig. 4.10

In any event, we can now see how group theory enters into chemistry by describing the lowest energy states of atoms: the chemical properties of an atom are determined by its electronic configuration.

The electron orbitals with a given value of n are said to form a *shell*. The reason for this terminology is that the 'radius' of such an orbital (in the sense of the highest density of $|\psi|^2$, for example) increases with n, as can be seen from Table 30. There is also a slight l dependence of this radius, and orbitals with a given value of n and l are said to form a *subshell*. For example, hydrogen's ground state is $1s$, just one electron in the $n = 1$ shell. Helium's ground configuration is $(1s)^2$, its $n = 1$ shell is filled. Lithium with $(1s)^2 2s$ has one electron in the $n = 2$ shell. The second shell, with $8 = (2 \cdot 1 + 2 \cdot 3)$ are successively filled until we reach neon with $(1s)^2(2s)^2(3s)^6$ for its ground state. Sodium, Na, with its configuration $(1s)^2(2s)^2(2p)^6(3s)$ has its first two shells filled and one 'electron' in the outer shell. This electron is attached by a charge of about $+ e$, the nuclear charge $+ 11e$ 'screened' by the ten electrons in the $n = 1$ and $n = 2$ shells. Thus the chemical properties of sodium are very similar to those of lithium. Similarly Argon, ^{18}Ar, has its $3s$ and $3p$ subshells filled, and its properties are very similar to those of neon. The point is that chemical behavior is determined by what happens in the outermost shell, or sometimes outside an s, p subshell. This accounts for the periodicity of the elements, which had been discovered in 1868 by Mendeleev some 60 years before its quantum mechanical explanation. We will not enter further into the details of the periodic table, which are presented in any elementary chemistry text. The point is that the 'magic numbers'

$$2, 6 = 2 \cdot 3, \quad 10 = 2 \cdot 5, \quad 14 = 2 \cdot 7$$

that enter all come from the representations of the group $SO(3)$.

We now explain briefly how group theory enters into the determination of the chemical properties of nitrogen. This involves studying the break up of the 20-fold degeneracy under the more refined Hamiltonian.

We may consider $H - H_0$ as a 'perturbation' where H is given by (6.1), H_0 by (6.2). Then apply perturbation theory to get a better approximation to values of H. This involves finding the eigenvalues of $T = P(H - H_0)P$, where P denotes projection of the Hilbert space of all states onto the 20-dimensional space we are studying. Now H commutes with $SO(3)$. Also, it does not involve the spins so commutes with the $SU(2)$ acting on $V_{\frac{1}{2}}$. Thus T commutes with $SO(3) \times SU(2)$ acting on $\wedge^3(V_1 \otimes V_{\frac{1}{2}})$. We claim that under $SO(3) \times SU(2)$ we have the decomposition

$$\wedge^3(V_1 \otimes V_{\frac{1}{2}}) = (V_0 \otimes V_{\frac{3}{2}}) \oplus (V_1 \otimes V_{\frac{1}{2}}) \oplus (V_2 \otimes V_{\frac{1}{2}})$$

and hence T has three (possibly) distinct eigenvalues.

The general method of decomposing tensor products under the symmetric group will be described in Chapter 5. Here we will illustrate the procedure 'by hand'. Let V and W be vector spaces. We may identify $(V \otimes W) \otimes (V \otimes W) \otimes (V \otimes W)$ with

$$(V \otimes V \otimes V) \otimes (W \otimes W \otimes W).$$

We are looking for completely anti-symmetric tensors in this space. One way of finding such a tensor is to take a tensor in $V \otimes V \otimes V$ which is completely anti-symmetric, and multiply it by a tensor in $W \otimes W \otimes W$ which is completely symmetric. The linear span of such tensors is just $\wedge^3(V) \otimes S^3(W)$. In other words,

$$\wedge^3(V) \otimes S^3(W) \subset \wedge^3(V \otimes W)$$

and is clearly an invariant subspace. In our present example, with $V = V_1$ and $W = V_{\frac{1}{2}}$, $\wedge^3(V) = \wedge^3(V_1)$ is clearly the trivial representation of $SO(3)$, i.e. $\wedge^3(V_1) = V_0$. If we identify $V_{\frac{1}{2}}$ as the space of linear polynomials on $\mathbb{C}^2$, then $S^3(V_{\frac{1}{2}})$ is clearly the space of cubic polynomials, so

$$\wedge^3(V_1) \otimes S^3(V_{\frac{1}{2}}) \sim V_0 \otimes V_{\frac{3}{2}}$$

as a representation space of $SO(3) \times SU(2)$. This space is four dimensional. Now *in general* we should also consider the space $S^3(V) \otimes \wedge^3(W)$. In our case, with $W = V_{\frac{1}{2}}$ two dimensional, $\wedge^3(V_{\frac{1}{2}}) = \{0\}$ so this space does not make its appearance. We must still look for a 16-dimensional complement to the four-dimensional subspace $V_0 \otimes V_{\frac{3}{2}}$ of $\wedge^3(V_1 \otimes V_{\frac{1}{2}})$.

Let e and f be elements of V and let u and v be elements of W. Consider the element

$$t = e \otimes e \otimes f \otimes u \otimes v \otimes u.$$

Let us completely anti-symmetrize t – that is, construct the element

$$A(t) = \frac{1}{6} \sum_{\pi \in S_3} (\text{sgn } \pi)(\pi t).$$

Explicitly,

$$6A(t) = e \otimes e \otimes f \otimes u \otimes v \otimes u + f \otimes e \otimes e \otimes u \otimes u \otimes v$$
$$+ e \otimes f \otimes e \otimes v \otimes u \otimes u - e \otimes e \otimes f \otimes v \otimes u \otimes u$$
$$- f \otimes e \otimes e \otimes u \otimes v \otimes u - e \otimes f \otimes e \otimes u \otimes u \otimes v.$$

Clearly $A(t)$ is completely anti-symmetric. Suppose we take e to be the eigenvector of

$$R^z_{2\theta} = \tau(U) = \tau\left(\begin{pmatrix} e^{i\theta} & 0 \\ 0 & e^{-i\theta} \end{pmatrix} \right)$$

with eigenvalue $\exp 2i\theta$ – the 'maximal eigenvalue' in V_1 – and take f to be the eigenvector with eigenvalue 1. Similarly, take $u = \uparrow$ the 'maximal eigenvector' in $V_{\frac{1}{2}}$ and $v = \downarrow$, the minimal eigenvector. Then the element t – and hence also $A(t)$ – is an eigenvalue of

$$R^z_{2\theta} \times U_\phi = \tau\left(\begin{pmatrix} e^{i\theta} & 0 \\ 0 & e^{-i\theta} \end{pmatrix} \right) \times \begin{pmatrix} e^{i\phi} & 0 \\ 0 & e^{-i\phi} \end{pmatrix}$$

with eigenvalue $\exp 4i\theta \cdot \exp i\phi$. As far as the θ dependence is concerned, this 4 is the highest exponent we can achieve. (We might try to take $e = f = $ maximal eigenvector for $R^z_{2\theta}$. Then t would be an eigenvector with eigenvalue $\exp 6i\theta \cdot \exp i\phi$. But because t is completely symmetric in its V_1 component, when we apply the anti-symmetrization, A to t, the effect is to completely anti-symmetrize the $V_{\frac{1}{2}}$ component:

$$t = e \otimes e \otimes e \otimes A(u \otimes v \otimes u) \in S^3(V_1) \otimes \wedge^3(V_{\frac{1}{2}}).$$

But since $V_{\frac{1}{2}}$ is two dimensional, $At = 0$.)

Thus, as far as the $SO(3)$ factor is concerned, the space $\wedge^3(V_1 \otimes V_{\frac{1}{2}})$ contains at least one copy of V_2. As far as the $SU(2)$ (the second factor in our group $SO(3) \times SU(2)$) is concerned, we have

$$V_{\frac{1}{2}} \otimes V_{\frac{1}{2}} = V_1 + V_0$$

hence

$$V_{\frac{1}{2}} \otimes V_{\frac{1}{2}} \otimes V_{\frac{1}{2}} = V_{\frac{3}{2}} + V_{\frac{1}{2}} + V_{\frac{1}{2}}.$$

The $V_{\frac{3}{2}} = S^3(V_{\frac{1}{2}})$ is completely symmetric and hence only occurs as a factor in $\wedge^3(V_1) \otimes S^3(V_{\frac{1}{2}})$. Thus, the space spanned by all linear combinations of $\rho(a \times b)A(t)$

must be isomorphic to
$$V_2 \otimes V_{\frac{1}{2}}$$
under $SO(3) \times SU(2)$. This space is $10 = 5 \times 2$ dimensional. Now consider the following two possibilities:

(a) $e_1 = $ 'maximal eigenvector' with eigenvalue $\exp 2i\theta$

         $f_1 = $ 'minimal eigenvector' with eigenvalue $\exp - 2i\theta$

and

(b) $e_2 = $ 'intermediate eigenvector' with eigenvalue 1

         $f_2 = $ 'maximal eigenvector' with eigenvalue $\exp 2i\theta$.

In both cases, $e \otimes e \otimes f$ has eigenvalue $\exp 2i\theta$ and clearly $A(t_1)$ and $A(t_2)$ are linearly independent for $u = \uparrow$ and $v = \downarrow$ and also for $u = \downarrow$ and $v = \uparrow$. More precisely, taking

$$t_1 = e_1 \otimes e_1 \otimes f_1 \otimes \uparrow \otimes \downarrow \otimes \uparrow$$
$$t_2 = e_2 \otimes e_2 \otimes f_2 \otimes \uparrow \otimes \downarrow \otimes \uparrow.$$

The elements $A(t_1)$ and $A(t_2)$ span a two-dimensional space of eigenvectors for $R_{2\theta}^z \times U_\phi$ with eigenvalue $\exp 2i\theta \cdot \exp i\phi$ while for

$$t_3 = e_1 \otimes e_1 \otimes f_1 \otimes \downarrow \otimes \uparrow \otimes \downarrow$$
$$t_4 = e_2 \otimes e_2 \otimes f_2 \otimes \uparrow \otimes \downarrow \otimes \uparrow$$

the elements $A(t_3)$ and $A(t_4)$ span a two-dimensional space with eigenvalue $\exp 2i\theta \cdot \exp - i\phi$. Now the space V_2 contains only a one-dimensional space of eigenvectors of any given eigenvalue, so $V_2 \otimes V_{\frac{1}{2}}$ also contains only a one-dimensional space of eigenvectors with eigenvalue $\exp 2i\theta \cdot \exp i\phi$. Thus, in addition to $V_2 \otimes V_{\frac{1}{2}}$, our 16-dimensional space must contain some other space of the form $V_p \otimes V_s$, where $p \geqslant 1$ and $s \geqslant \frac{1}{2}$; since $V_1 \otimes V_{\frac{1}{2}}$ is already six dimensional, we must have $p = 1$ and $s = \frac{1}{2}$. We have thus proved that

$$\wedge^3 (V_1 \otimes V_{\frac{1}{2}}) = V_0 \otimes V_{\frac{3}{2}} + V_2 \otimes V_{\frac{1}{2}} + V_1 \otimes V_{\frac{1}{2}}$$

under the group $SO(3) \times SU(2)$. We would therefore expect that the 20-fold degeneracy in $\wedge^3 (V_1 \otimes V_{\frac{1}{2}})$ should split into a four-fold, ten-fold, and six-fold degeneracy when we pass from H_0 to H.

So far, we have been considering the 'spin' of the electron as merely modifying the multiplicity of the eigenvalues of H, and affecting the Pauli exclusion principle. We have not taken into account that there might be some 'spin dependent forces'.

This approximation is known as the Russel–Saunders (or L–S) approximation. For the case we are studying, nitrogen, it turns out that in this approximation the

$$V_0 \otimes V_{\frac{3}{2}} = \wedge^3 (V_1) \otimes S^3 (V_{\frac{1}{2}})$$

space has the lowest energy.

More generally, the L–S approximation can be used to give a theoretical explanation for *Hund's rules* (1925) in spectroscopy which say that:

(a) in the decomposition under $SO(3) \times SU(2)$, the lowest energy state will be one for which the $SU(2)$ multiplicity is largest;

(b) among components with a given $SU(2)$ multiplicity, the subspace with a maximal $SO(3)$ multiplicity will have the smallest energy level.

To a large degree, it is the 'unpaired' electrons in the outer shell which determine the chemical properties of the element. Thus, for nitrogen, we might use the symbol

to indicate the element of $e \wedge f \wedge g \otimes \uparrow \otimes \uparrow \otimes \uparrow$ of $\wedge^3(V_1) \otimes S^3(V_{\frac{1}{2}}) \subseteq \wedge^3(V_1 \otimes V_{\frac{1}{2}})$ (where e, f and g is a basis of V_1). Thus the subspace $\wedge^3(V_1) \otimes S^3(V_{\frac{1}{2}})$ is sometimes specified in more elementary chemistry books by saying that (we quote from Coulson)

'(i) electrons tend to avoid being in the same space orbit, or cell, so far as possible, and

(ii) two electrons each singly occupying a given pair of equivalent orbitals (e.g., $2p_x, 2p_y$) tend to have their spins parallel in states of lowest energy.'

These rules show us at once that in nitrogen the $1s$ and $2s$ orbitals are doubly filled, and each of the $2p_x, 2p_y, 2p_z$ orbitals singly filled, the three electrons concerned having parallel spins; and that oxygen, with one more p-electron, has one p-orbital (say $2p_z$) doubly filled, the other two being singly occupied, with parallel spins, as shown. (See Fig. 4.11.) It is the two 'unpaired electrons' in oxygen that are largely responsible for its chemical valence. Our computations above show that the valence of nitrogen is three.

The L–S approximation is useful for the spectra of the lighter elements and for predicting chemical properties. More generally, one must take into account the effects of the electron spin on the Hamiltonian. The spinning electron, if thought of as a spinning extended body of electric charge, might be expected to have a magnetic moment. This magnetic moment would then interact with the electromagnetic field of the rest of the atom. This is known as spin–orbit coupling. If we include these forces, and so modify H, then our new Hamiltonian will no longer be invariant under the full group $SO(3) \times SU(2)$, for we can no longer 'rotate' the electron (about its center) independently of the rest of the atom without changing the energy. We can still apply an element of the form $\tau(A) \times A$ ($A \in SU(2)$) without changing the new Hamiltonian.

Thus, when including the spin dependent forces, we will apply a perturbation which is only invariant under the diagonal $SU(2)$ subgroup of $SO(3) \times SU(2)$ – that is, the

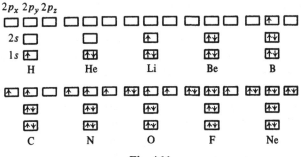

Fig. 4.11

subgroup consisting of all elements of the form $\tau(A) \times A$. Under this subgroup, the Clebsch–Gordan formula tells us that

$$V_0 \otimes V_{\frac{3}{2}} = V_{\frac{3}{2}}$$
$$V_1 \otimes V_{\frac{1}{2}} = V_{\frac{3}{2}} \oplus V_{\frac{1}{2}}$$

and

$$V_2 \otimes V_{\frac{1}{2}} = V_{\frac{5}{2}} \oplus V_{\frac{3}{2}}$$

so

$$\wedge^3(V_1 \otimes V_{\frac{1}{2}}) = V_{\frac{1}{2}} \oplus 3V_{\frac{3}{2}} \oplus V_{\frac{5}{2}},$$

under $SU(2)$. Then we would have to evaluate the perturbation matrix. Finding its eigenvalues will involve a third-degree polynomial. This method was developed by Slater in 1929. We refer the reader to his papers, or to the book by Condon and Shortley for the details.

This spin–orbit coupling becomes very important in the shell model of the nucleus which we briefly describe in the next section.

4.7 The shell model of the nucleus

By the 1920s it was recognized that the charge of an atomic nucleus was given by $Q = Ze$, where e is the charge of the electron and where Z, the atomic number, gives the position of the atom in the periodic table, a position which had been determined earlier by purely chemical properties. A given chemical element could exist in forms having differing atomic weights, forms which were called isotopes, and which could be separated from one another by various techniques. The mass of the nucleus of a given isotope was found to be approximately of the form

$$M = Am_p$$

where m_p denotes the mass of the proton. Thus the ground state of a given isotope is characterized by two integers, Z and A. Originally, it was thought that a nucleus consists of A protons and $A - Z$ electrons so that the total residual charge is Z. But this led to contradictions – principally with the observed spin values. After the discovery of the neutron by Chadwick in 1932, it became clear that the nucleus consists of Z protons and $N = A - Z$ neutrons. An isotope of some element – that is, a nuclear species with a given value of N and Z – is called a *nuclide*.

However, not all (N, Z) values are observed as neutron and proton numbers of nuclides. For lighter nuclei, the values of N and Z are approximately equal. For heavier nuclei, N tends to be larger than Z.

If we plot the relative abundance of nuclides, we see an uneven distribution. Thus the relative abundance of nuclides with N and Z both even are plotted in Fig. 4.12. There are peaks at N or Z values 50, 82, 126.

The *binding energy* of a nucleus is defined as follows. Consider a nucleus with Z protons and N neutrons. The total mass of the nucleus is slightly smaller than the sum

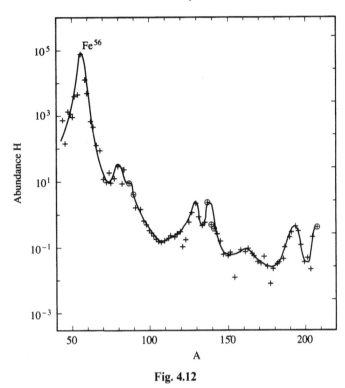

Fig. 4.12

of the masses of the constituents because of the binding energy, B, which holds the nucleus together. Using $E = mc^2$ we can convert this energy into a mass so that

$$\frac{B}{c^2} = Zm_p + Nm_n - m(Z, N)$$

where $m(Z, N)$ denotes the mass of the nuclide. Here $B = B(Z, N)$ is positive and represents the energy that is required to disintegrate the nucleus into its constituent protons and neutrons. Suppose, for example, that a neutron is removed from a nuclide (Z, N) so as to obtain a nuclide $(Z, N - 1)$. Then

$$S_n(Z, N) = B(Z, N) - B(Z, N - 1)$$

is the *separation energy* – the energy needed to remove one neutron from the (Z, N) nuclide.

Fig. 4.13 shows the separation energies of the isotopes of cerium, $Z = 58$. It is clear that the neutrons are more tightly bound for N even than for N odd. (This is part of a more general fact that even–even nuclei are more stable and tend to have spin zero in their ground states. This shows that two like particles tend to pair off to zero angular momentum and that there is a 'pairing interaction'.) If we consider only the N even nuclei, we see a break at $N = 82$. Neutrons after $N = 82$ are less tightly bound. Similar breaks (for other elements) are observed at the 'magic numbers'

$$2, \ 8, \ 20, \ 28, \ 50, \ 82, \ 126.$$

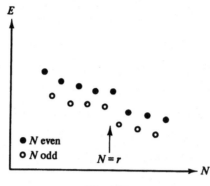

Fig. 4.13

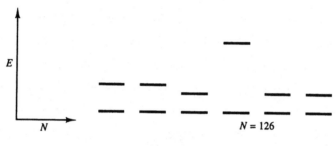

Fig. 4.14

There is additional evidence supporting the stability at these magic numbers, and that comes from the energy gap between the ground state and the first excited state. A large energy gap implies that it is hard to excite the nucleus to its first excited state. Presumably, this means extra stability in the sense that it will be hard to excite the nucleus to higher states – in particular, enough to dislodge a neutron. Fig. 4.14 shows the energy gaps of different isotopes of lead. Notice the extra large gap at $N = 126$.

The first function of the shell model of the nucleus is to explain the magic numbers. As in the study of the atomic spectrum, one begins with a self-consistent field – independent particle model. That is, we assume that each neutron or each proton responds (in the center of mass frame) to a potential $V(r)$, where V is some function obtained by averaging the nuclear forces. Now in our atomic model, we began with the assumption that for a simple electron, the potential was of the $1/r$ type. For nuclear forces, we know nothing about the 'single nucleon' potential other than it must be of quite short range, and strong enough to overcome the electric repulsion between protons. As a reasonable guess, we can expect that if the attraction between nucleons is of short range, then the smeared (self-consistent field) potential will be similar in form to the nuclear density. We might expect its shape to be similar to that shown in Fig. 4.15. Suppose we approximate $V(r)$ by a quadratic potential. Then the corresponding one-body quantum mechanical problems can be solved explicitly: the eigenvalues are evenly spaced and the kth level corresponds to $S^k(\mathbb{C}^3)$ and hence has multiplicity

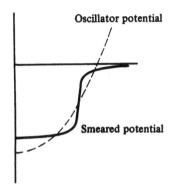

Oscillator potential

Smeared potential

Fig. 4.15

Table 31.

Level	Energy	Dimension of eigenspace
0	$\frac{3}{2}\omega$	1
1	$\frac{5}{2}\omega$	3
2	$\frac{7}{2}\omega$	6
3	$\frac{9}{2}\omega$	10

Table 32.

Degeneracy of kth level	Total number of levels
2	2
6	8
12	20
20	40
30	70
42	112
56	168

$(k + 1)(k + 2)/2$. Thus we get the multiplicities shown in Table 31. If we take into account that the nucleons have spin $\frac{1}{2}$, we have to double each of the number of multiplicities. Thus we have Table 32. Notice that the first three numbers on the right are the magic numbers; afterwards, they are not.

To remedy this, we might replace the harmonic oscillator potential by a more realistic potential – one with less degeneracy. We know from Section 4.5 that in terms of $SO(3)$, we have

$$S^{2k}(\mathbb{C}^3) = V_0 \oplus V_2 \oplus \cdots \oplus V_{2k}$$

and

$$S^{2k+1}(\mathbb{C}^3) = V_1 \oplus V_3 \oplus \cdots \oplus V_{2k+1}.$$

Now the higher angular momenta wave functions can be expected to be concentrated at larger radii. So if the potential we use lies below the harmonic oscillator potential for these values of r, we expect the eigenvalues to split as in Table 33.

We have not helped matters much if we simply double the multiplicities. In fact, we have made matters somewhat worse by introducing additional numbers such as 18, 34, etc.

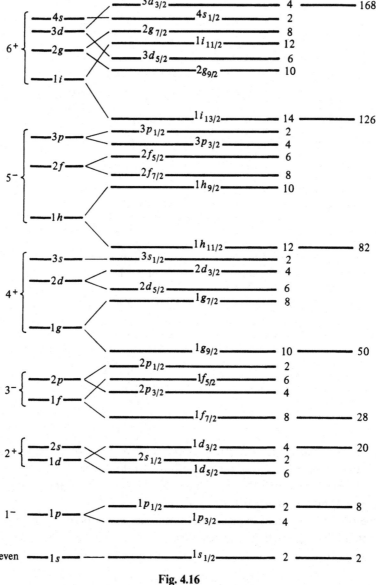

Fig. 4.16

Table 33.

Oscillator levels	$V(r)$ levels	Multiplicities	$2 \times$ levels up to this point
4	3p	1	70
	2d	5	68
	1g	9	58
3	2p	3	40
	1f	7	34
2	2s	1	20
	1d	5	18
1	1p	3	8
0	1s	1	2

All of this is posited on the assumption that there is no spin–orbit interaction; that the only function of the spin of the nucleon is to double the number of energy levels. Now let us assume that there is a spin–orbit interaction. Then each $V_l \otimes V_{\frac{1}{2}} = V_{l+\frac{1}{2}} + V_{l-\frac{1}{2}}$ as a representation of $SU(2)$. Let us assume that for given l, the energy of $V_{l+\frac{1}{2}}$ is lower than that for $V_{l-\frac{1}{2}}$ and that this difference becomes more pronounced as l increases. We can then expect energy levels as in Fig. 4.16, thus accounting for the magic numbers. Thus, the magic numbers tell us that the spin–orbit interaction is an important component of the nuclear force.

4.8   The Clebsch–Gordan coefficients and isospin

We know from Section 4.3 how $V_{s_1} \otimes V_{s_2}$ decomposes into irreducibles. For many applications we shall need more refined information: suppose we have chosen a standard orthonormal basis for V_s. In particular, we will denote our basis as

$$|s, s\rangle, |s, s-1\rangle, \ldots |s, -s\rangle$$

where $|s, j\rangle$ is an eigenvector of

$$\rho_s \begin{pmatrix} e^{i\theta} & 0 \\ 0 & e^{-i\theta} \end{pmatrix}$$

with eigenvalue $\exp 2ij\theta$. (We shall, in fact, choose a specific model for V_s and specific orthonormal basis in this model consisting of eigenvectors of

$$\rho_s\!\left(\!\begin{pmatrix} e^{i\theta} & 0 \\ 0 & e^{-i\theta} \end{pmatrix}\!\right)$$

If we are given any other representation (σ, W) equivalent to (ρ_s, W_s), then the image of $|s, s\rangle$ in W under any equivalence is determined up to a phase factor by requiring it to be of unit length. Fixing this phase factor then determines an orthonormal basis in W. Thus, for example, in the right-hand side of

$$V_1 \otimes V_1 = V_0 + V_1 + V_2,$$

fixing three phase factors, one each for V_0, V_1 and V_2, determines an orthonormal basis of the right-hand side. On the other hand, the orthonormal basis for each of the factors V_1 determines an orthonormal basis of the left-hand side and we can ask for the matrix which gives the change of basis.

Let us illustrate this computation for the case of $V_{\frac{1}{2}}$ and V_1, together with some mathematical consequences, before we describe the general case. We shall use some alternative notation whose physical meaning we will explain. But for the moment, everything we do is strictly mathematics. So, in addition to the $|s, j\rangle$ nomenclature, let us introduce names for the basis elements with $s = \frac{1}{2}$ and $s = 1$ as

$$\mathrm{p} = |\tfrac{1}{2}, \tfrac{1}{2}\rangle$$
$$\mathrm{n} = |\tfrac{1}{2}, -\tfrac{1}{2}\rangle$$

and

$$\pi^+ = |1, 1\rangle$$
$$\pi^0 = |1, 0\rangle$$
$$\pi^- = |1, -1\rangle$$

so that p, n form a basis $V_{\frac{1}{2}}$ and π^+, π^0, π^- form a basis of V_1. Thus

$$\mathrm{p} \otimes \pi^+, \mathrm{p} \otimes \pi^0, \mathrm{p} \otimes \pi^-, \mathrm{n} \otimes \pi^+, \mathrm{n} \otimes \pi^0, \mathrm{n} \otimes \pi^-$$

form an orthonormal basis of $V = V_{\frac{1}{2}} \otimes V_1$. For any linear operator A on $V = V_{\frac{1}{2}} \otimes V_1$ we can compute its matrix elements with respect to this basis, which we shall denote by the following notation: $A_{u \leftarrow v} = (Av, u)$ for any basis elements u, v, etc., where $(,)$ denotes the scalar product; in other words, $A_{u \leftarrow v}$ is the coefficient of u in the basis expansion of Av. So, for example,

$$A_{\mathrm{p} \otimes \pi^- \leftarrow \mathrm{p} \otimes \pi^0} = (A(\mathrm{p} \otimes \pi^0), \mathrm{p} \otimes \pi^-) \, \text{etc.}$$

Now if A is an $SU(2)$ morphism, then these 36 coefficients depend really on only two parameters. To see this explicitly, we must pass from the above basis to a basis adapted to the direct sum decomposition $V = V_{\frac{3}{2}} \oplus V_{\frac{1}{2}}$. Once we have chosen such a basis, we let C denote the change of basis matrix so that $A = CBC^{-1}$, where, by Schur's lemma, the 6×6 matrix B is of the form

$$\begin{pmatrix} \lambda I_4 & 0 \\ 0 & \mu I_2 \end{pmatrix},$$

where I_4 is the 4×4 identity matrix and I_2 is the 2×2 identity matrix. We can choose the new basis as follows. Notice that $\mathrm{p} \otimes \pi^+$ is an eigenvector of $\rho(U_\theta)$ with eigenvalue $\exp 3i\theta$. Hence it must lie in the subspace equivalent to $V_{\frac{3}{2}}$ and, under any identification of this subspace with our standard space, $V_{\frac{3}{2}}$ must be some multiple of the element $|\frac{3}{2}, \frac{3}{2}\rangle$. Let us choose our identification so that $\mathrm{p} \otimes \pi^+$ is in fact identified with this element. (At this point we have fixed one arbitrary phase factor.) This then fixes a basis for the subspace (isomorphic to) $V_{\frac{3}{2}}$. It is not hard to check that the element $\mathrm{p} \otimes \pi^0$ does not belong to the subspace $V_{\frac{3}{2}}$. It therefore has non-trivial projection onto the subspace $V_{\frac{1}{2}}$. On the other hand, $\mathrm{p} \otimes \pi^0$ is an eigenvector of $\rho(U_\theta)$ with eigenvalue $\exp i\theta$; and hence each of the projections of $\mathrm{p} \otimes \pi^0$ onto the subspaces $V_{\frac{3}{2}}$ and $V_{\frac{1}{2}}$ must be eigenvectors with the same eigenvalue (since the projections onto each component are $SU(2)$ morphisms). Thus the projection of $\mathrm{p} \otimes \pi^0$ onto $V_{\frac{1}{2}}$ must be a multiple of $|\frac{1}{2}, \frac{1}{2}\rangle$. We will fix our identification of this subspace with our standard $V_{\frac{1}{2}}$ by requiring this multiple to be positive. Thus

$$\mathrm{p} \otimes \pi^0 = x|\tfrac{3}{2}, \tfrac{1}{2}\rangle + y|\tfrac{1}{2}, \tfrac{1}{2}\rangle$$

and we fix a basis of $V_{\frac{1}{2}}$ by requiring y to be positive. We now have a new basis of V consisting of

$$|\tfrac{3}{2}, \tfrac{3}{2}\rangle, |\tfrac{3}{2}, \tfrac{1}{2}\rangle, |\tfrac{3}{2}, -\tfrac{1}{2}\rangle, |\tfrac{3}{2}, -\tfrac{3}{2}\rangle, |\tfrac{1}{2}, \tfrac{1}{2}\rangle, |\tfrac{1}{2}, -\tfrac{1}{2}\rangle.$$

The first problem is to express the old basis in terms of the new basis. Thus, for example,

$$\mathrm{p} \otimes \pi^+ = |\tfrac{3}{2}, \tfrac{3}{2}\rangle$$
$$\mathrm{n} \otimes \pi^- = |\tfrac{3}{2}, -\tfrac{3}{2}\rangle$$

and

$$\mathrm{n} \otimes \pi^+ = (\tfrac{1}{3})^{\frac{1}{2}}|\tfrac{3}{2}, \tfrac{1}{2}\rangle + (\tfrac{2}{3})^{\frac{1}{2}}|\tfrac{1}{2}, \tfrac{1}{2}\rangle.$$

In general, it is clear from the behavior of the elements on the left under $\rho(U_\theta)$ that there can be at most two non-zero coefficients on the right-hand side, and the problem is to compute these coefficients. Another way of putting the problem is that we are being asked to compute values of various scalar products such as $(\mathrm{p} \otimes \pi^0, |\frac{3}{2}, \frac{1}{2}\rangle)$. By using group theory we can reduce the computation of these scalar products to those already given.

To compute the coefficients, one way to proceed is to express all the basis elements in terms of homogeneous polynomials. To be explicit,

$$\mathrm{p} = z_1, \quad \mathrm{n} = z_2$$

and, similarly,

$$\pi^+ = \pi^+(y_1, y_2) = \frac{y_1^2}{2^{\frac{1}{2}}} = e_1^1(y)$$

$$\pi^0 = y_1 y_2 = e_1^0(y)$$

$$\pi^- = \frac{y_2^2}{2^{\frac{1}{2}}} = e_1^{-1}(y)$$

while

$$|\tfrac{3}{2},\tfrac{3}{2}\rangle = e_{3/2}^{3/2}(\bar{w}) = \frac{w_1^3}{6^{\frac{1}{2}}}, \text{etc.}$$

Since $|\tfrac{3}{2},\tfrac{3}{2}\rangle = p \otimes \pi^+$, it follows that

$$\rho_{\frac{3}{2}}(A)|\tfrac{3}{2},\tfrac{3}{2}\rangle = [\rho_{\frac{1}{2}}(A)p] \otimes [\rho_1(A)\pi^+].$$

Now recall that $\rho_1(A)e^n(z) = e^n(\bar{A}^{-1}z)$, where $\bar{A}^{-1}z = \begin{pmatrix} a & b \\ -\bar{b} & \bar{a} \end{pmatrix}\begin{pmatrix} z_1 \\ z_2 \end{pmatrix}$. All we have to do is to multiply out both sides and equate the coefficients of a^3, a^2b, ab^2 and b^3, thus identifying $|\tfrac{3}{2},\tfrac{3}{2}\rangle, |\tfrac{3}{2},\tfrac{1}{2}\rangle, |\tfrac{3}{2},-\tfrac{1}{2}\rangle$ and $|\tfrac{3}{2},-\tfrac{3}{2}\rangle$ as linear combinations of tensor products such as $p \otimes \pi^0, n \otimes \pi^+$, etc. Then choose $|\tfrac{1}{2},\tfrac{1}{2}\rangle$ to be a linear combination of $p \otimes \pi^0$ and $n \otimes \pi^+$ which is orthogonal to $|\tfrac{3}{2},\tfrac{1}{2}\rangle$, choosing the phase by making the coefficient of $p \otimes \pi^0$ as real and positive. Finally, transform $|\tfrac{1}{2},\tfrac{1}{2}\rangle$ to determine $|\tfrac{1}{2},-\tfrac{1}{2}\rangle$.

We first carrry out the expansion of $\rho_{\frac{3}{2}}(A)|\tfrac{3}{2},\tfrac{3}{2}\rangle$. Since $|\tfrac{3}{2},\tfrac{3}{2}\rangle = e_{3/2}^{3/2}(w) = w_1^3/6^{\frac{1}{2}}$, we have

$$\rho_{\frac{3}{2}}(A)|\tfrac{3}{2}\tfrac{3}{2}\rangle = e^{3/2}\left(\frac{aw_1 + bw_2}{-\bar{b}w_1 + \bar{a}w_2}\right) = \frac{(aw_1 + bw_2)^3}{6^{\frac{1}{2}}}$$

$$= \frac{a^3 w_1^3}{6^{\frac{1}{2}}} + (\tfrac{3}{2})^{\frac{1}{2}}a^2bw_1^2w_2 + (\tfrac{3}{2})^{\frac{1}{2}}ab^2w_1w_2^2 + \frac{b^3 w_2^3}{2^{\frac{1}{2}}}.$$

Recalling that $e_{3/2}^{1/2}(w) = (w_1^2 w_2)/2^{\frac{1}{2}}$, we see that

$$\rho_{(3/2)}(A)|\tfrac{3}{2}\tfrac{3}{2}\rangle = a^3|\tfrac{3}{2}\tfrac{3}{2}\rangle + 3^{\frac{1}{2}}a^2b|\tfrac{3}{2}\tfrac{1}{2}\rangle + 3^{\frac{1}{2}}ab^2|\tfrac{3}{2}-\tfrac{1}{2}\rangle + b^3|\tfrac{3}{2}-\tfrac{3}{2}\rangle.$$

On the other hand,

$$\rho_{(\frac{1}{2})}(A)p = e_{1/2}^{1/2}\left(\frac{az_1 + bz_2}{-\bar{b}z_1 + \bar{a}z_2}\right) = az_1 + bz_2$$

while

$$\rho_1(A)\pi^+ = e_1^1\left(\frac{ay_1 + by_2}{-\bar{b}y_1 + \bar{a}y_2}\right) = \frac{(ay_1 + by_2)^2}{2^{\frac{1}{2}}} = a^2\frac{y_1^2}{2^{\frac{1}{2}}} + 2^{\frac{1}{2}}aby_1y_2 + b^2\frac{y_2^2}{2^{\frac{1}{2}}}.$$

Expressing these results in terms of $p = z_1$, $n = z_2$, $\pi^+ = y_1^2/2^{\frac{1}{2}}$, $\pi^0 = y_1y_2$, $\pi^- = y_2^2/2^{\frac{1}{2}}$, we have

$$[\rho_{\frac{1}{2}}(A)p] \otimes [\rho_1(A)\pi^+] = (ap + bn) \otimes (a^2\pi^+ + 2^{\frac{1}{2}}ab\pi^0 + b^2\pi^-)$$

$$= a^3 p \otimes \pi^+ + a^2b(2^{\frac{1}{2}}p \otimes \pi^0 + n \otimes \pi^+) + ab^2(p \otimes \pi^- + 2^{\frac{1}{2}}n \otimes \pi^0) + b^3 n \otimes \pi^-.$$

Equating coefficients of the various monomials in a and b, we find

$$|\tfrac{3}{2}\tfrac{3}{2}\rangle = p \otimes \pi^+$$

$$|\tfrac{3}{2}\tfrac{1}{2}\rangle = \frac{1}{3^{\frac{1}{2}}}(2^{\frac{1}{2}}p \otimes \pi^0 + n \otimes \pi^+)$$

$$|\tfrac{3}{2} -\tfrac{1}{2}\rangle = \frac{1}{3^{\frac{1}{2}}}(p \otimes \pi^- + 2^{\frac{1}{2}}n \otimes \pi^0)$$

$$|\tfrac{3}{2} -\tfrac{3}{2}\rangle = n \otimes \pi^-.$$

We now determine the coefficients in $|\tfrac{1}{2}\tfrac{1}{2}\rangle = \alpha p \otimes \pi^0 + \beta n \otimes \pi^+$ by requiring that $|\tfrac{1}{2}\tfrac{1}{2}\rangle$ be orthogonal to $|\tfrac{3}{2}\tfrac{1}{2}\rangle$. This gives

$$2^{\frac{1}{2}}\alpha + \beta = 0$$

so that

$$\beta = -2^{\frac{1}{2}}\alpha.$$

The condition that $|\tfrac{1}{2}\tfrac{1}{2}\rangle$ has unit length gives $|\alpha|^2 + |\beta|^2 = 2|\alpha|^2 + |\alpha|^2 = 1$, so that $|\alpha| = (1/3)^{\frac{1}{2}}$. Finally, the convention that α is real and positive gives $\alpha = 1/3^{\frac{1}{2}}$, so that $\beta = -(2/3)^{\frac{1}{2}}$. Thus,

$$|\tfrac{1}{2}\tfrac{1}{2}\rangle = \frac{1}{3^{\frac{1}{2}}}(p \otimes \pi^0 - 2^{\frac{1}{2}}n \otimes \pi^+).$$

The easiest way to determine $|\tfrac{1}{2} -\tfrac{1}{2}\rangle$ is to apply a rotation through $180°$ about the y axis to $|\tfrac{1}{2}\tfrac{1}{2}\rangle$. For this rotation, $A = \begin{pmatrix} 0 & -1 \\ +1 & 0 \end{pmatrix}$, i.e. $a = 0$, $b = 1$. Hence

$$\rho_{\frac{1}{2}}(A)p = e^{\frac{1}{2}}\begin{pmatrix} z_2 \\ -z_1 \end{pmatrix} = z_2 = n,$$

$$\rho_{\frac{1}{2}}(A)n = e^{-\frac{1}{2}}\begin{pmatrix} z_2 \\ -z_1 \end{pmatrix} = -z_1 = -p,$$

$$\rho_1(A)\pi^0 = e^0\begin{pmatrix} y_2 \\ -y_1 \end{pmatrix} = -y_1 y_2 = -\pi^0,$$

$$\rho_1(A)\pi^1 = e^{-1}\begin{pmatrix} y_2 \\ -y_1 \end{pmatrix} = \frac{y_2^2}{2^{\frac{1}{2}}} = \pi^-.$$

It follows that

$$(\rho_{\frac{1}{2}} \otimes \rho_1)(A)\frac{1}{3^{\frac{1}{2}}}(p \otimes \pi^0 - 2^{\frac{1}{2}}n \otimes \pi^+) = \frac{1}{3^{\frac{1}{2}}}(-n \otimes \pi^0 + 2^{\frac{1}{2}}p \otimes \pi^-)$$

so that

$$|\tfrac{1}{2} -\tfrac{1}{2}\rangle = \frac{1}{3^{\frac{1}{2}}}(2^{\frac{1}{2}}p \otimes \pi^- - n \otimes \pi^0).$$

We can check that $|\tfrac{1}{2} -\tfrac{1}{2}\rangle$ is orthogonal to $|\tfrac{3}{2} -\tfrac{1}{2}\rangle$.

To summarize, we have

$$\pi^+ \otimes p = |\tfrac{3}{2}, \tfrac{3}{2}\rangle$$

$$\pi^- \otimes p = (\tfrac{1}{3})^{\frac{1}{2}}|\tfrac{3}{2}, -\tfrac{1}{2}\rangle - (\tfrac{2}{3})^{\frac{1}{2}}|\tfrac{1}{2}, -\tfrac{1}{2}\rangle$$

$$\pi^0 \otimes p = (\tfrac{2}{3})^{\frac{1}{2}} |\tfrac{3}{2}, \tfrac{1}{2}\rangle - (\tfrac{1}{3})^{\frac{1}{2}} |\tfrac{1}{2}, \tfrac{1}{2}\rangle$$

$$\pi^+ \otimes n = (\tfrac{1}{3})^{\frac{1}{2}} |\tfrac{3}{2}, \tfrac{1}{2}\rangle + (\tfrac{2}{3})^{\frac{1}{2}} |\tfrac{1}{2}, \tfrac{1}{2}\rangle$$

$$\pi^0 \otimes n = (\tfrac{2}{3})^{\frac{1}{2}} |\tfrac{3}{2}, \tfrac{1}{2}\rangle + (\tfrac{1}{3})^{\frac{1}{2}} |\tfrac{1}{2}, -\tfrac{1}{2}\rangle$$

$$\pi^- \otimes n = |\tfrac{3}{2}, -\tfrac{3}{2}\rangle$$

and, inverting this matrix, we get

$$|\tfrac{3}{2}, \tfrac{3}{2}\rangle = \pi^+ \otimes p$$

$$|\tfrac{3}{2}, \tfrac{1}{2}\rangle = (\tfrac{1}{3})^{\frac{1}{2}} \pi^+ \otimes n + (\tfrac{2}{3})^{\frac{1}{2}} \pi^0 \otimes p$$

$$|\tfrac{3}{2}, -\tfrac{1}{2}\rangle = (\tfrac{2}{3})^{\frac{1}{2}} \pi^0 \otimes n + (\tfrac{1}{3})^{\frac{1}{2}} \pi^- \otimes p$$

$$|\tfrac{3}{2}, -\tfrac{3}{2}\rangle = \pi^- \otimes n$$

and

$$|\tfrac{1}{2}, +\tfrac{1}{2}\rangle = (\tfrac{2}{3})^{\frac{1}{2}} \pi^+ \otimes n - (\tfrac{1}{3})^{\frac{1}{2}} \pi^0 \otimes p$$

$$|\tfrac{1}{2}, -\tfrac{1}{2}\rangle = (\tfrac{1}{3})^{\frac{1}{2}} \pi^0 \otimes n - (\tfrac{2}{3})^{\frac{1}{2}} \pi^- \otimes p.$$

Before proceeding to the mathematical discussion of the general method of calculating the Clebsch–Gordan coefficients, let us turn to the physical application of the special case worked out above.

It was observed in the early 1930s, soon after the discovery of the neutron, that in most properties the proton and the neutron were remarkably similar. The only difference between them consists in their electric charge. It was therefore suggested by Heisenberg that they be considered as different aspects of the 'same' particle. More specifically, it was suggested that if we could somehow 'turn off' or 'ignore' the electromagnetic interactions, then there should be a single particle – the nucleon – which is associated with the $V_{\frac{1}{2}}$ representation of the group $SU(2)$. The electromagnetic forces 'break' this symmetry and in fact give rise to the operator $\rho_{\frac{1}{2}} \begin{pmatrix} i & 0 \\ 0 & -i \end{pmatrix}$ acting on this space and the proton corresponds to the eigenvector $p = |\tfrac{1}{2}, \tfrac{1}{2}\rangle$ while the neutron corresponds to the eigenvector $n = |\tfrac{1}{2}, -\tfrac{1}{2}\rangle$. In the late 1940s various new elementary particles were discovered, among them the charged pions π^+ and π^- with masses 139.6 MeV/c^2 and the neutral pion π^0 with mass 135.0 MeV/c^2. Again, it was suggested that the observed mass difference be attributed to electromagnetic effects and that the pions 'would be' really all the same particles if we could 'turn off' the electromagnetic forces. This time, since there are three of them, the pions correspond to the V_1 representation $SU(2)$, with labeling as indicated above. Now all of this seems contrived and formal until we take into account scattering experiments between pions and nucleons. When a π^- pion is scattered off a proton, the end result can be (among others) another $p\pi^-$ pair or a $n\pi^0$ pair. Similarly, $p\pi^+ \to p\pi^+$ is a possible scattering process. The amplitudes of these various processes depend, of course, on the initial relative momenta and energies. For fixed such values, we want to compute the transition probabilities such as

$$\sigma(\pi^+ p \to \pi^+ p)$$

etc.

By the general principles of quantum mechanics we know that the transition probabilities are given as the square of the absolute values of the matrix entries

$$\sigma(u \to v) = |A_{v \leftarrow u}|^2$$

where u and v are any of the basis elements of $V_{\frac{1}{2}} \otimes V_1$ such as $p \otimes \rho^+$, etc. If we assume that A is an $SU(2)$ morphism, then

$$A = CBC^{-1}$$

where C is the change of basis matrix calculated above and where

$$B = \begin{pmatrix} A(\frac{3}{2})I_4 & 0 \\ 0 & A(\frac{1}{2})I \end{pmatrix}.$$

Thus the entries of A depend only on the two amplitudes $A(\frac{3}{2})$ and $A(\frac{1}{2})$. Explicitly, using the values for C that we have calculated, we see that, for example, we have the proportionalities

$$\sigma(\pi^+ p \to \pi^+ p) \propto |A(\tfrac{3}{2})|^2$$

$$\sigma(\pi^- p \to \pi^- p) \propto |(\tfrac{1}{3})A(\tfrac{3}{2}) + (\tfrac{2}{3})A(\tfrac{1}{2})|^2$$

$$\sigma(\pi^- p \to \pi^0 n) \propto |(\tfrac{2}{9})^{\frac{1}{2}}A(\tfrac{3}{2}) - (\tfrac{2}{9})^{\frac{1}{2}}A(\tfrac{1}{2})|^2.$$

Thus $SU(2)$ invariance predicts the ratios

$$\sigma(\pi^+ p \to \pi^+ p) : \sigma(\pi^- p \to \pi^- p) : \sigma(\pi^- p \to \pi^0 n)$$

$$= |A(\tfrac{3}{2})|^2 : |(\tfrac{1}{3})A(\tfrac{3}{2}) + (\tfrac{2}{3})A(\tfrac{1}{2})|^2 : |(\tfrac{2}{9})^{\frac{1}{2}}A(\tfrac{3}{2}) - (\tfrac{2}{9})^{\frac{1}{2}}A(\tfrac{1}{2})|^2.$$

A more precise prediction depends on the values of $A(\frac{3}{2})$ and $A(\frac{1}{2})$ which we expect to be energy dependent. Here are some ratios given by some special choice of these values:

  (a) $|A(\frac{3}{2})| \gg |A(\frac{1}{2})|$ then $\sigma(\pi^+ p \to \pi^+ p) : \sigma(\pi^- p \to \pi^- p) : \sigma(\pi^- p \to \pi^0 n)$

                                $= \qquad\quad 1 \quad : \quad \tfrac{1}{9} \quad : \quad \tfrac{2}{9}$

  (b) $|A(\frac{1}{2})| \gg |A(\frac{3}{2})|$ then $\sigma(\pi^+ p \to \pi^+ p) : \sigma(\pi^- p \to \pi^- p) : \sigma(\pi^- p \to \pi^0 n)$

                                $= \qquad\quad 0 \quad : \quad \tfrac{4}{9} \quad : \quad \tfrac{2}{9}$

  (c) $A(\frac{3}{2}) = A(\frac{1}{2})$ then $\sigma(\pi^+ p \to \pi^+ p) : \sigma(\pi^- p \to \pi^- p) : \sigma(\pi^- p \to \mu^0 n)$

                                $= \qquad\quad 1 \quad : \quad 1 \quad : \quad 0.$

The experimental evidence confirms that A does indeed have the desired form. Fig. 4.17 depicts the plot of the observed total scattering cross sections for $\sigma(\pi^+ p \to \pi^+ p)$ and $\sigma(\pi^- p \to \pi^- p)$ as a function of energy. An important point here is that there is a strong

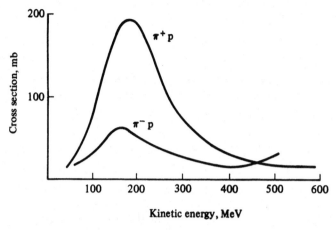

Fig. 4.17

peak at about 180 MeV. Around this range the scattering cross sections are given as

$$\sigma(\pi^+ p \to \pi^+ p) \sim 195\,\text{mb}$$

$$\sigma(\pi^- p \to \pi^- p) \sim \ \ 23\,\text{mb}$$

$$\sigma(\pi^- p \to \rho^0 n) \sim \ \ 45\,\text{mb}$$

so the ratios are quite close to 9:1:2. This suggests that $|A(\tfrac{3}{2})| \gg |A(\tfrac{1}{2})|$. We interpret this as saying that short lived particles corresponding to the $\tfrac{3}{2}$ representation of $SU(2)$ have been created, which then decay into a pion and a nucleon. These particles have a lifetime of about 10^{-23} s and were on this account sometimes called resonances rather than particles. We shall return to the significance of the lifetime 10^{-23} s later. These particles are called $\Delta^{++}, \Delta^+, \Delta^0$ and Δ^-, and are eigenvectors of $\rho_{\frac{3}{2}}(U_\theta)$ in $V_{\frac{3}{2}}$.

As a second example of the use of isospin we consider a weak interaction among hadrons, the decay of the lambda hyperon:

$$\Lambda^0 \to p + \pi^- \quad \text{and} \quad \Lambda^0 \to n + \pi^0$$

with a mean decay time of 2.5×10^{-10} s. (This relatively 'long' life for the Λ^0 is indicative of the fact that it is a weak interaction decay.) Now Λ^0 is an isosinglet while the representation corresponding to the decay product is $V_{\frac{1}{2}} \otimes V_1 = V_{\frac{1}{2}} + V_{\frac{3}{2}}$ so the decay cannot be given by an $SU(2)$ morphism. As it is usually phrased, isospin is not conserved by this weak interaction. However, the percentages of the two decay modes are measured to be around 64% for $p\pi^-$ and 36% for $n\pi^0$. These numbers are close to $\tfrac{2}{3}$ and $\tfrac{1}{3}$. So one is tempted to look for an explanation in terms of Clebsch–Gordan coefficients: Now the basis of the $V_{\frac{1}{2}}$ component of $V_{\frac{1}{2}} \otimes V_1$ is given, as we have seen, by

$$|\tfrac{1}{2}, +\tfrac{1}{2}\rangle = (\tfrac{2}{3})^{\frac{1}{2}} \pi^+ \otimes n - (\tfrac{1}{3})^{\frac{1}{2}} \pi^0 \otimes p$$

and

$$|\tfrac{1}{2}, -\tfrac{1}{2}\rangle = (\tfrac{1}{3})^{\frac{1}{2}} \pi^0 \otimes n - (\tfrac{2}{3})^{\frac{1}{2}} \pi^- \otimes p.$$

The first of these has electric charge $+1$ so is unavailable as a decay product. The squares of the coefficients of $|\tfrac{1}{2}, -\tfrac{1}{2}\rangle$ give the right decay percentages. Thus, we get the correct answer if we assume that the change, ΔI, in the isospin is $\tfrac{1}{2}$ under a weak interaction.

We now return to mathematics and illustrate a technique which allows the computation of the Clebsch–Gordan coefficients in general. The reader might not be interested in following all the details of the computation. But he should look at the general discussion of the method, since we shall have occasion to use it again in a more general context in Chapter 5 when we study the irreducible representation of general $SU(d)$.

The first observation is that the representations V_s are in fact representations of the group $Gl(2, \mathbb{C})$ or $Sl(2, \mathbb{C})$, as can be seen from its very construction. In terms of $Sl(2, \mathbb{C})$ we can give an abstract characterization of the 'maximal weight vectors' v_s.

Indeed, let $T_b = \begin{pmatrix} 1 & b \\ 0 & 1 \end{pmatrix} \in Sl(2, \mathbb{C})$. Then the line through $|s, s\rangle$ can be characterized in V_s as consisting of the only vectors which are left invariant under $\rho_s(T_b)$ (for all b). The elements in V_s on this line are also eigenvectors of $\rho_s(U_\theta)$ with eigenvalue $\exp 2is\theta$. Thus the line through v_s in $V_{s_1} \otimes V_{s_2}$ can be characterized as consisting of those vectors which satisfy the two sets of equations

$$(\rho_{s_1} \otimes \rho_{s_2})(T_b)v = v \quad \text{and} \quad (\rho_{s_1} \otimes \rho_{s_2})(U_\theta)(v) = (\exp 2is\theta)v \quad \text{(all } \theta \text{ all } b).$$

A possible v_s must satisfy these equations and also $\|v_s\| = 1$ (and so is determined up to phase). In order to make a convenient choice of the v_s, it is convenient for us to have still another model of the spaces V_s. Any homogeneous polynomial $f(z, w) = a_0 z^k + a_1 z^{k-1} w + \cdots + a_k w^k$ corresponds to the inhomogeneous polynomial $F(y) = a_0 + \cdots + a_k y^k$ by

$$F(y) = f(1, y) \quad \text{and} \quad f(z, w) = z^k F\left(\frac{w}{z}\right).$$

If $A = \begin{pmatrix} a & b \\ c & d \end{pmatrix} \in Sl(2, \mathbb{C})$, then $(\rho_s(A)f)(z, w) = f(dz - bw, -cz + aw)$ for any polynomial f of degree $2s$. The corresponding action on F is described by

$$(\rho_s(A)F)(y) = (-cy + d)^{2s} F\left(\frac{dy - b}{-cy + a}\right).$$

In this representation, the line through the element $|s, s\rangle$ clearly consists of the constants. In fact, it is clear that we have

$$|s, j\rangle = d(s, j) y^{s-j}, \quad \text{where} \quad d(s, j) = ((s - j)!(s + j)!)^{-\frac{1}{2}}.$$

Once we have identified V_s with the space of polynomials (of degree at most $2s$) in one variable under the above action, it is clear that we can identify $V_{s_1} \otimes V_{s_2}$ with the space of all polynomials in two variables y_1 and y_2 of degree at most s_1 in y_1 and s_2 in y_2 – we simply identify the element $y_1^j \otimes y_2^k$ with the polynomial $y_1^j y_2^k$. Under this identification, it is clear that

$$\rho_{s_1} \otimes \rho_{s_2}(A)F(y_1, y_2) = (-cy_1 + a)^{2s_1}(-cy_2 + a)^{2s_2}F\left(\frac{dy_1 - b}{-cy_1 + a}, \frac{dy_2 - b}{-cy_2 + a}\right).$$

In particular,

$$\rho_{s_1} \otimes \rho_{s_2}(T_b)F(y_1, y_2) = F(y_1 - b, y_2 - b).$$

Thus the vectors invariant under T_b are of the form $F(y_1, y_2) = G(y_1 - y_2)$, where G is a polynomial in one variable necessarily of degree at most $2 \min(s_1, s_2)$. The polynomial $(y_2 - y_1)^j$ is an eigenvector of $\rho_{s_1} \otimes \rho_{s_2}(U_\theta)$ with eigenvalue $\exp 2[(s_1 + s_2) - j]i\theta$. Thus v_s is a multiple of $(y_2 - y_1)^{s_1 + s_2 - s}$, and we shall fix the arbitrary phase by requiring that v_s be a real positive multiple of $(y_2 - y_1)^{s_1 + s_2 - s}$. To compute the actual v_s, we must simply find the length of $(y_2 - y_1)^{s_1 + s_2 - s}$. For example, if $s_1 = \frac{1}{2}$ and $s_2 = 1$, then

$$\| y_1 - y_2 \|^2 = \| y_1 \otimes 1 - 1 \otimes y_2 \|^2 = \| y_1 \otimes 1 \|^2 + \| 1 \otimes y_2 \|^2$$

since $y_1 \otimes 1$ and $1 \otimes y_2$ are orthogonal in $V_{s_1} \otimes V_{s_2}$. Now $\| y_1 \otimes 1 \|^2 = \| y_1 \|_{s_1}^2 \| 1 \|_{s_2}^2$ and $s_1 = \frac{1}{2}$ so $\| y_1 \|_{s_1}^2 = 2$. Thus $\| y_1 \otimes 1 \|^2 = 2$ and similarly $\| 1 \otimes y_2 \|^2 = 1$ so

$$\| y_2 - y_1 \|^2 = 3$$

and hence

$$v = 3^{-\frac{1}{2}}(y_1 \otimes 1 - 1 \otimes y_2).$$

Now, in $V_{\frac{1}{2}}$ we have $y = |\frac{1}{2}, -\frac{1}{2}\rangle = $ n and $1 = |\frac{1}{2}, \frac{1}{2}\rangle = $ p, while in V_1 we have $1 = 2^{\frac{1}{2}}|1, 1\rangle = 2^{\frac{1}{2}}\pi^+$ while $y_2 = |1, 0\rangle = \pi^0$. So the preceding equation can be written as

$$v_{\frac{1}{2}} = (\tfrac{2}{3})^{\frac{1}{2}}\text{n} \otimes \pi^+ - (\tfrac{1}{3})^{\frac{1}{2}}\text{p} \otimes \pi^0.$$

In general we will have

$$v_s = c(s_1, s_2, s)(y_2 - y_1)^{s_1 + s_2 - s}$$

where $c(s_1, s_2, s) = \| (y_2 - y_1)^{s_1 + s_2 + s} \|^{-1}$ as computed above. Having chosen this basis, it is now straightforward (but tedious) to compute all the Clebsch–Gordan coefficients. We first indicate how to find the matrix coefficients relative to the unnormalized bases $1, y, y^2, \dots$. Indeed, a direct computation shows that

$$\rho_{s_1} \otimes \rho_{s_2}(A)(y_2 - y_1)^j = (-cy_1 + a)^{2s_1 - j}(-cy_2 + a)^{2s_2 - j}(y_2 - y_1)^j$$

$$= (y_2 - y_1)^j \sum \binom{2s_1 - j}{r_1}\binom{2s_2 - j}{r_2}(-c)^{r_1 + r_2}$$

$$\times a^{2(s_1 + s_2 - j) - r_1 - r_2}y_1^{r_1}y_2^{r_2},$$

while in the space $V_{s_1 + s_2 - j}$ we have

$$\rho_{s_1 + s_2 - j}(A)(1) = (-cy + a)^{2(s_1 + s_2 - j)}$$

$$= \sum \binom{2(s_1 + s_2 - j)}{r}(-cy)^r a^{2(s_1 - s_2 - j) - r}.$$

Table 34. $j_1 = j_2 = \frac{1}{2}$

| | | J | 1 | 1 | 0 | 1 |
		M	1	0	0	-1
m_1	m_2					
1/2	1/2		1			
1/2	-1/2			$(1/2)^{\frac{1}{2}}$	$(1/2)^{\frac{1}{2}}$	
-1/2	1/2			$(1/2)^{\frac{1}{2}}$	$-(1/2)^{\frac{1}{2}}$	
-1/2	-1/2					1

Table 35. $j_1 = 1, j_2 = \frac{1}{2}$

| | | J | 3/2 | 3/2 | 1/2 | 3/2 | 1/2 | 3/2 |
		M	3/2	1/2	1/2	-1/2	-1/2	-3/2
m_1	m_2							
1	1/2		1					
1	-1/2			$(1/3)^{\frac{1}{2}}$	$(2/3)^{\frac{1}{2}}$			
0	1/2			$(2/3)^{\frac{1}{2}}$	$-(1/3)^{\frac{1}{2}}$			
0	-1/2					$(2/3)^{\frac{1}{2}}$	$(1/3)^{\frac{1}{2}}$	
-1	1/2					$(1/3)^{\frac{1}{2}}$	$-(2/3)^{\frac{1}{2}}$	
-1	-1/2							1

Comparing coefficients of various powers of $(-cy)$ and a we see that the vector y^r in $V_{s_1+s_2-j}$ corresponds to the vector

$$(y_2 - y_1)^j \sum p(r_1, r_2, j) y_1^{r_1} y_2^{r_2} \quad \text{in} \quad V_{s_1} \otimes V_{s_2}$$

where

$$p(r_1, r_2, j) = \frac{\binom{2s_1 - j}{r_1}\binom{2s_2 - j}{r_2}}{\binom{2(s_1 + s_2 - j)}{r_2 + r_2}}$$

is a quotient of binomial coefficients. To get the actual Clebsch–Gordan coefficients, we must still put in the appropriate normalization factors. For small values they are given below.

Clebsch–Gordan coefficients

$$j = s, \quad \phi(j, m) = |j, m\rangle$$

$$\phi(J, M) = \sum_{m_1, m_2} C(J, M, j_1, m_1, j_2, m_2)\phi(j_1, m_1)\phi(j_2, m_2)$$

For each pair of value of j_1 and j_2 Tables 34–37 give:

Table 36. $j_1 = j_2 = 1$

		J 2	2	1	2	1	0	2	1
		M 2	1	1	0	0	0	-1	-1
m_1	m_2								
1	1	1							
1	0		$(1/2)^{\frac{1}{2}}$	$(1/2)^{\frac{1}{2}}$					
0	1		$(1/2)^{\frac{1}{2}}$	$-(1/2)^{\frac{1}{2}}$					
1	-1				$(1/6)^{\frac{1}{2}}$	$(1/2)^{\frac{1}{2}}$	$(1/3)^{\frac{1}{2}}$		
0	0				$(2/3)^{\frac{1}{2}}$	0	$-(1/3)^{\frac{1}{2}}$		
-1	1				$(1/6)^{\frac{1}{2}}$	$-(1/2)^{\frac{1}{2}}$	$(1/3)^{\frac{1}{2}}$		
0	-1							$(1/2)^{\frac{1}{2}}$	$(1/2)^{\frac{1}{2}}$
-1	0							$(1/2)^{\frac{1}{2}}$	$-(1/2)^{\frac{1}{2}}$
-1	-1								

Table 37. $j_1 = \frac{3}{2}$, $j_2 = \frac{1}{2}$

		J 2	2	1	2	1	2	1	2
		M 2	1	1	0	0	-1	-1	-2
m_1	m_2								
3/2	1/2	1							
3/2	-1/2		1/2	$\sqrt{3}/2$					
1/2	1/2		$\sqrt{3}/2$	-1/2					
1/2	-1/2				$(1/2)^{\frac{1}{2}}$	$(1/2)^{\frac{1}{2}}$			
-1/2	1/2				$(1/2)^{\frac{1}{2}}$	$-(1/2)^{\frac{1}{2}}$			
-1/2	-1/2						$\sqrt{3}/2$	1/2	
-3/2	1/2						1/2	$-\sqrt{3}/2$	
-3/2	-1/2								1

	J	J	$\cdots$
	M	M	$\cdots$
m_1	m_2		
m_1	m_2		
$\vdots$	$\vdots$		

Blank spaces indicate zeros.

4.9 Relativistic wave equations

In Section 3.9 we classified all the irreducible representations of the Poincaré group (assuming Mackey's theorem about the irreducible representations of semidirect products of topological groups). The most familiar of these representations were known before Wigner's general classification. They were realized on spaces of solutions of certain partial differential equations. We wish to explain these realizations. In this section we will make free use of the Fourier transform and of basic notions in the theory of distributions (generalized functions).

Let us begin with the simplest case – the representations with $m^2 > 0$ and $s = 0$. The representation is on the space of all (complex-valued) functions on M_m^+, where M_m^+ is the 'positive mass shell' by

$$M_m^+ = \{p \mid \|p\|^2 > 0, p_0 > 0\}.$$

The functions must be square integrable with respect to the invariant measure $\mathrm{d}\mu_m$. If we use p_1, p_2, p_3 as coordinates on M_m^+, then (up to an arbitrary multiplicative constant)

$$\mathrm{d}\mu_m = \frac{1}{p_0} \mathrm{d}p_1 \wedge \mathrm{d}p_2 \wedge \mathrm{d}p_3. \tag{9.1}$$

This can be intuitively seen as follows. Consider the small region between M_m^+ and $M_{m+\varepsilon}^+$. The four-dimensional volumes between these two mass shells over any small region is conserved. The height of such a region is given by

$$\mathrm{d}\|p\|^2 = 2(p_0 \, \mathrm{d}p_0 - p_1 \, \mathrm{d}p_1 - p_2 \, \mathrm{d}p_2 - p_3 \, \mathrm{d}p_3)$$

so that the invariant three-dimensional volume will be given up to a constant by a three-form ω such that

$$\omega \wedge (p_0 \mathrm{d}p_0 - p_1 \, \mathrm{d}p_1 - p_2 \, \mathrm{d}p_2 - p_3 \mathrm{d}p_3) = \mathrm{d}p_0 \wedge \mathrm{d}p_1 \wedge \mathrm{d}p_2 \wedge \mathrm{d}p_3.$$

Clearly

$$\omega = \frac{1}{p_0} \mathrm{d}p_1 \wedge \mathrm{d}p_2 \wedge \mathrm{d}p_3$$

does the job.

In terms of the theory of distributions, we can rephrase the above argument by saying that we are looking at the δ-measure given by

$$\delta(\|p\|^2 - m^2). \tag{9.2}$$

(see Appendix G for a discussion of the notion of δ-measure.)

The Hilbert space structure is given by

$$(f_1, f_2) = \int_{M_m} f_1 \bar{f}_2 \, \mathrm{d}\mu_m. \tag{9.3}$$

Now instead of thinking of f as a function defined on M_m^+, we can think of

$$u = f\delta(\|p\|^2 - m^2) \tag{9.4}$$

as a function defined on all of momentum space, V^*. Since the support of $\delta(\|p\|^2 - m^2)$ is $M_m^+ \cup M_m^-$, we assume that f is defined to be zero on M_m^-. Now a necessary and sufficient condition for a distribution to be of the form (9.4) is

$$(\|p\|^2 - m^2)u = 0. \tag{9.5}$$

Let us consider the inverse Fourier transform of u,

$$w = \frac{1}{(2\pi)^2} \int (\exp ip \cdot x) u(p) \, d^4p, \quad w = F^{-1}u \tag{9.6}$$

$$u = \frac{1}{(2\pi)^2} \int (\exp - ip \cdot x) w(x) \, d^4x, \quad u = Fw. \tag{9.7}$$

Here $x \in V$, where V denotes Minkowski space and $p \in V^*$ is an energy momentum vector. Then the Fourier transform of

$$\frac{\partial w}{\partial x_j} \quad \text{is} \quad ip_j u$$

so $(\|p\|^2 - m^2)$ is the Fourier transform of $-(\Box^2 + m^2)w$, where

$$\Box^2 = \frac{\partial^2}{\partial x_0^2} - \frac{\partial}{\partial x_1^2} - \frac{\partial}{\partial x_2^2} - \frac{\partial}{\partial x_3}. \tag{9.8}$$

Thus u satisfies (9.5) if and only if w is a solution of the Klein–Gordon equation

$$(\Box^2 + m^2)w = 0. \tag{9.9}$$

Let $\rho_{m,0}^+$ denote the positive mass spin zero irreducible representation of the Poincaré group that we are studying. Let ρ denote the usual representation of the Poincaré group function on Minkowski space, V. Let K–G_m denote the space of solutions of the Klein–Gordon. Then it is easy to check that

$$F^{-1}\rho_{m,0}^+(a)u = \rho(a)F^{-1}u. \tag{9.10}$$

Thus the Fourier transform is a morphism for the Poincaré group.

Thus we can regard $\rho_{m,0}$ as acting on the space of solutions of the Klein–Gordon equation. Of course we do not want *all* solutions. First of all, we are only interested in the *positive energy* solutions, i.e. those w for which Fw has its support in M_m^+. Secondly, we want to only consider those $w \in$ K–G_m for which the corresponding $u = Fw$ is given by (9.4) with F square integrable. For such w, one defines the scalar product (w_1, w_2)

$$(w_1, w_2) = (f_1, f_2) = \int_{M_m^+} f_1 f_2 \, d\mu_m \tag{9.11}$$

where

$$f_i \delta(\|p\|^2 - m^2) = Fw_i.$$

It is slightly awkward to express the scalar product (9.11) directly in terms of w_1 and w_2.

The situation for the representations with $s > 0$ is more subtle. In this case the f's are not functions, but sections of a vector bundle, $E \to M$. Since $f(m) \in E_m$, and the vector space E_m varies with m, one cannot directly take the Fourier transform. What we shall do is embed E as a sub-bundle of a trivial bundle $E \to F$, where $F = M \times W$ and W is a fixed vector space. We can then consider the trivial bundle $V^* \times W$ (where V^* is the full energy momentum space, so $M \subset V^*$). We can then take the Fourier transform. The conditions that a (generalized) section u of $V^* \times W$ have its support in M and take its values in the sub-bundle E will be expressed as certain algebraic conditions on u which translate into a system of partial differential equations for its Fourier transform. We shall illustrate this method for two examples: we shall show that $m^2 > 0$, $s = \frac{1}{2}$ representation can be realized as the space of solutions of the Dirac equation and that the $m^2 = 0$, $s = \pm 1$ representations come from Maxwell's equations.

Let us begin with Maxwell's equations. In free space, these equations are

$$\mathrm{d}\omega = 0, \quad \mathrm{d}{*}\omega = 0 \tag{9.12}$$

where ω is a two-form (the electromagnetic field). We can think of ω as a function from V (Minkowski space) to $W = \Lambda^2(V^*)^{\mathbb{C}}$. Then the Fourier transform

$$u = \frac{1}{(2\pi)^2} \int (\exp - \mathrm{i} p \cdot x) \omega(x) \, \mathrm{d}x$$

satisfies

$$p \wedge u(p) = 0 \tag{9.13}$$

and

$$p \wedge {*}u(p) = 0. \tag{9.14}$$

If $p \neq 0$, then (9.13) implies that

$$u(p) = p \wedge a$$

where $a = a(p)$ is determined up to adding some multiple of p. We first claim that if $u(p) \neq 0$, then (9.14) implies that $\|p\|^2 = 0$. Indeed, if $\|p\|^2 \neq 0$, then by replacing a by

$$a - \frac{(a, p)}{(p, p)} p,$$

we can arrange that $a \perp p$. Then, by $\|p\|^2 \neq 0$, we can choose an 'orthonormal' basis of V^* such that

$$p = re_i \quad \text{and} \quad a = se_j$$

for some $i \neq j = 0, 1, 2, 3$. Then

$$p \wedge a = rse_i \wedge e_j$$

so

$$*(p \wedge a) = \pm rse_k \wedge e_l \quad k \neq i, j, \quad l \neq i, j.$$

and

$$p \wedge {*}(p \wedge a) = \pm r^2 se_i \wedge e_k \wedge e_l \neq 0.$$

Thus we must have $\|p\|^2 = 0$. So u must be supported on $M_0 = \{p|\|p\|^2 = 0\}$. Strictly speaking, of course, we must regard u as a generalized (or distributional) section of W and interpret (9.13) and (9.14) in the weak sense, i.e. by integrating against a 'test section'. See Appendix G for details. This argument (which is similar to that above) we leave to the reader. It gives the stronger result that

$$u = f\delta(\|p\|^2)$$

where now f is a section of $M_0 \times W \to M_0$. But not all sections will do. Equations (9.13) and (9.14) imply that f must be a section of the sub-bundle $E \to M_0 \times W$, where, for each $p \in M_0$, $E_p \subset W = \Lambda^2(V^*)$ is the subspace consisting of all α satisfying

$$p \wedge \alpha = 0 \tag{9.15}$$

and

$$p \wedge *\alpha = 0. \tag{9.16}$$

Since (9.15) and (9.16) are conditions which involve only linear algebra and the metric on V^* (which determines the $*$ operator) they are invariant under the Lorentz group, so the vector bundle E is in fact a Lorentz (and hence Poincaré) invariant sub-bundle. To see what bundle this is, it is enough to examine the action of the isotropy group of some p on E_p. Let us choose $p = e_0 + e_3 = \begin{pmatrix} 2 & 0 \\ 0 & 0 \end{pmatrix}$, so the isotropy group consist of all matrices of the form

$$\begin{pmatrix} e^{i\theta} & b \\ 0 & e^{-i\theta} \end{pmatrix}.$$

Condition (9.15) implies that

$$\alpha = (e_0 + e_3) \wedge [r(e_0 - e_3) + x_1 e_1 + x_2 e_2]$$
$$= 2re_0 \wedge e_3 + (e_0 + e_3) \wedge (x_1 e_1 + x_2 e_2).$$

Then

$$*\alpha = 2re_1 \wedge e_2 + (e_0 + e_3) \wedge (x_1 e_2 - x_2 e_1)$$

so

$$(e_0 + e_3) \wedge *\alpha = 2r(e_0 + e_3) \wedge e_1 \wedge e_2.$$

So condition (9.16) implies that $r = 0$ so

$$\alpha = (e_0 + e_3) \wedge (x_1 e_1 + x_2 e_2). \tag{9.17}$$

Now $x_1 e_1 + x_2 e_2$ corresponds to the self-adjoint matrix

$$\begin{pmatrix} 0 & x_1 + ix_2 \\ x_1 - ix_2 & 0 \end{pmatrix}$$

and

$$\begin{pmatrix} e^{i\theta} & b \\ 0 & e^{-i\theta} \end{pmatrix} \begin{pmatrix} 0 & x_1 + ix_2 \\ x_1 - ix_2 & 0 \end{pmatrix} \begin{pmatrix} e^{-i\theta} & 0 \\ \bar{b} & e^{i\theta} \end{pmatrix}$$
$$\begin{pmatrix} y & e^{2i\theta}(x_1 + ix_2) \\ e^{-2i\theta}(x_1 - ix_2) & 0 \end{pmatrix}$$

where $y = 2Re[(\exp i\theta)(x_1 + ix_2)\bar{b}]$. So the vector $x_1e_1 + x_2e_2$ gets sent into $y(e_0 + e_3) + R_{2\theta}(x_1e_1 + x_2e_2)$, where $R_{2\theta}$ denotes rotation through angle 2θ about the e_3 axis. But adding the term $y(e_0 + e_3)$ has no effect when we multiply by $e_0 + e_3$; i.e.

$$\begin{pmatrix} e^{i\theta} & b \\ 0 & e^{-i\theta} \end{pmatrix} \text{ sends } (e_0 + e_3) \wedge (x_1e_1 + x_2e_2) \text{ into } (e_0 + e_3) \wedge R_{2\theta}(x_1e_1 + x_2e_2). \quad (9.18)$$

Thus the vector bundle E is a two plane bundle corresponding to the two-dimensional representation

$$\begin{pmatrix} e^{i\theta} & b \\ 0 & e^{-i\theta} \end{pmatrix} \rightsquigarrow R_{2\theta}.$$

(In physics language this says that the electromagnetic field is a transverse vector field.)

This representation is not irreducible (over the complex numbers). The matrix $R_{2\theta}$ has the eigenvalues $\exp 2i\theta$ and $\exp -2i\theta$ (The corresponding sub-bundles are called right-handed and left-handed circularly polarized light.) The vector bundles corresponding to the two eigenspaces are exactly the ones associated to the representations $s = 1$ and $s = -1$ of $E(2)_+$. Thus Maxwell's equations, photons, correspond to $m = 0$, $s = \pm 1$.

Let us now turn to the case $m^2 > 0$, $s = \frac{1}{2}$. Here, for variety, we will start with the momentum space description. We will construct a representation of the group $Sl(2, \mathbb{C})$ on $\mathbb{C}^4$. This will give us a trivial bundle $V^* \times \mathbb{C}^4 \to V^*$. We shall use the results of the end of Section 3 of Chapter 3 to show that this vector bundle, when restricted to any 'mass shell' M_m^+, $m^2 > 0$, is equivalent to a direct sum of two copies of the vector bundle $E_{\frac{1}{2}}^m$ corresponding to $s = \frac{1}{2}$. We will then write down an algebraic equation (involving m) for (generalized) reactions of this trivial bundle which has two effects: (1) it forces the section to be of the form $f\delta(\|p\|^2 - m^2)$ with f a section of the bundle restricted to M_m, and (2) the section of $E_{\frac{1}{2}}^m \oplus E_{\frac{1}{2}}^m$ corresponding to f must lie in the diagonal sub-bundle. Thus the set of all (generalized) sections of $V^* \times \mathbb{C}^4$ satisfying our equations will, in fact, be the same as the space of all (ordinary) sections of $E_{\frac{1}{2}}^m$. Taking the Fourier transform will convert our algebraic equation into a partial differential equation – the Dirac equation.

Recall that for any 2×2 matrix

$$X = \begin{pmatrix} x & y \\ z & w \end{pmatrix}$$

we let X^a denote the matrix

$$X^a = \begin{pmatrix} w & -y \\ -z & x \end{pmatrix}$$

so

$$XX^a = \det X \begin{pmatrix} 1 & 0 \\ 0 & 1 \end{pmatrix}.$$

If

$$X = \begin{pmatrix} x_0 + x_3 & x_1 + ix_2 \\ x_1 - ix_2 & x_0 - x_3 \end{pmatrix}$$

then

$$X^a = \begin{pmatrix} x_0 - x_3 & -(x_1 + ix_2) \\ -(x_1 - ix_2) & x_0 + x_3 \end{pmatrix}$$

so the operation $X \to X^a$ reduces to spatial inversion in this case, as we have seen.

For any self-adjoint 2×2 matrix

$$P = \begin{pmatrix} P_0 + iP_3 & P_1 + iP_2 \\ P_1 - iP_2 & P_0 - P_3 \end{pmatrix}$$

define the 4×4 matrix $\gamma(P)$ by

$$\gamma(P) = \begin{pmatrix} 0 & P \\ P^a & 0 \end{pmatrix}. \tag{9.19}$$

Then

$$\gamma(P)^2 = \det P I \tag{9.20}$$
$$= \|P\|^2 I.$$

Given any $A \in Sl(2, \mathbb{C})$, define the 4×4 matrix $\tau(A)$ by

$$\tau(A) = \begin{pmatrix} A & 0 \\ 0 & A^{*-1} \end{pmatrix}. \tag{9.21}$$

It is clear that τ is a homomorphism, and it is easy to check that

$$\gamma(APA^*) = \tau(A)\gamma(P)\tau(A)^{-1}. \tag{9.22}$$

Let $U: V^* \to \mathbb{C}^4$ be a (generalized) function. We want to consider the equation

$$\gamma(p)u(p) = mu(p). \tag{9.23}$$

Multiplying both sides of this equation by $\gamma(p)$ and applying (9.20) shows that we must have supp $u \subset M_{m^2}$ and a more refined argument, as usual, shows that

$$u = f\delta(\|p\|^2 - m^2)$$

where $f: M_m \to \mathbb{C}^4$ is an ordinary function which must satisfy

$$\gamma(p)f(p) = mf(p). \tag{9.24}$$

Let ρ denote the restriction of τ to $SU(2)$. Thus

$$\rho(A) = \begin{pmatrix} A & 0 \\ 0 & A \end{pmatrix}$$

so ρ is the direct sum of two copies of the $s = \frac{1}{2}$ representation of $SU(2)$. Let $E_\rho \to M_m$ be the vector bundle induced by the representation ρ, so that $E_\rho = E_{\frac{1}{2}} \oplus E_{\frac{1}{2}}$, with the obvious notation. We are now in the situation described at the end of Section 3.3. The representation, ρ, of $SU(2)$ giving rise to the vector bundle E_ρ, extends to the representation τ of $Sl(2, \mathbb{C})$. Thus a section, s, of E_ρ corresponds to a function,

$f: M_m \to \mathbb{C}^4$, where if

$$s\left(A\begin{pmatrix} m & 0 \\ 0 & m \end{pmatrix} A^* \right) = [(A, \phi(A))],$$

then

$$f\left(A\begin{pmatrix} m & 0 \\ 0 & m \end{pmatrix} A^* \right) = \tau(A)\phi(A).$$

We can now translate (9.23) into a condition on ϕ: by (9.22)

$$\gamma\left(A\begin{pmatrix} m & 0 \\ 0 & m \end{pmatrix} A^* \right) f\left(A\begin{pmatrix} m & 0 \\ 0 & m \end{pmatrix} A^* \right) = \tau(A)\gamma\begin{pmatrix} m & 0 \\ 0 & m \end{pmatrix}\tau(A^{-1})\tau(A)\phi(A)$$

$$= \tau(A)\gamma\begin{pmatrix} m & 0 \\ 0 & m \end{pmatrix}\phi(A).$$

So (9.23) becomes (after multiplying on the left by $\tau(A)^{-1}$)

$$\gamma\begin{pmatrix} m & 0 \\ 0 & m \end{pmatrix}\phi(A) = m\phi(A)$$

or

$$\begin{pmatrix} 0 & 0 & m & 0 \\ 0 & 0 & 0 & m \\ m & 0 & 0 & 0 \\ 0 & m & 0 & 0 \end{pmatrix}\begin{pmatrix} \phi_1 \\ \phi_2 \\ \phi_3 \\ \phi_4 \end{pmatrix}(A) = m\begin{pmatrix} \phi_1 \\ \phi_2 \\ \phi_3 \\ \phi_4 \end{pmatrix}(A),$$

i.e. $\phi_1(A) = \phi_3(A)$ and $\phi_2(A) = \phi_4(A)$. Thus $\phi(A)$ lies in the diagonal sub-bundle $\phi = \begin{pmatrix} \Psi \\ \Psi \end{pmatrix}$, where Ψ is a section of $E_{\frac{1}{2}}$. We have thus identified solutions of (9.23) with sections of the bundle $E_{\frac{1}{2}} \to M_m$ as promised.

Taking the Fourier transform of (9.23) gives the celebrated Dirac equation

$$\frac{h}{i}\gamma\left(\frac{\partial}{\partial x} \right)w = mw. \tag{9.25}$$

Let us consider the (limiting) case of $m = 0$ of the Dirac equation (9.23). Thus

$$\gamma(p)u = 0. \tag{9.26}$$

Again, multiplying by $\gamma(p)$ we see that u must be supported on M_0, and a more careful argument shows that

$$u = f\delta(\|p\|^2)$$

where f is a function from M_0 to $\mathbb{C}^4$ which satisfies

$$\gamma(p)f(p) = 0. \tag{9.27}$$

To see what this amounts to, let us pick our usual representative point on M_0^+

$$p_0 = \begin{pmatrix} 2 & 0 \\ 0 & 0 \end{pmatrix} = e_0 + e_3.$$

Then (9.27) becomes

$$\begin{pmatrix} 0 & 0 & 2 & 0 \\ 0 & 0 & 0 & 0 \\ 0 & 0 & 0 & 0 \\ 0 & 2 & 0 & 0 \end{pmatrix} \begin{pmatrix} a_1 \\ a_2 \\ a_3 \\ a_4 \end{pmatrix} = \begin{pmatrix} 0 \\ 0 \\ 0 \\ 0 \end{pmatrix}, \quad \text{where } f(p_0) = \begin{pmatrix} a_1 \\ a_2 \\ a_3 \\ a_4 \end{pmatrix}$$

or

$$a_2 = 0, \quad a_3 = 0. \tag{9.28}$$

The isotropy group fixing p_0 is $\tilde{E}(2)$, consisting of all matrices of the form

$$\begin{pmatrix} e^{i\theta} & b \\ 0 & e^{-i\theta} \end{pmatrix}.$$

Let $Y \subset \mathbb{C}^4$ denote the set of vectors satisfying (9.28). Thus Y is the two-dimensional space of all sections of the form

$$\begin{pmatrix} a_1 \\ 0 \\ 0 \\ a_4 \end{pmatrix}.$$

The representation τ restricted to $\tilde{E}(2)$ leaves Y invariant. Explicitly,

$$\tau\begin{pmatrix} e^{i\theta} & b \\ 0 & e^{-i\theta} \end{pmatrix} = \begin{pmatrix} e^{i\theta} & b & 0 & 0 \\ 0 & e^{-i\theta} & 0 & 0 \\ 0 & 0 & e^{i\theta} & 0 \\ 0 & 0 & -\bar{b} & e^{-i\theta} \end{pmatrix}$$

so

$$\tau\begin{pmatrix} e^{i\theta} & b \\ 0 & e^{-i\theta} \end{pmatrix} \begin{pmatrix} a_1 \\ 0 \\ 0 \\ a_4 \end{pmatrix} = \begin{pmatrix} e^{i\theta}a_1 \\ 0 \\ 0 \\ e^{-i\theta}a_4 \end{pmatrix}.$$

Thus, on Y restriction of τ to $E(2)$ decomposes as a direct sum of the two one-dimensional representations corresponding to $s = +\frac{1}{2}$ and $s = -\frac{1}{2}$. So, by the same arguments as before, the set of all functions from M_0 to $\mathbb{C}^4$ which satisfy (9.26) can be identified with the space of all sections of

$$E_{\frac{1}{2}}^0 \oplus E_{-\frac{1}{2}}^0$$

and so coincides with the direct sum of the two irreducible representations of the Poincaré group corresponding to $m = 0$, $s = \frac{1}{2}$ and $m = 0$, $s = -\frac{1}{2}$.

Notice that in the $\frac{1}{2}$ and $-\frac{1}{2}$ entirely decouple. We can in fact get separate wave equations for each: indeed, consider the space of generalized functions from V^* to $\mathbb{C}^2$ satisfying the equation

$$P^a u(p) = 0 \qquad\qquad (9.29)_+$$

i.e.

$$\begin{pmatrix} p_0 - p_3 & -(p_1 + ip_2) \\ -(p_1 - ip_2) & p_0 + p_3 \end{pmatrix} \begin{pmatrix} u_1(p) \\ u_2(p) \end{pmatrix} = \begin{pmatrix} 0 \\ 0 \end{pmatrix}.$$

This equation can have non-trivial solutions if and only if

$$\det P = \|p\|^2 = 0.$$

As before

$$u = f \delta(\|p\|^2)$$

with $f : M_0 \to \mathbb{C}^2$ satisfying

$$P^a f = 0.$$

Consider the representation of $Sl(2, \mathbb{C})$ on $\mathbb{C}^2$ given by

$$\tau_+(A) = A.$$

Then the corresponding action on functions to $\mathbb{C}^2$ sends $u(P)$ into

$$u_A(p) = Au(A^{-1}pA^{*-1}).$$

If u satisfies $(9.29)_+$ then so does $u_A(p)$ since

$$\begin{aligned} P^a u_A(p) &= P^a u(A^{-1}PA^{*-1}) \\ &= A^{*-1}(A^*P^a A)u(A^{-1}PA^{*-1}) \\ &= 0 \end{aligned}$$

since

$$A^a = A^{-1} \quad \text{for } A \subset Sl(2, \mathbb{C})$$

and

$$(APA^*)^a = A^{*-1}P^a A^{-1}.$$

Taking $P = P_0 = \begin{pmatrix} 2 & 0 \\ 0 & 0 \end{pmatrix}$ in $(9.29)_+$ shows that at P_0, f must be of the form

$$\begin{pmatrix} a_1 \\ 0 \end{pmatrix}.$$

Let Y_+ denote the one-dimensional space of such vectors. Then

$$\tau_+ \begin{pmatrix} e^{i\theta} & b \\ 0 & e^{-i\theta} \end{pmatrix} \begin{pmatrix} a \\ 0 \end{pmatrix} = \begin{pmatrix} e^{i\theta}a \\ 0 \end{pmatrix}.$$

So the restriction of τ_T to $E(2)_{c=0}$ acts as the $\frac{1}{2}$ representation of Y. Thus the set of functions from M_0^+ to $\mathbb{C}^2$ satisfying $(9.29)_+$ is the same as the set of all sections of $E_{\frac{1}{2}}$. Taking the Fourier transform of $(9.29)_+$ gives a partial differential equation known as the Weyl equation.

Similarly, we can consider the equation

$$Pu = 0, \tag{9.29}$$

i.e.

$$\begin{pmatrix} p_0 + p_3 & p_1 + ip_2 \\ p_1 - ip_2 & p_0 - p_3 \end{pmatrix} \begin{pmatrix} u_1 \\ u_2 \end{pmatrix} = \begin{pmatrix} 0 \\ 0 \end{pmatrix}.$$

This time we let $Sl(2, \mathbb{C})$ act on $\mathbb{C}^2$ by

$$\tau_-(A) = A^{*-1}$$

and the same results hold as before with $+$ replaced by $-$.

Weyl introduced his equations in 1933. They were immediately rejected by the physics community on the grounds that they were not invariant under parity transformations. It was not until 1956, with the discovery of the violation of parity conservation in weak interactions, that Weyl's equation was accepted as the equation of the neutrino. We have discussed the issues involved in parity conservation or non-conservation in Chapter 3.

4.10   Lie algebras

Let $G = O(n)$, the group of $n \times n$ orthogonal matrices. Thus G consists of all matrices which satisfy

$$AA^\dagger = I \tag{10.1}$$

where $A^\dagger$ denotes matrix transpose. Equation (10.1) defines G for us as a subgroup of the set of all $n \times n$ real matrices. Thus G is a linear Lie group as defined in the introduction to this chapter. Let $A(t)$ be a differentiable curve of matrices. (This means that each of the matrix entries of A is a differentiable function of t.) Suppose that

$$A(t) \in G \quad \text{for all} \quad t \quad \text{and} \quad A(0) = I. \tag{10.2}$$

If we differentiate (10.1) with respect to t and set $t = 0$ we get

$$A'(0) + A'(0)^\dagger = 0.$$

In other words,

$$A'(0) \in g \tag{10.3}$$

where g denotes the vector space consisting of all anti-symmetric matrices. Conversely, suppose that $X \in g$. Set

$$A(t) = \exp tX.$$

Then

$$A'(t) = X \exp tX = (\exp tX)X$$

and

$$\frac{d}{dt}(A(t)A^\dagger(t)) = A'(t)A^\dagger(t) + A(t)A^\dagger(t)'$$

$$= (\exp tX)X(\exp tX)^\dagger + (\exp tX)X^\dagger(\exp tX)^\dagger$$
$$= (\exp tX)(X + X^\dagger)(\exp tX)^\dagger$$
$$= 0.$$

Combined with the initial condition

$$A(0)A^\dagger(0) = I$$

we get (10.1) for all t. Thus

$$\text{if } X \in g \text{ then } \exp tX \in G \text{ for all } t. \tag{10.4}$$

Similarly, consider the group $U(n)$ of all $n \times n$ unitary matrices. Thus a complex matrix A belongs to $U(n)$ if and only if

$$AA^* = I.$$

If $A(t)$ is a curve of such matrices then differentiating at $t = 0$ as before gives

$$X + X^* = 0 \quad \text{for} \quad X = A'(0).$$

So if g now denotes the (real) vector space of all complex skew Hermitian matrices then (10.3) holds. An argument similar to the one given above shows that (10.4) holds as well.

Consider the group $Sl(n)$ of all $n \times n$ matrices of determinant unity. If $A(t)$ is the curve of matrices with $A(0) = I$ and $A'(0) = X$, we have

$$A(t) = \begin{pmatrix} 1 + tx_{11} & tx_{12} & \cdots & tx_{1n} \\ tx_{21} & 1 + tx_{22} & & \\ & & \ddots & \\ & & & 1 + tx_{nn} \end{pmatrix} + \cdots$$

where the $\cdots$ involve terms of higher order in t. Upon expanding the determinant of the matrix on the right-hand side, any non-diagonal term will contribute a term of order at least t^2 so only the diagonal terms matter. In expanding the product of the diagonal terms, we get

$$\det A(t) = 1 + t \sum x_{ii} + \cdots$$

so

$$\frac{d}{dt}(\det A(t))|_{t=0} = \operatorname{tr} X.$$

On the other hand,

$$\det \exp(t + s)X - \det \exp tX = \det(\exp sX \cdot \exp tX) - \det \exp tX$$
$$= (\det \exp tX)[\det(\exp sX) - 1]$$

so dividing by s and letting $s \to 0$ gives

$$\frac{d}{dt}[\det \exp tX] = \operatorname{tr} X[\det A(t)].$$

Thus, if $A(t)$ is a curve in $Sl(n, \mathbb{R})$ and $A(0) = I$ then $A'(0) = X$, where

$$\operatorname{tr} X = 0.$$

Conversely, if $\operatorname{tr} X = 0$ then $\det \exp tX \equiv 1$. So again (10.3) and (10.4) hold with $G = Sl(n, \mathbb{R})$ and $g = $ set of matrices with trace zero.

The general pattern is clear. To each linear Lie group G we associate a vector space g given by all $A'(0)$, where $A(t) \in G$ and $A(0) = I$. This space is called *the Lie algebra of G*. (The reason for the word 'algebra' will soon become clear, and we will give an abstract definition of the term Lie algebra.) Then (10.4) holds. We will not give the proof in general. The reader can check it for all the examples that he will encounter. (In fact, there is a more general definition that applies to all (not only linear) Lie groups; one can define g to be TG_e, the tangent space of G at the identity. A more sophisticated definition of the exponential map is required, but then (10.4) still holds. We refer the reader to any standard text on Lie groups or differential geometry.)

Suppose that G is a linear Lie group and g its Lie algebra. Let $X \in g$ and $A \in G$. Then

$$A(\exp tX)A^{-1} \in G$$

for all t and, hence, differentiating at the origin shows that

$$AXA^{-1} \in g.$$

In other words, g is stable under conjugation by $A \in G$, i.e. $AgA^{-1} = g$. We thus get a representation of G on g known as the *adjoint representation*.

Let X and Y be two elements of g. Then $\exp tX \in G$ for each fixed t, and hence

$$(\exp tX)Y(\exp - tX) \in g$$

for all t. But

$$(\exp tX)Y(\exp - tX) = (I + tX + (t^2/2!)X^2 + \cdots)Y(I - tX + (t^2/2)X^2 - \cdots)$$
$$= Y + t(XY - YX) + O(t^2).$$

This clearly implies that the coefficient of t, which is

$$[X, Y] = XY - YX,$$

must lie in g. Thus g is closed under the bracket operation sending X, Y into $[X, Y]$. The commutator bracket, as defined above, is a bilinear product which satisfies

$$[X, Y] = -[Y, X] \qquad \text{(anti-symmetry)} \qquad (10.5)$$

and

$$[X, [Y, Z]] = [[X, Y], Z] + [Y, [X, Z]] \qquad \text{(Jacobi's identity).} \qquad (10.6)$$

Any vector space with a bilinear product satisfying the above two identities is known as a *Lie algebra*. It is true that the tangent space of any Lie group at the identity has the structure of a Lie algebra. We have demonstrated this for the case of linear Lie groups, assuming the property (10.4), a property that we have verified for the case of $U(n)$, $O(n)$ and $Sl(n)$ (and hence also $SU(n)$).

Let ρ be any continuous linear representation of the linear Lie group G on some finite-dimensional vector space W. If $A(t) = \exp tX$ is a one parameter subgroup of G, then $\rho(A(t))$ is a one parameter subgroup of linear transformations of W, and, since W is finite dimensional, there is some linear transformation, which we shall denote by $\dot\rho(X)$ such that

$$\rho(\exp tX) = \exp t\dot\rho(X). \tag{10.7}$$

Of course it follows from (10.7) that we have

$$\dot\rho(X) = \frac{\mathrm{d}}{\mathrm{d}t}\rho(\exp tX)|_{t=0}. \tag{10.8}$$

Let $Z = \alpha X + \beta Y$, where α and β are real numbers. Then

$$A(t) = \exp t\alpha X \cdot \exp t\beta Y \in G$$

for all t. Now $\rho(A(t))$ is differentiable and is tangent to $\rho(B(t)) = \rho(\exp tZ)$. On the other hand,

$$\rho(A(t)) = \rho(\exp t\alpha X)\rho(\exp t\beta Y).$$

So, by (10.8),

$$\dot\rho(\alpha X + \beta Y) = \alpha\dot\rho(X) + \beta\dot\rho(Y). \tag{10.9}$$

In other words the map $X \rightsquigarrow \dot\rho(X)$ is linear. Also for $A \in G$ and $Y \in g$ we have

$$\rho(A)\rho(\exp tY)\rho(A^{-1}) = \rho(\exp tAYA^{-1})$$

so differentiating at $t = 0$ gives

$$\dot\rho(AYA^{-1}) = \rho(A)\dot\rho(Y)\rho(A^{-1}). \tag{10.10}$$

Taking $A = \exp tX$ in (10.10) and differentiating at $t = 0$ gives

$$\dot\rho([X, Y]) = [\dot\rho(X), \dot\rho(Y)]. \tag{10.11}$$

Let g be any Lie algebra and W any vector space. A linear map $r: g \to \mathrm{Hom}\,(W, W)$ which satisfies

$$r([X, Y]) = [r(X), r(Y)] \tag{10.12}$$

is called a representation of the Lie algebra. We have shown that every finite-dimensional representation ρ of a Lie group G determines a representation $\dot\rho$ of its Lie algebra. The converse is not necessarily true.

In case we are given representation of G on an infinite-dimensional topological vector space, W, the situation is somewhat more complicated because the one parameter groups do not necessarily have the simple form $\exp At$, where A is a continuous linear operator. Let us call a vector, $\mathbf{w}$, in W *differentiable*, if the function $a \rightsquigarrow \rho(a)\mathbf{w}$ is a differentiable function from G to W. We claim that the differentiable vectors are dense in W. Indeed, let f be any function on G which is differentiable and has compact support, and consider the vector

$$\mathbf{w}_f = \int_G f(g)\rho(g)\mathbf{w}\,\mathrm{d}g.$$

Then

$$\rho(a)\mathbf{w}_f = \int_G f(a^{-1}g)\rho(g)\mathbf{w}\,dg = \mathbf{w}_{af}.$$

Thus $\mathbf{w}_f$ is a differentiable vector. If we let the f's have smaller and smaller support around e, and have $\int f(g)\,dg = 1$, then it is clear that $\mathbf{w}_f$ converges weakly to $\mathbf{w}$. Thus the set of all $\mathbf{w}_f$, and hence of all differentiable vectors, is dense. The definition and properties of 'analytic vectors' are a little more complicated and will not be discussed here.

Many of the notions that we developed for the representations of groups have rather straightforward translations into corresponding notions for the representations of Lie algebras. For instance, if $\rho(\exp tX)v = v$ for all t, then differentiating at at $t = 0$ yields the equation $\dot\rho(X)v = 0$. So the condition that a vector v is invariant under a representation, r, of a Lie algebra, L, is that $r(x)v = 0$ for all $X \in L$. If $(,)$ is a scalar product which is invariant under the group G, then differentiating the equation

$$(\rho(\exp tX)u, \rho(\exp tX)v) = (u, v)$$

yields

$$(\dot\rho(X)u, v) + (u, \dot\rho(X)v) = 0.$$

The notion of an invariant subspace remains the same, as does the concept of complete reducibility and irreducible representation. It need not be true that a finite-dimensional representation of a Lie algebra be completely reducible; indeed the same assertion need not be true for general Lie groups. We have only proved it for finite (and in Appendix E for compact Lie) groups. If ρ_1 and ρ_2 are two representations of the group G on the finite-dimensional vector spaces W_1 and W_2 then we defined their tensor product, $\rho_1 \otimes \rho_2$ on $W_1 \otimes W_2$ by

$$(\rho_1 \otimes \rho_2)(a)v_1 \otimes v_2 = \rho_1(a)v_1 \otimes \rho_2(a)v_2.$$

Setting $a = \exp tX$ and differentiating with respect to t at $t = 0$ yields

$$(\rho_1 \otimes \rho_2)'(X)(v_1 \otimes v_2) = \dot\rho_1(X)v_1 \otimes v_2 + v_1 \otimes \dot\rho_2(X)v_2.$$

So we define the tensor product of two representations, r_1 and r_2, of a Lie algebra by the rule

$$(r_1 \otimes r_2)(X)(v_1 \otimes v_2) = r_1(X)v_1 \otimes v_2 + v_1 \otimes r_2(X)v_2.$$

The idea in translating the various definitions from groups to algebras is always the same: one differentiates the appropriate equation at the origin. We will not state all the necessary translations here.

4.11 Representations of $su(2)$

In this section we illustrate some Lie algebra techniques in the course of classifying all the finite-dimensional irreducible representations of $su(2)$. As a result, we will get another proof of the classification of the irreducible representations of $SU(2)$.

The algebra, $su(2)$, of all 2×2 skew Hermitian matrices is three dimensional, and has as a basis the matrices

$$X_1 = (i/2)\sigma_1 = (i/2)\begin{pmatrix} 0 & 1 \\ 1 & 0 \end{pmatrix}$$

$$X_2 = -(i/2)\sigma_2 = (i/2)\begin{pmatrix} 0 & i \\ -i & 0 \end{pmatrix}$$

and

$$X_3 = (i/2)\sigma_3 = (i/2)\begin{pmatrix} 1 & 0 \\ 0 & -1 \end{pmatrix}.$$

A direct computations shows that the following bracket relations hold:

$$[X_1, X_2] = X_3, [X_2, X_3] = X_1 \quad \text{and} \quad [X_3, X_1] = X_2. \tag{11.1}$$

Now we are interested in representations of $su(2)$ on complex vector spaces. We can thus consider complex linear combinations of the linear transformations $r(X)$, which can be regarded as a representation of the 'complexification' of the Lie algebra $su(2)$: for any real Lie algebra, we construct the complex vector space $g^{\mathbb{C}} = g \otimes \mathbb{C}$. Thus every element of $g^{\mathbb{C}}$ can be written as $X + iY$, where X and Y are elements of g. The bracket of two such elements is given, as expected, by

$$[X_1 + iY_1, X_2 + iY_2] = [X_1, X_2] - [Y_1, Y_2] + i([X_1, Y_2] + [Y_1, X_2]).$$

Any representation of $su(2)$ on a complex vector space clearly extends to a representation of $su(2)^{\mathbb{C}}$. Let us set

$$E_+ = -i(X_1 - iX_2) = \begin{pmatrix} 0 & 1 \\ 0 & 0 \end{pmatrix},$$

$$E_- = -i(X_1 + iX_2) = \begin{pmatrix} 0 & 0 \\ 1 & 0 \end{pmatrix}$$

and

$$H = -iX_3 = \tfrac{1}{2}\begin{pmatrix} 1 & 0 \\ 0 & -1 \end{pmatrix}.$$

A direct computation shows that

$$[H, E_+] = E_+, \quad [H, E_-] = -E_- \quad \text{and} \quad [E_+, E_-] = 2H. \tag{11.2}$$

We shall use these bracket relations to determine all the finite-dimensional irreducible representations. Notice that the matrices E_+, E_- and H are real 2×2 matrices forming a basis of the Lie algebra $sl(2, \mathbb{R})$. This should not be surprising since $sl(2, \mathbb{R}) \otimes \mathbb{C} = su(2) \otimes \mathbb{C} = sl(2, \mathbb{C})$: any complex traceless matrix can be written as $A + iB$, where A and B are real traceless matrices, or as $C + iD$, where C and D are in $su(n)$. We will thus, in the process of determining all the finite-dimensional representations of $SU(2)$, also determine all the finite-dimensional representations of $Sl(2, \mathbb{R})$. However, the group $Sl(2, \mathbb{R})$ is not compact and its representation theory is considerably more complicated. For instance, none of its non-trivial finite-dimensional representations is unitary, and it has a continuous family of infinite-dimensional irreducible unitary representations.

The key use of the bracket relations (11.2) is via the following simple observation. Let r be a representation of $sl(2, \mathbb{C})$ on some vector space, V.

> Let $v_\lambda \in V$ be an eigenvector of $r(H)$ with eigenvalue λ. Then $r(E_+)v_\lambda$ is   (11.3)
> an eigenvector of $r(H)$ with eigenvalue $\lambda + 1$ and $r(E_-)v_\lambda$ is an
> eigenvector of $r(H)$ with eigenvalue $\lambda - 1$.

Proof By (11.2) we know that $r(H)r(E_+) = r(E_+)r(H) + r([H, E_+]) = r(E_+)r(H) + r(E_+)$. Applying this to the vector v_λ shows that $r(E_+)v_\lambda$ is an eigenvector of $r(H)$ with eigenvalue $\lambda + 1$, and a similar argument works for $r(E_-)v$.

Let V be a finite-dimensional complex vector space on which we are given a representation of the Lie algebra $su(2)$. Then H has at least one eigenvalue. Let v_m be an eigenvector corresponding to the maximum eigenvalue, m. Then we must have $r(E_+)v_m = 0$. Let us set $v_{m-1} = r(E_-)v_m$, $v_{m-2} = r(E_-)v_{m-1}$, etc. so that $v_{m-k} = r(E_-)^k v_m$. By the finite dimensionality of V, there will be some value of k such that $r(E_-)^k v_m \neq 0$ but $r(E_-)^{k+1} v_m = 0$. The set of vectors $v_m, v_{m-1}, \ldots, v_{m-k}$ is clearly stable under H and E_-. Also $r(E_+)v_{m-i-1} = r(E_+)r(E_-)v_{m-i} = r(E_-)r(E_+)v_{m-i} + 2Hv_{m-i}$, which proves, by induction, that $r(E_+)v_{m-i-1}$ is a multiple of v_{m-i}. In particular, the space spanned by the v_{m-i} is an invariant subspace for $su(2)$. If we assume that V is irreducible, this must be all of V.

Relative to this basis we know that

$$r(E_-)v_{m-i} = v_{m-i-1} \quad 0 < i < k \quad r(E_-)v_{m-k} = 0$$
$$r(H)v_{m-i} = (m - i)v_{m-i}$$
$$r(E_+)v_{m-i} = a_{m-i}v_{m-i+1} \quad 0 < i < k \quad r(E_+)v_m = 0. \tag{11.4}$$

We must determine the possible values of m and k and of the a_{m-i}, and this will fix the possible representations. For this purpose, it is convenient to introduce the 'Casimir operator'

$$\Delta = \tfrac{1}{2}(E_- E_+ + E_+ E_-) + H^2.$$

When we write such an expression involving products of elements of a Lie algebra, the rules for commuting a product are given by the bracket relations; for example, $E_- E_+ = E_+ E_- - [E_+, E_-]$ and so on. In any representation, r, of the algebra, the expression $r(\Delta)$ is obtained by forming the corresponding product and sum of the $r(x)$ occurring on the right, so that, for example,

$$r(\Delta) = \tfrac{1}{2}(r(E_-)r(E_+) + r(E_+)r(E_-)) + r(H)^2.$$

The general theory of such expressions is treated in texts on Lie algebras. They are elements of the 'universal enveloping algebra' of a Lie algebra. For the present, it is clear enough how to compute with the specific operator, Δ:

$$\Delta = \tfrac{1}{2}(E_-E_+ + E_+E_-) + H^2$$
$$= E_-E_+ + H(H+1)$$
$$= E_+E_- + H(H-1)$$

and

$$\Delta H = H\Delta, \quad \Delta E_+ = E_+\Delta \text{ and } \Delta E_- = E_-\Delta.$$

The last three equations say that $r(\Delta)$ commutes with all the operators $r(X)$ for X in $su(2)$. If V is irreducible, this implies that $r(\Delta)$ must be a scalar matrix. (If we go back and examine the proof of Schur's lemma, we see that it works just as well for irreducible representations of a Lie algebra as for irreducible representations of a Lie group. In fact, all we needed was some family of operators acting irreducibly on a finite-dimensional vector space V, which commute with some linear transformation, ϕ, to conclude that ϕ must be a scalar matrix, if we are over the complex numbers.) Now

$$r(\Delta)v_m = [r(E_-)r(E_+) + r(H)(r(H)+1)]v = m(m+1)v_m$$

showing that

$$r(\Delta) = m(m+1)I.$$

On the other hand,

$$E_+E_- = \Delta - H(H-1)$$

so that

$$r(E_+)v_{m-i-1} = r(E_+)r(E_-)v_{m-i} = [m(m+1) - (m-i)(m-i-1)]v_{m-i}.$$

If we take $i = k$ in the above equation, so that $r(E_-)v_{m-k} = 0$, we see that

$$2m = k$$

so the possible values of m are half integers:

$$m = 0, \tfrac{1}{2}, 1, \tfrac{3}{2}, 2, \dots.$$

We shall now drop the r's and write E_+v instead of $r(E_+)v_j$. Also to obtain a more symmetrical form for the operators E_+ and E_-, it is more convenient to replace the vectors v_j by w_j, where the w_j are scalar multiples of the v_j (so that we still have $Hv_j = jv_j$, $j = m, m-1, m-2, \dots, -m$) and where the w_j are so chosen that

$$E_-w_j = (m+j)w_{j-1},$$

so that we may take

$$w_m = v_m, \quad w_{m-1} = (2m)^{-1}v_{m-1}, \text{ etc.}$$

Then

$$E_+w_j = (m+j+1)^{-1}E_+E_-w_{j+1} = (m+j+1)^{-1}[\Delta - H(H-1)]w_{j+1}$$
$$= (m+j+1)^{-1}[m(m+1) - j(j+1)]w_{j+1}$$
$$= (m-j)w_{j+1}.$$

To summarize, we have proved:

Proposition 11.1

The finite-dimensional irreducible representations of the algebra $su(2)$ are parametrized by m as m ranges over the non-negative half integers. For each value of m, there is a unique irreducible representaton on the vector space V_m of dimension $2m + 1$. The operator Δ takes on the scalar value $m(m + 1)$ on V_m. We can choose a basis, $w_m, w_{m-1}, \ldots, w_{-m}$ of V_m so that

$$Hw_j = jw_j, \quad E_-w_j = (m + j)w_{j-1}, \quad E_+w_j = (m - j)w_{j+1}.$$

We can easily check that these representations are precisely of the form $\dot\rho$, where ρ ranges over the irreducible representations of $SU(2)$. Recall that these are actually representations of $Sl(2, \mathbb{C})$ acting on the space W_m of homogeneous polynomials of degree $2m$. Any $A \in Sl(2, \mathbb{C})$ sends the polynomial, p, into $\rho(A)p$, where

$$\rho(A)p(u) = p(A^{-1}u) \quad \text{for any } u \in \mathbb{C}^2.$$

If we write $u = (x, y)$, it is clear that W_m is spanned by the polynomials

$$x^{2m}, x^{2m-1}y, x^{2m-2}y^2, \ldots, y^{2m}$$

and thus has dimension $2m + 1$. Now

$$\exp tE_+ = \begin{pmatrix} 1 & t \\ 0 & 1 \end{pmatrix}$$

so that

$$\rho(\exp tE_+)p(x, y) = p(x - ty, y)$$

and therefore

$$\dot\rho(E_+)p = -y(\partial p/\partial x).$$

Also

$$\exp tH = \begin{pmatrix} e^{t/2} & 0 \\ 0 & e^{-t/2} \end{pmatrix}$$

so that

$$\rho(\exp tH)p(x, y) = p((\exp - t/2)x, (\exp t/2)y)$$

and

$$\dot\rho(H)p = \tfrac{1}{2}(y(\partial p/\partial y) - x(\partial p/\partial x)).$$

In particular, we see that

$$\dot\rho(E_+)(y^{2m}) = 0$$

and

$$\dot\rho(H)(y^{2m}) = my^{2m},$$

where m is the maximum eigenvalue of H. These were the properties we used to characterize the irreducible representations on V_m in Proposition 11.1.

Now let us consider the group $SO(3)$. The homomorphism $\tau : SU(2) \to SO(3)$ is two-to-one, but is *locally* one-to-one near the identity. That is, given any smooth curve $B(t)$ in $SO(3)$, with

$$B(0) = \begin{pmatrix} 1 & 0 & 0 \\ 0 & 1 & 0 \\ 0 & 0 & 1 \end{pmatrix},$$

there exists a unique smooth curve $A(t)$ in $SU(2)$ with $A(0) = \begin{pmatrix} 1 & 0 \\ 0 & 1 \end{pmatrix}$ and $\tau(A(t)) = B(t)$.

Thus τ induces an *isomorphism* between the Lie algebra of $SU(2)$ and the Lie algebra of $SO(3)$. From this we see that the Lie algebra does not determine the Lie group. We also see that not every representation of the Lie algebra g of a group G comes from a representation of the group; the half integral spin representations of the Lie algebra of $SO(3)$ do not give rise to representations of $SO(3)$.

Let us consider the representation ρ of $SU(2)$ on (smooth) functions on $\mathbb{R}^3$:

$$[\rho(A)f](v) = f(\tau(A)^{-1}v).$$

Taking

$$X_1 = i \begin{pmatrix} \frac{1}{2} & 0 \\ 0 & -\frac{1}{2} \end{pmatrix}$$

$$\exp \theta X = \begin{pmatrix} e^{i\theta/2} & 0 \\ 0 & e^{-i\theta/2} \end{pmatrix}$$

and

$$\tau(\exp \theta X) = R^z_\theta,$$

rotation through angle θ about the z axis. Thus

$$- \dot{\rho}(X_1) = x \frac{\partial}{\partial y} - y \frac{\partial}{\partial x}$$

with similar formulas for $\dot{\rho}(X_2)$ and $\dot{\rho}(X_3)$. Thus

$$\dot{\rho}(\Delta) = \dot{\rho}(X_1^2) + \dot{\rho}(X_2^2) + \dot{\rho}(X_3^2)$$

or

$$\dot{\rho}(\Delta) = - \Delta_{S^2}.$$

From this formula we see immediately that the possible eigenvalues of Δ_{S^2} are $- m(m + 1)$. (The fact that all such eigenvalues actually occur, and the decomposition of $L^2(S^2)$ into irreducibles, each occurring with multiplicity one, require some more refined argument such as the one we presented in Section 4.4.)

In quantum mechanics, the self-adjoint operator

$$\frac{h}{i} \dot{\rho}(X_1)$$

is called the angular momentum about the z axis, and

$$- h^2 \dot{\rho}(\Delta)$$

is called the total angular momentum.

In much of the physical literature, the various facts about the representations of the rotation group are presented and used in Lie algebra form, where they are called the 'theory of angular momentum'.

We close this section by giving a sketch of the group theoretical explanation of the degeneracy of the hydrogen atom spectrum. (Actually, this explanation, using the language of angular momenta, was discovered by Pauli in 1925, one year before Schrödinger obtained his equation.) In the classical central force problem, the bound states are ellipses. Thus in addition to the conservation of angular momentum,

$$\mathbf{L} = \mathbf{r} \times \mathbf{p},$$

which guarantees motion in a plane, and Kepler's area law, the parameters describing the ellipse are also constants of motion. In particular, a direct computation shows that the vector

$$\mathbf{a} = -\frac{\mathbf{p} \times \mathbf{L}}{m} + \frac{e^3}{\|\mathbf{r}\|}\mathbf{r},$$

called the Runge–Lenz vector, is conserved. It can be checked that $\mathbf{a}$ points along the direction of the major axis of the ellipse and has length $\|a\| = e^2\varepsilon$. where ε is the eccentricity of the ellipse.

Pauli's idea was to consider a quantum mechanical version of A. Let $E < 0$ be an eigenvalue of the hydrogen atom Hamiltonian and W_E the corresponding eigenspace. On W_E define the operator $\mathbf{A} = \mathbf{A}_E$ by

$$\mathbf{A} = \frac{1}{2(-2E)^{\frac{1}{2}}}\left(\mathbf{L} \times \mathbf{p} - \mathbf{p} \times \mathbf{L} + \frac{2\mathbf{r}}{r}\right),$$

where $\mathbf{p} = (p_x, p_y, p_z) = -\mathrm{i}h(\partial/\partial x, \partial/\partial y, \partial/\partial z)$ is the vector whose components are the linear momentum operators, etc. Then one checks that the components of $\mathbf{L}$ and of $\mathbf{A}$ all commute with H on W_E. Furthermore, the six component operators satisfy

$$[L_1, L_2] = \mathrm{i}L_3 + \cdots, \text{ etc.}$$
$$[L_1, A_2] = \mathrm{i}A_3 + \cdots, \text{ etc.}$$
$$[A_1, A_2] = \mathrm{i}L_3 + \cdots, \text{ etc.}$$

Setting

$$C_j = -\tfrac{1}{2}\mathrm{i}(L_j + A_j) \quad \text{and} \quad D_j = -\tfrac{1}{2}\mathrm{i}(L_j - A_j)$$

satisfy

$$[C_1, C_2] = C_3, \text{ etc.}$$
$$[D_1, D_2] = D_3, \text{ etc.}$$

and

$$[C_i, D_j] = 0.$$

Thus the Lie algebra spanned by the C's and D's is just $su(2) \times su(2)$. Its irreducible representations are just the tensor product of one irreducible for each $su(2)$ factor, hence are parametrized by pairs of integers, (s, t), each giving the value $s(s + 1)$ and $t(t + 1)$ to the operators

$$C^2 = C_1^2 + C_2^2 + C_3^2 \quad \text{and} \quad D^2 = D_1^2 + D_2^2 + D_3^2.$$

It follows from the definition of A that

$$\mathbf{A} \cdot \mathbf{L} = \mathbf{L} \cdot \mathbf{A} = 0$$

and

$$L^2 + A^2 = \mathbf{L} \cdot \mathbf{L} + \mathbf{A} \cdot \mathbf{A} = -\left[1 + \frac{1}{2E} \right].$$

Thus

$$C^2 = D^2 = \tfrac{1}{4}(L^2 + A^2) = -\tfrac{1}{4}\left[1 + \frac{1}{2E} \right].$$

From this we see that the only irreducibles are ones where $s = t$. The dimension of this irreducible is

$$(2s + 1) \times (2s + 1) = n^2, n = 2s + 1.$$

The corresponding energy is given by solving

$$-\left[1 + \frac{1}{2E} \right] = s(s + 1)$$

for E yielding

$$E = \frac{-1}{2n^2}.$$

Thus we get the energy levels and multiplicities. The subalgebra generated by L acts as the diagonal in $su(2) \times su(2)$. Thus the restriction of $V_s \otimes V_s$ to the diagonal is given by the Clebsch–Gordan formula

$$V_s \otimes V_s = V_0 \oplus V_1 \oplus \cdots \oplus V_{2s}$$

and this is how the $(2s + 1)$st energy level decomposes under $SO(3)$. For more details see Guillemin and Sternberg, *Geometric Asymptotics*, and Guillemin and Sternberg, *Variations on a Theme by Kepler*, AMS Publications.

THE IRREDUCIBLE

REPRESENTATIONS OF $SU(n)$

In this chapter we will construct all the irreducible representations of the groups $SU(n)$ for any n, and also describe the Clebsch–Gordan decomposition for the tensor product of two such irreducibles. In the course of doing so, we will call on a large variety of techniques and describe many applications. Before embarking on the details, it is best to describe in outline how we intend to proceed.

We will begin by establishing the remarkable connection between the irreducible representations of the symmetric group, described at the end of Chapter 2, and certain (actually all polynomial) irreducible representations of the general linear group, $Gl(n)$. This connection was first discovered by Issai Schur and exploited by Hermann Weyl, and the methods involved are purely algebraic. Since $SU(n) \subset Sl(n, \mathbb{C}) \subset Gl(n, \mathbb{C})$, we will have constructed representations of the groups $SU(n)$. We will see that these representations are irreducible, and, in fact, are all the irreducible representations of $SU(n)$. To prove this fact, we will need some detailed information about the structure of the group $Sl(n, \mathbb{C})$, a little bit of complex analysis and some information about representations of Lie algebras. Many of the theorems have a broad range of validity. We shall try to adhere to the policy of formulating the results as theorems about general Lie groups, but proving them in the special cases needed. Having collected our results, we will then describe some of the applications to nuclear and elementary particle physics.

5.1 The representation of $Gl(V)$ on the r-fold tensor product

Let V be a finite-dimensional vector space. The vector space $V \otimes V$ is spanned by decomposable tensors of the forms $\mathbf{x} \otimes \mathbf{y}$, where $\mathbf{x}$ and $\mathbf{y}$ are vectors in V. The symmetric group on two letters, S_2, acts on $V \otimes V$ by the formula

$$(12)\mathbf{x} \otimes \mathbf{y} = \mathbf{y} \otimes \mathbf{x}$$

on decomposable tensors and extending linearly to all tensors. The group S_2 has two irreducible (one-dimensional) representations: the trivial representation and the sgn representation in which (12) is sent into -1. Correspondingly, the space $V \otimes V$ decomposes into the direct sum $V \otimes V = S^2(V) \oplus \wedge^2 V$ where $S^2 V$ consists of symmetric tensors and the space $\wedge^2 V$ consists of anti-symmetric tensors. The space $S^2 V$ is spanned by vectors of the form $\frac{1}{2}(\mathbf{x} \otimes \mathbf{y} + \mathbf{y} \otimes \mathbf{x})$ and is a direct sum of $\frac{1}{2}n(n+1)$ copies of the trivial irreducible representation of S_2 while $\wedge^2 V$ is spanned by tensors of the form $\frac{1}{2}(\mathbf{x} \otimes \mathbf{y} - \mathbf{y} \otimes \mathbf{x})$ and is thus a direct sum of $\frac{1}{2}n(n-1)$ copies of the sgn representation of S_2. Here $n = \dim V$. Notice that if $\dim V = 1$ then there are no anti-symmetric tensors, i.e. $\wedge^2 V = 0$. If A is a linear transformation on V then A acts on $V \otimes V$ by sending $\mathbf{x} \otimes \mathbf{y}$ into $A\mathbf{x} \otimes A\mathbf{y}$ (and extending linearly to all tensors). This action commutes with the action of S_2. In fact,

$$A(12)\mathbf{x} \otimes \mathbf{y} = A(\mathbf{y} \otimes \mathbf{x}) = A\mathbf{y} \otimes A\mathbf{x} = (12)A\mathbf{x} \otimes A\mathbf{y} = (12)A(\mathbf{x} \otimes \mathbf{y}).$$

Thus the group, $Gl(V)$, of all linear transformations of V, acts on $V \otimes V$ in such a way that all of its elements commute with the action of S_2. In particular, each of the spaces $S^2 V$ and $\wedge^2 V$ is invariant under the action of $Gl(V)$. It is not hard to check directly that the representation of $Gl(V)$ on each of these subspaces is irreducible. The purpose of the present section is to develop the corresponding facts when the two-fold tensor product, $V \otimes V$, is replaced by the r-fold tensor product $V \otimes \cdots \otimes V$.

Let $T_r V$ denote the space $V \otimes \cdots \otimes V$ (r times). Thus $T_r V$ is spanned by the decomposable tensors $\mathbf{v}_1 \otimes \cdots \otimes \mathbf{v}_r$. The group S_r acts on the set of such decomposable tensors by

$$s(\mathbf{v}_1 \otimes \cdots \otimes \mathbf{v}_r) = \mathbf{v}_{s^{-1}(1)} \otimes \cdots \otimes \mathbf{v}_{s^{-1}(r)}, \quad s \in S_r.$$

In other words, regard the set of entries $\mathbf{v}_1, \ldots, \mathbf{v}_r$ occurring in the decomposable tensor as a function from the r-element set $\{1, \ldots, r\}$ to V and let S_r act as it usually does on functions. From this point of view it is clear that we do indeed have a group action, but we repeat the proof:

$$\begin{aligned} s(t(\mathbf{v}_1 \otimes \cdots \otimes \mathbf{v}_r)) &= s(\mathbf{v}_{t^{-1}(1)} \otimes \cdots \otimes \mathbf{v}_{t^{-1}(r)}) \\ &= \mathbf{v}_{t^{-1}s^{-1}(1)} \otimes \cdots \otimes \mathbf{v}_{t^{-1}s^{-1}(r)} \\ &= (st)(\mathbf{v}_1 \otimes \cdots \otimes \mathbf{v}_r). \end{aligned}$$

We emphasize that the elements of S_r permute the positions of the vectors occurring in the tensor product but do not change the vectors themselves. For example, when $r = 3$,

$$(123)(\mathbf{x} \otimes \mathbf{y} \otimes \mathbf{z}) = \mathbf{z} \otimes \mathbf{x} \otimes \mathbf{y}.$$

It is clear that this action of S_r on decomposable tensors extends by linearity to give a representation of S_r on $T_r V$.

If A is a linear transformation on V we define the transformation $T_r A$ on $T_r V$ by

$$T_r A(\mathbf{v}_1 \otimes \cdots \otimes \mathbf{v}_r) = A\mathbf{v}_1 \otimes \cdots \otimes A\mathbf{v}_r,$$

for decomposable tensors and extending by linearity to all of $T_r V$. It is clear that every transformation of the form $T_r A$ commutes with the action of any $s \in S_r$.

It is also clear that

$$T_r AB = T_r A T_r B$$

so we get a representation of $Gl(V)$ on $T_r V$. The main task of the next two sections is to show that the representation of $Gl(V)$ on T_rV is completely reducible and to describe its decomposition into irreducibles.

5.2 $Gl(V)$ spans $\mathrm{Hom}_{S_r}(T_rV, T_rV)$

Let S^rV denote the subspace of T_rV consisting of symmetric tensors, i.e. those $w \in T_rV$ which satisfy $sw = w$ for all $s \in S_r$. There is an obvious projection operator from T_rV to S^rV given by averaging over S_r:

Namely, set

$$\&w = \frac{1}{r!}\sum sw$$

where the sum ranges over all s in S_r. Let us denote $\&(v_1 \otimes \cdots \otimes v_r)$ by $v_1 \cdots v_r$, and, if some of the v's are repeated, use the power notation; so, for example, $\&(e \otimes e \otimes e \otimes f \otimes f) = e^3 f^2$. With this notation, if $e_1, \ldots, e_n$ is a basis of V then the elements

$$e_1^{k_1} \cdot e_2^{k_2} \cdots e_n^{k_n}$$

form a basis of S^rV as the $k = (k_1, \ldots, k_n)$ range over all n-tuplets of non-negative integers k_i with $k_1 + \cdots + k_n = r$. The dimension of S^rV is equal to the number of such possible k. We can count them as follows. For each k_i write a sequence of k_i X's followed by a slash, /, to separate the sequence of X's from the next k_{i+1} X's. Thus, for example, corresponding to 5-tuple $(7, 2, 0, 1, 3)$ we write

$$XXXXXXX/XX//X/XXX.$$

We can recover the original n-tuple from a knowledge of the position of the slashes. The total number of symbols, X's and slashes, is $n + r - 1$, and there are $n - 1$ slashes which can occupy any positions. Thus the number of k's is the same as the number of ways of selecting $n - 1$ objects from $n + r - 1$ so

$$\dim S^rV = \binom{n+r-1}{n-1} = \frac{n(n+1)\cdots(n+r-1)}{r!}.$$

Thus, for example, if $\dim V = 3$ then

$$\dim S^3V = 10.$$

Let $a_1, \ldots, a_n$ be numbers and set $v = a_1 e_1 + \cdots + a_n e_n$. Then

$$v^r = \&(v \otimes \cdots \otimes v) = \sum k_1! \cdots k_n! a_1^{k_1} \cdots a_n^{k_n} e_1^{k_1} \cdots e_n^{k_n}.$$

As the a's vary, the v's range over all of V. We claim that the elements of the form v^r span all of S^rV; in other words that

every element of S^rV can be written as a linear combination of elements of the form v^r.

Indeed, let W denote the subspace spanned by all the $\mathbf{v}^r$. We must show that $W = S^r V$. For this it suffices to show that all the elements $\mathbf{e}_1^{k_1} \cdots \mathbf{e}_r^{k_r}$ lie in W. But the limit of elements in W lie in W since W is a subspace. So if we apply the differential operator $(\partial/\partial a_1)^{k_1} \cdots (\partial/\partial a_n)^{k_n}$ to the above formula for $\mathbf{v}^r$ we find that a non-zero multiple of $\mathbf{e}_1^{k_1} \cdots \mathbf{e}_n^{k_n}$ and hence $\mathbf{e}_1^{k_1} \cdots \mathbf{e}_n^{k_n}$ itself lies in W.

We now turn to the question of describing all elements of $\text{Hom}(T_rV, T_rV)$ which commute with S_r. For this purpose it is convenient to make the following identification. The space $\text{Hom}(V, V)$ of all linear transformations of V is itself a vector space so we can form its 'tensor power'

$$T_r(\text{Hom}(V, V)) = \text{Hom}(V, V) \otimes \text{Hom}(V, V) \otimes \cdots \otimes \text{Hom}(V, V) \quad (r \text{ times}).$$

We can define the map $\phi: T_r(\text{Hom}(V, V)) \rightarrow \text{Hom}(T_rV, T_rV)$ by

$$\phi(\mathbf{A}_1 \otimes \cdots \otimes \mathbf{A}_r)(\mathbf{v}_1 \otimes \cdots \otimes \mathbf{v}_r) = \mathbf{A}_1\mathbf{v}_1 \otimes \cdots \otimes \mathbf{A}_r\mathbf{v}_r$$

on decomposable elements, and extending by linearity (in the $\mathbf{A}$'s and in the $\mathbf{v}$'s). It is clear that

$$\phi(\mathbf{A}_1\mathbf{B}_1 \otimes \cdots \otimes \mathbf{A}_r\mathbf{B}_r) = \phi(\mathbf{A}_1 \otimes \cdots \otimes \mathbf{A}_r)\phi(\mathbf{B}_1 \otimes \cdots \otimes \mathbf{B}_r)$$

where the multiplications on the left are in $\text{Hom}(V, V)$ and the multiplication on the right is in $\text{Hom}(T_rV, T_rV)$. It is easy to see that the map ϕ is injective, and since $\text{Hom}(T_rV, T_rV)$ and $T_r\text{Hom}(V, V)$ both have the same dimension, n^{2r}, we see that ϕ is an isomorphism. The symmetric group acts on T_rV. Thus each element of S_r defines an element of $\text{Hom}(T_rV, T_rV)$; and in particular each s in S_r acts on $\text{Hom}(T_rV, T_rV)$ by conjugation, sending $Y \in \text{Hom}(T_rV, T_rV)$ into sYs^{-1}. The symmetric group also acts on $T_r\text{Hom}(V, V)$ when we regard $\text{Hom}(V, V)$ as a vector space. We claim that

$$\phi(sX) = s\phi(X)s^{-1} \quad \text{for any } X \in T_r\text{Hom}(V, V) \text{ any } s \in S_r.$$

To prove this, it suffices to check it on decomposable $X = \mathbf{A}_1 \otimes \cdots \otimes \mathbf{A}_r$ and evaluate both sides on decomposable $\mathbf{v}_1 \otimes \cdots \otimes \mathbf{v}_r$. Now

$$sX = \mathbf{A}_{s^{-1}(1)} \otimes \cdots \otimes \mathbf{A}_{s^{-1}(r)}$$

so

$$\phi(sX)\mathbf{v}_1 \otimes \cdots \otimes \mathbf{v}_r = \mathbf{A}_{s^{-1}(1)}\mathbf{v}_1 \otimes \cdots \otimes \mathbf{A}_{s^{-1}(r)}\mathbf{v}_r$$

while

$$s^{-1}(\mathbf{v}_1 \otimes \cdots \otimes \mathbf{v}_r) = \mathbf{v}_{s(1)} \otimes \cdots \otimes \mathbf{v}_{s(r)}$$

so

$$\phi(X)s^{-1}(\mathbf{v}_1 \otimes \cdots \otimes \mathbf{v}_r) = \mathbf{A}_1\mathbf{v}_{s(1)} \otimes \cdots \otimes \mathbf{A}_r\mathbf{v}_{s(r)}$$

so that

$$s\phi(X)s^{-1}(\mathbf{v}_1 \otimes \cdots \otimes \mathbf{v}_r) = \mathbf{A}_{s^{-1}(1)}\mathbf{v}_1 \otimes \cdots \otimes \mathbf{A}_{s^{-1}(r)}\mathbf{v}_r$$

as was to be proved.

Now an element Y of $\text{Hom}(T_rV, T_rV)$ commutes with all $s \in S_r$, i.e. belongs to $\text{Hom}_{S_r}(T_rV, T_rV)$ if and only if $sYs^{-1} = Y$ for all $s \in S_r$. If $Y = \phi(X)$ this means that $sX = X$ for all $s \in S_r$; in other words that X lies in $S^r(\text{Hom}(V, V))$. We thus see that

$$\text{Hom}_{S_r}(T_rV, T_rV) = \phi(S^r(\text{Hom}(V, V))).$$

Now $S^rHom(V, V)$ is spanned by elements of the form $\mathbf{A} \otimes \mathbf{A} \otimes \cdots \otimes \mathbf{A}$ and the image under φ of such an element is just T_rA. Thus the set of elements of the form T_rA, as A ranges over $Hom(V, V)$ spans $Hom_{S_r}(T_rV, T_rV)$. The group $Gl(V)$ consists of those A in $Hom(V, V)$ which are invertible. The set of linear combinations of T_rA, as A ranges over $Gl(V)$, will span a subspace of $Hom_{S_r}(T_rV, T_rV)$. On the other hand, any linear transformation can be approximated as closely as we like by an invertible transformation. Hence every element of $Hom_{S_r}(T_rV, T_rV)$ can be approximated by a linear combination of T_rA as the A range over $Gl(V)$ and hence the subspace spanned by these linear combinations must already be the whole space. We have thus proved

the space of linear combinations of T_rA, $A \in Gl(V)$ is all of $Hom_{S_r}(T_rV, T_rV)$.

5.3 Decomposition of T_rV into irreducibles

Let $F_1, \ldots, F_p$ be the various inequivalent irreducible representations of S_r. We can decompose T_rV into a direct sum of irreducibles under S_r and collect together all the irreducibles equivalent to a given irreducible representation. That is, we can write

$$T_rV = W_1 \oplus W_2 \oplus \cdots \oplus W_p$$

where W_1 is a direct sum of m_1 irreducible subspaces all equivalent to F_1, W_2 is a direct sum of m_2 irreducible subspaces all equivalent to F_2, etc., with possibly some of the $m_i = 0$, in which case the corresponding W_i do not occur. No element of $Hom_{S_r}(T_rV, T_rV)$ can carry any vector in W_i out of W_i so that every element of $Hom_{S_r}(T_rV, T_rV)$ is a direct sum of operators on each W_i, and commuting with S_r there. Conversely, given an element of $Hom_{S_r}(W_i, W_i)$ for each i, we can take the direct sum of these operators to get an operator of T_rV which commutes with S_r. Thus $Hom_{S_r}(T_rV, T_rV) = \oplus Hom_{S_r}(W_i, W_i)$.

Let us write

$$W_i = U_i \otimes F_i$$

where $U_i = \mathbb{C}^{m_i}$. By equation (6.6) of Chapter 2 we know that the linear combinations of the linear transformations coming from elements of S_r on F_i span all of $Hom(F_i, F_i)$. Thus, if $L \in Hom_{S_r}(W_i, W_i)$, then L must commute with all transformations of the form $I \otimes Q$, where Q is any transformation of F_i. (Let us drop the subscript i temporarily in the discussion, since the argument holds for every i.) Now any transformation of the form $P \otimes I$ commutes with all $I \otimes Q$, since

$$(P \otimes I)(I \otimes Q)(u \otimes f) = (P \otimes I)(u \otimes Qf) = Pu \otimes Qf$$
$$= (I \otimes Q)(Pu \otimes f)$$
$$= (I \otimes Q)(P \otimes I)(u \otimes f).$$

Conversely, if $f_1, \ldots, f_N$ is a basis of F, then we may write

$$L(u \otimes f_i) = \sum_j L_{ij} u_i \otimes f_j$$

as the most general operator on $U \otimes F$, where $L_{ij} \in Hom(U, U)$. Let π_1 be the projection operator of F onto the subspace spanned by f_1 given by $\pi_1(f_j) = 0$, $j \neq 1$, $\pi_1(f_1) = f_1$. Then $(I \otimes \pi_1)(u \otimes f_1) = u \otimes f_1$, so the condition $(I \otimes \pi_1)L = L(I \otimes \pi_1)$ implies that

$$L(u \otimes f_1) = L_{11} u \otimes f_1$$

and similarly $L(u \otimes f_i) = L_{ii} u \otimes f_i$ for each i. Let $\pi_{ij} f_i = f_j$, $\pi_{ij} f_j = f_i$. Then we see that $L_{ii} = L_{jj}$ is the same operator, say P. Thus $L = P \otimes I$.

We thus see that $\text{Hom}_{S_r}(W, W) \sim \text{Hom}(U, U)$. We thus write

$$\text{Hom}_{S_r}(T_rV, T_rV) = \oplus \text{Hom}(U_i, U_i).$$

Now suppose that $R \in \text{Hom}(T_rV, T_rV)$ commutes with all elements of $\text{Hom}_{S_r}(T_rV, T_rV)$. In terms of the decomposition

$$T_r(V) = \oplus W_i = \oplus U_i \otimes F_i$$

let us write R in 'block form' as

$$R = (R_{ij})$$

where $R_{ij} \in \text{Hom}(W_j, W_i)$. The assumption that R commutes with all of $\text{Hom}_{S_r}(T_rV, T_rV)$ says that

$$R_{ij} = (P_j \otimes I) = (P_i \otimes I)R_{ij},$$

where the P_i and P_j can be chosen independently for $i \neq j$. Taking $P_i = I$ and $P_j = 0$ shows that $R_{ij} = 0$ if $i \neq j$ and R_{ii} commutes with all the $P_i \otimes I$ on W_i. But then the argument that we gave above, this time with the roles of U and F interchanged, shows that R must be of the form $R = R_1 + \cdots + R_r$ with $R_i = I \otimes Q_i$ on W_i. But all such operators are linear combinations of elements coming from S_r. We thus have proved that

> if R is a linear transformation which commutes with all of $\text{Hom}_{S_r}(T_rV, T_rV)$, then R is a linear combination of elements coming from S_r.

Each $A \in Gl(V)$ determines T_rA which then gives a linear operator on each U_i. Since $T_rAB = T_rAT_rB$, we see that we get a linear representation, ρ_i, of $Gl(V)$ on U_i. Since the linear combinations of the T_rA span all of $\text{Hom}_{S_r}(T_rV, T_rV)$, we see that the $\rho_i(A)$ span all of $\text{Hom}(U_i, U_i)$. In particular, the representation ρ_i of $Gl(V)$ on U_i is irreducible.

We claim that all the different (ρ_i, U_i) are inequivalent. Indeed, if $L: U_i \to U_j$ commutes with all of $Gl(V)$, then for any $B \in \text{Hom}(F_i, F_j)$, $L \otimes B: W_i \to W_j$ commutes with $\text{Hom}_{S_r}(T_rV, T_rV)$. But, by the preceding, this implies that $L \otimes B$ is a linear combination of the elements of S_r. But each such element leaves the spaces W_i invariant. Hence $L = 0$. To summarize:

There exist distinct irrducible representations (ρ_i, U_i) of $Gl(V)$, each associated with a different representation of (σ_i, F_i) of S_r. We have the decomposition

$$T_rV = U_1 \otimes F_1 \oplus \cdots \oplus U_p \otimes F_p.$$

If $f_i = \dim F_i$ and $s_i = \dim U_i$, we see that each T_rV decomposes, under $Gl(V)$ into a direct sum of f_i copies of each U_i and that T_rV decomposes under S_r into a direct sum of s_i copies of each F_i. Some of the s_i may equal zero, meaning that the corresponding U_i do not occur.

We can describe the Clebsch–Gordan decomposition for these irreducible representations of $Gl(V)$ in terms of the corresponding representations of S_r

Let μ parametrize the various irreducible representations of S_m and let ν parametrize the various irreducible representations of S_n. Let F_λ be a representation of S_{m+n} and regard $S_m \times S_n$ as a subgroup of S_{m+n}. Suppose that for each λ

$$F_\lambda | S_m \times S_n = \sum a_{\mu\nu}^\lambda F_\mu \otimes F_\nu.$$

We claim that then

$$U_\mu \otimes U_\nu = \sum a_{\mu\nu}^\lambda U_\lambda$$

is the Clebsch–Gordan decomposition for $Gl(V)$. Indeed, let $l = m + n$ and write

$$T_l V = \oplus W_\lambda = \oplus U_\lambda \otimes F_\lambda \quad \text{under } S_l$$
$$= \oplus U_\lambda \otimes (\oplus a_{\mu\nu}^\lambda F_\mu \otimes F_\nu) \quad \text{under } S_m \times S_n.$$

On the other hand, $T_l V = T_m V \otimes T_n V$ so, under $S_m \times S_n$, we have

$$T_l V = (\oplus W_\mu) \otimes (\oplus W_\nu) = (\oplus U_\mu \otimes F_\mu) \otimes (\oplus U_\nu \otimes F_\nu)$$
$$= \oplus U_\mu \otimes U_\nu \otimes F_\mu \otimes F_\nu.$$

Comparing the 'coefficients' of the unique irreducible representation $F_\mu \otimes F_\nu$ of $S_m \times S_n$ gives the desired result.

We thus see that the Clebsch–Gordan decomposition for these representations of $Gl(V)$ can be completely determined from facts about the representation theory of the various symmetric groups.

5.4 Computational rules

The following computational rules are easy to state but relatively hard to prove. Proofs will be given in Appendix C. We will provide rules for

(i) dim F_λ, where F_λ is the irreducible representation of S_n associated with the diagram λ (we have already described this in Chapter 2); and

(ii) dim U_λ, where U_λ is the irreducible representation of $Gl(V)$ associated with λ and dim $V = d$.

(iii) The Clebsch–Gordan decomposition:

$$U_\mu \otimes U_\nu = \sum a_{\mu\nu}^\lambda U_\lambda.$$

Recall that for any node in a Young diagram its 'hook length' is the sum of the nodes to the right of it plus the nodes underneath it plus one. Here is a Young diagram with the hook length printed at each node:

$$
\begin{array}{llll}
6 & 4 & 3 & 1 \\
4 & 2 & 1 & \\
1 & & &
\end{array}
$$

Then for a Young diagram λ with n nodes

$$\dim F_\lambda = \frac{n!}{\prod (\text{all hook lengths})}.$$

Thus in the above example $n = 8$ and

(i) $$\dim F_\lambda = \frac{8!}{6 \cdot 4 \cdot 3 \cdot 4 \cdot 2} = 70.$$

We shall see in the next section that if $\dim V = d$, no U_λ can appear if λ has more than d rows. At the i, j node of λ, put $d + j - i$, where i is the row and j is the column. Thus in the above diagram we would have

$$\begin{array}{cccc} d & d+1 & d+2 & d+3 \\ d-1 & d & d+1 & \\ d-2 & & & \end{array}.$$

Multiply all these numbers and divide by the product of the hook lengths. This gives $\dim U_\lambda$. Thus, if $d = 3$,

(ii) $$\dim U_\lambda = \frac{3 \cdot 4 \cdot 5 \cdot 6 \cdot 2 \cdot 3 \cdot 4}{6 \cdot 4 \cdot 3 \cdot 4 \cdot 2} = 15.$$

Thus, if $\dim V = 3$ and

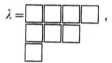

the 15-dimensional representation U_λ occurs 70 times in $T_8 V$.

(iii) The Clebsch–Gordan decomposition. Let λ and μ be diagrams. Write down μ_1 1's, μ_2 2's, etc:

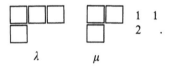

Add the 1's to λ so as to be a legitimate diagram, with no two 1's in the same column:

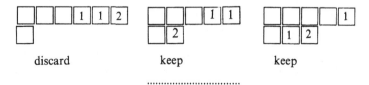

Now add the 2's so as to be a legitimate diagram, with no two 2's in the same column.

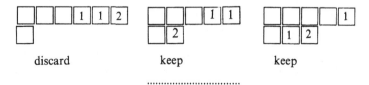

Continue. Discard any diagram which, when reading from right to left in successive rows, has the jth 2 appearing before the jth 1, or the jth 3 appearing before the jth 2, etc., for all integers and any j. Erase the numbers in the added nodes. The diagrams so obtained, counted with multiplicity, give the Clebsch–Gordan coefficients. Example:

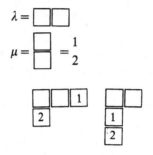

so

$$U_{\square\square} \otimes U_{\substack{\square\\\square}} = U_{\substack{\square\square\square}} \otimes U_{\substack{\square\square\\\square}} \,.$$

(Check the dimensions on both sides of this equation when $d = 3$.)

5.5 Description of tensors belonging to W_λ

In this section we make use of the notations and arguments of Section 2.8. We advise the reader to review that section before proceeding.

Let λ be a Young diagram and F_λ the corresponding irreducible representation of S_r. We know that $T_r V$ has a component $W_\lambda = U_\lambda \otimes F_\lambda$. We now wish to describe which tensors actually belong to W_λ. More precisely, let t be a tableau of type λ. For any $\phi \in \mathrm{Hom}_{S_r}(F_\lambda, T_r V)$, the element $\phi(e_t)$ will be in W_λ. We would like to describe the set of all such tensors as ϕ ranges over all of $\mathrm{Hom}_{S_r}(F_\lambda, T_r V)$. In terms of the decomposition $T_r V = \oplus W_\lambda$ and $W_\lambda = U_\lambda \otimes F_\lambda$, these elements will, of course, all lie in W_λ and be of the form $u \otimes e_t$. But we would like a more transparent description in terms of the symmetry properties of the tensor. We claim the following: let R_t denote the subgroup of S_r which is the isotropy subgroup of $\{t\}$ (so that $\pi \in R_t$ if and only if π permutes the elements in the various rows of t). Let C_t be the subgroup permuting the elements of the columns of t. Set

$$B_t = \sum_{\pi \in R_t} \pi$$

$$A_t = \sum_{\sigma \in C_t} (\mathrm{sgn}\,\sigma)\sigma$$

and

$$E_t = A_t B_t$$

so

$$E_t = \sum_{\substack{\sigma \in C_t \\ \pi \in R_t}} (\text{sgn } \sigma)\sigma\pi.$$

(The elements A_t, B_t and E_t are in the 'group algebra' of S_r – the space of linear combinations of group elements. We can think of this space as $\mathscr{F}(S_r)$ – the space of functions on S_r, if we identify the group element τ with the delta function δ_τ.)

The element E_t acts on $T_r V$ and we claim that

$$U_\lambda \otimes \mathbf{e}_t = \{\phi(\mathbf{e}_t), \phi \in \text{Hom}_{S_r}(F_\lambda, T_r V)\} = E_t(T_r V).$$

In other words, we claim that the desired set of tensors is precisely the image of $T_r V$ under E_t. For example, if $t = {}^{12}_{\ 3}$, then we claim that the corresponding space $\{\phi(\mathbf{e}_t)\}$ is spanned by all tensors of the form

$$\mathbf{x} \otimes \mathbf{y} \otimes \mathbf{z} + \mathbf{y} \otimes \mathbf{x} \otimes \mathbf{z} - \mathbf{z} \otimes \mathbf{y} \otimes \mathbf{x} - \mathbf{z} \otimes \mathbf{x} \otimes \mathbf{y}$$

while if $t = {}^1_2{}^3$ we would get the space spanned by all tensors of the form

$$\mathbf{x} \otimes \mathbf{y} \otimes \mathbf{z} + \mathbf{z} \otimes \mathbf{y} \otimes \mathbf{x} - \mathbf{y} \otimes \mathbf{x} \otimes \mathbf{z} - \mathbf{z} \otimes \mathbf{y} \otimes \mathbf{x}.$$

To prove the claim, we first point out that for any $\pi \in S_r$

$$\pi B_t = B_t \text{ if and only if } \pi \in R_t, \text{ i.e. if and only if } \pi\{t\} = \{t\}.$$

Thus the map of $S_r \cdot B_t \to M_\lambda$ sending πB_t into $\{\pi t\} = \pi\{t\}$ is well defined. This extends to an identification of the space $\{(S_r \cdot B_t)\}$ of all expressions $(\sum_{\pi \in S_r} a_\pi \pi) \cdot B_t \to \mathscr{F}(M_\lambda)$ sending $(\sum a_\pi \pi) \cdot B_t$ into $\sum a_\pi \delta_{\pi\{t\}}$. Under this map, $E_t = A_t \cdot B_t$ goes into $\mathbf{e}_t$. Now F_λ is spanned by all linear combinations of $\tau \mathbf{e}_t$ as τ ranges over all of S_r. Since F_λ is irreducible, this means that we can identify F_λ with the space of all left multiples of E_t in the 'group ring' $F(S_r)$ of S_r, i.e. with the set of all expressions $f \cdot E_t$, where $f = \sum a_\tau \tau$.

We let L_t denote this space. In Chapter 2 we proved that multiplication by E_t on F_λ carries every element into some multiple of $\mathbf{e}_t$, hence left multiplication by E_t on L_t carries every element into some multiple of E_t. In particular,

$$E_t^2 = c_\lambda E_t$$

where c_λ is some constant. We claim that $c_\lambda \neq 0$. In fact, let $f_\lambda = \dim L_t = \dim F_\lambda$. We claim that

$$c_\lambda = \frac{r!}{f_\lambda}.$$

Indeed, consider the operation α of *right* multiplication by E_t on the group ring, i.e.

$$\alpha(f) = f E_t.$$

Then $\alpha(\mathscr{F}(S_r)) \subset L_t$ and, for each $g E_t \in L_t$,

$$\alpha(g E_t) = g E_t E_t = c_\lambda g E_t.$$

Thus, if we choose a basis of $\mathscr{F}(S_r)$ whose first f_λ elements form a basis of L_t, we see that

$$\operatorname{tr}\alpha = f_\lambda c_\lambda,$$

since the matrix of α with respect to this basis has c_λ on the first f_λ diagonal elements and zero elsewhere. On the other hand, if we write

$$E_t = \sum (\operatorname{sgn}\sigma)\sigma\pi,$$

the only term in this sum which is a multiple of the identity e is the term corresponding to $\sigma = e$ and $\pi = e$. Indeed, if $\sigma\pi = e$, then $\sigma^{-1} = \pi$ is a permutation of t which preserves both the rows and the columns of t and hence must be the identity. Let us now choose the basis of $\mathscr{F}(S_r)$ consisting of the elements τ, as τ ranges over S_r. Then the τ component of τE_t is exactly τ. Thus α has 1 at all the diagonal elements relative to this basis. Thus

$$\operatorname{tr}\alpha = r!,$$

proving that

$$c_\lambda = r!/f_\lambda.$$

Thus

$$E_t e_t = c_\lambda e_t, \quad \text{with } c_\lambda = r!/f_\lambda \neq 0.$$

Now let $\phi \in \operatorname{Hom}_{S_r}(F_\lambda, Z)$ for any representation space, Z, of S_r. Let

$$x = \phi(e_t).$$

Then

$$E_t x = E_t \phi(e_t) = \phi(E_t e_t) = c_\lambda \phi(e_t) = c_\lambda x$$

so

$$x = E_t z, \quad \text{where } z = \frac{1}{c_\lambda}x.$$

Conversely, for any z, the map $L_t \to Z$ sending $fE_t \to fE_t z$ is clearly an S_r morphism from L_t (and hence from F_λ) to $T_r V$. This proves our claim that the elements of $U_\lambda \otimes e_t$ in $T_r E$ is spanned by all tensors of the form $E_t z$.

Let p be the number of rows in the Young diagram λ. The operation A_t will involve complete anti-symmetrization with respect to certain tensor positions, namely those which correspond to p entries of t all lying in the same column. In particular,

$$E_t z = A_t B_t z$$

will be completely anti-symmetric in this position. For example, if $t = {1 \atop 3}^2 \cdot$

$$E_t(\mathbf{x}_1 \otimes \mathbf{x}_2 \otimes \mathbf{x}_3) = A_t(\mathbf{x}_1 \otimes \mathbf{x}_2 \otimes \mathbf{x}_3 + \mathbf{x}_2 \otimes \mathbf{x}_1 \otimes \mathbf{x}_3)$$
$$= \mathbf{x}_1 \otimes \mathbf{x}_2 \otimes \mathbf{x}_3 - \mathbf{x}_3 \otimes \mathbf{x}_2 \otimes \mathbf{x}_1$$
$$+ \mathbf{x}_2 \otimes \mathbf{x}_1 \otimes \mathbf{x}_3 - \mathbf{x}_3 \otimes \mathbf{x}_1 \otimes \mathbf{x}_2$$

is completely anti-asymmetric in the first and third positions.

If dim $V < p$, this cannot happen unless $E_t z = 0$. In fact, if $\mathbf{v}_1, \ldots, \mathbf{v}_d$ is a basis of V, every tensor in $T_r V$ is a linear combination of tensors of the form $\mathbf{v}_{i_1} \otimes \cdots \otimes \mathbf{v}_{i_r}$, where the $\mathbf{v}_{i_j}$ are chosen from $\mathbf{v}_1, \ldots, \mathbf{v}_d$. So it suffices to show that $E_t z = 0$ for a tensor of this form. But if $p > d$, then at least two of the p anti-symmetric positions, say i and j, must be occupied by the same $\mathbf{v}_i = \mathbf{v}_j$. Then $(ij)z = z$, since $\mathbf{v}_i = \mathbf{v}_j$, but $(ij)z = -z$, since z is anti-symmetric in this position. Thus $z = 0$.

On the other hand, suppose that $d \geqslant p$. Choose z to be given as a product of the $\mathbf{v}_i$, where $\mathbf{v}_1$ occurs at all positions given by the entries in the first row of t, where $\mathbf{v}_2$ occurs at all positions corresponding to the entries in the second row of t, etc. Thus, for example, if

$$t = \begin{array}{ccc} 3 & 1 & 6 \\ 2 & 4 & \\ 5 & & \end{array} \quad ,$$

choose

$$z = \mathbf{v}_1 \otimes \mathbf{v}_2 \otimes \mathbf{v}_1 \otimes \mathbf{v}_2 \otimes \mathbf{v}_3 \otimes \mathbf{v}_1.$$

Then $\pi z = z$ for any $\pi \in R_t$, and hence $B_t \cdot z$ is a positive multiple of z. But then $E_t Z = A_t B_t z$ is a positive multiple of $A_t z \neq 0$, since $v_1 \wedge \cdots \wedge v_p \neq 0$.

Thus the non-zero W_λ are those that correspond to Young diagrams having at most d rows, where $d = \dim V$.

Let λ be a Young diagram. Let us choose a single Young tableau t with this diagram. We have just proved that the space $E_t(T_r V)$ gives one copy of the irreducible representation U_λ. In particular the elements

$$E_t(\mathbf{v}_{i_1} \otimes \cdots \otimes \mathbf{v}_{i_r})$$

will span the space $E_t(T_r V)$ when $\mathbf{v}_1, \ldots, \mathbf{v}_n$ is a basis of V and the i_j range from 1 to n. Of course these elements will not be linearly independent. We must select from among them to get a basis of $E_t(T_r V)$.

Let us choose t to be the tableau given by writing the integers $1, \ldots, n$ in natural order from left to right in successive rows of λ. Thus for

we take t to be

$$\begin{array}{l} 1234 \\ 56 \\ 7 \end{array} \quad .$$

We then will express the element

$$E_t(\mathbf{v}_{i_1} \otimes \cdots \otimes \mathbf{v}_{i_r})$$

by putting the $\mathbf{v}_{i_j}$ in order in the successive rows of λ. So, in the above example we

would write

$$E_t(\mathbf{v}_1 \otimes \mathbf{v}_2 \otimes \mathbf{v}_3 \otimes \mathbf{v}_4 \otimes \mathbf{v}_5 \otimes \mathbf{v}_6 \otimes \mathbf{v}_7)$$

as

and

$$E_t(\mathbf{v}_1 \otimes \mathbf{v}_1 \otimes \mathbf{v}_1 \otimes \mathbf{v}_2 \otimes \mathbf{v}_2 \otimes \mathbf{v}_2 \otimes \mathbf{v}_3)$$

as

It is not hard to prove that the following elements form a basis. Arrange the vectors so that the subscripts i_j are non-decreasing along the rows and strictly increasing on the columns of λ. Then the set of all such elements give a basis of U_λ. For example, if $\dim V = 3$ and

$$\lambda = \boxed{}\boxed{} \atop \boxed{} \;\; ,$$

then

$$\begin{array}{cccc}
\boxed{\mathbf{v}_1}\boxed{\mathbf{v}_1} & \boxed{\mathbf{v}_1}\boxed{\mathbf{v}_1} & \boxed{\mathbf{v}_1}\boxed{\mathbf{v}_2} & \boxed{\mathbf{v}_1}\boxed{\mathbf{v}_2} \\
\boxed{\mathbf{v}_2} & \boxed{\mathbf{v}_3} & \boxed{\mathbf{v}_2} & \boxed{\mathbf{v}_3} \\[4pt]
\boxed{\mathbf{v}_1}\boxed{\mathbf{v}_3} & \boxed{\mathbf{v}_1}\boxed{\mathbf{v}_2} & \boxed{\mathbf{v}_2}\boxed{\mathbf{v}_2} & \boxed{\mathbf{v}_2}\boxed{\mathbf{v}_3} \\
\boxed{\mathbf{v}_2} & \boxed{\mathbf{v}_3} & \boxed{\mathbf{v}_3} & \boxed{\mathbf{v}_3}
\end{array}$$

form a basis for the eight-dimensional space

$$E_{\boxed{}\boxed{} \atop \boxed{}}(T_3 V),$$

which we can take as a copy of

$$U_{\boxed{}\boxed{} \atop \boxed{}}.$$

5.6 Representations of $Gl(V)$ and $Sl(V)$ on U_λ

We now wish to examine the representations of $Gl(V)$ and $Sl(V)$ on U_λ more closely. The main result, to be proved in Section 5.8, is that the U_λ are all the irreducible finite-dimensional representations of *Sl*(V). To prove this, we need some structural facts about *Sl*(V). We may take $V = \mathbb{C}^d$ and so we study the groups

$$Gl(d, \mathbb{C}) = Gl(\mathbb{C}^d) = \text{all invertible } d \times d \text{ matrices}$$

and

$$Sl(d, \mathbb{C}) = Sl(\mathbb{C}^d) = \text{all } d \times d \text{ matrices of determinant 1.}$$

We let N_+ denote the subgroup consisting of all upper triangular matrices with ones on the diagonal. So

$$N_+ = \left\{ \begin{pmatrix} 1 & - & - & - & - & - & - \\ & 1 & - & - & - & - & - \\ & & 1 & - & - & - & - \\ & & & \cdot & - & - & - \\ & & & & \cdot & - & - \\ & & & & & \cdot & - \\ 0 & & & & & & 1 \end{pmatrix} \right\}$$

Strictly speaking, we should write $N_+(d)$ if we wish to be precise. But d will be held fixed throughout this section, so we will not need to make it explicit. Similarly, we let N_- denote the group of lower triangular matrices wth ones on the diagonal:

$$N_- = \left\{ \begin{pmatrix} 1 & & & & & & \quad 0 \\ - & 1 & & & & & \\ - & - & 1 & & & & \\ - & - & - & \cdot & & & \\ - & - & - & - & \cdot & & \\ - & - & - & - & - & \cdot & \\ - & - & - & - & - & - & 1 \end{pmatrix} \right\}$$

Finally, let $\hat{H}$ denote the subgroup of all non-singular diagonal matrices, so $\delta \in \hat{H}$ looks like

$$\delta = \left\{ \begin{pmatrix} \delta_1 & & \\ & \ddots & \\ & & \delta_d \end{pmatrix} \right\}$$

For any matrix A, let $\Delta_i(A)$ denote the determinant of the $i \times i$ submatrix in the upper left corner, so if

$$A = \begin{pmatrix} a_{11} & a_{12} & a_{13} & \cdots \\ a_{21} & a_{22} & a_{23} & \cdots \\ a_{31} & a_{32} & a_{33} & \cdots \\ \vdots & \vdots & \vdots & \end{pmatrix}$$

then

$$\Delta_1(A) = a_{11}$$

$$\Delta_2(A) = a_{11}a_{22} - a_{12}a_{21} = \det\begin{pmatrix} a_{11} & a_{12} \\ a_{21} & a_{22} \end{pmatrix}$$

$$\Delta_3(A) = \det\begin{pmatrix} a_{11} & a_{12} & a_{13} \\ a_{21} & a_{22} & a_{23} \\ a_{31} & a_{32} & a_{33} \end{pmatrix}$$

etc. The main fact we will need is the so-called 'Gauss decomposition': if A is a matrix with $\Delta_i(A) \neq 0$, $i = 1, \ldots, d$, then there exist unique matrices

$$C_+ \in N_+, \quad C_- \in N_- \quad \text{and} \quad \delta \in \hat{H},$$

such that

$$A = C_- \delta C_+.$$

Proof   The uniqueness is easy: if

$$C'_- \delta' C'_+ = A = C_- \delta C_+$$

then

$$C'^{-1}_- C_- = \delta' C'_+ C_+^{-1} \delta^{-1}.$$

But the left-hand side of this equation is a lower triangular matrix with ones on the diagonal, while the right-hand side is upper triangular. The only way this can happen is for both sides to equal the identity. Thus

$$C_- = C'_- \quad \text{and} \quad \delta' C'_+ C'_+ \delta^{-1} = I.$$

This second equation implies that $\delta'^{-1}\delta = C'_+ C_+^{-1}$. Again, the left-hand side is diagonal and the right-hand side is upper triangular with 1's on the diagonal. This can only happen if $\delta = \delta'$ and $C_+ = C'_+$. This proves the uniqueness.

To prove the existence, let us write a matrix $C \in N_+$ as

$$C = \begin{pmatrix} 1 & x_{12} & x_{13} & \cdots \\ 0 & 1 & x_{23} & \cdots \\ \vdots & \vdots & \vdots & \end{pmatrix}$$

and write out the equations which say that

$$AC \text{ is lower triangular,}$$

i.e. that

$$(AC)_{ij} = 0 \text{ for } i < j.$$

For example

$$(AC)_{12} = a_{11}x_{12} + a_{12}$$

and the equation

$$a_{11}x_{12} + a_{12} = 0$$

can be solved for x_{12} since $a_{11} \neq 0$. Similarly

$$(AC)_{13} = a_{11}x_{13} + a_{12}x_{23} + a_{13}$$
$$(AC)_{23} = a_{21}x_{13} + a_{22}x_{23} + a_{23}$$

so the conditions that $(AC)_{13} = 0$ and $(AC)_{23} = 0$ are a pair of linear equations for x_{13} and x_{23} which can be solved since $\Delta_2(A) \neq 0$. Proceeding this way it is clear that the conditions that the entries in the kth column of AC above the diagonal all vanish is a system of linear equations for $x_{1k}, \ldots, x_{k-1,k}$ which can be solved since $\Delta_{k-1}(A) \neq 0$. Thus AC is a lower triangular matrix. Also, $\det(AC) = \det A = \Delta_d(A) \neq 0$, so that none of the diagonal entries of AC can vanish. Thus we can write

$$AC = C_-\delta \quad C_- \in N_-, \delta \in \hat{H}.$$

Letting $C_+ = C^{-1}$ we get the desired decomposition.

Notice that each of the conditions $\Delta_i(A) = 0$ defines a lower-dimensional set in $Sl(s, \mathbb{C})$ or $Gl(d, \mathbb{C})$. Thus, given any matrix A, we can always find a matrix A^1 arbitrarily close to A and for which the conditions $\Delta_i(A^1) \neq 0$ are all satisfied. In particular, a continuous function on $Sl(d, \mathbb{C})$ or $Gl(d, \mathbb{C})$ is completely determined by its values on all matrices of the form $C_-\delta C_+$.

We let B (or B_+) denote the subgroup of all matrices of the form δC_+, i.e. of all nonsingular upper triangular matrices. If A_1 and A_2 are two elements of B, then $A_1 A_2$ and $A_2 A_1$ have the same diagonal elements. Thus $A_1 A_2 A_1^{-1} A_2^{-1} \in N_+$. If C_1 and C_2 are two elements of N_+, then the entries in $C_1 C_2$ along the diagonal immediately above the main diagonal (i.e. at all the positions i, $i+1$) are obtained by adding the corresponding entries in C_1 and C_2. Thus $C_1 C_2$ and $C_2 C_1$ have the same entries in that diagonal, and $C_1 C_2 C_1^{-1} C_2^{-1}$ has zero there. Let N'_+ denote the group of all such matrices, i.e. all matrices of the form

$$\begin{pmatrix} 1 & 0 & - & - & - & - \\ & 1 & - & - & - & - \\ & & \cdot & - & - & - \\ & & & \cdot & - & - \\ & & & & \cdot & - \\ 0 & & & & & 1 \end{pmatrix}$$

If D_1 and D_2 are two matrices in N'_+, then the same argument shows that $D_1 D_2 D_1^{-1} D_2^{-1}$ has all zeros in the two diagonals above the main diagonal. That is, $D_1 D_2 D_1^{-1} D_2^{-1}$ belong to the subgroup N''_+ of all matrices of the form

$$\begin{pmatrix} 1 & 0 & 0 & - & - & - & - & - \\ & 1 & - & - & - & - & - & - \\ & & \cdot & - & - & - & - & - \\ & & & \cdot & - & - & - & - \\ & & & & \cdot & - & - & - \\ & & & & & \cdot & 0 & 0 \\ & & & & & & 1 & 0 \\ 0 & & & & & & & 1 \end{pmatrix}$$

Continuing this eventually gets us to the trivial, identity subgroup. We claim that this fact implies that the only irreducible finite-dimensional representations of B are one dimensional.

Indeed, this is a consequence of a general group theoretical result known as Lie's theorem. In order to formulate this result, we first need to make some general group theoretical definitions. Let G be any group. The subgroup $[G, G]$ generated by all elements of the form $aba^{-1}b^{-1}$ is called the commutator subgroup of G. If G is a topological group, the closure of $[G, G]$ is called the (first) derived subgroup of G and is denoted by G'. Since

$$c(aba^{-1}b^{-1})c^{-1} = cac^{-1}cbc^{-1}ca^{-1}c^{-1}cb^{-1}c^{-1}$$

we see that $[G, G]$ and hence G' is a normal subgroup of G. Notice that G/G' is commutative.

The kth derived subgroup, $G^{(k)}$, is defined inductively as

$$G^{(k)} = (G^{k-1})'.$$

A group G is called *solvable* if $G^{(k)} = \{e\}$ for some k. The smallest k for which this holds is called the rank of G. Thus a solvable group of rank 1 is Abelian (and hence we know that its irreducible finite-dimensional representations are one dimensional). Lie's theorem states

> Let G be a connected solvable group and (ρ, V) a finite-dimensional irreducible representation of G. Then V is one dimensional.

(The reader may prefer to skip or postpone the proof of this basic theorem until after he sees how we will use it.)

Proof By induction on the rank. We know the theorem for all solvable groups of rank 1. Let us assume the theorem for all solvable groups of rank $k - 1$. Let G be a connected solvable group of rank k. Then G' is connected and solvable of rank $k - 1$. Choose some subspace V_0 of V irreducible for G'. By induction, V_0 is one dimensional and hence there is some function $\lambda: G' \to \mathbb{C}$ such that

$$\rho(b)v = \lambda(b)v \quad \text{for any } v \neq 0 \text{ in } V_0,$$
$$\text{and all } b \in G'.$$

Now for any $a \in G$ and $b \in G'$, we have $a^{-1}ba \in G'$. Therefore

$$\rho(b)\rho(a)v = \rho(a)\rho(a^{-1}ba)v$$
$$= \rho(a)\lambda_a(b)v$$
$$= \lambda_a(b)\rho(a)v$$

where $\lambda_a(b) = \lambda(a^{-1}ba)$. Thus $\rho(a)v$ is another simultaneous eigenvector of G' with eigenvalues $\lambda_a(b)$. Since V is finite dimensional, there are only finitely many possibilities for the values $\lambda_a(b)$, hence these values form a discrete set. Since G is connected, they must all be equal, i.e. $\lambda_a(b) = \lambda(b)$ for all a and b. This implies that the set of simultaneous eigenvectors, i.e. all v, such that

$$\rho(b)v = \lambda(b)v$$

is an invariant subspace, hence all of V.

Now $\rho(b) = \lambda(b)I$ so $\det \rho(b) = \lambda(b)^n$, where $n = \dim V$. On the other hand,

$$\det \rho(a_1 a_2 a_1^{-1} a_2^{-1}) = \det \rho(a_1)\rho(a_2)\rho(a_1)^{-1}\rho(a_2)^{-1} = 1.$$

Thus $\det \rho(b) = 1$ for all $b \in [G, G]$ and hence for all $b \in G'$. Thus $\lambda(b)^n = 1$. Since G' is connected, this implies that $\lambda(b) \equiv 1$.

Thus $\rho(b) = I$ for all $b \in G'$. This implies that ρ gives a representation of the commutative group G/G'. By Schur's lemma, V must be one dimensional.

5.7 Weight vectors

Let (ρ, V) be a finite-dimensional representation of the group $Sl(d, \mathbb{C})$. A vector $v \in V$ is called a *weight vector* if it is a simultaneous eigenvector for all the elements of the subgroup H, where H denotes the subgroup of $\hat{H}$ consisting of those elements with determinant 1, i.e. $\delta_1 \ldots \delta_d = 1$. Thus $\mathbf{i}$ is a weight vector if there exists some complex-valued function μ on H such that

$$\rho(\delta)v = \mu(\delta)v$$

for all $\delta \in H$. This implies that

$$\mu(\delta\delta^1) = \mu(\delta)\mu(\delta^1). \tag{7.1}$$

Let us write

$$\delta = \begin{pmatrix} \delta_1 & & \mathbf{0} \\ & \ddots & \\ \mathbf{0} & & \delta_d \end{pmatrix}$$

Since $\delta_1 \cdots \delta_d = 1$, we can use $\delta_1, \ldots, \delta_{d-1}$ as coordinates on H and (7.1) clearly implies that

$$\mu(\delta) = \delta_1^{n_1} \cdots \delta_{d-1}^{n_{d-1}} \tag{7.2}$$

where the exponents $n_1, \ldots, n_{d-1}$ are integers. It is more convenient not to single out the last coordinate and to write

$$\mu(\delta) = \delta_1^{m_1} \cdots \delta_d^{m_d} \tag{7.3}$$

where the vector $\mathbf{m} = (m_1, \ldots, m_d)$ is only determined up to adding a vector $\mathbf{r} = (r, \ldots, r)$, i.e. $\mathbf{m}$ and $\mathbf{m} + \mathbf{r}$ determine the same μ. The relation between (7.2) and (7.3) is

$$n_i = m_i - m_d \quad \text{for} \quad i = 1, \ldots, d - 1.$$

For example, take $V = \mathbb{C}^d$ and let $\mathbf{e}_1, \ldots, \mathbf{e}_d$ be the standard basis, and take the standard representation of $Sl(\mathbb{C}^d)$ on $\mathbb{C}^d$. Then

$$\rho(\delta)\mathbf{e}_i = \delta_i \mathbf{e}_i$$

so $\mathbf{e}_i$ is a weight vector with $\mathbf{m} = (0, \ldots, 1, \ldots, 0)$ with 1 in the ith position and zeros elsewhere. The $\mathbf{e}_i$ are clearly the only weight vectors in $\mathbb{C}^d$. If we take $V = T_3(\mathbb{C}^d)$ with the representation of $Sl(d, \mathbb{C})$ we have been considering above, then

$$\begin{array}{|c|c|} \hline \mathbf{e}_1 & \mathbf{e}_2 \\ \hline \mathbf{e}_3 \\ \cline{1-1} \end{array}$$

is a weight vector with $\mathbf{m} = (1, 1, 1, 0 \ldots 0)$, and

$$\begin{array}{|c|c|} \hline \mathbf{e}_1 & \mathbf{e}_1 \\ \hline \mathbf{e}_2 \\ \cline{1-1} \end{array}$$

is a weight vector with $\mathbf{m} = (2, 1, 0 \ldots 0)$. If $\lambda = (\lambda_1, \ldots, \lambda_p)$ is an rth order diagram, then in the representation of $Sl(d, \mathbb{C})$ on $T_r(\mathbb{C}^d)$

$$\begin{array}{|c|c|c|c|} \hline \mathbf{e}_1 & \mathbf{e}_1 & \cdots & \mathbf{e}_1 \\ \hline \mathbf{e}_2 & \cdots & \mathbf{e}_2 \\ \cline{1-2} \vdots \end{array}$$

is a weight vector with $\mathbf{m} = (\lambda_1, \lambda_2, \ldots, \lambda_p, 0 \ldots 0)$.

A vector v is called a *maximal weight vector* if it is a weight vector and if it is left fixed by all the elements of N_+.

Thus, in $\mathbb{C}^d$, the vector $\mathbf{e}_1$ is the only maximal weight vector since $\mathbf{e}_1$ is left fixed by all elements of N_+, while elements of N_+ move the remaining $\mathbf{e}_i$. For example

$$\begin{pmatrix} 1 & 1 \\ 0 & 1 \end{pmatrix} \mathbf{e}_2 = \mathbf{e}_2 + \mathbf{e}_1.$$

The element

$$\begin{array}{ccc} \boxed{\mathbf{e}_1} \cdots & & \boxed{\mathbf{e}_1} \\ \boxed{\mathbf{e}_2} \cdots & \boxed{\mathbf{e}_2} & \\ \vdots & & \\ \boxed{\mathbf{e}_k} \cdots \boxed{\mathbf{e}_k} & & \end{array}$$

is a maximal weight vector. Indeed, any element of N_+ sends each $\mathbf{e}_i$ into $\mathbf{e}_i +$ (a combination of $\mathbf{e}_j$ with $j < i$). Now replacing any $\mathbf{e}_i$ by an $\mathbf{e}_j$ with $j < i$ in the above expression will give zero, since we will get a column with a repeated vector.

If v is a maximal weight vector, it is a simultaneous eigenvector for all of $B_+ = H \cdot N_+$. Conversely, let v be a simultaneous eigenvector for B_+. Thus $\rho(n)v = \lambda(n)v$ for all $n \in N_+$. But every element of N_+ can be written as a commutator of elements of B_+. This implies that $\lambda(n) = 1$. Thus

> v is a maximal weight vector if and only if it is a simultaneous eigenvector of B_+.

If (ρ, W) is a finite-dimensional representation of $Sl(d, \mathbb{C})$, choose a subspace W_0 which is irreducible under B_+. Since B_+ is solvable, by Lie's theorem, this is one dimensional. Thus we have proved

> every finite-dimensional representation of $Sl(d, \mathbb{C})$ has a maximal weight vector.

We will now show that an irreducible representation (σ, W) of $Sl(d, \mathbb{C})$ possesses a unique maximal weight vector (up to multiples); in other words, that there is a unique one-dimensional subspace of vectors w in W satisfying

$$\sigma(\delta C_+)v = \mu(\delta)v$$

for all $\delta \in H, C_+ \in N_+$. Furthermore, we will prove that the function μ completely determines the representation (σ, W) up to equivalence.

For this we recall from Chapter 2 that we have an $Sl(d, \mathbb{C})$ morphism from $W \otimes W^* \to \mathcal{F}(Sl(d, \mathbb{C}))$ sending $w \otimes l$ into f^l_w, where

$$f^l_w(a) = \langle \rho(a^{-1})w, l \rangle$$

for $w \in W, l \in W$ and $a \in Sl(d, \mathbb{C})$. Fix some $l \in W^*$. Let $\mathcal{F}^l_w$ denote the space of functions f^l_w as w ranges over W. We know that $\mathcal{F}^l_w$ is a subspace of $\mathcal{F}(Sl(d, \mathbb{C}))$ which is invariant, and the representation of $Sl(d, \mathbb{C})$ on $\mathcal{F}^l_w$ is equivalent to its representation on W. Furthermore, each of the functions f^l_w is continuous, and hence determined by its values on elements a of the form $a = C_- \delta C_+$.

Now choose $l \in W^*$ to be a 'minimal weight vector' for the representation on W^*, i.e.

$$\rho^{*-1}(C_- \delta)l = v(\delta)l = v(\delta)l$$

for $C_- \in N_-$ and $\delta \in H$. By Lie's theorem, we know that such an $l \neq 0$ exists. Then

$$f^l_w(C_+ \delta C_-) = \langle \rho(C_+ \delta C_-)^{-1}w, l \rangle = \langle \rho(C_+^{-1})w, \rho^{*-1}(C_- \delta)l \rangle$$
$$= v(\delta)f^l_w(C_+).$$

If we choose $w = v$ to be a maximal weight vector, we have

$$f^l_v(C_+ \delta C_-) = \langle \rho(C_+ \delta)^{-1}v, f(C_-)^{*-1}l \rangle = \mu(\delta^{-1})\langle v, l \rangle.$$

The function f^l_v cannot vanish on all $C_+ \delta C_-$ for then it would vanish on all of $Sl(d, \mathbb{C})$ by continuity, contradicting the fact that $f^l_v \neq 0$. So $\langle v, l \rangle \neq 0$. We may choose v so that $\langle v, l \rangle = 1$. Then

$$f^l_v(\delta) = \mu(\delta^{-1}) = v(\delta).$$

This shows that the function μ is completely determined – having chosen a minimal weight vector l, with corresponding function v, we must have $\mu(\delta) = v(\delta^{-1})$ which also shows that v is unique. Now consider the function $f^l_v \in \mathcal{F}^l_w$. It is completely determined by the formula

$$f^l_v(C_+ \delta C_-) = \mu(\delta^{-1}).$$

Let r denote the regular representation of $G = Sl(d, \mathbb{C})$ on $\mathscr{F}(G)$. Thus

$$(r(b)f)(a) = f(b^{-1}a).$$

The set of all linear combinations of elements of the form

$$\rho(b)f_v^l$$

is an invariant subspace of $\mathscr{F}_w^l$ and hence must coincide with $\mathscr{F}_w^l$ as $\mathscr{F}_w^l$ is irreducible. It is also equivalent to W. Since any maximal weight vector gives rise to a function which must be a multiple of f_v^l, we see that there is a unique one-dimensional space of maximal weight vectors. We have thus proved the following.

Let (ρ, W) be an irreducible finite-dimensional representation of the group $G = Sl(d, \mathbb{C})$. Then it has a unique one-dimensional space of maximal weight vectors and a corresponding function μ. Define the function $f \in \mathscr{F}(G)$ by the formula

$$f(C_+ \delta C_-) = \mu(\delta^{-1})$$

and extend by continuity to all of G. The space $\mathscr{F}^\mu$ of linear combinations of the $r(a)f$ forms a finite-dimensional subspace of $\mathscr{F}(G)$ and the representation of G on $\mathscr{F}^\mu$ is equivalent to (ρ, W). In particular, every irreducible representation of G is determined, up to equivalence, by its function μ.

5.8   Determination of the irreducible finite-dimensional representations of $Sl(d, \mathbb{C})$

We have seen that any finite-dimensional irreducible representation of $Sl(d, \mathbb{C})$ is completely determined by its function

$$\mu(\delta) = \delta_1^{m_1} \cdots \delta_d^{m_d} \tag{8.1}$$

where the m_i are integers, determined up to replacing the m_i by $m_i + \mathbf{r}$, i.e. by

$$m = (m_1, \ldots, m_d) \to m + r(1, \ldots, 1).$$

A Young diagram gives rise to an irreducible representation with m satisfying

$$m_1 \geqslant m_2 \geqslant \cdots \geqslant m_d. \tag{8.2}$$

Here the m_i are the number of boxes in the ith row of the Young diagram. (Since we are allowed to subtract $m_d(1, \ldots, 1)$ from m without changing the representation, we may assume that $m_d = 0$, i.e. that the Young diagram has at most $d - 1$ rows.)

If we can show that any m occurring in the μ of an irreducible representation must satisfy (8.2), then we will have proved that the representations associated to Young diagrams give all the irreducible finite-dimensional representations of $Sl(d, \mathbb{C})$. This is precisely what we shall prove in this section. It will be convenient and instructive to prove this by passing to the Lie algebra.

The Lie algebra, $sl(d, \mathbb{C})$ of $Sl(d, \mathbb{C})$, consists of all complex $d \times d$ matrices of trace zero. Let E_{ij} denote the matrix which has 1 in the (i,j)th position and 0 elsewhere. Thus $E_{ij} \in sl(d, \mathbb{C})$ if $i \neq j$ while

$$e_i = E_{ii} - E_{i+1,i+1} \in sl(d, \mathbb{C}).$$

It is clear that the E_{ij}, all $i \neq j$, and e_i $(i = 1, \ldots, d-1)$ form a basis of $sl(d, \mathbb{C})$.

Now matrix multiplication yields

$$E_{ij}E_{kl} = \delta_{jk}E_{il}$$

so

$$[E_{ij}, E_{kl}] = \delta_{jk}E_{il} - \delta_{li}E_{kj}. \tag{8.3}$$

In particular,

$$[\textstyle\sum X_i E_{ii}, E_{kl}] = (X_k - X_l)E_{kl}. \tag{8.4}$$

Let n_+ denote the space of strictly upper triangular matrices, i.e. matrices whose non-zero entries all lie above the main diagonal. It is clear that n_+ is a subalgebra of $sl(d, \mathbb{C})$. If $X \in n_+$, then $\exp X = I + X + \frac{1}{2}X^2 \cdots$ is in N_+. This sum is finite since $X^d = 0$. Conversely, suppose that $C = I + L \in N_+$. Then

$$\log C = L - \tfrac{1}{2}L^2 + \cdots \text{(finite sum)}$$

is in n_+ and $\exp(\log C) = C$. Thus n_+ is the Lie algebra of N_+. Similarly, we let h denote the set of diagonal matrices in $sl(d, \mathbb{C})$. It is the Lie algebra of H. Finally, we let n_- consist of all lower triangular matrices. It is the Lie algebra of N_-. Notice that h is a maximal commutative subalgebra of $g = sl(d, \mathbb{C})$, in the sense that any two elements of h commute and any $X \in g$ commuting with all elements of h must belong to h. Also, each of the spaces $\{E_{ij}\}$ is a one-dimensional simultaneous eigenspace under bracket by $X \in h$. This is the content of (8.4). The subalgebra h is called a *Cartan subalgebra* of g.

Recall that a representation (R, V) of the Lie algebra g is a linear map $R: g \to \text{Hom}(V, V)$ such that

$$R(X)R(Y) - R(Y)R(X) = R([X, Y])$$

for all X and Y in g. Any representation (ρ, V) of G gives rise to a representation $(\dot\rho, V)$ of g by the formula

$$\dot\rho(X) = \frac{d}{dt} \rho(\exp tX)|_{t=0}.$$

Not every representation of g comes from a representation of G. A subspace $W \subset V$ is called invariant if $R(X)W \subset W$ for all $X \in g$. The representation is called irreducible if it has no proper invariant subspaces.

Suppose that (ρ, V) is a representation of G and W is an invariant subspace for ρ. Then clearly W is an invariant subspace for the representation $\dot\rho$ of g. Conversely, suppose that W is an invariant subspace for $\dot\rho$. In particular, $\dot\rho(L)W \subset W$ for all $L \in n_+$ and so $\rho(C_+)W \subset W$ for all $C_+ \in N_+$. Similarly, W is invariant under all of H and under all of N_-. Hence $\rho(C_- \delta C_+)W \subset W$ and hence by continuity $\rho(a)W \subset W$ for all $a \in G$. Thus

(ρ, V) is irreducible under G if and only if ($\dot\rho$, V) is irreducible as a representation of g.

Let $v \in V$. The cyclic subspace of v is defined to be the set of all linear combinations of v and all the vectors

$$R(X_n) \cdots R(X_1)v \qquad (8.5)$$

where the $X_1, \ldots, X_n$ are arbitrary elements of g and where $n = 1, 2, 3, \ldots$. This is an invariant subspace since $R(X)$ applied to an element of the form (8.5) is an expression of the same form, with n replaced by $n + 1$ and $X_{n+1} = X$.

A vector $v \in V$ is called a weight vector for the representation R if it is a simultaneous eigenvector for all the elements of h, i.e. if

$$R(X)v = \alpha(X)v$$

for all $X \in h$. Here α is clearly a linear function on h called the *weight* of the vector v.

If (ρ, V) is a representation of G and v is a weight vector for ρ, so

$$\rho(\delta)v = \mu(\delta)v$$
$$\mu(\delta) = \delta_1^{m_1} \cdots \delta_d^{m_d},$$

then v is a weight vector for $\dot\rho$ with $\alpha = \alpha_m$ given by

$$\alpha_m(X) = m_1 x_1 + \cdots + m_d x_d$$

$$\text{if } X = \begin{pmatrix} x_1 & & 0 \\ & \ddots & \\ 0 & & x_d \end{pmatrix}.$$

(Notice that the linear function α_m is unambiguously defined since $\operatorname{tr} X = 0$; adding a constant to all the m_i does not change α_m.)

For example, consider the adjoint representation of G on g given by

$$\rho(a)Y = aYa^{-1}.$$

The corresponding representation, $\dot\rho$, of g on itself is denoted by *ad* and given by

$$ad(X)Y = [X, Y].$$

Equation (8.4) says that each of the elements E_{ij} is a weight vector for the adjoint representation with weight α_{ij} given by

$$\alpha_{ij}(X) = x_i - x_j.$$

The weights of the adjoint representation are called *roots* and the corresponding eigenvectors, E_{ij}, are called *root vectors*.

Let (R, V) be a representation of g and suppose that $v \in V$ is a weight vector with weight α. Then

$$R(X)R(E_{ij})v = R(X)R(E_{ij}) - R(E_{ij})R(X) + R(E_{ij})R(X)v$$

$$= R([X, E_{ij}])v + R(E_{ij})R(X)v$$
$$= (x_i - x_j)R(E_{ij})v + \alpha(X)R(E_{ij})v.$$

In other words, $R(E_{ij})v$ is again a weight vector, this time with weight $\alpha + \alpha_{ij}$.

Since the E_{ij} together with h span all of g, this shows that the cyclic subspace of h is spanned by weight vectors. In particular, if an irreducible representation space contains a weight vector, then it is spanned by weight vectors.

We now restrict attention to the representations (ρ, V) of $G = Sl(V)$ and the corresponding representations $\dot\rho$ of g. Let v be a weight vector with weight α and let S be a permutation matrix, i.e. S is the matrix obtained by applying a permutation $\sigma \in S_d$ to the columns of the identity matrix. Then $S \in U(d)$ and, for any diagonal matrix X,

$$S^{-1}XS = X_\sigma$$

where X_σ is the diagonal matrix with its entries permuted by σ. (The group S_d acting on h is called the Weyl group.)

Now

$$\dot\rho(X)\rho(S)v = \rho(S)\rho(S^{-1})\dot\rho(X)\rho(S)v$$
$$= \rho(S)\dot\rho(S^{-1}XS)v$$
$$= \rho(S)\dot\rho(X_\sigma)v$$
$$= \alpha(X_\sigma)\rho(S)v.$$

So $\rho(S)v$ is again a weight vector whose weight is obtained by applying the permutation σ to α. In other words,

if $(m_1, \ldots, m_d)$ gives a weight, so does $(m_{\sigma(1)}, \ldots, m_{\sigma(d)})$ for any permutation $\sigma \in S_d$.

Now all the weights of a representation of G are given by $\mathbf{m} = (m_1, \ldots, m_d)$, where the m_i can be taken to be integers, in particular real numbers.

Let us (temporarily) agree to choose our representatives m of weights σ such that $m_d = 0$. We then order the weights lexicographically:

$$\mathbf{m} = (m_1, \ldots, m_{d-1}, 0) > \mathbf{n} = (n_1, \ldots, n_{d-1}, 0)$$

if

$$m_1 > n_1, \text{ or, if } m_1 = n_1$$

if

$$m_2 > n_2, \text{ or, if } m_1 = n_1 \text{ and } m_2 = n_2$$

if

$$m_3 > n_3, \text{ etc.}$$

This puts a total order on the space of all weights of a given representation.

Suppose that v is a weight vector with weight α given by $\mathbf{m}$ and $E_{ij} \in n_+$, so $i < j$. Then

$$\dot\rho(E_{ij})v$$

is a weight vector with weight $\alpha + \alpha_{ij}$, where the $\mathbf{m}'$ representing this weight is obtained from **m** by increasing the ith entry by one and decreasing the jth entry by one. Thus

$$\mathbf{m}' > \mathbf{m}.$$

In particular, if we choose v to be a weight vector whose weight is highest among all weights of the representation, we must have

$$\dot{\rho}(E_{ij})v = 0 \quad \text{for all } i < j$$

or

$$\dot{\rho}(C)v = 0 \quad \text{for all } C \in n_+.$$

This implies that $\rho(\exp C)v = v$, i.e. that v is a maximal weight vector in the sense of the preceding paragraph.

Finally, we come to the punch line: if **m** is the highest weight, then

$$m_1 \geqslant m_2 \geqslant \cdots \geqslant m_{d-1} \geqslant m_d. \tag{8.6}$$

Indeed, given any weight **m** we can apply a permutation σ to its entries so that the new weight $\sigma\mathbf{m}$ has the same entries as **m**, but in decreasing order. But then $\sigma\mathbf{m} \geqslant \mathbf{m}$ by the lexicographical ordering. Thus if **m** is the highest weight, its entries must be in decreasing order.

Now we are done! We know that any irreducible representation of $G = Sl(d, \mathbb{C})$ has a unique highest weight vector and is determined up to equivalence by the corresponding highest weight. We have just shown that these highest weights must all satisfy (8.2). But a set of integers satisfying (8.2) (with $m_d = 0$) is just a Young diagram with $d - 1$ rows, and the maximal weight of the corresponding representation is given by **m**. Thus the representation of $Sl(d, \mathbb{C})$ constructed from the Young diagrams give all the irreducible finite-dimensional representations.

In particular, we obtain all the irreducible representations of $SU(d)$. Now we claim that two distinct Young diagrams with at most $d - 1$ rows give rise to two inequivalent representations of $SU(d)$. In fact, if two representations of $SU(d)$ are equivalent, then so are the corresponding representations of the Lie algebra $su(d)$. But any representation of $su(d)$ extends to a representation of its complexification, $su(d) \otimes \mathbb{C}$ which is $sl(d, \mathbb{C})$. Equivalent representations of $su(d)$ give equivalent representations of $sl(d, \mathbb{C})$. The representations of $sl(d, \mathbb{C})$ that we would get would be the ones given by the Young diagrams. But we know that two distinct Young diagrams with at most $d - 1$ rows have different highest weight vectors and so are inequivalent.

For $SU(2)$ we have only the representations $\square, \square\square, \square\square\square$, etc. We know that these correspond to the symmetric powers $S^k(\mathbb{C}^2)$. This gives still another proof of the classification of the irreducible representations for $SU(2)$.

Let us graph some of the weight diagrams for the low-dimensional representations of $SU(3)$. We first must choose a basis for the subalgebra h. Let

$$I_3 = \begin{pmatrix} \frac{1}{2} & 0 & 0 \\ 0 & -\frac{1}{2} & 0 \\ 0 & 0 & 0 \end{pmatrix}$$

Table 38.

Weight vector	**m**	$\alpha(I_3)$	$\alpha(Y)$	$\alpha(T_8)$
$\boxed{e_1}$	$(1, 0, 0)$	$\dfrac{1}{2}$	$\dfrac{1}{3}$	$\dfrac{1}{2 \cdot 3^{\frac{1}{2}}}$
$\boxed{e_2}$	$(0, 1, 0)$	$-\dfrac{1}{2}$	$\dfrac{1}{3}$	$\dfrac{1}{2 \cdot 3^{\frac{1}{2}}}$
$\boxed{e_3}$	$(0, 0, 1)$	0	$-\dfrac{2}{3}$	$-\dfrac{1}{3^{\frac{1}{2}}}$

and

$$Y = \begin{pmatrix} \frac{1}{3} & 0 & 0 \\ 0 & \frac{1}{3} & 0 \\ 0 & 0 & -\frac{2}{3} \end{pmatrix}.$$

These matrices clearly span h, and, as we shall see, play a crucial role in the theory of elementary particles. These matrices also have the desirable property that they are orthogonal relative to the scalar product on h given by the trace:

$$\operatorname{tr} I_3 Y = 0.$$

Unfortunately, they do not have the same length relative to the trace form:

$$\operatorname{tr} I_3^2 = \tfrac{1}{2}, \quad \operatorname{tr} Y^2 = \tfrac{2}{3}.$$

To exhibit the symmetry of the figures under the Weyl group $(= S_3)$, it is best to use a basis with elements of the same length, so we will replace Y by

$$T_8 = \frac{3^{\frac{1}{2}}}{2} Y.$$

Next to each weight we shall, however, write down its values on I_3 and Y since these are of physical interest. So, for example, in the fundamental representation of $SU(3)$ on $\mathbb{C}^3$, the three weight vectors, corresponding m, and values on I_3 and Y and T_8 are tabulated in Table 38, and plotted in Fig. 5.1. For the Young diagram $\boxed{}$ we would have Table 39,

and graph as in Fig. 5.2. Again, we repeat that the symbol $(\frac{1}{2}, -\frac{1}{3})$ next to the lower right-hand point means that the I_3 value is $\frac{1}{2}$ and the Y value is $-\frac{1}{3}$. The actual coordinates of this point in terms of the orthogonal basis I_3, T_8 are $(\frac{1}{2}, -\frac{1}{2 \cdot 3^{1/2}})$.

For the Young diagram $\boxed{}$ we have Table 40, and graph as in Fig. 5.3. For the Young diagram $\boxed{}$ we get Table 41. Notice that the weight 0 occurs with multiplicity 2.

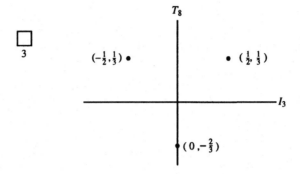

Fig. 5.1

Table 39.

Weight vector	**m**	$\alpha(I_3)$	$\alpha(Y)$	$\alpha(T_8)$
$\boxed{e_1}$ $\boxed{e_2}$	$(1, 1, 0)$	0	$\dfrac{2}{3}$	$\dfrac{1}{3^{\frac{1}{2}}}$
$\boxed{e_1}$ $\boxed{e_3}$	$(1, 0, 1)$	$\dfrac{1}{2}$	$-\dfrac{1}{3}$	$-\dfrac{1}{2\cdot3^{\frac{1}{2}}}$
$\boxed{e_2}$ $\boxed{e_3}$	$(0, 1, 1)$	$-\dfrac{1}{2}$	$-\dfrac{1}{3}$	$-\dfrac{1}{2\cdot3^{\frac{1}{2}}}$

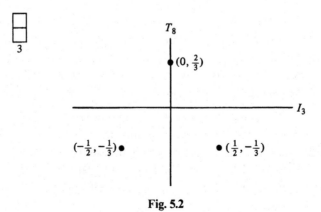

Fig. 5.2

Table 40.

Weight vector	$\mathbf{m}$	$\alpha(I_3)$	$\alpha(Y)$	$\alpha(T_8)$
$\boxed{e_1}\boxed{e_1}$	$(2, 0, 0)$	1	$\dfrac{2}{3}$	$\dfrac{1}{3^{\frac{1}{2}}}$
$\boxed{e_1}\boxed{e_2}$	$(1, 1, 0)$	0	$\dfrac{2}{3}$	$\dfrac{1}{3^{\frac{1}{2}}}$
$\boxed{e_2}\boxed{e_2}$	$(0, 2, 0)$	-1	$\dfrac{2}{3}$	$\dfrac{1}{3^{\frac{1}{2}}}$
$\boxed{e_1}\boxed{e_3}$	$(1, 0, 1)$	$\dfrac{1}{2}$	$-\dfrac{1}{3}$	$-\dfrac{1}{2 \cdot 3^{\frac{1}{2}}}$
$\boxed{e_2}\boxed{e_3}$	$(0, 1, 1)$	$-\dfrac{1}{2}$	$-\dfrac{1}{3}$	$-\dfrac{1}{2 \cdot 3^{\frac{1}{2}}}$
$\boxed{e_3}\boxed{e_3}$	$(0, 0, 2)$	0	$-\dfrac{4}{3}$	$-\dfrac{4}{3^{\frac{1}{2}}}$

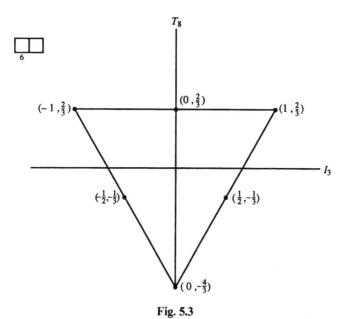

Fig. 5.3

Table 41.

Weight vector	$\mathbf{m}$	$\alpha(I_3)$	$\alpha(Y)$	$\alpha(T_8)$
$\boxed{e_1}\boxed{e_1}$ $\boxed{e_2}$	$(2,1,0)$	$\dfrac{1}{2}$	1	$\dfrac{3^{\frac{1}{2}}}{2}$
$\boxed{e_1}\boxed{e_2}$ $\boxed{e_2}$	$(1,2,0)$	$-\dfrac{1}{2}$	1	$\dfrac{3^{\frac{1}{2}}}{2}$
$\boxed{e_1}\boxed{e_1}$ $\boxed{e_3}$	$(2,0,1)$	1	0	0
$\boxed{e_1}\boxed{e_2}$ $\boxed{e_3}$	$(1,1,1)$	0	0	0
$\boxed{e_1}\boxed{e_3}$ $\boxed{e_2}$	$(1,1,1)$	0	0	0
$\boxed{e_2}\boxed{e_2}$ $\boxed{e_3}$	$(0,2,1)$	-1	0	0
$\boxed{e_1}\boxed{e_3}$ $\boxed{e_3}$	$(1,0,2)$	$\dfrac{1}{2}$	-1	$-\dfrac{3^{\frac{1}{2}}}{2}$
$\boxed{e_2}\boxed{e_3}$ $\boxed{e_3}$	$(0,1,2)$	$-\dfrac{1}{2}$	-1	$-\dfrac{3^{\frac{1}{2}}}{2}$

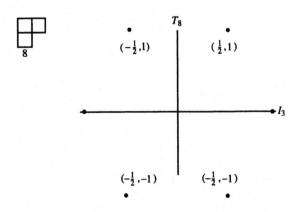

Fig. 5.4

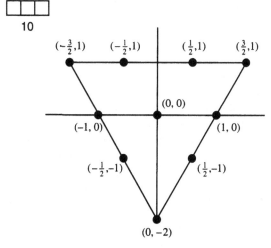

Fig.5.5

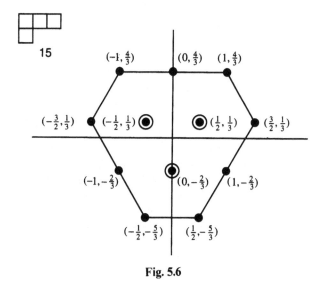

Fig. 5.6

This is indicated in the graph (Fig. 5.4) by a circle around the dot at the origin. Some more diagrams are shown in Figs. 5.5 and 5.6.

5.9   Strangeness

It is now time to return to the study of elementary particles. We begin with a rapid review of what we have presented so far. By the end of the 1930s, it was recognized that there are four types of forces of nature:

The strong interaction, which, for example, is responsible for the forces holding the nucleus together. It is of short range (10^{-13} cm). A particle decaying via the strong interaction typically has a lieftime of about 10^{-23} s.

The electromagnetic interaction. Of long range. About 10^{-2} as strong as the strong interaction.

The weak interaction. Of short range and about 10^{-12} of the strength of the electromagnetic interactions. The weak interaction is responsible for the instability of the neutron, and other forms of β decay.

Gravitation. Of long range, but only 10^{-37} of the strength of the strong interaction. Usually ignored in elementary particle physics.

The particles were also classified by various schemes:

Classification by interaction: hadrons and leptons

Not all particles seem to participate in the strong interaction. Examples are electrons, muons and (their) neutrinos. Such particles are called *leptons*. Particles which do participate in the strong interaction are called *hadrons*. Examples are the nucleons – neutrons and protons – and the pions.

Classification by end product of decays: mesons and baryons

Any hadron will eventually decay into some collection of protons, electrons, photons and neutrinos (and their anti-particles). There may be various decay modes. But the total number of protons – the total number of anti-protons in the final product – is independent of the mode of decay. This integer, called the *baryon number*, and denoted by B, is thus an intrinsic property of the particle. A particle with $B \neq 0$ is called a *baryon*. The earliest examples were the neutron and proton, both with $B = 1$. The pion π^+ has a decay mode

$$\pi^+ \to \mu^+ + \nu$$
$$ \hookrightarrow e^+ + \nu + \bar{\nu}$$

so $B(\pi^+) = 0$. Hadrons with $B = 0$ are called *mesons*.

Electric charge

This is conserved in all interactions. Thus the charge, Q, of a particle is an invariant.

Classification by mass and spin

Each elementary particle corresponds to an irreducible representation of $P = Sl(2, \mathbb{C}) \circledS \mathbb{R}^{1,3}$. It thus is labeled by m and s, where, for $m^2 > 0$, $s = 0, \frac{1}{2}, 1, \frac{3}{2}, \dots$ and for $m = 0$, $s = 0, \pm\frac{1}{2}, \pm 1, \pm\frac{3}{2}$, etc. In elementary particle physics it is customary to use J as the symbol for the spin. Baryons have half-integer spins and mesons have integer spins.

(We have also seen that as far as strong interactions are concerned, P seemed to

Table 42. *Hadrons*

		I	I_3	J	Q	Mass (MeV/c^2)	Mean lifetime (s)
Baryons	p	$\frac{1}{2}$	$\frac{1}{2}$	$\frac{1}{2}$	1	938.3	stable
$B=1$	n	$\frac{1}{2}$	$-\frac{1}{2}$	$\frac{1}{2}$	0	939.6	9.6×10^2
Mesons	π^+	1	1	0	1	139.6	2.6×10^{-8}
$B=0$	π^0	1	0	0	0	135.0	0.89×10^{-16}
	π^-	1	-1	0	-1	139.6	2.6×10^{-8}

Table 43. *Leptons*

	J	Q	Mass (MeV/c^2)	Mean lifetime (s)
ν_e	$\frac{1}{2}$	0	0?	stable
ν_μ	$\frac{1}{2}$	0	0?	stable
e^+	$\frac{1}{2}$	1	0.51	stable
μ^+	$\frac{1}{2}$	1	105.6	2.2×10^{-6}

be a symmetry of nature. Thus, in the fast world of strong interactions, hadrons with positive mass would have an additional invariant – the $\pm$ of intrinsic parity.)

Classification by statistics: fermions and bosons
The spin-statistics theorem asserts that particles with integer spin are *bosons* (the k particle states lie in $S^k(W)$, where W is the space of one particle states) while the particles of non-integer spin are *fermions* (whose k particle states transform as $\Lambda^k(W)$).

Classification by isospin
As we explained in Chapter 4, Heisenberg suggested that, as far as strong interactions are concerned, the proton and neutrons should be regarded as different eigenstates of the operator $I_3 = \left(\begin{smallmatrix} \frac{1}{2} & 0 \\ 0 & -\frac{1}{2} \end{smallmatrix}\right)$ in a two-dimensional representation of a group $SU(2)$. This was fully justified, as we explained, by the pion–nucleon scattering experiments if we group the three pions into the V_1 representations of $SU(2)$. Thus, for hadrons, we have two additional labels – the isospin I labelling the representation of $SU(2)$, and a label of the eigenvalue of $\dot{\rho}_I\left(\begin{smallmatrix} \frac{1}{2} & 0 \\ 0 & -\frac{1}{2} \end{smallmatrix}\right)$ (corresponding to the particle which is an eigenvector) which is conventionally written as I_3.

Thus, the known particles in 1947 had the labels shown in Tables 42 and 43, and, in a class by itself, the photon with $m=1$, $J=1$, $Q=0$ (and it is stable).

Each of the particles in Tables 42 and 43 comes equipped with an anti-particle with opposite charge (and opposite value of I_3 where applicable). Notice that for the hadrons we have

$$Q = I_3 + \tfrac{1}{2}B.$$

In 1947 a number of new particles were discovered in cloud chamber photographs of 'showers' produced by penetrating cosmic ray particles. These new particles were studied intensely over the next few years, at sea level in cloud chambers (where one had to wait a long time between events), and by sending stacks of layers of photographic emulsion up in balloons to great altitudes. There was a great advance when accelerators were able to produce these particles in the laboratory (Brookhaven, 1954).

The new particles fell into two main groups. One consisted of particles heavier than the nucleon which yielded nucleons as parts of their decay products. They thus had baryon number $B = 1$ (or $B = -1$ for their anti-particles). At the time they were called hyperons and denoted by the symbols Λ, Σ and Ξ. The other group were bosons with spin 0 and baryon number 0. They are called K-mesons.

A typical hyperon is the Λ^0, an uncharged particle which decays with a mean lifetime of 2.5×10^{-10} s, the main mode of decay being into a proton and a pion: $\Lambda^0 \to p + \pi^-$. The 'interaction time' for the strong interaction is around 10^{-23} s, the time it takes for a particle moving with the velocity of light to traverse the range of the nuclear forces. Thus the Λ^0 lives a long time on the strong interaction scale – it decays by the weak interaction. On the other hand, the rate at which the Λ^0 and the other new particles were produced was typical of the strong interactions. So these particles can be produced in the strong interaction but decay via the weak interaction. In 1952, Pais made a suggestion of how to explain this phenomenon. Perhaps the new particles were produced in pairs, via the strong interaction, and could only mutually annihilate one another in pairs. If the two particles moved apart, they could no longer individually decay by the strong interaction, and so decayed by the much slower weak interaction. The idea was that the process would be similar to the production

$$\gamma \to e^+ + e^-$$

of an electron position pair from a high energy photon. Once produced (via the electromagnetic interaction), the individual electrons are stable particles. They can only disappear by mutual annihilation. This notion of 'associated production' was confirmed in the laboratory. Thus, for example, a 1.5×10^9 eV pion upon collision with a proton produces a K^0 and a Λ^0 (via the strong interaction). Each subsequently decays (via the weak interaction):

$$\pi^- + p \to K^0 \quad + \quad \Lambda^0$$
$$\to \pi^+ + \pi^- \qquad \quad \;\;\; \vert$$
$$\hookrightarrow \pi^- + p.$$

Fig. 5.7 is a diagram drawn from a cloud chamber photograph of this interaction.

Now the fact that electrons are stable – that they only disappear by mutual annihilation – is usually expressed by saying that (or by 'attributing' the stability to) the conservation of electric charge. Thus, in 1953, Gell-Mann and Nishijima suggested that

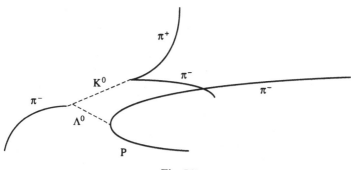

Fig. 5.7

there is a new quantum number (unfortunately) called strangeness, S, which is conserved in strong interactions. Thus, the familiar particles were assigned $S = 0$. Assigning $S = -1$ to Λ^0 and $S = 1$ to K^0 explains why the reaction

$$\pi^- + p \to \Lambda^0 + K^0$$

could take place. The decays $\Lambda^0 \to p + \pi^-$ and $K^0 \to \pi^+ + \pi^-$ could not take place via the strong interaction. They do proceed via the weak interaction with $|\Delta S| = 1$. The Ξ particles (there are two of them, Ξ^0 and Ξ^- (with their anti-particles)) do not decay, even via the weak interaction, directly into nucleons and pions. Rather, they decay via a 'cascade'

$$\Xi^- \to \Lambda^0 + \pi^- \text{ or } \Xi^0 \to \Lambda^0 + \pi^0$$

and

$$\Lambda^0 \to \quad \begin{array}{c} p + \pi^- \\ \text{or} \\ n + \pi^0. \end{array}$$

This could be explained if we assume that $|\Delta S| \leqslant 1$ in the weak interaction. That is, that although the weak interaction does not preserve strangeness, the change in strangeness is at most one unit at a time. Recall that we were led to a similar hypothesis concerning isospin at the end of Section 4.8. The list of known hadrons in 1957 was as in Table 44, and, of course, their anti-particles with the values of S, B and Q reversed.

Before we give the corresponding table for the mesons, we have to discuss the neutral kaon, K^0, a little more closely. As we have already indicated in our discussion of parity, as far as decay modes are concerned, there seem to be two types of kaons, one which decays into two pions and one which decays into three. Those which decay into two pions have a relatively short lifetime of 8.9×10^{-11} s and those which decay into three pions have a longer lifetime of about 5.2×10^{-8} s. This effect can be readily observed since a beam consisting of a mixture of the two types will exhibit the two pion decay mode within a distance of at most $c \cdot (5.2 \times 10^{-8} \text{ s}) = 2.59$ cm, from the source. All decays at a further distance will be into three pions. A careful experiment showed that there was a minute mass difference: the 'long lived' kaon K_L^0 was about 3.6×10^{-12} MeV heavier than the 'short lived' kaon K_S^0. Thus K_L^0 and K_S^0 are not the anti-

Table 44. *Baryons* $B = 1$

	S	I	I_3	Q	J	Mass (MeV/c^2)	Lifetime (s)
Ξ^-	-2	$\frac{1}{2}$	$-\frac{1}{2}$	-1	$\frac{1}{2}$	1321	1.7×10^{-10}
Ξ^0	-2	$\frac{1}{2}$	$\frac{1}{2}$	0	$\frac{1}{2}$	1314	3×10^{-10}
Σ^-	-1	1	-1	-1	$\frac{1}{2}$	1197	1.4×10^{-10}
Σ^0	-1	1	0	0	$\frac{1}{2}$	1192	$< 1 \times 10^{-14}$
Σ^+	-1	1	$+1$	1	$\frac{1}{2}$	1189	0.8×10^{-10}
Λ^0	-1	0	0	0	$\frac{1}{2}$	1115	2.5×10^{-10}
n	0	$\frac{1}{2}$	$-\frac{1}{2}$	0	$\frac{1}{2}$	939	10^3
p	0	$\frac{1}{2}$	$\frac{1}{2}$	1	$\frac{1}{2}$	938	stable

particles of one another. On the other hand, as far as the strong interaction is concerned, the indications were that the K^0 particle has strangeness $S = 1$, and so its anti-particle should have strangeness $S = -1$. Thus certain reactions would proceed strongly for K^0 and not for its anti-particle $CK^0 = \overline{K^0}$. (We use the symbol $C = $ charge conjugation to denote the operation which takes a particle into its anti-particle.) Thus

$$K^0 + p \rightarrow K^+ + n \qquad \overline{K^0} + p \not\rightarrow K^+ + n$$
$$S = 1 \quad +0 \quad 1 \quad +0 \qquad -1 +0 \quad 1 \quad +0$$

and

$$K^0 + n \not\rightarrow \quad K^- + n \qquad \overline{K^0} + p \rightarrow K^- + n$$
$$S = 1 \quad +0 \quad -1 \quad +0 \qquad -1 +0 \quad -1 +0.$$

Thus the K^0 and $\overline{K^0}$ would be produced in different reactions: for example

$$p + n \rightarrow p + \Lambda^0 + K^0$$
$$S = 0 + 0 \quad 0 + \; -1 + 1$$

$$K^- + p \rightarrow \overline{K^0} + n$$
$$S = -1 \quad +0 \quad -1 + 0.$$

The explanation for this behavior of the neutral kaons was given by a brilliant insight of Pais and Gell-Mann in 1955. (In fact, the Gell-Mann–Pais theory was created before many of the experiments were performed, and served as a guide for future experimentation.) Their proposal was that what was at issue here was the symmetry CP – supposed to be conserved in weak interactions. Suppose we assign to the kaons the intrinsic parity -1. As far as the strong interaction is concerned, we have already mentioned that the pions have intrinsic parity -1. Now by our parity assignments to the kaon

$$PK^0 = -K^0$$

and

$$CK^0 = \exp i\theta \overline{K^0}$$

where $\exp i\theta$ is some phase factor. The convention in the choice of this phase factor is 1 so that

$$CPK^0 = \overline{K^0}$$

and

$$CP\overline{K^0} = K^0.$$

Thus K^0 and $\overline{K^0}$ are not eigenstates of CP but

$$\frac{1}{2^{\frac{1}{2}}}(K^0 + \overline{K^0})$$

and

$$\frac{1}{2^{\frac{1}{2}}}(K^0 - \overline{K^0})$$

are, with eigenvalues $+1$ and -1, respectively.

For the pions, let us assume that $C\pi^0 = \pi^0$ and $C\pi^+ = \pi^-$, $C\pi^- = \pi^+$. Thus (with all the pions eigenvectors of P with eigenvalue -1) we have

$$CP(\pi^0 \otimes \pi^0) = \pi^0 \otimes \pi^0$$

and

$$CP(\pi^+ \otimes \pi^- + \pi^- \otimes \pi^+) = \pi^+ \otimes \pi^- + \pi^- \otimes \pi^+$$

while

$$CP(\pi^0 \otimes \pi^0 \otimes \pi^0) = -\pi^0 \otimes \pi^0 \otimes \pi^0$$

and

$$CP(\pi^+ \otimes \pi^- \otimes \pi^0 + \pi^- \otimes {}^+ \otimes \pi^0)$$
$$= -(\pi^+ \otimes \pi^- \otimes \pi^0 + \pi^- \otimes \pi^+ \otimes \pi^0).$$

Thus we can assume that

$$K_s = \frac{1}{2^{\frac{1}{2}}}(K^0 + \overline{K^0})$$

and

$$K_L = \frac{1}{2^{\frac{1}{2}}}(K^0 - \overline{K^0})$$

Then, assuming CP is conserved in weak interaction, K_S^0 can decay into two pions but not into three, while K_L^0 can decompose into three pions, but not into two. The two seemingly different particles K_S and K_L are just (quantum mechanical) superpositions of the K^0 and $\overline{K^0}$. As far as the strong interaction is concerned, where S is a conserved quantum number, the K^0 and $\overline{K^0}$ arise differently. As far as the weak decay is concerned, it is the operator CP which controls. This was strikingly confirmed by the

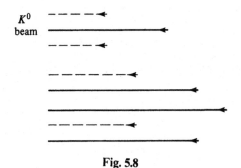

Fig. 5.8

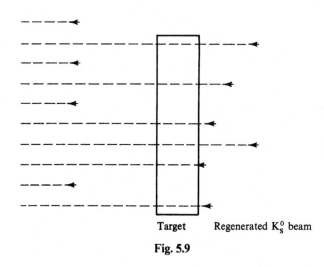

Target   Regenerated K_S^0 beam

Fig. 5.9

following type of experiment. Consider a beam of K^0 particles produced by the reaction

$$p + n \rightarrow p + \pi^0 + K^0.$$

Half the beam decays quickly as K_S^0 into two pions. The other half travels further to decay into three pions as shown in Fig. 5.8. However, if we place a target in the remaining K_L^0 beam (Fig. 5.9), the K^0 and $\overline{K^0}$ components will interact differently because of the conservation of S in strong interactions. As a result, when the beam emerges from the target the relative amplitudes of the K^0 and $\overline{K^0}$ components will have changed. The emerging beam will no longer be pure K_L^0. Some K_S^0 particles will reappear and two pion decays will have been 'regenerated'.

The mesons known in 1957 are given in Table 45. The K^+ and K^0 have been grouped into a single $I = \frac{1}{2}$ representation of the isospin $SU(3)$, while K^0 and K^-, their anti-particles, form another $SU(2)$ representation with $I = \frac{1}{2}$. The anti-particle of the π^+ is the π^- and π^0 is its own anti-particle. The three pions group together to form an $I = 1$ representation of $SU(2)$. There are several important lessons to be

Table 45. *Mesons B = 0*

	S	I	I_3	Q	J	Mass (MeV/c^2)	Lifetime (s)
K^+	1	$\frac{1}{2}$	$\frac{1}{2}$	1	0	493.89	1.2×10^{-8}
K^0	1	$\frac{1}{2}$	$-\frac{1}{2}$	0	0	497.0	$K_S = \frac{1}{2^{1/2}}(K^0 + \overline{K^0})$
$\overline{K^0}$	-1	$\frac{1}{2}$	$\frac{1}{2}$	0	0	497.0	$K_L = \frac{1}{2^{1/2}}(K^0 - \overline{K^0})$
							$\rightarrow 5.2 \times 10^{-5}$
K^-	-1	$\frac{1}{2}$	$-\frac{1}{2}$	-1	0	493.84	1.2×10^{-8}
π^+	0	1	1	1	0	139.5	2.6×10^{-8}
π^0	0	1	0	0	0	139.9	8×10^{-11}
π^-	0	1	-1	-1	0	139.5	2.6×10^{-8}

learned from an examination of the table of baryons and mesons. First of all, for all particles, the charge Q is related to I_3, B and S by the formula

$$Q = I_3 + \tfrac{1}{2}(B + S).$$

This is remarkable, since both B and S were introduced as conserved quantities to explain the non-occurrence of certain types of strong interactions. The combination $B + S$ was denoted by Y and called the 'hypercharge'. Thus the preceding equation would be written as

$$Q = I_3 + \tfrac{1}{2}Y.$$

In fact, at the time that Gell-Mann and Nishijima defined strangeness, the short lived Σ^0 had not been discovered, although on the basis of the preceding formula they had predicted its existence. Similarly, the Ξ^- with its anti-particle was known and the Ξ^0 was predicted and then discovered.

Another important point (whose importance will become clear in terms of the quark model) has to do with relation between particles and anti-particles. For the baryons (with half integer spin), the particles are all different from their corresponding anti-particles. For the mesons (with integral spin) certain neutral particles (the π^0) are identical with their anti-particles, while others (the K^0) are distinct from their anti-particles. This could be explained if we assume that all hadrons are composite – built up from more fundamental spin $\frac{1}{2}$ fermions. The bosons would be made up of an odd number of these particles, such as abc or aab or aaa, etc. So no such composite could be identical to its anti-particle. The mesons would be made up of an even combination of these fundamental particles with their anti-particles, for example, $a\bar{a}$ or $a\bar{b}$. Then $a\bar{a}$ would be identical to its anti-particle $\overline{a\bar{a}} = \bar{a}a = a\bar{a}$, while $a\bar{b}$ would not: $\overline{a\bar{b}} = \bar{a}b \neq a\bar{b}$. We shall see in detail how this is implemented when we study the quark model.

5.10   The eight-fold way

Towards the end of the 1950s, with several major particle accelerators in action, many new particles were discovered. Most of them decayed by the strong interaction, and so had a lifetime of about 10^{-23} s. At the time, there was some reluctance to call them particles because of their ephemeral existence. At the time they were called 'resonances'. Since they were short lived, they could not be detected in the bubble chamber or on any photographic emulsion. There were two principal means of detection:

(i)   Resonances in scattering experiments

We have already briefly described this method in our discussion of the Δ resonance in pion–nucleon scattering. A blip in the (total) scattering cross section as a function of energy (Fig. 5.10) indicates the formations of an intermediate particle at this energy which subsequently decays. Thus, we interpreted the blip at 180 MeV in the $\pi^+ p \to \pi^+ p$ scattering as being caused by the formation of a short lived Δ^{++} which subsequently decays back into a π^+ and a proton, as shown in Fig. 5.11. The protons were targets, i.e. at rest. At 180 MeV for the pion, we find that its (total) momentum P_π is given by $P_\pi = 288$ MeV/c, so at collision we have the total energy (in GeV) as

$$E = 0.14 + 0.94 + 0.18$$

where 0.14 is the rest mass of the pion and 0.94 is the rest mass of the proton. Thus the

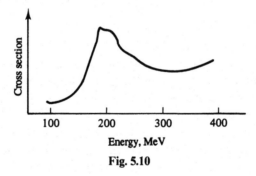

Energy, MeV

Fig. 5.10

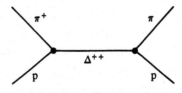

Fig. 5.11

mass of the Δ^{++} is

$$m(\Delta^{++}) = (E^2 - P^2)^{\frac{1}{2}}$$
$$= \{(0.14 + 0.94 + 0.18)^2 - (0.288)^2\}^{\frac{1}{2}} = 1.23.$$

An analysis of the various angles of scattering – the differential scattering cross section together with the principle of conservation of angular momentum – can be used to determine the spin of the short lived particle. (We refer to any text on elementary particle physics for details.) For the Δ particles, the spin was found to be $\frac{3}{2}$.

(ii) Detection by energy momentum correlations

Very few particles live long enough to serve as targets in scattering experiments. So the number of different resonances detected in scattering experiments was rather limited. We now go on to describe the simplest version of another procedure, although the idea extends to more complicated interactions. Suppose a particle A decays at rest into two known particles B and C and we only choose to measure the kinetic energy of B. This kinetic energy is completely determined by the (known) rest masses of A, B and C via the law of conservation of energy–momentum. Indeed, an elementary computation will show that the kinetic energy T_B is given by

$$\frac{(m(A) - m(B))^2 - m(C)^2}{2m(A)}.$$

Thus the distribution of the observed value of the kinetic energy will be a sharp spike at this value of T_B. If particle A decays into three particles

$$A \to B + D + E,$$

then the kinetic energy of B is not uniquely determined by the law conservation of energy–momentum. In fact, it can take any value on the interval

$$0 \leqslant T_B \leqslant \frac{(m(A) - m(B))^2 - (m_D + m_E)^2}{2m_A}.$$

Furthermore, quantum statistics predicts a definite probability distribution for T_B in this range. Thus we expect a probability curve of the shape shown in Fig. 5.12.

Now if, on occasion, we have

$$A \to B + C$$
$$\lfloor\!\!-\!\!\to D + E$$

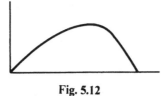

Fig. 5.12

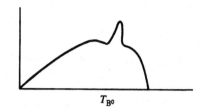

Fig. 5.13

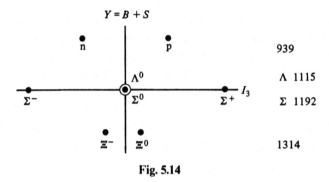

Fig. 5.14

in addition to the direct decay $A \rightarrow B + D + E$, we would get a mix of these two distributions, and so a probability curve of the form shown in Fig. 5.13 would result. The blip in this curve indicates the formation of an intermediate particle whose mass we can compute from T_{B^0}. In practice, the situation is more complicated in that, instead of a decay of a single particle from rest, one studies collisions and subsequent decays. But the principle is the same.

Thus, in 1961, in pion–deuteron scattering experiments

$$\pi^+ + d \rightarrow p + p + \pi^+ + \pi^0 + \pi^-,$$

two peaks were observed. One, corresponding to a mass of 549 MeV, was called the η and the other at 794 MeV the ω.

A large proliferation of particles emerged, and a search was on for finding some organizational principle. The idea was to find some group larger than $SU(2)$ whose representations would give some order to the growing list of particles. After various attempts and false starts, this was achieved by Gell-Mann and Ne'eman working independently. Consider the collection of eight baryons listed at the end of Section 5:9. Suppose we plot this collection on a graph with coordinates Y and I_3 (Fig. 5.14). Notice that they fit the same pattern as the weights for the eight-dimensional (adjoint) representation of $SU(3)$. On the right of the figure we have listed the average mass in MeV for the given isospin family. Similarly, if we add the η to the list of seven mesons in Table 45, we can plot these mesons on the same weight diagram (Fig. 5.15). A number of meson resonances also fit into this octet pattern; for example, see Fig. 5.16. It therefore became reasonable to search for other representations whose weight

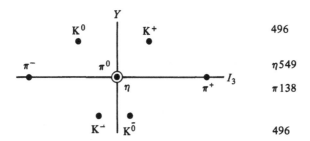

Fig. 5.15

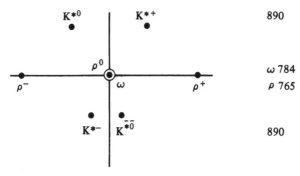

Fig. 5.16

diagrams could be associated to particles. In particular, a scheme was derived which included the four Δ particles which correspond to the $\frac{3}{2}$ representation of $SU(2)$. Remember that we are considering $SU(2)$ as the subgroup of $SU(3)$ which fixes a vector, say

$$\begin{pmatrix} 0 \\ 0 \\ 1 \end{pmatrix}$$

so it sits as

$$\begin{pmatrix} SU(2) & 0 \\ 0 & 1 \end{pmatrix}$$

in $SU(3)$. Now the representation $V_{\frac{3}{2}}$ is $S^3(\mathbb{C}^2)$, so the simplest representation of $SU(3)$ which contains it is the ten-dimensional $S^3(\mathbb{C}^3)$. Known particles which fit part of this scheme were Z and Ξ resonances at 1385 and 1530 MeV, respectively. Thus one obtained Fig. 5.17. Notice the difference of about 150 MeV between the two levels. This suggested that the apex of the triangle should be completed at about 1680 MeV. Notice that since $B = 1$ and $Y = -2$, we would have to have $S = -3$ for this particle. Now the only way this particle could decay via the strong interaction so as to preserve

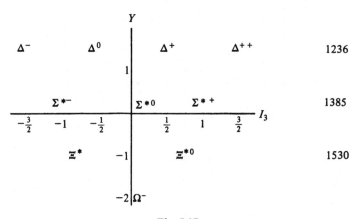

Fig. 5.17

strangeness would be into an $S = -2$ particle and an $S = -1$ particle. The lightest such known combination would be into $\Xi^- + \overline{K^0}$. For this to be energetically possible, the mass of the missing particle would have to be at least the sum of the masses of Ξ and K, i.e. $1321 + 497 = 1818$. As this is more than 1680, the missing particle could not decay via the strong interaction. Thus, in contrast to the other members of the decouplet which are resonances, the missing particle would be observable via its tracks. Furthermore, as $|\Delta S| \leqslant 1$ in the weak interaction, this particle would have to decay in three stages. This particle was named the Ω^-, and its possible decay modes were predicted by Gell-Mann and Ne'eman at a conference at CERN in 1962. In fact, two years later the Ω^- was discovered in the bubble chamber with the decay mode

$$\Omega^- \to \Xi^0 + \pi^-$$
$$\qquad\quad \hookrightarrow \Lambda^0 + \pi^0$$
$$\qquad\qquad\quad \hookrightarrow p + \pi^-.$$

This was justly regarded as a triumph of the $SU(3)$ theory. Indeed, many baryon and meson resonances fit into either the octets, decouplets or singlets. It thus appeared that $SU(3)$ was an approximate symmetry of nature.

5.11 Quarks

What was puzzling in the $SU(3)$ classification scheme was that the only observed multiplets were singlets, octets and decouplets. No other representations of $SU(3)$ seemed to enter. The most obvious representation would be the basic representation on $\mathbb{C}^3$. The three basic vectors (relative to which I_3 and Y are diagonal) were called 'quarks' by Gell-Mann. Assume that such particles exist and call them u, d and s. We wish to regard the baryons and the mesons as composite particles, built up out of the

Table 46.

Name of quark	Symbol	J	I	I_3	B	S	Y	Q
up	u	$\frac{1}{2}$	$\frac{1}{2}$	$\frac{1}{2}$	$\frac{1}{3}$	0	$\frac{1}{3}$	$\frac{2}{3}$
down	d	$\frac{1}{2}$	$\frac{1}{2}$	$-\frac{1}{2}$	$\frac{1}{3}$	0	$\frac{1}{3}$	$-\frac{1}{3}$
strange	s	$\frac{1}{2}$	0	0	$\frac{1}{3}$	-1	$-\frac{2}{3}$	$-\frac{1}{3}$
(anti) $\overline{\text{up}}$	$\bar{\text{u}}$	$\frac{1}{2}$	$\frac{1}{2}$	$\frac{1}{2}$	$-\frac{1}{3}$	0	$-\frac{1}{3}$	$-\frac{2}{3}$
(anti) $\overline{\text{down}}$	$\bar{\text{d}}$	$\frac{1}{2}$	$\frac{1}{2}$	$-\frac{1}{2}$	$-\frac{1}{3}$	0	$-\frac{1}{3}$	$\frac{1}{3}$
(anti) strange	$\bar{\text{s}}$	$\frac{1}{2}$	$\frac{1}{2}$	0	$-\frac{1}{3}$	1	$\frac{2}{3}$	$\frac{1}{3}$

quarks. As the baryons have half integral spin, so must the quarks. Thus the simplest assumption is that the quarks have spin $\frac{1}{2}$ and that the baryons are built up out of three quarks. Then since

$$T_3(V) = S^3(V) + 2T_{\boxminus\boxminus}(V) + \Lambda^3(V) \quad V = \mathbb{C}^3$$

$$27 \;=\; 10 \;+\; 2 \times 8 \;+\; 1$$

we can understand why only 10's, 8's and 1's occur in the number of elements in baryon multiplets. This suggests that the quarks be assigned spin $\frac{1}{2}$ and baryon number $\frac{1}{3}$. We would then identify the anti-particles of the quarks with the dual space $V^* \sim \Lambda^2(V)$, corresponding to the diagram

$$\boxminus \; .$$

Then

$$V \otimes V^* = sl(V) + \text{trivial}$$

$$= \boxminus\boxminus + \boxminus$$

$$8 + 1.$$

Thus, if we assume that mesons are built out of a quark–anti-quark pair, we can understand why only octets and singlets of mesons were observed. We also understand why the mesons all have spin 0 or spin 1. Also, as explained at the end of Section 5.9, we understand why some neutral mesons are their own anti-particles while this does not occur for baryons. Thus, we can construct Table 46 of the quarks and anti-quarks. Then, for example, the baryon octet containing the nucleons would be given (in terms of the basis we have indicated in Section 5.10) as in Fig. 5.18, and the mesons would

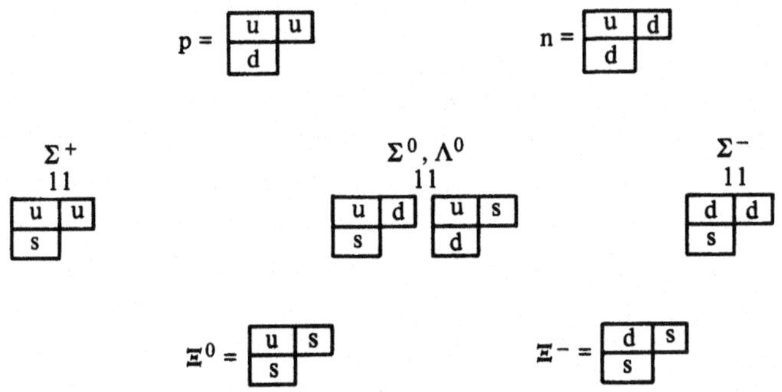

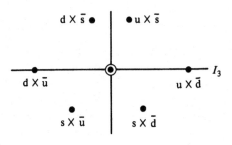

Fig. 5.18

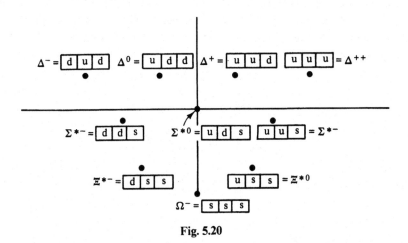

Fig. 5.19

Fig. 5.20

be described in terms of quark–anti-quark pairs as in Fig. 5.19. The two weight zero particles will be traceless diagonal matrices, for example

$$\frac{1}{6^{\frac{1}{2}}}(-d\otimes\bar{d}-u\otimes\bar{u}+2s\otimes\bar{s}) \quad \text{and} \quad \frac{1}{2^{\frac{1}{2}}}(u\otimes\bar{u}-d\otimes\bar{d})$$

if we want two normalized elements. The decouplet of baryon resonances would then be described by Fig. 5.20.

Let us now take the spin into account. We are assuming that the quarks are spin $\frac{1}{2}$ particles. Thus we are regarding a quark state as lying in $\mathbb{C}^3 \otimes \mathbb{C}^2$ under $SU(3) \times SU(2)$. Now we have already mentioned that the Δ's are spin $\frac{3}{2}$ particles. Thus, if the Δ is to be regarded as an element of $T_3(\mathbb{C}^3 \otimes \mathbb{C}^2) = T_3(\mathbb{C}^3) \otimes T_3(\mathbb{C}^2)$, the $T_3(\mathbb{C}^2)$ component is $S^3(\mathbb{C}^2)$ since this is the $J = \frac{3}{2}$ representation of $SU(2)$. Thus, the decouplet will correspond to the irreducible representation of $SU(3) \times SU(2)$ on $S^3(\mathbb{C}^3) \otimes S^3(\mathbb{C}^2)$. Notice that

$$ S^3(\mathbb{C}^3) \otimes S^3(\mathbb{C}^2) \subset S^3(\mathbb{C}^3 \otimes \mathbb{C}^2). $$

Also notice that $\mathbb{C}^3 \otimes \mathbb{C}^2$ is six dimensional so that $\dim S^3(\mathbb{C}^3 \otimes \mathbb{C}^2) = 56$. We claim that

$$ S^3(\mathbb{C}^3 \otimes \mathbb{C}^2) = S^3(\mathbb{C}^3) \otimes S^3(\mathbb{C}^2) \oplus U^{\square\square}(Sl(3)) \otimes U^{\square\square}(Sl(2)) $$

$$
\begin{aligned}
56 &= 40 + 16 \\
&= 10 \times 4 + 8 \times 2
\end{aligned}
$$

as representations of $SU(3) \times SU(2)$. A dimension count will show that to prove the above decomposition, it is enough to prove that

$$ T_{\square\square}(\mathbb{C}^3) \otimes T_{\square\square}(\mathbb{C}^3) $$

contains some symmetric tensors. Here we use

$$ T_{\square\square}(\mathbb{C}^3) $$

to denote the subspace of $T_3(\mathbb{C}^3)$ transforming as the representation $\square\square$ of S_3. That is, it is the space we denoted $W^{\square\square}$ for the space $V = \mathbb{C}^3$, and similarly for $\mathbb{C}^2$. Indeed, the space $T_{\square\square}(V)$ is a direct sum of copies of $U_{\square\square}(Sl(V))$ for $V = \mathbb{C}^2$ or $\mathbb{C}^3$. So $T_{\square\square}(\mathbb{C}^3) \otimes T_{\square\square}(\mathbb{C}^2)$ is a direct sum of the irreducible representations $U_{\square\square}(SL(3)) \otimes U_{\square\square}(Sl(2))$. If we think of $T_{\square\square}(\mathbb{C}^3) \otimes T_{\square\square}(\mathbb{C}^2)$ as a subspace of $T_3(\mathbb{C}^3) \otimes T_3(\mathbb{C}^2) = T_3(\mathbb{C}^3 \otimes \mathbb{C}^2)$, the set of symmetric tensors in $T_{\square\square}(\mathbb{C}^3) \otimes T_{\square\square}(\mathbb{C}^2)$ is an invariant subspace, and so must be a direct sum of copies of $U_{\square\square}(SL(3)) \otimes U_{\square\square}(Sl(2))$. However, this space is already 16 dimensional. So if we prove that

$$ S^3(\mathbb{C}^3 \otimes \mathbb{C}^2) \cap T_{\square\square}(\mathbb{C}^3) \otimes T_{\square\square}(\mathbb{C}^2) \neq \{0\}, $$

then, since $56 - 40 = 16$, we will know that the above intersection consists precisely of one copy of $U_{\square\square}(SL(3)) \otimes U_{\square\square}(Sl(2))$. Now we know from the representation theory of S_3 (indeed it is a trivial computation with characters) that

$$ F^{\square\square} \otimes F^{\square\square} = F^{\square\square\square} \oplus F^{\square\square} \oplus F^{\square}_{\square\square}. $$

Thus

$$T_{\boxminus}(\mathbb{C}^3) \otimes T_{\boxminus}(\mathbb{C}^2)$$

contains a non-trivial subspace which transforms as the

(trivial) representation under S_3, i.e. contains symmetric tensors. Thus

$$S^3(\mathbb{C}^3 \otimes \mathbb{C}^2) = S^3(\mathbb{C}^3) \otimes S^3(\mathbb{C}^2) + T_{\boxminus}(\mathbb{C}^3) \otimes T_{\boxminus}(\mathbb{C}^2).$$

Now

$$T_{\boxminus}(\mathbb{C}^2) \sim V_{\frac{1}{2}}$$

as a representation of $SU(2)$, and this is the spin of the nucleon. This suggests that the baryon decouplet and the octet both fit together as $S^3(\mathbb{C}^3 \otimes \mathbb{C}^2)$ according to the above decomposition. This is given striking confirmation from the study of nucleon magnetic moments, which will also allow us to compute quark masses. We first need to make some preliminary remarks.

In (classical, non-relativistic) electromagnetic theory, the magnetic energy of a small magnet in the presence of magnetic field B is given by

$$E_{\mathrm{mag}} = -\boldsymbol{\mu} \cdot \mathbf{B}$$

where $\boldsymbol{\mu}$ is the magnetic (dipole) moment. Now the magnetic field B (at a given point) is an element of $\Lambda^2(\mathbb{R}^3) \sim su(2) = o(3)$. So, in quantum mechanics, $\boldsymbol{\mu}$ should be considered as a representation operator corresponding to the (coadjoint) representation of $SU(2)$ on $su(2)^* \sim su(2)$. Since the angular momentum is also such a representation operator, we expect that

$$\boldsymbol{\mu} = \mu \mathbf{J}$$

where μ is a scalar and $\mathbf{J}$ is the angular momentum. Indeed, in classical physics, a magnet is considered as coming from a small current loop: if a particle of charge q moves with velocity v in a circular orbit of radius r, the particle revolves $v/2\pi r$ times/s and hence produces a current $qv/2\pi r$. The magnetic moment of such a magnet is given by

$$\frac{1}{c} \, \mathrm{current} \times \mathrm{area} = \frac{1}{2c} qvr$$

(pointing in the direction perpendicular to the plane). The angular momentum of a particle moving in such a circle is

$$mvr$$

(pointing in the same direction). So classically we have

$$\boldsymbol{\mu} = \frac{q}{2mc} \mathbf{J}$$

or $\mu = q/2mc (= q/2m$ in units where $c = 1$). If this simple-minded model is correct, we would expect that the magnetic moment of the proton would be the nuclear magneton defined as

$$\mu_N = \frac{q\hbar}{2m_p c}$$

(where q is the proton charge and m_p the proton mass and $\hbar = h/2\pi$, where h is Planck's constant). It would also predict that the neutron magnetic moment should be zero. The observed values are

$$\mu_p = 2.78\,\mu_N$$
$$\mu_n = -1.92\,\mu_N.$$

These values can be explained in terms of our picture of the nucleons being composite – in particular the assumption that the nucleon octet sits as inside $T_3(\mathbb{C}^3 \otimes \mathbb{C}^2)$ as the completely symmetric component of

$$T_{\boxminus}\mathbb{C}^3 \otimes T_{\boxminus}\mathbb{C}^2.$$

Indeed, we shall see that it follows from this hypothesis about the nucleon octet that $\mu_p = 3\mu_N$ and $\mu_n = -2\mu_N$, which are reasonably close to the observed values.

Let us make the following two assumptions:

(i) That there is no orbital component to the magnetic moment of the 'composite' nucleons, so the magnetic dipole moment of the member of the nucleon octet is the sum of the magnetic dipole moments of the component quarks.

(ii) For the individual quarks, we have the formula

$$\mu = \frac{q}{2m}\mathbf{J}.$$

Let us explain in more detail what we mean by condition (i). We assume that each quark has its own magnetic moment, so we have

$$\boldsymbol{\mu}_u = \mu_u \mathbf{J}$$
$$\boldsymbol{\mu}_d = \mu_d \mathbf{J}$$
$$\boldsymbol{\mu}_s = \mu_s \mathbf{J}$$

for each of the three quarks. Then, in $T_3(\mathbb{C}^3 \otimes \mathbb{C}^2) = T_3(\mathbb{C}^2) \otimes T_3(\mathbb{C}^2)$, we would have an $SU(2)$ representation operator $\boldsymbol{\mu}_3$ in

$$\mathrm{Hom}_{SU(2)}(su(2)), \mathrm{Hom}(T_3(\mathbb{C}^3 \otimes \mathbb{C}^2), T_3(\mathbb{C}^3 \otimes \mathbb{C}^2))$$

defined by

$$\boldsymbol{\mu}_3(\xi)(u \otimes u \otimes d \otimes \alpha \otimes \beta \otimes \gamma)$$
$$= u \otimes u \otimes d \otimes (\mu_u \mathbf{J}(\xi)\alpha \otimes \beta \otimes \gamma + \alpha \otimes \mu_u \mathbf{J}(\xi)\beta \otimes \gamma + \alpha \otimes \beta \otimes \mu_d \mathbf{J}(\xi)\gamma)$$

for any $\xi \in su(2)$, where $\alpha, \beta, \gamma \in \mathbb{C}^2$, and with similar expressions for all the basis elements. Let P denote the projection of $T_3(\mathbb{C}^3 \otimes \mathbb{C}^2)$ onto the subspace *Bar* corresponding to the 16-dimensional baryon subspace of $S^3(\mathbb{C}^3 \otimes \mathbb{C}^2)$. Thus

$$P = R \circ \left(E^3_{\scriptsize\yng(2,1)} \otimes E^2_{\scriptsize\yng(2,1)} \right)$$

where

$E^3_{\scriptsize\yng(2,1)}$ is the projection of $T_3(\mathbb{C}^3)$ onto $T_{\scriptsize\yng(2,1)}(\mathbb{C}^3)$

and

$E^2_{\scriptsize\yng(2,1)}$ is the projection of $T_3(\mathbb{C}^2)$ onto $T_{\scriptsize\yng(2,1)}(\mathbb{C}^2)$

(both corresponding to the character

$$\chi_{\scriptsize\yng(2,1)}$$

of S^3), and where R is the projection of

$$T_{\scriptsize\yng(2,1)}(\mathbb{C}^3) \otimes T_{\scriptsize\yng(2,1)}(\mathbb{C}^2)$$

onto the symmetric subspace under S_3 (in the decomposition

$$\yng(2,1) \otimes \yng(2,1) = \yng(3) \otimes \yng(2,1) + \yng(1,1,1)$$

under S^3). Thus R is given by complete symmetrization. So assumption (i) says that

$$P\mu_3 P \in \mathrm{Hom}_{su(2)}(su(2), \mathrm{Hom}(Bar, Bar)).$$

Since

$$Bar = U_{\scriptsize\yng(2,1)}(\mathbb{C}^3) \otimes U_{\scriptsize\yng(2,1)}(\mathbb{C}^2)$$

$$= U_{\scriptsize\yng(2,1)}(\mathbb{C}^3) \otimes V_{\frac{1}{2}},$$

we conclude that

$$P\mu P = A \otimes \mathbf{J}_{\frac{1}{2}}$$

where

$$A \in \mathrm{Hom}\left(U_{\scriptsize\yng(2,1)}(\mathbb{C}^3) \otimes U_{\scriptsize\yng(2,1)}(\mathbb{C}^3) \right)$$

and $\mathbf{J}_{\frac{1}{2}}$ is the vector operator which assigns to each $\xi \in su(2)$ its image in the representation of $su(2)$ on $V_{\frac{1}{2}}$, i.e.

$$\mathbf{J}(\xi) = \dot{\rho}_{\frac{1}{2}}(\xi).$$

The baryons form a basis, p, n, Σ^+, Σ^0. Σ^-, Λ^0, Ξ^0, Ξ^- of $U_{\square\!\square}(\mathbb{C}^3)$. We can write the

operator A as a matrix relative to this basis. For any one of the baryons, y, the corresponding diagonal entry A_{yy} will give the magnetic moment coefficient. In other words, the magnetic moment coefficient of the proton for example will be

$$\mu_p = A_{pp}$$

so that the magnetic dipole moment of the proton in a definite spin state α will be the vector $\boldsymbol{\mu}_{p,\alpha} \in \mathbb{R}^{3*}$ given by

$$\mu_{p,\alpha}(\xi) = A_{pp}\langle \alpha, \dot{\rho}_{\frac{1}{2}}(\xi)\alpha \rangle.$$

The non-zero weight vectors, that is p, n, Σ^+, Σ^-, Ξ^0 and Ξ^- are determined (up to a phase factor), because these weight spaces are one dimensional. The zero weight space is two dimensional and there is no preferred way of choosing a basis. So further information will be needed to specify the magnetic moments of Σ^0 and Λ^0.

To compute the coefficient A_{pp} (and similar coefficients), we can proceed as follows. Choose the element α in some definite spin state, say in the eigenstate $+\frac{1}{2}$ for the operator $\dot{\rho}_{\frac{1}{2}}(\xi)$, where

$$\xi = \begin{pmatrix} \frac{1}{2} & 0 \\ 0 & -\frac{1}{2} \end{pmatrix}.$$

Then choose the $z \in T_3(\mathbb{C}^3) \otimes T_3(\mathbb{C}^2)$ which corresponds to this element (and has unit length). Then

$$\tfrac{1}{2}A_{pp} = A_{pp}\langle \alpha, \dot{\rho}_{\frac{1}{2}}(\xi)\alpha \rangle = \langle z, \mathbf{J}_3(\xi)z \rangle.$$

Our problem, therefore, is to find the elements z corresponding to the different baryons. There will be two types of baryons: an

$$\boxed{a}\boxed{b}$$
$$\boxed{c}$$
type: $\Sigma^0, \Lambda^0 = \boxed{u}\boxed{d}$ and $\boxed{u}\boxed{s}$
$$\qquad\qquad\quad \boxed{s} \qquad\qquad \boxed{d}$$

and the

aab type: $p = \boxed{u}\boxed{u}$, $n = \boxed{u}\boxed{d}$, $\Sigma^+ = \boxed{u}\boxed{u}$
$$\qquad\qquad\quad \boxed{d} \qquad\qquad\quad \boxed{d} \qquad\qquad\quad \boxed{s}$$

$$\Xi^0 = \boxed{u}\boxed{s} \quad , \quad \Xi^- = \boxed{d}\boxed{s} \quad , \quad \Xi^- = \boxed{d}\boxed{d}$$
$$\qquad\quad \boxed{s} \qquad\qquad\quad \boxed{s} \qquad\qquad\quad \boxed{s}$$

It is immediate to check that

$$\frac{1}{2\cdot3^{\frac{1}{2}}}\{(u\otimes d\otimes s - d\otimes u\otimes s)\otimes(\uparrow\otimes\downarrow\otimes\uparrow - \downarrow\otimes\uparrow\otimes\uparrow) + \text{cyclic permutation}\}$$

and

$$\frac{1}{2\cdot3^{\frac{1}{2}}}\{(u\otimes s\otimes d - s\otimes u\otimes d)\otimes(\uparrow\otimes\downarrow\otimes\uparrow - \downarrow\otimes\uparrow\otimes\uparrow) + \text{cyclic permutation}\}$$

lie in $S^3(\mathbb{C}^3 \otimes \mathbb{C}^2)$ and in

$$T_{\boxed{}}(\mathbb{C}^3) \otimes T_{\boxed{}}(\mathbb{C}^2)$$

and hence span the two-dimensional space spanned by Λ^0 and Σ^0. Similarly, any state of the form aab will have its spin $+\frac{1}{2}$ given by z of the form

$$\frac{1}{3\cdot 2^{\frac{1}{2}}} \{a \otimes a \otimes b \otimes (2\uparrow \otimes \uparrow \otimes \downarrow - \uparrow \otimes \downarrow \otimes \uparrow - \downarrow \otimes \uparrow \otimes \uparrow) + \text{cyclic permutation}\}.$$

Now for a state of type aab and $\xi = \begin{pmatrix} 1 & 0 \\ 0 & -1 \end{pmatrix}$, we have

$$\boldsymbol{\mu}_3(\xi)z = \frac{1}{3\cdot 2^{\frac{1}{2}}} \{a \otimes a \otimes b \otimes (\mu_a[2\uparrow \otimes \uparrow \otimes \downarrow - \uparrow \otimes \downarrow \otimes \uparrow + \downarrow \otimes \uparrow \otimes \uparrow]$$

$$+ \mu_a[2\uparrow \otimes \uparrow \otimes \downarrow + \uparrow \otimes \downarrow \otimes \uparrow - \downarrow \otimes \uparrow \otimes \uparrow]$$

$$+ \mu_b[-2\uparrow \otimes \uparrow \otimes \downarrow - \uparrow \otimes \downarrow \otimes \uparrow - \downarrow \otimes \uparrow \otimes \uparrow])$$

$$+ \text{cyclic permutations}\}$$

$$= \frac{1}{3\cdot 2^{\frac{1}{2}}} \{a \otimes a \otimes b \otimes [(4\mu_a - 2\mu_b)\uparrow \otimes \uparrow \otimes \downarrow - \mu_b \uparrow \otimes \downarrow \otimes \uparrow - \mu_b \downarrow \otimes \uparrow \otimes \uparrow]$$

$$+ \text{cyclic permutations}\}.$$

Thus

$$\langle z, \boldsymbol{\mu}_3(\xi)z \rangle = \frac{1}{9\cdot 2} \cdot 3[8\mu_a - 4\mu_b + \mu_b + \mu_b]$$

$$= \frac{1}{3}[4\mu_a - \mu_b].$$

Thus we obtain

$$\mu_p = \tfrac{1}{3}[4\mu_u - \mu_d], \quad \mu_n = \tfrac{1}{3}[4\mu_d - \mu_u], \quad \mu_{\Sigma^+} = \tfrac{1}{3}(4\mu_u - \mu_s)$$

$$\mu_{\Xi^0} = \tfrac{1}{3}[4\mu_s - \mu_u], \quad \mu_{\Xi^-} = \tfrac{1}{3}[4\mu_s - \mu_d], \quad \mu_{\Xi^-} = \tfrac{1}{3}(4\mu_d - \mu_s).$$

So far we have only used assumption (i), that the baryon magnetic moments are sums of the quark magnetic moments. Let us now assume that the individual quark magnetic moment coefficients are given by

$$\mu_a = \frac{q(a)}{2m_a}.$$

Thus

$$\mu_u = \frac{1}{3m_u}, \mu_d = \frac{-1}{6m_d} \quad \text{and} \quad \mu_s = \frac{-1}{6m_s}.$$

Now, since the proton and neutron have approximately the same mass, it is reasonable to assume, to first approximation, that $m_u = m_d = \tfrac{1}{3}m_p = \tfrac{1}{3}m_n$. Then

$$\mu_u = \frac{1}{m_p} \quad \text{and} \quad \mu_d = \frac{-1}{2m_p}$$

so, in terms of the nuclear magneton $\mu_n = 1/2m_p$, we have

$$\mu_p = \frac{1}{3m_p}[4 + \tfrac{1}{2}] = \frac{9}{3 \cdot 2m_p} = 3\mu_N$$

and

$$\mu_n = \frac{1}{3m_p}[-4 \cdot \tfrac{1}{2} - 1] = -2\mu_N$$

which compares reasonably well with the observed values of

$$\mu_p = 2.79\mu_{N'}$$

and

$$\mu_n = -1.9\mu_N.$$

We have taken the mass of the u and d quarks to be $\tfrac{1}{3}$ of the proton mass – and so equal to about 313 MeV. The observed magnetic moment of the Λ^0 is about -0.614 nuclear magnetons. If we take

$$\Lambda^0 = \boxed{u}\,\boxed{d}$$
$$\boxed{s}$$

then the computation using the expression for z given about yields $\mu(\Lambda^0) = \mu_s$. This gives the mass of the strange quark at about 500 MeV.

Another way of getting an estimate for the quark masses is to assume that the binding energies in the formation of baryons are small in comparison with the quark masses. Assuming $m_u = m_d$, we would have

$$m_u = \tfrac{1}{6}(m_p + m_\Delta) = \tfrac{1}{6}(938 + 1232) \sim 360 \text{ MeV}$$

and

$$m_s = \tfrac{1}{3}m_{\Omega^-} = \tfrac{1}{3}1672 = 557.$$

These do not differ too much from the masses as estimated from the magnetic moment calculations. Furthermore,

$$m_{1115.6}(\Omega^0) \approx m_{1190}(\Sigma) \approx m_{1380}(\Sigma^*) \approx 2m_u + m_s = 1280 \text{ MeV}$$

and

$$m_{1314}(\Xi) \approx m_{1532}(\Xi^*) \approx m_u + 2m_s = 1480 \text{ MeV}$$

are not too far from the mark.

5.12 Color and beyond

Despite the successes of the quark model – only some of which were described in the preceding section – a number of puzzling features were recognized. Immediately after the proposal of the quark model in 1964, an intensive search was begun for the quarks. Various procedures were attempted all over the world. The number of ions produced per centimeter of track by a particle of charge q is proportional to q^2. Thus a particle of charge $\frac{2}{3}$ should produce a track whose droplet density is $\frac{4}{9}$, and a particle of charge $\frac{1}{3}$ should produce a track whose droplet density is $\frac{1}{9}$ of the density of a normal particle. Despite intensive searches in cloud and bubble chambers, no such tracks were found. Similarly, accelerator experiments, which should detect quarks by the radius of their circular orbit in a uniform magnetic field, gave negative results. Only one experiment, the modern analog of the Millikan oil drop experiment using supercooled niobium pellets instead of oil drops, gave any indications of the existence of particles of fractional charge. This experiment, performed by Fairbank at Stanford from 1964–77 has not been reproduced elsewhere, and its status remains unclear.

From a theoretical point of view the following questions arose: Why are only the ten-dimensional

and the eight-dimensional

representations observed as baryon multiplets; why are only the eight-dimensional representations consisting of quark–anti-quark pairs and the corresponding singlet observed for the mesons? Our discussion of the spin of the decouplets and the magnetic moments of the baryons showed that quarks enter symmetrically: the correct tensor space was the third *symmetric* power $S^3(\mathbb{C}^3 \otimes \mathbb{C}^2)$. (Furthermore the magnetic moment computation posited no orbital contribution, so the quarks were all in the same symmetrical ground state.) But the quarks are spin $\frac{1}{2}$ particles, so the Pauli exclusion principle should apply, forbidding formation of such symmetric states. It was therefore suggested (Greenberg, Gell-Mann) that the quarks possess an additional internal degree of freedom which is three dimensional. Thus a quark (at rest) lies in a space of the form

$$V \otimes \mathbb{C}^2 \otimes W$$

where $V (= \mathbb{C}^3)$ is spanned by u, d, s, where $\mathbb{C}^2 = V_{\frac{1}{2}}$ gives the spin and the three-dimensional space W describes the additional degree of freedom which, unfortunately, is called color. Thus, if we call a basis of W 'red', 'blue' and 'green', the u quark can exist in a linear combination of these three color states. This introduces an entirely new group, the group $SU(W) \sim SU(3)$ which is called $SU(3)_{\text{color}}$. Assume that only color singlets, i.e. the trivial representation of $SU(3)_{\text{color}}$, can arise. If we combine this with the Pauli exclusion principle, we see that the baryon states are the totally anti-symmetric states given by

$$S^3(V \otimes \mathbb{C}^2) \otimes \Lambda^3(W).$$

The meson states would then be given by $[V \otimes \mathbb{C}^2 \otimes W] \otimes [V \otimes \mathbb{C}^2 \otimes W]^*$.

At the time, this appeared as a rather outrageous *ad hoc* assumption made in order to reconcile the Pauli exclusion principle with the quark model. However, it has gained a central role in recent theories of elementary particles. First of all, it was soon realized that there were experiments which could be interpreted as counting the number of quarks, and these only gave the right answer if the three-fold multiplicity implied by color was taken into account. For example, the rate of decay of π^0 into two photons depends on the number of quark–anti-quark states available, and the factor of nine resulting in the three internal color states is just what is needed to make the computations in the quark model agree with experiment. Similarly, computations for electron–positron annihilation involved the creation of quark–anti-quark pairs, and the factor of three coming from the color states made its contribution. Indeed, the fact that observed particles must be 'color neutral' suggested that the forces between quarks involve 'color charges' just as the electrical charges are involved in electromagnetic forces. This led to a theory called 'quantum chromodynamics' in analogy to quantum electrodynamics. There would then be 'particles which carry the color force' in analog to photons which 'carry the electromagnetic forces'. These particles correspond to elements of the Lie algebra of $SU(3)$. There would thus be eight linearly independent particles which were called *gluons*. No free gluons have been observed, just as there is no confirmed observation of free quarks. However, indirect evidence for their existence developed in terms of electron–proton scattering experiments and electron–positron annihilation.

Additional experimental support for the existence of color came from electron–positron annihilations in the GeV range. From the energy blob created by this annihilation various new particles can be formed: a new electron–positron pair, a muon–anti-muon pair, or a quark–anti-quark pair (which is converted into a jet of hadrons). Quantum electrodynamics predicts the hadrons and a muon–anti-muon pair as being equal to the sum over all quarks of the squares of their charges. This factor is called R. For the u, d, s quarks we get contributions of $(2/3)^2$, $(-1/3)^2$ and $(-1/3)^2$, respectively, which add up to $2/3$. If we multiply by 3 for the colors, we get $R = 2$. Experiments in 1974 showed that up to 3 GeV the value of R is between 2 and 2.5, agreeing with this prediction. In the 1960s, Glashow had predicted the existence of a fourth quark with charge $2/3$. If this also came in three colors, this would add $3 \cdot 4/9$ to the R factor giving a value of $R = 4/3$. In the range from 3 to 4 GeV the R value did move up. But instead of stabilizing around 3-1/3, there was a sharp peak in hadron formation at about 3.1 GeV. This was interpreted as being due to the formation of a new meson resonance with a lifetime of about 10^{-20} s, an unusually long lifetime for such a heavy particle. This particle, discovered independently by Ting and Richter, was interpreted as being of the form $c\bar{c}$, where c is the fourth quark, a particle whose existence had been predicted earlier by Appelquist and Politzer.

So, by the mid 1970s, the original $SU(3)$ – the one proposed by Gell-Mann and Ne'eman – no longer was regarded as fundamental. There were four quarks, u, d, s and c, instead of the original three. But they still come in three colors. Indeed, with the advent of the charmed quark, a more natural grouping appeared to be

$$
\begin{array}{cc}
\text{d} & -\tfrac{1}{3} \\
\text{u} & \tfrac{2}{3}
\end{array}
$$

and

$$
\begin{array}{cc}
\text{s} & -\frac{1}{3} \\
\text{c} & \frac{2}{3},
\end{array}
$$

where we have listed the electric charge next to each quark.

Up until 1975 the only known leptons were the electron and its neutrino and the muon with its neutrino. In 1975 a new lepton (named τ) was discovered by Perl. In 1977 a new heavy resonance was discovered, which was interpreted as being a quark–anti-quark meson built out of a new quark labelled b, of charge $-1/3$. This led to the conjecture of the existence of still another quark, t, of charge 2/3 so that we would have the 'families'

$$
\begin{array}{c}
e, v_e, u, d \\
\mu, v_e, c, s \\
\tau, v_\tau, t, b
\end{array}
$$

of leptons and hadrons, with each of the quarks coming in three colors. (At present, the existence of the t has not yet been established.)

This suggests that there should be some more fundamental symmetry which underlies and unites the leptons and hadrons, and also the various forces of nature. At the moment, several such theories have been proposed, but must still be considered speculative.

5.13   Where do we stand?

In this section we summarize the results discussed so far in the application of group theory to elementary particle physics. We will allow ourselves to use a little differential geometry, as the current 'standard model' postulates a gauge theory of forces. We refer to any standard text for background.

As we have seen in Section 3.9, an elementary particle 'is' an irreducible representation of the group G of nature, and so a primary task is to determine G and its 'physically admissible' representations. One of the requirements on G is that it be related to the geometry of space time. For example, if we have a theory which does not include gravitational interaction, we would demand that G act on space time so as to include all the isometries, or some connected components. In a flat space time this amounts to the requirement that we be given a homomorphism of G onto the Poincaré group. (One might want to replace the Poincaré group by one of the other isometry groups for a homogeneous space time, such as the deSitter or anti-deSitter groups.) For example, in Wigner's original paper, G is taken to be the Poincaré group itself, i.e. $G = Sl(2, \mathbb{C}) \circledS \mathbb{R}^{1,3}$, the semidirect product of $G = Sl(2, \mathbb{C})$ with the group of translations, $\mathbb{R}^{1,3}$. His celebrated result, as we saw in Section 3.9, was that the physically relevant irreducible representations of the Poincaré group are parametrized by a non-negative parameter, m, the mass, and a half-integer parameter, s, the spin, where $s = 0, \frac{1}{2}, 1, \ldots,$ if $m > 0$, and $s = 0 \pm \frac{1}{2}, \pm 1, \ldots,$ if $m = 0$.

Since the mass and the spin are not the only physical parameters describing particles, and since the observed values of the masses cannot be explained by the Poincaré group alone, this implies the need for a larger group.

An important issue is that of localizability, that is the existence of a 'current'. Let M denote space time, and let $Z = Z^3(M)$ denote the space of closed three-forms on M (which vanish sufficiently rapidly at spacelike infinity if we take M to be ordinary Minkowski space). Suppose that we have a unitary representation of G on a Hilbert space H. Let V be some representation (usually finite dimensional) of G. A V-valued current on H is a sesquilinear form

$$J : H \times H \to V \otimes Z,$$

which is equivariant for the action of G. The fact that $J(\phi, \psi)$ is closed implies that its integral over any spacelike surface, $\int_S J(\phi, \psi)$, is a V-valued sesquilinear form on H which is independent of the choice of S. If V is the trivial one-dimensional representation of G then J is simply called a current. In the case that H is irreducible, then $\int_S J(\phi, \psi)$ is an invariant scalar valued sesquilinear form which then must be some multiple of the scalar product (ϕ, ψ). For example, if we take G to be the Poincaré group, and consider mass-zero representations, then the representations with $s = 0$ and $s = \frac{1}{2}$ have currents, but the representations with $s \geq 1$ do not. For $s = 1$ this expresses the well known fact that the photon cannot be localized. On the other hand, if we take $V = \mathbb{R}^{1,3*} = M^*$ and let G act on V via the homomorphism from G to $Sl(2, \mathbb{C})$ then a V-valued current exists. Thus, although the photon cannot be localized, its energy or momentum *can* be localized, a fact that was known to Poincaré. For $s > 1$ such a V-valued current does not exist either. The mass-zero representations have the property that they all extend to irreducible unitary representations of $SU(2, 2)$ the (four-fold cover of the connected component of) the group of conformal transformations of conformally completed Minkowski space. The $s = 0$ representation does not have an $SU(2, 2)$ current, but the $s = \pm\frac{1}{2}$ representations do. This suggests that 'matter' is built up out of spin-$\frac{1}{2}$ particles. One of the reasons why the currents are of importance is that, to lowest order in perturbation theory, many of the standard interactions involve the currents associated to the in and out states of the scattered particles.

In order to describe interactions, it is assumed that the irreducible representations are realized on a space of sections of some vector bundle on space time. That is, it is assumed that H is the Hilbert space completion of a space $\mathbf{E}$ of smooth sections of a vector bundle $E \to M$, where $\mathbf{E}$ is characterized as being the space of solutions of a differential operator expressed in terms of a flat connection on E. For example, for particles whose behavior under the Poincaré group is that of a spin-$\frac{1}{2}$ particle, the vector bundle E would be of the form $W \otimes S$, where S is the spin bundle and W is a vector bundle with a flat connection. Then the *gauge theory of forces* says that interactions are manifested by the presence of a non-flat connection on W. More specifically, suppose that W is an associated bundle to a principal bundle P with structure group K. Let $\mathscr{C}$ denote the space of all connections on P and $\mathscr{A}^0(M, E)$ the space of sections of E. Then the interactions are expressed in terms of a Lagrangian which is a local map from $\mathscr{C} \times \mathscr{A}^0(M, E)$ to functions on M. For example, the Yang–Mills Lagrangian is given by

$$L(A, s) = \tfrac{1}{2}(\parallel \mathbf{F} \parallel^2 + (\mathbf{D}_{A}s - ms, s)) \tag{13.1}$$

where $\mathbf{F} = \mathbf{F}(A)$ denotes the curvature of the connection A, and where $\mathbf{D}_{A}s$ denotes the Dirac operator associated to the connection A. The metric and scalar product on the right-hand side of (13.1) are determined by the Lorentzian metric on M, by an invariant metric on W and by an invariant bilinear form on the Lie algebra of K, and m denotes the mass matrix. The relation with the representation point of view is then the following. We consider the Lagrangian L in the neighborhood of the trivial connection on P and the zero section of E. Then the quadratic terms in (13.1) will give a quadratic Lagrangian,

$$L_0 = L_{01} + L_{02}. \tag{13.2}$$

The terms on the right-hand side of (13.2) correspond to the two terms on the right-hand side of (13.1). Here L_{02} does not involve the connection A and so it is a quadratic Lagrangian on E. The Euler–Lagrange equations of L_{02} then define the representation space E. Upon restriction to the Poincaré group, E decomposes into a direct sum of spin-$\tfrac{1}{2}$ representations with masses given by the eigenvalues of the mass matrix m. Similarly, L_{01} does not involve s; it is purely a function of A. In fact, the space, $\mathscr{C}$, of all connections is an affine space whose associated linear space is $\mathscr{A}^1(k)$ the space of one-forms on M with values in k, where k is the Lie algebra of K. Hence L_{01} is a quadratic Lagrangian on $\mathscr{A}^1(k)$. The corresponding Euler–Lagrange equations define a subspace, $\mathbf{C}$, of the space of connections, which is a representation space for G. Upon restriction to the Poincaré group this decomposes into a direct sum of dim k copies of the $m = 0$, $s = \pm 1$ representations. These are regarded as the 'force carriers' in that the forces giving rise to bound states and to the scattering matrix are determined from the higher order terms in (13.1) according to the Feynman rules. A detailed discussion of the Feynman rules from the point of view of quantum field theory can be found in any text on the subject, for example. For a clear and elementary presentation of these rules without invoking the entire machinery of quantum field theory (but essentially dealing with the lowest order in perturbation theory, the tree diagrams) see Griffiths, *Int. to Elem. Particles*.

The above described theory, although beautiful from a mathematical point of view, must be modified in order to conform to physical reality. The trouble is that it predicts too many (dim k) massless vector bosons, i.e. particles with $m = 0$, $s = \pm 1$. For example, in the theory of strong interactions, where $k = su(3)$, the theory would predict $8 = \dim su(3)$ such particles, and in electroweak interactions, where $k = su(2) \times u(1)$, it predicts four. *But only the photon is observed.* This problem is dealt with differently in the two domains. In the theory of strong interactions, it is believed that these particles are not observed as free particles because of 'confinement'. In the electroweak theory, the three extra vector bosons are 'given a mass' by the Higgs mechanism of spontaneous symmetry breaking. In addition, the Higgs mechanism gives masses to some of the massless fermions. That is, one starts with a theory with

$m = 0$ in (13.1), but the Higgs mechanism ends up with some of the fermions having mass. The mathematics of confinement is not yet understood. However, the Higgs mechanism, and the associated notion of spontaneous symmetry breaking, has a differential geometric meaning in terms of the 'reduction of the principal bundle'.

Let $P_K \to M$ be a principal bundle with structure group K over the base manifold M. There are various geometrical objects that we can associate with P_K. For example, if the vector space F is a K module, then we can construct the associated bundle to F and then consider the space of j-forms on the base with values in this vector bundle. Let us denote the space of all such j-forms by $\mathscr{A}^j(M, F)$ or $\mathscr{A}^j(F)$. Let us denote the direct sum (over j) of all the $\mathscr{A}^j(M, F)$ spaces by $\mathscr{A}^*(M, F)$. We may want to consider various different modules, F. Summing over various different F's that come into consideration gives us a space which we shall denote by $\mathscr{A}^*(P_K)$. We can also consider the space of all connections on P_K which we have denoted by $\mathscr{C}$. By a *field* we shall mean an element of the space $\mathscr{A}^*(P_K) \times \mathscr{C}$. We shall denote the general field by ϕ.

Let $\operatorname{Aut} P_K$ denote the group of all diffeomorphisms of P_K which commute with the action of K. Then $\operatorname{Aut} P_K$ acts on the space of all fields, $\mathscr{A}^*(P_K) \times \mathscr{C}$. We can construct Lagrangian densities

$$L: \mathscr{A}^*(P_K) \times \mathscr{C} \to C^\infty(M)$$

out of geometric data which will be equivariant for the action of $\operatorname{Aut} P_K$, or at least for the action of a large subgroup of $\operatorname{Aut} P_K$. For example, the Lagrangian density given in (13.1) will be equivariant under the subgroup of $\operatorname{Aut} P_K$ which induces a transformation of M, which is an isometry. More generally, suppose that M has a (pseudo-) Riemann metric and that the Lie algebra of K is given a K invariant scalar product as is each of the representation spaces, F, that enter into our definition of the geometrical fields. Then we can consider various scalar valued functions on M built out of the geometrical fields such as

$$\begin{aligned}
&\|\omega\|^2 &&\omega \in \mathscr{A}^k(F) \\
&\|d_A \omega\|^2 &&\omega \in \mathscr{A}^k(F), A \in \mathscr{C} \\
&\|curv(A)\|^2 &&A \in \mathscr{C} \\
&V(f) &&f \in \mathscr{A}^0(F), V: F \to \mathbb{R},
\end{aligned}$$

where in the last item V is a K invariant function on F. If we have a projection of P_K onto a sub-bundle of the bundle of orthonormal frames of M, so that $T(M)$ and $T^*(M)$ are associated bundles to P_K, then we might also have expressions such as

$$W(f, \omega),$$

where W is some K invariant polynomial.

In any event, we can build some Lagrangian density, L, out of these objects and then integrate $L(\phi)$ with respect to some volume density, $dvol$, to obtain

$$\mathbf{L}(\phi) = \int L(\phi) dvol. \tag{13.3}$$

In quantum field theory one starts from the 'functional integral'

$$\int e^{iL(\phi)} \mathcal{D}(\phi) = \int e^{i\int L(\phi) dvol} \mathcal{D}(\phi). \tag{13.4}$$

Of course, these functional integrals are not very well defined in a mathematical sense, but via perturbation theory they give rise to the Feynman rules of physics. However, even within the framework of perturbation theory there is a problem of interpretation due to the fact that the integrand in (13.4) is invariant under a very large group of symmetries. Indeed, let $\mathrm{Gau}(P_K)$ denote the subgroup of $\mathrm{Aut}(P_K)$ consisting of those automorphisms which induce the identity transformation on M. It is what the mathematicians call the *gauge group*, and what the physicists call the *group of local gauge transformations*. Then (13.3) is invariant under $\mathrm{Gau}(P_K)$. One of the ideas of spontaneous symmetry breaking is to remove some of the redundancy in (13.4) by providing a partial cross section for the action of $\mathrm{Gau}(P_K)$.

In order to describe how this works, we first recall an alternative description of the group $\mathrm{Gau}(P_K)$. Let K act on itself by conjugation. Then the associated bundle of P_K relative to this conjugation action is a fiber bundle over M, each of whose fibers is a group. This makes the space of smooth sections of this associated bundle into a group. This group of sections can be identified with $\mathrm{Gau}(P_K)$ as follows. A section of the bundle associated to the conjugation action can be thought of as a function $\tau: P_K \to K$, which satisfies the identity

$$\tau(pb) = b^{-1}\tau(p)b. \tag{13.5}$$

To each such τ one associates the diffeomorphism ψ_τ of P_K defined by

$$\psi_\tau(p) = p\tau(p). \tag{13.6}$$

It is easy to check that the map $\tau \mapsto \psi_\tau$ is a homomorphism from the group of sections τ onto $\mathrm{Gau}(P_K)$.

Next, one singles out a particular space, U, on which K acts, a section of the associated bundle to the space U being called a *Higgs field*. (In the Weinberg–Salam model the group K is taken to be $SU(2) \times U(1)$ and the space U is then taken to be $\mathbb{C}^2$.)

We assume the following about the action of K on U: that outside of a small G invariant singular set $S \subset U$ there is a cross section, Γ, for the K action. Thus we assume that

$$U - S = \Gamma \times (K/H) \tag{13.7}$$

as K spaces, where H is the common isotropy group of all points of Γ. For example, in the case of the Weinberg–Salam model with $K = SU(2) \times U(1)$ and $U = \mathbb{C}^2$ we could take $S = 0$ and Γ to consist of all vectors of the form $\begin{pmatrix} 0 \\ \rho \end{pmatrix}$ with ρ real and positive, and H is isomorphic to $U(1)$.

Let $\mathscr{B}(U)$ denote the space of all sections of the bundle associated to U which do not intersect the sub-bundle determined by S. Thus $\mathscr{B}(U)$ consists of those smooth maps $f: P_K \to U$ which satisfy the equivariance condition

$$f(pa) = a^{-1} f(p), \quad \forall a \in K, \quad p \in P_K \tag{13.8}$$

and such that

$$f(P_K) \subset U - S. \tag{13.9}$$

In view of the decomposition (13.7), condition (13.9) implies that we can write f as

$$f = (g, r),$$

where $g: P_K \to \Gamma$ and $r: P_K \to K/H$, where each of the components g and r satisfies the equivariance condition (13.8). Since the action of K on Γ is trivial, condition (13.8) simply says that g is a Γ-valued function on M. For r, condition (13.8) says that r is a section of the bundle associated to the homogeneous space K/H. Now a section of this bundle is the same as a reduction of the principal bundle, P_K, to an H sub-bundle. Indeed, define the sub-bundle P_H by

$$P_H^r = r^{-1}(H) = \{p \in P_K | r(p) \in H \in K/H\}. \tag{13.10}$$

Note that $P_K = P_H \times {}_H K$. This explains why we say that we have reduced the structure group of P_K to H. This observation is the starting point of the theory of 'Cartan connections'. The group $\text{Gau}(P_K)$ acts on all geometric objects, in particular on the space of all smooth sections of the bundle associated to K/H, i.e. on the space of all reductions to an H sub-bundle. We will assume that this space is non-empty, and that $\text{Gau}(P_K)$ acts transitively on it. For example, in the case that $H = e$ is the trivial subgroup, this action is always transitive. Indeed, let r_i, $i = 1, 2$, be two maps r_i: $P_K \to K$ satisfying (13.8). If we then define $\tau: P_K \to K$ by

$$\tau(p) = r_1(p) r_2(p)^{-1}$$

then τ satisfies (13.5) and hence defines an element of $\text{Gau}(P_K)$. Also

$$r_1(p\tau(p)) = r_2(p) r_1(p)^{-1} r_1(p) = r_2(p)$$

and so ψ_τ carries r_1 into r_2.

Under our assumptions of the existence of a cross section, r, and the transitivity of the action of $\text{Gau}(P_K)$, all of the bundles P_H^r are carried into one another by an element of $\text{Gau}(P_K)$. In particular, they define the same abstract principal H bundle which we will denote by P_H. From this point of view, the choice of an $f \in \mathcal{B}(U)$ and the decomposition $f = (g, r)$ then determines a specific embedding

$$r: P_H \to P_K \tag{13.11}$$

with $r(P_H) = P_H^r$. If F is any representation space of K, it becomes a representation space of H by restriction. Hence r determines a pullback map

$$r^*: \mathcal{A}^* P_K \to \mathcal{A}^* P_H, \tag{13.12}$$

which is consistent with the various geometrical Lagrangians, so that, for example,

$$\|r^*\omega\|^2(q) = \|\omega\|^2(r(q)), \quad q \in P_H. \tag{13.13}$$

Notice that r^* is one-to-one since $P_K \times {}_K F = P_H \times {}_H K \times {}_K F = P_H \times {}_H F$. Now let us examine how connections pull back under r. Let h denote the Lie algebra of H so that h is a subalgebra of k, the Lie algebra of K. Let us assume that h has an H invariant vector space complement, l in k, so that

$$k = h + l. \tag{13.14}$$

It is always possible to choose such a complement if H is compact or semisimple. Now a connection, A, on P_K is a k-valued one-form on P_K satisfying an equivariance condition

$$R_b^* A = Ad_b^* A \tag{13.15}$$

together with the interior product condition

$$\iota(\xi_{P_K}) A = \xi, \xi \in k, \tag{13.16}$$

where ξ_{P_K} is the vector field on P_K coming from the right action of K. Now on $r(P_H) = P'_H$ only the subgroup H acts. The decomposition (13.14) implies that A decomposes as $A = A_H + \theta_l$; i.e., if we restrict to $\xi \in h$, the decomposition (13.14) implies that the restriction of A decomposes as a sum of a connection on P_H and an element of $\mathscr{A}^1(P_H, l)$, the space of one-forms on M with values in the vector bundle over M associated to the principal bundle P_H and the representation, l, of H. In other words, if $\mathscr{C}(P_K)$ denotes the space of connections on P_K and $\mathscr{C}(P_H)$ the space of connections on P_H, the map r defines a pullback map

$$r^* : \mathscr{C}(P_K) \to \mathscr{C}(P_H) \times \mathscr{A}^1(P_H, l). \tag{13.17}$$

For example, if $K = SU(2) \times U(1)$ and $H = U(1)$, a K connection gives rise to a $U(1)$ connection A_H which represents an electromagnetic field together with a one-form with values in a three-dimensional bundle giving the W_+-, W_--, and Z_0-particles. The relation between the curvature of the connection A on P_K at the point $p = r(q)$ and the two components of its pullback under (13.17) is given by

$$F_A(p) = F_{A_H}(q) + d_{A_H} \theta_l(q) - \tfrac{1}{2}[\theta_l, \theta_l]. \tag{13.18}$$

Thus we can write down a geometric Lagrangian $L' : \mathscr{C}(P_H) \times \mathscr{A}^1(P_H, l) \to C^\infty(M)$ such that

$$L'(r^*(A)) = \| F_A \|^2.$$

Proposition 13.1 (Reduction of the principal bundle)

Let r be a section of the bundle associated to K/H. Then r induces a pullback map of the fields:

$$r^* : \mathscr{C}(P_K) \times \mathscr{A}^*(P_K) \to \mathscr{C}(P_H) \times \mathscr{A}^*(P_H). \tag{13.19}$$

For any geometrical Lagrangian, L, on P_K there is a Lagrangian, L', on P_H such that

$$L' \circ r^* = L.$$

In particular this holds for the r component of a Higgs field $f = (g, r)$.

In this way the $\text{Gau}(P_K)$ redundancy inherent in L has been reduced to a $\text{Gau}(P_H)$ redundancy inherent in L'.

The 'mass acquisition' occurs in Proposition 13.1 as follows. Consider the function

$$\mathscr{C}(P_K) \times \mathscr{A}^0(U) \to C^\infty(M)$$

given by

$$(A, f) \mapsto \|d_A f\|^2.$$

Suppose that f avoids S, so $f \in \mathscr{B}(U)$ can be written as $f = (g, r)$. Then

$$r^* f = (g, H),$$

where the second component, H, is constant, and $g: M \to \Gamma$. Thus along the sub-bundle $r(P_H)$ we can identify df with $dg: TM \to T\Gamma$. Since h acts trivially on $T(K/H)_H$, we have

$$A \cdot f(p) = \theta_l \cdot f \in T(O_{g(m)}),$$

where $m \in M$ is the image of p under the projection of P_K to M and $O_{g(m)}$ denotes the K orbit through the point $g(m) \in \Gamma$ in U. Suppose that U carries a K invariant Riemann metric and that the cross section, Γ, has been chosen so that the tangent space $T\Gamma_{g(m)}$ and $TO_{g(m)}$ are perpendicular. For example, in the Weinberg–Salam model, Γ is a radial ray passing through the 'north pole' of the three spheres $O_{g(m)}$ for all m. Then the orthogonal decomposition

$$TU_{g(m)} = T\Gamma_{g(m)} \oplus T(O_{g(m)})_{g(m)}$$

gives

$$\|d_A f\|^2(p) = \|dg(p)\|^2 + \|\theta_l \cdot g(m)\|^2.$$

Now the action of K gives us an identification of l with the tangent space $T(O_{g(m)})_{g(m)}$ and hence the metric on $T(O_{g(m)})_{g(m)}$ pulls back to a metric on L (depending on the point $g(m)$). Let us denote this metric by $\|\cdot\|^2_{g(m)}$. Then we can write the preceding equation as

$$\|d_A f\|^2(p) = \|dg(p)\|^2 + \|\theta_l\|^2_{g(m)}. \tag{13.20}$$

Let us assume that h and l are perpendicular for the scalar product on k. Then substituting into (13.18) gives

$$\|F_A\|^2 = \|dA_H\|^2 + \|d\theta_l\|^2 + \cdots \tag{13.21}$$

along $r(P_H)$, where $\cdots$ consist of terms cubic or higher in the fields. Let us combine (13.20) with (13.21). Suppose that our original Lagrangian on P_K contains (as a summand) a term of the form

$$L = \|F_A\|^2 + a\|d_A f\|^2 + V(f), \tag{13.22}$$

where V is some K invariant function defined on U. (Here a is some constant.) If we expand (13.22) about zero in f and A (about a flat connection and the zero section of U), then the quadratic terms in A decouple as $\dim k$ copies of the Lagrangian for Maxwell's equations. In other words, we get $\dim k$ particles of mass 0 and spin ± 1. But suppose that V takes on a minimum along some non-zero orbit $O_{\min}$, where

$O_{\min} \subset U - S$ and we choose f_0 to be a section of $O_{\min}$. According to our assumptions, such an f_0 exists, and the group $\mathrm{Gau}(P_K)$ acts transitively on the space of such sections. Let

$$\|\theta_l\|_0^2 = \|\theta_l\|_{g_0(m)}^2,$$

where $f_0 = (g_0, r_0)$ so that g_0 is the constant function which maps P_K to $O_{\min} \cap \Gamma$. By the transitivity of the $\mathrm{Gau}(P_K)$ action and the invariance of our Lagrangian, we may assume that we have chosen $f_{(0)}$ with $r_0 \equiv H$. If we then write $L = L' \circ r^*$ we find that, ignoring overall constants,

$$L' = \|d\theta_l\|^2 + a\|\theta_l\|_0^2 + \|F_{A_H}\|^2 + \cdots. \tag{13.23}$$

The quadratic terms in (13.23) are those of the Maxwell equations in the A_H component but the Proca Lagrangian (positive mass, spin 1) in the l component. To summarize, we have proved Proposition 13.2.

Proposition 13.2 (Mass acquisition for vector bosons)

Suppose that we expand (13.22) about a non-zero minimum section under our assumptions as to the existence of the cross section and the action of $\mathrm{Gau}(P_K)$. This then leads to a reduction of the principal bundle as in Proposition 13.1, and the replacement of (13.22) by (13.23) so that the l components of A acquire mass.

Similarly, the additional term in the Lagrangian for the fermions leads to mass acquisition for them as well.

The above describes, in schematic terms, the methodology of the Higgs mechanism. However, the detailed implementation, including a detailed prediction of the mass spectrum and of the generational phenomenon, remains a desired goal.

Appendix A

THE BRAVAIS LATTICES AND THE
ARITHMETICAL CRYSTAL CLASSES

The purpose of this appendix is to take the study of crystallography a bit further than in the main text. In particular, we wish to explain the work of Bravais and indicate some of the mathematical considerations that go into the study of the crystallographic groups. A thorough presentation of the classification of the crystallographic groups in two and three dimensions can be found in the book Burckhardt (1966). Recall that a *lattice* is a discrete subgroup of the group $E(3)$. A subgroup G of the group of Euclidean motions, $E(n)$, is called cystallographic if it satisfies the following two conditions:

(1) Discreteness: for any $x \in \mathbb{R}^n$ there is an $\varepsilon > 0$ such that $\| ax - bx \| < \varepsilon, a, b \in G$ implies that $ax = bx$, and

(2) Space filling: there is a compact subset K of $\mathbb{R}^n$ such that $\bigcup_{a \in G} aK = \mathbb{R}^n$.

These two conditions reflect the idea that the group G represents the symmetry of a discrete repetitive pattern that fills up all of space. A subgroup L of the group of *translations* which satisfies these two conditions is called a *lattice*. It is true, but surprisingly tricky to prove, that any crystallographic group contains a lattice. In other words, that there are enough *translational* symmetries of the discrete repetitive pattern to fill up all of space. Rather than go into the details of this proof, we shall simply take its conclusion as our starting point. We will be interested in discrete subgroups of $E(n)$ which contain lattices. We must begin with some basic facts about lattices.

A.1   The lattice basis and the primitive cell

The purpose of this section is to prove that every lattice has a basis. Let L be a discrete subgroup of the group of translations of $\mathbb{R}^n$.

We wish to show that there exist vectors $\mathbf{v}_1, \ldots, \mathbf{v}_k$ which are linearly independent, which are elements of L, and such that every element of L is a combination of the $\mathbf{v}_i$ with *integer* coefficients. By general vector space theory, we know that we can always choose a maximum set $\mathbf{u}_1, \ldots, \mathbf{u}_k$ of linearly independent vectors in L, and then every element of L is a linear combination of the $\mathbf{u}$'s with *real* coefficients. Our strategy is to show that for any such choice of u's each element of L can be expressed as a *rational* combination of the $\mathbf{u}$'s and to estimate the common denominator of these coefficients in terms of a property of the parallelepiped spanned by the $\mathbf{u}$'s. We let $P(\mathbf{u}_1, \ldots, \mathbf{u}_k)$ denote the half open parallelepiped spanned by the **u**'s, so that

$$P(\mathbf{u}_1, \ldots, \mathbf{u}_k) = x_1\mathbf{u}_1 + \cdots + x_k\mathbf{u}_k \quad 0 \leqslant x_i < 1, \text{ all } i.$$

By the discreteness of L, any bounded set can contain only finitely many elements of L. Thus the above parallelepiped contains finitely many, say N, elements, $\mathbf{w}_1, \ldots, \mathbf{w}_N$ of L. The coefficients of each of the **w**'s, when expressed in terms of the **u**'s all lie between zero and one and are strictly less than one. Thus no two **w**'s can differ from one another by a linear combination of the **u**'s with integer coefficients. Let us introduce the following notation. For any real number, x, the symbol $[x]$ denotes the largest integer less than or equal to x. For any vector

$$\mathbf{c} = x_1\mathbf{u}_1 + \cdots x_k\mathbf{u}_k$$

we set

$$[\mathbf{c}] = [x_1]\mathbf{u}_1 + \cdots + [x_k]\mathbf{u}_k.$$

Thus, for any $\mathbf{c} \in L$,

$$\mathbf{c} - [\mathbf{c}] \in P(\mathbf{u}_1, \ldots, \mathbf{u}_k),$$

and also

$$\mathbf{c} + \mathbf{w}_j - [\mathbf{c} + \mathbf{w}_j] \in P(\mathbf{u}_1, \ldots, \mathbf{u}_k) \quad j = 1, \ldots, N.$$

We can thus write

$$\mathbf{c} + \mathbf{w}_j - [\mathbf{c} + \mathbf{w}_j] = \mathbf{w}_{\phi(j)} \tag{$*$}_j$$

for some $\phi(j)$, where the assignment $j \to \phi(j)$ depends on **c**. We claim that ϕ is one-to-one: if $\phi(i) = \phi(j)$ then subtracting $(*)_i$ from $(*)_j$ will express $\mathbf{w}_j - \mathbf{w}_i$ as a linear combination of the **u**'s with integer coefficients, which is impossible. If we add all the equations $(*)_j$, the **w**'s cancel since each **w** occurs exactly once on both sides. We get

$$N\mathbf{c} = [\mathbf{c} + \mathbf{w}_1] + \cdots + [\mathbf{c} + \mathbf{w}_N].$$

Thus every $\mathbf{c} \in L$ can be written as

$$\mathbf{c} = r_1\mathbf{u}_1 + \cdots + r_k\mathbf{u}_k$$

where each r is a rational number with denominator N, where N is the number of elements of L in $P(\mathbf{u}_1, \ldots, \mathbf{u}_k)$. We would like to choose the **u**'s so that $N = 1$, i.e. that 0 is the only element of L lying in the parallelepiped. Any other choice of k linearly independent **v**'s in L can be written as

$$\mathbf{v}_i = r_{ij}\mathbf{u}_j$$

where the matrix (r_{ij}) has rational entries. Let Vol P denote the (k-dimensional) volume of the parallelepiped, P. Then

$$\text{Vol } P(\mathbf{v}_1, \ldots, \mathbf{v}_k) = \det(r_{ij})\text{Vol } P(\mathbf{u}_1, \ldots, \mathbf{u}_k).$$

Now $\det(r_{ij})$ is a positive rational number with denominator N^k. Thus we can choose the **v**'s so that their parallelepiped has minimal volume. We claim that $P(\mathbf{v}_1, \ldots, \mathbf{v}_k)$

contains no non-zero element of L. Suppose $\mathbf{w} = x_1\mathbf{v}_1 + \cdots + x_k\mathbf{v}_k$ lies in $P(\mathbf{v})$ with, say $x_1 \neq 0$. Then $\mathbf{w}, \mathbf{v}_2, \ldots, \mathbf{v}_k$ are linearly independent and $\operatorname{Vol} P(\mathbf{w}, \mathbf{v}_2, \ldots, \mathbf{v}_k) = x_1 \operatorname{Vol} P(\mathbf{v}_1, \ldots, \mathbf{v}_k)$ with $0 \leqslant x_1 < 1$, contradicting our choice of $\mathbf{v}$'s.

For simplicity in discussion, let us restrict attention to the case where the lattice spans the whole vector space V, i.e. that there are k linearly independent vectors in L, where k is the dimension if V. Let us call any half open parallelepiped whose vertices are lattice points a *cell*. The translates of any cell cover all of the space V. A cell is called a *primitive cell* if its translates cover V in a one-to-one fashion, i.e. that no point lies in two translates of the cell, or, what amounts to the same thing, that no translate of the cell under any non-zero element of L has any non-empty intersection with the original cell. We have seen that there are cells of minimal volume, and these cells are precisely the primitive cells. Since the cell is half open, in any primitive cell exactly one vertex actually lies in the cell. If we choose a primitive cell whose vertex is the origin, then the vectors spanning this parallelepiped form an integral basis for L.

The notion of a primitive cell of a lattice is related to a more general notion, that of a cross section for a group action. Let the group G act on the set M. We say that a subset, C, of M is a *cross section* if C contains exactly one point from each orbit. Thus $G(C) = M$ and ax does not lie in C for any x in C if $ax \neq x$. In our present case, we can think of L as a group of transformations of V, acting by translations, and a primitive cell is a particular kind of cross section for the action of L on V, namely a parallelepiped. (Sometimes the word primitive cell is used to describe any (half open) polyhedron which is a cross section. Notice that the volume of all such polyhedra must be equal, so that one can still talk about *the* volume of a primitive cell.)

Frequently we would like our cross sections to have additional nice properties, and these are usually hard to come by. For instance, consider the group $\mathbb{R}$ acting on $\mathbb{R}^n$, the action being replaced by a vector field, i.e. a system of first order time independent ordinary differential equations. Here the orbits are all curves in n-dimensional space, and we would like to choose our cross section to be a nice $(n-1)$-dimensional surface. The fact is that in general we are unable to do this, and, in a sense, this is why there is a subject called the qualitative theory of ordinary differential equations. We sometimes have to make do with something less than a cross section in order to have a set, C, with reasonable geometric properties.

A.2 The 14 Bravais lattices

Let G be a subgroup of the group of Euclidean motions of $\mathbb{R}^n$. Remember that we have agreed to call a group G *crystallographic* if

   (i) G is discrete, and

   (ii) L_G contains n linearly independent translations, where L_G denotes the translation subgroup of G.

In particular, L_G is a lattice which spans all of $\mathbb{R}^n$.

When should two crystallographic groups be regarded as the 'same'? Well, suppose that G is the symmetry group of some subset, T, of $\mathbb{R}^n$, i.e.

$$G = E(n)_T$$

where $E(n)$ denotes the group of Euclidean motions. Let b denote any Euclidean motion. We do not want to regard the patterns T and bT as being essentially different. From equaton (6.1), Chapter 1, we know that

$$E(n)_{bT} = b(E(n)_T)b^{-1}$$

so that we should regard G and bGb^{-1} as being the same, where b is any Euclidean motion. Suppose that b is not a Euclidean motion, but is a similarity transformation. For instance, suppose that b were a dilatation by a factor of two. Then bT and T look more or less the same, except that the spacing between the points has been doubled. Thus we should regard G and bGb^{-1} as equivalent if b is any similarity transformation. But even with this notion of equivalence we still are in some trouble. Consider the case where G is itself a lattice, i.e. $G = L_G$, and that we are interested in classifying lattices under similarity transformations. It is clear that the set of angles between the vectors joining the origin to its nearest neighbors is invariant under similarity transformations, and thus we will get a continuum of different lattices. It is more convenient to use a rougher classification, which would have the effect of identifying all purely lattice groups, and have the advantage of giving a finite list. Further refinements, such as reintroducing the angles, could then be introduced as additional parameters used in the description of each class. The definition that is used is that

> G and bGb^{-1} are regarded as equivalent, where b is any orientation preserving affine motion, i.e. $bx = Bx + v$, where $\det B > 0$.

The result of Fedorov, Schoenflies and Barrow was that there are 230 different crystallographic groups in three-space. Hilbert, in his list of outstanding mathematical problems, presented in 1900, raised the question as to whether there are only a finite number of different crystallographic groups in any dimension. This was proved by Bieberbach in 1910.

Let G be a crystallographic group and let L be its lattice subgroup of translations. If $a: x \to Ax + v$ lies in G, then a acts on L by conjugation, and $ata^{-1} = At$. If we compose a with any translation, we get the same action on L. Thus the quotient group

$$F = G/L$$

acts as orthogonal transformations preserving L. (In general, suppose that G is a group and N a commutative normal subgroup. Then $a: n \to ana^{-1}$ defines an action of G on N, since N is normal, and this action is trivial for $a \in N$, and so gives a well defined action of G/N on N.)

The elements of F act *faithfully* on L, i.e. if $a \in G$ acts trivially on L, then a must be a translation. Since F is a discrete subgroup of the orthogonal group, it is finite. Thus each crystallographic group gives rise to a pair, (F, L), consisting of a finite subgroup of the orthogonal group which acts on the lattice L. If $G' = bGb^{-1}$ then we must have

$$L' = BL \text{ and } F' = BFB^{-1}$$

where B is the linear part of b. This introduces the appropriate notion of equivalence of the pairs (F, L). We can analyse the procedure for classifying the crystallographic groups in three steps:

(1) find all finite subgroups of the orthogonal group that can preserve a lattice;

(2) for each such finite group, F, find the possible actions of F on the lattices, i.e. find the possible (F, L) (up to equivalence);

(3) for a given pair (F, L), find the various crystallographic groups with L as lattice subgroup and $F = G/L$.

Step (1) has already been accomplished with our classification of the 32 point groups in Chapter 1. We shall discuss step (2) in the present section.

Suppose that the group F acts on the lattice, L. If we choose a basis for the lattice, then each element of F is represented by a matrix with integral entries. If we change the basis, then each matrix, B, is replaced by ABA^{-1}, where A is again a matrix with integer coefficients, since A is the matrix expressing one basis of the lattice in terms of another. Thus the pair (F, L) determines a representation of the elements of F by integer matrices, this representation being determined up to conjugacy by integer matrices. Now conjugacy by integer matrices is a more stringent condition than conjugacy by real matrices. Consider, for example, the two matrices

$$\begin{pmatrix} 1 & 0 \\ 0 & -1 \end{pmatrix} \text{ and } \begin{pmatrix} 0 & 1 \\ 1 & 0 \end{pmatrix}$$

in the plane. They both represent reflections, the first through the x axis, and the second through the line $x = y$. Since all reflections are conjugate to one another by an appropriate rotation, we conclude that these two matrices are indeed conjugate if we allow conjugation by real matrices. On the other hand, these two matrices are not conjugate over the integers, i.e. that there is no invertible matrix

$$\begin{pmatrix} p & r \\ q & s \end{pmatrix}$$

with integer entries so that

$$\begin{pmatrix} p & r \\ q & s \end{pmatrix}\begin{pmatrix} 1 & 0 \\ 0 & -1 \end{pmatrix} = \begin{pmatrix} 0 & 1 \\ 1 & 0 \end{pmatrix}\begin{pmatrix} p & r \\ q & s \end{pmatrix}.$$

The slickest way to see this is to observe that if such an equation were to hold, then it would hold if we reduced all the integers mod 2. But $1 \equiv -1 \bmod 2$, and the matrix $\begin{pmatrix} 1 & 0 \\ 0 & -1 \end{pmatrix}$ becomes the identity matrix which is conjugate to no other matrix. In more pedestrian fashion, we can multiply out and see that the above equation implies $p = q$

(a)   (b)

Fig. A.1

and $r = -s$ which implies

$$\det\begin{pmatrix} p & r \\ q & s \end{pmatrix} = 2ps \neq 1$$

so that the matrix will not be invertible over the integers. (The method of reducing modulo a prime is a very useful technique in studying equations over the integers.)

This example shows that the two lattices in the plane (Fig. A.1) are different. For both lattices, the group of all automorphisms consists of $I, -I, \tau$ and $-\tau$, where τ is a reflection. In Fig. A.1(a) the matrix of τ is $\begin{pmatrix} 1 & 0 \\ 0 & -1 \end{pmatrix}$ and in (b) it is $\begin{pmatrix} 0 & 1 \\ 1 & 0 \end{pmatrix}$ in terms of the bases we have drawn. (In both cases the axis of reflection is the x axis.)

Thus, in three-space, each of the 32 groups may give rise to several different integral representations. Each of these is known as an arithmetical crystal class. It turns out that there are 73 of them in all. In order to classify the arithmetical crystal classes, it is more convenient to classify the maximal ones first. That is, we look for pairs (F, L), where F is the group of *all* orthogonal transformations preserving the lattice. Now not every group on our list of 32 can actually arise as the group of *all* transformations preserving a lattice. For instance, every lattice is invariant under $-I$. Thus, only a group appearing in the second column of Table 4 of Chapter 1 can be the group of all automorphisms of a lattice. Not all of these can occur either. It turns out, as shall emerge from the ensuing discussion, that only seven of the eleven groups can so arise, namely

$$S_2, C_{2h}, D_{2h}, D_{4h}, D_{3d}, D_{6d} \text{ and } O_h$$

These correspond to the 'seven crystal systems'. Some of these groups have several inequivalent integral representations. If we list all the possibilities for (F, L) with F the group of all automorphisms of L, it turns out that there are 14 possibilities altogether. These are known as the 14 *Bravais lattices*. We proceed with the computations.

(1) The generic lattic will admit only I and $-I$ as automorphisms. It is called the *triclinic* lattice. If we choose the three shortest vectors as basis then they will all have different lengths and the angles between them will be unequal, hence the name triclinic. The group, of course, is S_2. In all bases I has the same matrix representation and so does $-I$.

(2) Suppose that the group of automorphisms of the lattice L (i.e. the group of orthogonal transformations preserving L) contains a rotation through 180°, about

some axis, which we might as well take to be the z axis. Let r be this rotation and let u' be some vector in L. Then $r(u' - ru') = -(u' - ru')$ so that $u = u' - ru'$ is orthogonal to the z axis. Since we can always choose u' to lie off the z axis, we can arrange that $u \neq 0$. We can choose some $v' \in L$ which is independent of u and the z direction. Then $v = v' - rv'$ is also orthogonal to the z axis and is independent of u. Thus u and v span the xy plane, P. We conclude that $L \cap P$ is a lattice which spans the plane P. We now modify our choice of u and v, if necessary, so that they form a basis for the two-dimensional lattice, $L \cap P$. Let w be some lattice vector not in the xy plane, whose z component is smallest in absolute value. There always exists such a w in view of the discreteness of the lattice. If w' is any other lattice vector not in the xy plane we can subtract suitable multiples of w from w', i.e. consider $w' - kw$, such that the z component of $w' - kw$ is smaller than the z component of w. Hence $w' - kw \in L \cap P$ and is hence an integral combination of u and v. We have thus established that u, v and w form a basis of L. Let us normalize our choice of w somewhat further. Let $\bar{w}$ be the projection of w onto the plane P. Notice that $w - rw = 2\bar{w}$ is a lattice vector. There are thus two possibilities:

(i) $\bar{w}$ is a lattice vector. In this case, by subtracting off suitable multiples of u and v, we can arrange that w lies on the z axis

(ii) $\bar{w}$ is not a lattice vector. In this case there is no way of choosing a basis u, v, w', with w' on the z axis, since there is no way of writing $\bar{w}$ as an integral combination of such vectors. By subtracting off suitable multiples of u and v we can arrange that $\bar{w}$ is either $\frac{1}{2}u, \frac{1}{2}v$ or $\frac{1}{2}(u + v)$. Replacing the basis u, v by $u, (u + v)$ or $(u + v), v$, if necessary, we can arrange that $\bar{w} = \frac{1}{2}(u + v)$.

Suppose that the group of automorphisms of L contains a rotation through $180°$ and that in the preceding analysis the lattice $L \cap P$ is the generic plane lattice. In particular, assume that we have chosen u and v to be a basis in $L \cap P$ where u is the vector of shortest length in $L \cap P$ and v is the vector of next shortest length, that the angle between u and v is not $\pi/2$ or $\pi/4$ and that u, v and w all have unequal lengths. Any automorphism of the lattice must then carry $u \to \pm u, v \to \pm v$ and $w \to \pm w$. A *rotation* sending $w \to -w$ is impossible, since the induced transformation in the xy plane would be a reflection about some axis in that plane, implying that the angle between u and v is either $\pi/2$ or $\pi/4$. Thus the only rotation is the original one we started with and the group in question is C_{2h}. The lattice in this case is called *monoclinic*, since a single angle, that between u and v, is at our disposal. The two cases, (i) and (ii) give rise to two Bravais lattices, the primitive monoclinic lattice (P) and the face centered monoclinic lattice (C) (Fig. A.2).

(3) Suppose that F contains a rotation, r_4, through $90°$. We can apply the preceding analysis to the rotation r_4^2, but now we can specify that u and v are orthogonal and have equal length. This limits our choice so that we must consider now four distinct alternatives for $\bar{w}$, namely $\bar{w} = 0, \frac{1}{2}u, \frac{1}{2}v$, or $\frac{1}{2}(u + v)$. If $\bar{w} = \frac{1}{2}u$, then

$$w - r_4 w = \tfrac{1}{2}(u - v)$$

This vector lies in $L \cap P$, but is shorter in length than u, which is impossible. Thus $\bar{w} = \frac{1}{2}u$ or $\frac{1}{2}v$ cannot occur. There are thus two possibilities:

$$\bar{w} = 0 \quad \text{or} \quad \bar{w} = \tfrac{1}{2}(u + v)$$

(a) Monoclinic (P)   (b) Monoclinic (C)

Fig. A.2

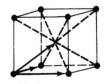

(a) Tetragonal (P)   (b) Tetragonal (I)

Fig. A.3

In either case, the group of the lattice clearly contains rotations about the x and y axes, i.e. the group contains D_4 (so that C_{4h} cannot arise as the group of all automorphisms of a lattice). If w has a different length from u and v, then these, together with powers of r_4, are clearly the only rotations and the group of automorphisms is D_{4h}. The two cases above give rise to two Bravais lattices, the primitive tetragonal lattice (P) and the body centered tetragonal lattice (I) (Fig. A.3).

(4) Suppose that F contains D_2. We can apply the above analysis to one of the non-trivial rotations, with the additional conclusion that the plane lattice, $P \cap L$ is invariant under reflection about some line in the plane. We need a lemma about plane lattices.

Lemma

If the plane lattice $P \cap L$ is invariant under reflection through a line, then the lattice is either rectangular or diamond shaped, i.e. we can choose basis vectors so that the fundamental cell is either a rectangle or a rhombus.

Proof We may assume that the x axis is the axis of the reflection, s. If u'' is any lattice vector, then $u'' + su''$ lies on the x axis and $u'' - su''$ lies on the y axis. By linear independence we conclude that there exist non-zero lattice vectors on both the x and y axes. Let u' be a shortest lattice vector on the x axis and v' be a shortest lattice vector on the y axis. Then for any $l \in P \cap L$ we have $l + sl = mu'$ and $l - sl = nv'$, where m and n are integers; thus $2l = mu' + nv'$, i.e. every l can be written as a half integral combination of u' and v'. There are now two possibilities: u' and v' form a basis, i.e. all the m's and n's are even. In this case the lattice is rectangular. If not, an element of the form $\frac{1}{2}u' + (n/2)v'$ is in the lattice (or $(m/2)u' + \frac{1}{2}v'$, interchanging u' and v' reduces one case to the other).

Then n cannot be even, since $\frac{1}{2}u'$ does not belong to the lattice. Subtracting off suitable multiples of v' gives that $u = \frac{1}{2}(u' + v')$ and hence $v = su = \frac{1}{2}(u' + v')$ belong to the lattice, and it is easy to see that they generate $P \cap L$. This $P \cap L$ is diamond shaped in this case. QED.

We return to the three-dimensional considerations. Suppose that u, v and w all have different lengths. Then D_2 contains all the rotations and the group is D_{2h}. Suppose that the lattice $P \cap L$ is rectangular. Then there are three possibilities:

$$\bar{w} = 0, \ \bar{w} = \tfrac{1}{2}u \text{ or } \bar{w} = \tfrac{1}{2}(u + v)$$

(the case $w = \frac{1}{2}v$ becomes $\bar{w} = \frac{1}{2}u$ upon interchanging the x and y axis.) Suppose that $P \cap L$ is diamond shaped. Then $\bar{w} = \frac{1}{2}u$ is impossible because there is a rotation r'' which interchanges u and v about an axis in the xy plane, yielding $w + r''w = \frac{1}{2}(u + v)$ and this is not a lattice vector. Thus either $\bar{w} = 0$ or $\bar{w} = \frac{1}{2}(u + v)$. If $\bar{w} = 0$, then w is perpendicular to u and v, and hence u and w span a rectangular plane lattice and we are back into the previous case under a rotation of the axes. Thus the only new case is $\bar{w} = \frac{1}{2}(u + v)$. The four alternatives are called primitive orthorhombic (P), base centered orthorhombic (C) (which is case (2), but usually drawn rotated so as to correspond to diamond shaped $P \cap L$ with $\bar{w} = 0$) body centered orthorhombic (I) and face centered orthorhombic (F) (see Fig. A.4).

(5) Suppose F contains a rotation, r_3, through $2\pi/3$. We may assume the rotation is about the z axis. Choose some lattice vector u' off the z axis. Then u', $r_3 u'$ and $r_3^2 u'$ form an equilateral triangle in the plane P. The lattice $P \cap L$ is a planar lattice invariant under rotations through $2\pi/3$. Let u be a vector of shortest length in this plane lattice. Then $r_3 u$ has the same length as u, and we claim that the parallelogram spanned by u and $r_3 u$ contains no other vector in $P \cap L$. Indeed, the triangle whose vertices consist of the origin, u, and $r_3 u$ is an equilateral triangle, and hence can contain no other lattice vector since such a lattice vector would have to be shorter than u. Similarly for the triangle. Thus u and $r_3 u$ span $P \cap L$. Let w be a lattice vector not in P, and $\bar{w}$ its projection onto P. As before, we may assume that $\bar{w}$ lies in the parallelogram spanned by u and $v = r_3 u$ and has shorter length. Thus $\|\bar{w}\| \leqslant \|u\|$. Now $r_3 w - w$ lies in P and is a lattice vector. Furthermore,

$$\|r_3 w - w\| = \|r_3 \bar{w} - \bar{w}\| = 3^{\frac{1}{2}} \|\bar{w}\| < 3^{\frac{1}{2}} \|u\|.$$

Now the only lattice vectors in $P \cap L$ whose length is $< 3^{\frac{1}{2}}u$ are 0, u, $u + v$, v, $-u$, $-u - v$ and $u - v$, the last six all having the same length. Thus either $r_3 w - w = 0$ or is

(a) Orthorhombic (P) (b) Orthorhombic (C) (c) Orthorhombic (I) (d) Orthorhombic (F)

Fig. A.4

Table 47.

System	Group	Number of lattices in system	Bravais lattices	Number of arithmetical crystal classes in system
triclinic	S_2	1	P	2
monoclinic	C_{2h}	2	P, C	6
orthorhombic	D_{2h}	4	P, C, I, F	13
tetragonal	D_{4h}	2	P, I	16
cubic	O_h	3	P, I, F	15
trigonal	D_{3d}	1	P	5
hexagonal	D_{6h}	1	P	16

one of the above six vectors. Any one of the six can serve as part of a basis so that there are two alternatives

$$\bar{w} = 0 \text{ or } r_3\bar{w} - \bar{w} = -u,$$

which solves to

$$\bar{w} = \tfrac{2}{3}u + \tfrac{1}{3}v$$

In the first case the lattice is invariant under the group D_{6d} and the lattice is called hexagonal (Fig. A.5(a)). In the second case the symmetry group is D_{3d} and the lattice is called trigonal (Fig. A.5(b)).

(6) Finally, suppose that the group of automorphisms of L contains T. Since T contains D_2 we can apply the analysis of case (4). Let u be the shortest of the three vectors, u, v and w. Since T contains rotations, r_3, of order 3, we conclude that $r_3 u \neq \pm u$ for a suitable choice of r_3, and hence that $r_3 u = \pm v$ and $r_3^2 u = \pm w$. Hence u, v and w all have the same length. Thus we are in the case of the orthorhombic situations except the rectangular parallelepiped is actually a cube. The base centered lattice is not invariant under the elements of order 3. There are thus three possibilities, simple (primitive) cubic (P) body centered cubic (I) and face centered cubic (F). The figures are the same as for

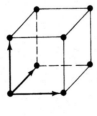

(a) Hexagonal

(b) Trigonal

Fig. A.5

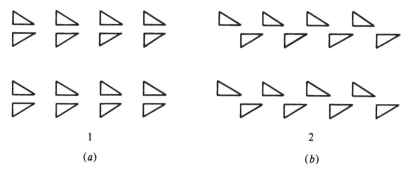

Fig. A.6

the orthorhombic case drawn on p. 317, with the understanding that the outside figure is meant to now be a cube.

We summarize our results in the Table 47. We have not derived the results in the last column. They can be easily obtained by examining the representations of the various subgroups of the seven listed groups in terms of the Bravais lattices. The detailed list of arithmetical classes can be found in the book by Burkhardt.

We close this section with a brief discussion of why the knowledge of (F, L) still does not determine the crystallographic group. Consider the patterns shown in Fig. A.6 in the plane. Both patterns admit the same set of translations as symmetries. The first pattern (a) also admits the linear transformation, τ, as a symmetry, where τ is a reflection about the x axis, $\tau = \begin{pmatrix} 1 & 0 \\ 0 & -1 \end{pmatrix}$. The reflection, τ, does not occur as a symmetry of the second pattern (b), but only the glide reflection

$$\begin{pmatrix} x \\ y \end{pmatrix} \rightarrow \begin{pmatrix} x + \frac{1}{2} \\ -y \end{pmatrix} = \begin{pmatrix} 1 & 0 \\ 0 & -1 \end{pmatrix}\begin{pmatrix} x \\ y \end{pmatrix} + \begin{pmatrix} \frac{1}{2} \\ 0 \end{pmatrix}$$

For both cases, the group F consists of I and τ and it acts in the same way on L. But in the first case, the group of all symmetries actually contains a subgroup isomorphic to F. In the second case, the group, G, of all symmetries does not contain a subgroup isomorphic to F, even though $F = G/L$.

From an abstract point of view, it turns out that the question boils down to the following: given a group F which acts as automorphisms of some commutative group, N, in how many different (i.e. non-isomorphic) ways can we find a group G which has N as a normal commutative subgroup with $G/N \sim F$ and such that the induced action of G/N on N is the specified action of F on N. Such a group, G, is called an *extension* of the group F by N. The general analysis of the problem of group extensions, together with some of the key tools in the theory, was developed by Frobenius as an outgrowth of his study of the crystallographic groups. The subject is now part of the theory of cohomology of groups, a branch of homol[gial algebra.

Appendix B

TENSOR

PRODUCT

The notion of the tensor product of two (or more) vector spaces is essential to much of what we do in this book. It is a basic notion in the theory of vector spaces, but as it does not appear in many of the elementary linear algebra texts, we give a brief introduction to the subject here. For a full development, we refer to the book *Multilinear algebra* by Graeb.

Let V and W be vector spaces, and let U be a third vector space. A map

$$f : V \times W \to U$$

is called *bilinear* if

$$f(v_1 + v_2, w) = f(v_1, w) + f(v_2, w)$$
$$f(v, w_1 + w_2) = f(v, w_1) + f(v, w_2)$$

and

$$f(av, w) = af(v, w) = f(v, aw).$$

In other words, f is called bilinear if it is linear in v when w is held fixed, and is linear in w when v is held fixed.

A familiar example, where $V = W = U$ is the 'vector product' in ordinary three-dimensional space.

Starting with V and W we wish to construct a vector space Z and a bilinear map $b : V \times W \to Z$ which is 'universal' in the following sense. Suppose that $f : V \times W \to U$ is any bilinear map. Then there exists a *unique* linear map $l : Z \to U$ such that

$$f = l \circ b.$$

Diagrammatically,

$$V \times W \xrightarrow{\;b\;} Z$$

for any f there is a unique linear l making the diagram commute.

Notice that *if* b and Z exist, they are unique up to isomorphism. Indeed, suppose b' and Z' were a second 'universal' bilinear map. Then taking $U = Z'$ and $f = b'$ in the

above diagram, we get

so $l_{b'}: Z \to Z'$ with

$$b' = l_{b'} \circ b.$$

Similarly we get a linear map $l_b: Z' \to Z$ with

$$b = l_b \circ b'.$$

But then

$$b = (l_b \circ l_{b'}) \circ b.$$

But

$$
\begin{array}{ccc}
V \times W & \xrightarrow{b} & Z \\
b \downarrow & \text{id} & {\Large\diagup}\, l_b \circ l_{b'} \\
Z & &
\end{array}
$$

then id and $l_b \circ l_{b'}$ both satisfy the equation

$$b = l \circ b.$$

By the uniqueness, we conclude that

$$l_b \circ l_{b'} = \text{id};$$

similarly

$$l_{b'} \circ l_b = \text{id}.$$

Thus $l_{b'}$ gives an isomorphism of Z with Z' and $b' = l_{b'} \circ b$. So, up to isomorphism, (b, Z) are unique.

Our problem is thus to show that such a (b, Z) actually exists. We shall give two, rather different looking, constructions. Of course, we know by uniqueness that they must give the same final answer.

Let Y denote the space of all scalar valued bilinear functions on $V \times W$. Thus $\alpha \in Y$ is a bilinear map of $V \times W \to \mathbb{C}$. It is clear that the set of all bilinear functions, α, forms a vector space and this is Y. Now take

$$Z = Y^*.$$

So a vector in Z is a linear function on Y. Define the map $b: V \times W \to Z$ by

$$[b(v, w)](\alpha) = \alpha(v, w).$$

That is, $b(v, w)$ is that linear function which assigns to each bilinear function α its value of α at (v, w). It is clear that $b(v, w)$ is a linear function of α and that it depends bilinearly on v and w. We claim that b and Z give a solution to our universal problem.

Indeed, suppose that $f: V \times W \to U$ is a bilinear map, where U is some vector space. For each $v \in U^*$ we get a bilinear function

$$v \circ f \in Y,$$

We have thus defined a linear map, $l^*: U^* \to Y$:

$$l^*(v) = v \circ f.$$

Therefore, we get its adjoint, $l^{**}: Z = Y^* \to U^{**}$. By definition if we evaluate l^{**} on the particular element $b(v, w) \in Y^*$ we get

$$
\begin{aligned}
l^{**}(b(v, w))(v) &= b(v, w)(l^*(v)) \\
&= (l^*(v))(v, w) \\
&= (v \circ f)(v, w) \\
&= v(f(v, w)).
\end{aligned}
$$

In other words, $l^{**}(b(v, w))$ is that linear function on U^* which assigns to each v the number $v(f(v, w))$. Now recall that there is a canonical injection

$$U \xrightarrow{\ i\ } U^{**}$$

where each $u \in U$ is identified with the linear function $i(u)$ on U^* given by

$$i(u)(v) = v(u) \quad u \in U$$
$$v \in U^*.$$

We can thus write

$$l^{**}(b(v, w)) = i(f(v, w)).$$

Using i, we can write $l^{**} = i \circ l$, where l is well defined as a map of $Z \to U$. Thus we have found a map $l: Z \to U$ satisfying

$$l \circ b = f.$$

Suppose there were two such maps. Then the difference, m, would satisfy

$$(m \circ b)(v, w) = 0, \quad \text{for all } v, w \in V.$$

So

$$m^*(v)(b(v, w)) = 0 \quad \text{all } v \in U^*.$$

Thus $m^*(v) \in Y$ is that bilinear function on $V \times W$ which assigns 0 to all pairs (v, w). Thus $m^*(v)$ is the zero bilinear function. Thus m^* sends all of U^* into 0 so is the zero linear map; thus $m = 0$. This proves the uniqueness of l.

The space Z is called the *tensor product* of the spaces V and W and is denoted by $V \otimes W$. We shall also use the notation

$$b(v, w) = v \otimes w.$$

Suppose that V and W are finite dimensional. Let $e_1, \ldots, e_m$ be a basis of V and $f_1, \ldots, f_n$ be a basis of W. Then a bilinear function β on $V \times W$ is completely determined by its values $\beta(e_i, f_j)$: we can write

$$\beta = \sum \beta(e_i, f_j) \varepsilon_{ij}$$

where ε_{ij} is the bilinear function determined by

$$\varepsilon_{ij}(v, w) = v_i w_j \quad \text{if} \quad v = v_i e_i + \cdots + v_m e_m$$
$$w = w_1 f_1 + \cdots + w_n f_n.$$

The ε_{ij} are clearly linearly independent and span Y; they form a basis of Y. The dual basis of $Z = Y^*$ is just $b(e_i, f_j) = e_i \otimes f_j$. Thus,

if $e_1, \ldots, e_m$ is a basis of V and $f_1, \ldots, f_n$ is a basis of W, then

$$\{e_i \otimes f_j\}_{\substack{i = 1, \ldots, m \\ j = 1, \ldots, n}}$$

gives a basis in $V \otimes W$. In particular

$$\dim(V \otimes W) = (\dim V)(\dim W).$$

Here is another construction of $V \otimes W$. It is more direct – and abstract – but has the disadvantage of invoking some infinite-dimensional spaces. For any set M, let $F(M)$ denote the vector space of all formal linear combinations of elements of M. Thus an element of $F(M)$ is a finite expression of the form

$$a_{m_1} m_1 + \cdots + a_{m_k} m_k$$

where the m_i are elements of M and the a_{m_i} are scalars. We add two such expressions by adding the coefficients. If M were a finite set, $F(M)$ would simply be the space of *all* functions on M, the function f corresponds to the formal expression

$$\sum_{m \in M} f(m) m.$$

If M is not finite, then $F(M)$ can be thought of as the space of all functions on M which vanish except on a finite number of points. (Here we are identifying $m \in M$ with the function δ_m, where $\delta_m(n) = 0$ if $n \neq m$ and $\delta_m(m) = 1$. Thus the most general element of $F(M)$ is a finite linear combination of the δ_m. The map $M \to F(M)$ sending $m \to \delta_m$ gives M as a subspace of $F(M)$.)

The space $F(M)$ is 'universal' with respect to maps of M into vector spaces: given any map $\phi : M \to U$, where U is a vector space, there is a unique linear map, $L_\phi : F(M) \to U$ such that

$$L_\phi \circ \delta_m = \phi(m).$$

Indeed, this last equation defines L_ϕ on the elements δ_m and hence L_ϕ extends by linearity to all of $F(M)$.

Now let us get back to our problem. Take $M = V \times W$. If ϕ is any map of the *set* $V \times W \to U$, there is an $L_\phi : F(V \times W) \to U$ which is linear and satisfies $L_\phi \circ \delta_m = \phi(m)$ for all $m \in V \times W$. If $f : V \times W \to U$ is bilinear, then L_f must vanish on all the elements

$$\delta_{(v_1 + v_2, w)} - \delta_{(v_1, w)} - \delta_{(v_2, w)}$$
$$\delta_{(v, w_1 + w_2)} - \delta_{(v, w_1)} - \delta_{(v, w_2)}$$
$$\delta_{(rv, w)} - r \delta_{(v, w)}$$
$$\delta_{(v, rw)} - r \delta_{(v, w)}.$$

Thus, L_f must vanish on the subspace B spanned by these elements. We can thus define $V \otimes W$ to be the quotient space

$$V \otimes W = F(V \times W)/B$$

and the map $b: V \times W \to V \otimes W$ by

$$b(v, w) = [\delta_{v,w}]$$

where $[\cdot]$ denotes the equivalence class mod B. It is now clear that this definition of b and $V \otimes W$ fulfils the universal properties.

Let us now draw some consequences of the universal property. Let V' and W' be another pair of vector spaces. Let $A: V \to V'$ and $B: W \to W'$ be linear transformations. Consider the map sending

$$(v, w) \text{ into } Av \otimes Bw.$$

This is clearly bilinear. Hence there exists a unique *linear* map from $V \otimes W \to V' \otimes W'$, which we shall denote by $A \otimes B$ such that

$$(A \otimes B)(v \otimes w) = Av \otimes Bw.$$

Suppose that $A': V' \to V''$ and $B': W' \to W''$ be a second pair of vector spaces and linear transformations. Then the map sending (v, w) into $A'AV \otimes B'Bw$ gives

$$A'A \otimes B'B: V \otimes W \to V'' \otimes W''.$$

But we also have the maps

$$A \otimes B: V \otimes W \to V' \otimes W'$$

and

$$A' \otimes B': V' \otimes W' \to V'' \otimes W''$$

so we get

$$(A' \otimes B') \cdot (A \otimes B): V \otimes W \to V'' \otimes W''.$$

By construction $(A' \otimes B')(Av \otimes Bw) = AA'v \otimes BB'w$. Hence we conclude (from the uniqueness part of the universal property) that

$$(A' \otimes B') \cdot (A \otimes B) = A'A \otimes B'B.$$

In particular, suppose that G and H are groups, that $V = V' = V''$ and we are given a representation, r, of G on V. Similarly, suppose that $W = W' = W''$ and that we are given a representation, s, of H on W. Then we get a representation $r \boxtimes s$ of $G \times H$ on $V \otimes W$ by

$$(r \boxtimes s)(a, b) = r(a) \otimes s(b).$$

Here is another consequence of the defining property of tensor products. Suppose that V and W are equipped with scalar products, $(,)_V$ and $(,)_W$. Consider the scalar-valued function f defined on $(V \times W) \times (V \times W)$ by

$$f((v, w), (v', w')) = (v, v')_V (w, w')_W.$$

For fixed v' and w', f is bilinear in v and w, hence defines a linear function $l_{(v',w')}$ on $V \otimes W$ satisfing

$$l_{(v',w')}(v \otimes w) = (v, v')_V(w, w')_W.$$

For fixed v and w, this expression is (anti) bilinear in v' and w'. Since every $\alpha \in V \otimes W$ is a finite sum of elements of the form $v \otimes w$, we conclude that

$$l_{v',w'}(\alpha)$$

is (anti) bilinear in v' and w', for any fixed α in $V \otimes W$. Thus there is an (anti) linear function l^α on $V \otimes W$ such that

$$l^\alpha(v' \otimes w') = l_{v',w'}(\alpha).$$

Then

$$l^\alpha(\beta)$$

is linear in α and (anti) linear in β. It is easy to check that

$$(\alpha, \beta)_{V \otimes W} = l^\alpha(\beta)$$

defines a scalar product on $V \otimes W$. To summarize:

there is a unique scalar product $(,)_{V \otimes W}$ defined on $V \otimes W$ which has the property that

$$(v \otimes w, v' \otimes w')_{V \otimes W} = (v, v')_V(w, w')_W.$$

Here is one further identification involving tensor products which is very useful. Let V and W be vector spaces. For each $v \in V$ and $\mu \in W^*$, consider the rank 1 linear transformation $T_v^u : W \to V$ defined by

$$T_v^u(w) = \langle \mu, w \rangle v.$$

Here $\langle \mu, w \rangle$ denotes the value of the linear function $\mu \in W^*$ on the vector $w \in W$. The map T_v^u clearly depends linearly on v for fixed μ and linearly on μ for fixed v. Thus we have a unique linear map

$$i : V \otimes W^* \to \operatorname{Hom}(W, V)$$

determined by

$$i(v \otimes \mu)w = \langle \mu, w \rangle v.$$

It is easy to check that the map i is an injection. It maps $V \otimes W^*$ onto the subspace of $\operatorname{Hom}(W, V)$ consisting ot those linear transformations of finite rank. If W of V is finite dimensional, this is all of $\operatorname{Hom}(W, V)$. If V and W are both infinite dimensional, this is a proper subspace.

There are a number of extensions and modifications of the notion of tensor product which we now briefly describe. Suppose that instead of just two vector spaces V and W we had k vector spaces $V_1, \ldots, V_k$. We can then define a k-linear (or multilinear) map $f : V_1 \times \cdots \times V_k \to U$ to be a map which is linear in any one of the variables when all the others are held fixed. Then there is a 'universal' space $V_1 \otimes \cdots \otimes V_k$ and multilinear map

$m: V_1 \times \cdots \times V_k \to V_1 \otimes V_2 \cdots \otimes V_k$ just as in the case. It follows from the universal properties that there is an isomorphism of

$$(V_1 \otimes \cdots \otimes V_k) \otimes (V_{k+1} \otimes \cdots \otimes V_{k+l})$$

with

$$V_1 \otimes \cdots \otimes V_{k+l}$$

since they both satisfy the universal property for $(k + l)$-linear maps.

As in the case $k = 2$, if we are given linear maps $A_1: V_1 \to W_1, A_2: V_2 \to W_2$, etc., we get a linear map

$$A_1 \otimes \cdots \otimes A_k: V_1 \otimes \cdots \otimes V_k \to W_1 \otimes \cdots \otimes W_k.$$

The generalization of the rule for composition holds as well. Also, under the identification of $(V_1 \otimes \cdots \otimes V_k) \otimes (V_{k+1} \otimes \cdots \otimes V_{k+l})$ with $V^1 \otimes \cdots \otimes V_{k+l}$, we get an identification of $(A_1 \otimes \cdots \otimes A_k) \otimes (A_{k+1} \otimes \cdots \otimes A_{k+l})$ with $A_1 \otimes \cdots \otimes A_{k+l}$.

Appendix C

INTEGRAL GEOMETRY AND THE
REPRESENTATIONS OF THE SYMMETRIC GROUP

Integral geometry, as formulated by Gel'fand, deals with the following situation. Let X and Y be spaces, and let Z be a subset of $X \times Y$. We wish to think of Z as describing an 'incidence relation'. For example, we might take X to be the ordinary three-dimensional Euclidean space, and Y to be the space of all lines in X. We can then let Z consist of all pairs (x, l) such that $x \in l$. Thus, in this case, the subset Z describes the incidence relation 'the point x lies on the line l'. Let $\pi : Z \to X$ denote the (restriction to Z of) projection onto the first factor, so $\pi(x, l) = x$, and, similarly, let $\rho : Z \to Y$ denote projection onto the second factor. So we have the diagram

Thus, for example, the set $\pi(\rho^{-1}(y))$ for $y \in Y$ consists of 'all x that are incident to y'. In our example, if y is a line, then $\pi(\rho^{-1}(y))$ consists of all points x lying on the line y. Let f be a (continuous) function on X. Then its pull back, $\pi^* f$ is a function on Z given by $\pi^* f(x, y) = f(x)$. Suppose that we are given a notion of 'integral' or 'measure' on each of the sets $\rho^{-1}(y)$. (For instance, in our example one would use the ordinary one-dimensional (Lebesgue or Riemann) integral on each line that derives from the Euclidean geometry of X.) Then we can integrate the function $\pi^* f$ over each fiber $\rho^{-1}(y)$ to obtain a function Rf on Y. In the example, R is known as the *Radon transform*, first studied by Radon in 1917. It assigns to each function on Euclidean three-space a function on the space of lines given by integrating each function over the line. The problem of recovering f from Rf is the central issue in X-ray tomography.

In this appendix we shall be concerned with the case that X and Y are *finite sets* and the measure that we use is the 'counting measure' that assigns weight 1 to each point. Thus

$$(Rf)(y) = \sum_{x \mid (x,y) \in Z} f(x)$$

Following a remarkable paper of James (1977) we shall see that many of the basic facts about the representations of the symmetric group can be derived from a careful study of the various Radon transforms associated to incidence relations between the various spaces M_λ where λ is a Young diagram. To see what is involved, let us examine the representations of S_4. Let us take

$$Y = M_{\square\square\square\square} = M_{(4)}$$

so Y is a one-point set and

$$X = M_{\square\square\atop\square} = M_{(3,1)}.$$

Let us take Z to be the trivial incidence relation, so all points (x, y) belong to Z. The R is just the operator

$$Rf(y) = \sum f(x).$$

We have seen in Sections 2.5 and 2.8 that the representation of S_4 on $F_{(3,1)}$ is exactly the kernel of R. In fact we proved the corresponding fact for all S_n. Next let us take

$$X = M_{(2,2)} \quad \text{and} \quad Y = M_{(3,1)}.$$

A point, x, of $M_{(2,2)}$ is a tabloid

$$x = \begin{Bmatrix} a & b \\ c & d \end{Bmatrix}$$

and a point y of $M_{(3,1)}$ is a tabloid

$$y = \begin{Bmatrix} e & f & g \\ h & & \end{Bmatrix}.$$

Let us say that x and y are incident if $h = c$ or $h = d$; in other words, if the set on the second row of y is a subset of the set of the second row of x. In all of this appendix we shall use the shorthand t for the function δ_t. Thus

$$R\begin{Bmatrix} a & b \\ c & d \end{Bmatrix} = \begin{Bmatrix} a & b & c \\ d & & \end{Bmatrix} + \begin{Bmatrix} a & b & d \\ c & & \end{Bmatrix}.$$

In particular, R carries constants into constants, and the above equation shows that the image of R contains non-constant elements. Since R is an S_4 morphism, the image of R must be an invariant subspace of $\mathscr{F}(M_{(3,1)})$ and hence all of $\mathscr{F}(M_{(3,1)})$.

Since $\mathscr{F}(M_{(2,2)})$ is six dimensional and $\mathscr{F}(M_{(3,1)})$ is four dimensional, we conclude that $\ker R$ is an invariant two-dimensional subspace of $\mathscr{F}(M_{(2,2)})$ and hence must coincide with $F_{(2,2)}$. (It is instructive to check directly from the definitions that the elements e_t, which span $F_{(2,2)}$, for tableaux of type $(2, 2)$ do indeed lie in $\ker R$.)

Next we will consider the 12-dimensional space

$$\mathscr{F}(M_{(2,1,1)}).$$

Let us first take

$$X = M_{(2,1,1)} \quad \text{and} \quad Y = M_{(2,2)}.$$

Declare the elements

$$Rf \begin{Bmatrix} a & b \\ c & \\ d & \end{Bmatrix} \quad \text{and} \quad \begin{Bmatrix} a & b \\ c & d \end{Bmatrix}$$

to be incident. In other words, x and y are incident if their top rows define the same subset of $\{1,2,3,4\}$. Thus

$$\left(\begin{Bmatrix} a & b \\ c & d \end{Bmatrix} \right) = f\left(\begin{Bmatrix} a & b \\ c & \\ d & \end{Bmatrix} \right) + f\left(\begin{Bmatrix} a & b \\ d & \\ c & \end{Bmatrix} \right).$$

Applied to δ functions (remember we are writing x for δ_x) we have

$$R \begin{Bmatrix} a & b \\ c & \\ d & \end{Bmatrix} = \begin{Bmatrix} a & b \\ c & d \end{Bmatrix}$$

so the map R is surjective. Thus $\ker R$ is an invariant six-dimensional subspace of $\mathscr{F}(M_{(2,1,1)})$. The elements

$$\begin{Bmatrix} a & b \\ c & \\ d & \end{Bmatrix} - \begin{Bmatrix} a & b \\ d & \\ c & \end{Bmatrix}$$

clearly lie in $\ker R$ and are linearly independent for the various choices of the subset a, b. There are six such choices so these elements span $\ker R$. Now take

$$Y' = M_{(3,1)}$$

and declare the elements

$$\begin{Bmatrix} a & b \\ c & \\ d & \end{Bmatrix} \quad \text{of} \quad M_{(2,1,1)} \quad \text{and} \quad \begin{Bmatrix} a & b & c \\ d & & \end{Bmatrix} \quad \text{of} \quad M_{(3,1)}$$

to be incident. In other words, x and y' are incident if they have the same last rows. Let R' denote the corresponding Radon transform. Then

$$R'\left(\begin{Bmatrix} a & b \\ c & \\ d & \end{Bmatrix} - \begin{Bmatrix} a & b \\ d & \\ c & \end{Bmatrix} \right) = \begin{Bmatrix} a & b & c \\ d & & \end{Bmatrix} - \begin{Bmatrix} a & b & d \\ c & & \end{Bmatrix} = e_{\substack{cab \\ d}} \in F_{(3,1)}.$$

Thus $R'(\ker R) = F_{(3,1)}$ and hence $\ker R \cap \ker R'$ is an invariant three-dimensional subspace of $\mathscr{F}(M_{(2,1,1)})$. It is easy to check directly from the definitions that the elements e_t, where $t \in M_{(2,1,1)}$ lie in the intersection of the two kernels and hence that

$$\mathscr{F}_{(2,1,1)} = \ker R \cap \ker R'.$$

One of the main results of James's analysis is that the irreducible subspaces $F_\lambda \subset \mathscr{F}(M_\lambda)$ can be characterized as the intersection of kernels of Radon transforms – this for the general symmetric group S_n. In all of what follows we will be following James (1977) very closely, except for minor rearrangements and changes in notation.

C.1 Partition pairs

In analyzing the various Radon transforms we will be considering various intermediate spaces between F_λ and $\mathscr{F}(M_\lambda)$. For this purpose it will be convenient to make a change in notation. By a partition, λ, of n we mean a sequence of non-negative integers, $\lambda_1, \ldots, \lambda_n$ which add up to n: $\lambda_1 + \cdots + \lambda_n = n$. We drop the requirement that $\lambda_1 \geqslant \lambda_2 \geqslant \cdots \geqslant \lambda_n$. We shall sometimes write a partition λ as before as a Young diagram consisting of rows of boxes. Thus the Young diagram corresponding to the partition $(2,5,3)$ of 10 would be written as

A Young tableau corresponding to a given diagram consists of placing the numbers from 1 to n in the corresponding positions. Thus

$$
\begin{array}{l}
5\ 2 \\
3\ 9\ 1\ 6\ 4 \\
10\ 7
\end{array}
$$

is a Young tableau corresponding to the above diagram. If t is a Young tableau, the corresponding tabloid $\{t\}$ is the partition of $1, \ldots, n$ into subsets given by the entries of the rows of t. The space of all tabloids for a given diagram λ will be denoted by M_λ. All of this is as in Chapter 2 except that we have dropped the non-increasing requirement on λ. In order to avoid having to draw large parentheses we shall also denote the tabloid corresponding to a given tableau by underlining the rows in the tableau. Thus

$$
\begin{array}{l}
\underline{5\ 2} \\
\underline{3\ 9\ 1\ 6\ 4} \\
\underline{10\ 7}
\end{array}
$$

denotes the partition of $1, \ldots, 10$ into three subsets $\{5, 2\} \{3, 9, 1, 6, 4\}$ and $\{10, 7\}$. We shall use this same symbol to denote the associated delta function in $\mathscr{F}(M_\lambda)$. Let μ and ν be partitions which agree at all positions except the ith and $(i+1)$st and suppose that $\nu_i > \mu_i$ so that $\nu_{i+1} < \mu_{i+1}$. Let $x \in M_\mu$ so that x is a partition of $\{1, \ldots, n\}$ into subsets $x_1 \cup \cdots \cup x_p$ of sizes $\mu_1, \ldots, \mu_p$, and let $y \in M_\nu$. In particular, x_{i+1} is a subset of size μ_{i+1} and y_{i+1} is a subset of size ν_{i+1}. We will now define the 'incidence relation' i.e. the subset $Z_{\mu,\nu} \subset M_\mu \times M_\nu$ by

$$(x, y) \in Z_{\mu,\nu} \quad \text{if and only if} \quad x_j = y_j \quad \text{for} \quad j \neq i \quad \text{or} \quad i+1 \quad \text{and} \quad y_{i+1} \subset x_{i+1}.$$

We let $R_{\mu,\nu} \colon \mathscr{F}(M_\mu) \to \mathscr{F}(M_\nu)$ denote the corresponding Radon transform. These will be the Radon transforms which we shall be using. In order to analyze the structure of these Radon transforms it is convenient to introduce the notion of a partition pair. By this we mean a pair (λ', λ), where λ is a partition of n (with $\lambda_1 \neq 0$) and λ' is a partition of some $n' \leqslant n$ such that.

$$\lambda_i' \leqslant \lambda_i$$

and

$$\lambda_i' \geqslant \lambda_{i+1}' \quad \text{for all } i.$$

If we draw our partition λ as rows of boxes or X's we can draw a partition pair by enclosing a boundary around the first λ_i' boxes or X's in each row. Thus

$$
\begin{array}{llll}
\text{X} & \text{X} & \text{X} & \text{X}| \\
\text{X} & \text{X}| & \text{X} \\
\text{X} & | \text{X} \\
\text{X} & | \text{X}
\end{array}
$$

is how we would draw the partition pair $((4, 2, 1, 1); (4, 3, 2, 2))$. We will be proving theorems by a kind of induction on partition pairs. For this purpose it will be convenient to introduce the following two operators a and r on the space of all partition pairs. Let c be the largest number such that for all $i < c$ $\lambda_i' = \lambda_i$. Assume that $c > 1$ and that $\lambda' \neq \lambda$. The operator a acts on λ' by changing λ_c' into $\lambda_c' + 1$ (unless $\lambda_{c-1}' = \lambda_c'$ in which case we would not get a partition pair – in this case we let the operator a send (λ', λ) into $(0, 0)$). The operator r acts on λ by changing λ_c to λ_c' and λ_{c-1} to $\lambda_{c-1} + \lambda_c - \lambda_c'$. Intuitively speaking, a encloses an additional node on the cth row, while r moves all the unenclosed nodes from the cth row to the $(c-1)$st row. So $a((3, 1); (3, 3)) = ((3, 2); (3, 3))$ and $r((3, 1); (3, 3)) = ((3, 1); (5, 1))$;

$$
\begin{array}{lll}
\text{X} & \text{X} & \text{X}| \\
\text{X} & | \text{X} & \text{X}
\end{array}
\xrightarrow{a}
\begin{array}{lll}
\text{X} & \text{X} & \text{X}| \\
\text{X} & \text{X} & | \text{X}
\end{array}
$$

$$r\downarrow$$

$$
\begin{array}{lllll}
\text{X} & \text{X} & \text{X}| & \text{X} & \text{X} \\
\text{X}|
\end{array}
$$

It will be convenient in carrying out induction arguments to always automatically replace λ_1' by λ_1, that is automatically enclose all boxes on the first row. So we would replace the last partition pair $((3, 1); (5, 1))$ by $((5, 1); (5, 1))$ for example.

We introduce an ordering on partitions by saying that $\mu \geqslant \lambda$ if and only if $\sum_1^j \mu_i \geqslant \sum_1^j \lambda_i$ for all j. This is a partial order on the set of all partitions. We order the set of partition pairs by setting $(\mu', \mu) \geqslant (\lambda', \lambda)$ if $\mu \geqslant \lambda$, and, if $\mu = \lambda$, then $\mu_i' \geqslant \lambda_i'$ for all i. The operations r and a increase the order of a partition pair (or map into $(0; 0)$). For example, starting with $((3, 0); (3, 3))$ and applying the operations r and a successively we obtain

$$
\begin{array}{lll}
\text{X} & \text{X} & \text{X}| \\
\text{X} & \text{X} & \text{X}
\end{array}
\xrightarrow{a}
\begin{array}{lll}
\text{X} & \text{X} & \text{X}| \\
\text{X}| & \text{X} & \text{X}
\end{array}
\xrightarrow{a}
\begin{array}{lll}
\text{X} & \text{X} & \text{X}| \\
\text{X} & \text{X}|
\end{array}
\xrightarrow{a}
\begin{array}{lll}
\text{X} & \text{X} & \text{X}| \\
\text{X} & \text{X} & \text{X}|
\end{array}
$$

$$
\begin{array}{c} r\downarrow \\ \begin{array}{llllll} \text{X} & \text{X} & \text{X} & \text{X} & \text{X} & \text{X}| \end{array} \end{array}
\qquad
\begin{array}{c} r\downarrow \\ \begin{array}{llllll} \text{X} & \text{X} & \text{X} & \text{X} & \text{X}| \\ \text{X}| \end{array} \end{array}
\qquad
\begin{array}{c} r\downarrow \\ \begin{array}{llll} \text{X} & \text{X} & \text{X} & \text{X}| \\ \text{X} & \text{X}| \end{array} \end{array}
$$

It is clear that the end results of the sequence of r and a operations when applied to any (λ', λ) gives a set of (μ', μ) with $\mu \geqslant \lambda$.

Let (λ', λ) be a partition pair and let t be a λ tableau. The group $C_t^{\lambda', \lambda}$ is the subgroup of S_n consisting of those permutations which do not move the numbers of t which lie outside of the positions corresponding to λ' and permute the numbers lying in the same column of λ'. Thus if

$$t = \begin{array}{ccc} 1 & 2 & 3 \\ 4 & 5 & 6 \end{array} \quad \text{and} \quad \lambda' = (3, 1)$$

then

$$C_t^{\lambda', \lambda} = S_{\{1,4\}} = \{e, (14)\}$$

while if $\lambda' = (3, 2)$ then

$$C_t^{\lambda', \lambda} = S_{\{1,4\}} \times S_{\{2,5\}} = \{e, (14), (25), (14)(25)\}.$$

We let $e_t^{\lambda', \lambda} \in \mathscr{F}(M_\lambda)$ be defined by

$$e_t^{\lambda', \lambda} = \sum_{\sigma \in C_t^{\lambda', \lambda}} (\operatorname{sgn} \sigma) \sigma \underline{t}.$$

It is easy to check that $\pi e_t^{\lambda', \lambda} = e_{\pi t}^{\lambda', \lambda}$ for any $\pi = S_n$. We let $F_{\lambda', \lambda} \subset \mathscr{F}(M_\lambda)$ denote the space spanned by the $e_t^{\lambda', \lambda}$ as t ranges over all tableaux. When $\lambda' = 0$ then $F_{0, \lambda} = \mathscr{F}(M_\lambda)$. When $\lambda' = \lambda$ (so that the λ_i are non-increasing: we say that λ is a *proper* partition or diagram) then the space $F_{\lambda, \lambda}$ is just the space F_λ that gives the irreducible representations of S_n. So the spaces $F_{\lambda', \lambda}$ are intermediate between $\mathscr{F}(M_\lambda)$ and F_λ.

If $\mu_i' \leqslant \mu_i''$ (for all i) then it is easy to see that $C_t^{\mu', \mu} \subset C_t^{\mu'', \mu}$ and hence that $e_t^{\mu'', \mu}$ is a linear combination of some $e_s^{\mu', \mu}$. Thus $F_{\mu'', \mu} \subset F_{\mu', \mu}$.

Let (μ', μ) be a partition pair and $(\mu', \nu) = r(\mu', \mu)$ so that ν differs from μ only in the $(c-1)$st and cth rows. We can therefore form the Radon transform $R = R_{\mu, \nu}$: $\mathscr{F}(M_\mu) \to \mathscr{F}(M_\nu)$. Recall that $F_{\mu', \mu}$ is a subspace of $\mathscr{F}(M_\mu)$, that $F_{a(\mu', \mu)} \subset F_{\mu', \mu}$ and that $F_{r(\mu', \mu)}$ is a subspace of $\mathscr{F}(M_\nu)$. We claim that

$$R(F_{\mu', \mu}) = F_{r(\mu', \mu)}$$

and

$$R(F_{a(\mu', \mu)}) = 0.$$

Proof   Let t be any μ tableau and let E_t be the element of the group algebra defined by

$$E_t = \sum_{\sigma \in C_t^{\mu', \mu}} (\operatorname{sgn} \sigma) \sigma$$

so that

$$E_t \underline{t} = e_t^{\mu', \mu}.$$

Now

$$R\underline{t} = \sum \underline{s}$$

where the sum extends over various tabloids obtained by leaving a subset of the cth row of size μ_c' on the cth row and moving the rest up to the $(c-1)$st row. One of these tabloids comes from leaving the first μ'_c elements on the cth row. Let z denote the v tableau obtained by moving the last $\mu_c - \mu'_c$ numbers up to the $(c-1)$st row, so that z is the above described tabloid. None of the permutations in $C_t^{\mu',\mu}$ affect any of the numbers we have moved so that

$$E_t(z) = e_z^{\mu',v}$$

For each of the other $\underline{s}$ that arise in the sum for $R\underline{t}$ we have moved at least one number x lying among the first μ'_i. Let y be the number immediately above x in t. Then the transposition (xy) lies in $C_t^{\mu',\mu}$ and $(xy)\underline{s} = \underline{s}$ since x and y lie in the same row of $\underline{s}$. But $(1 - (xy))$ is a factor of E_t and hence

$$E_t \underline{s} = 0.$$

Thus, since R commutes with the action of S_n, $RE_t\underline{t} = E_t R\underline{t} = E_t(\underline{z})$ so

$$Re_t^{\mu',\mu} = e_z^{\mu',v}.$$

As t ranges over all tableaux of μ, the tableau z ranges over all v tableaux, proving that

$$R(F_{\mu',\mu}) = F_{r(\mu',\mu)} = F_{\mu',v}.$$

Now $a(\mu', \mu)$ has one more node enclosed on the cth row. Hence if t' is any tableau and $E_{t'}^{a(\mu',\mu)} = E'$ is the corresponding group algebra element, we have $E'\underline{s} = 0$ for all $\underline{s}$ occurring in the expression for $R\underline{t}$ for any $\underline{t}$ since we must have moved at least one element x up to the $(c-1)$st row where x was enclosed in t'. This shows that

$$RF_{a(\mu',\mu)} = 0.$$

We shall, in fact, prove that the sequence

$$0 \to F_{a(\mu',\mu)} \to F_{\mu',\mu} \xrightarrow{R} F_{r(\mu',\mu)} \to 0 \qquad (C.1)$$

is exact. We shall do so by constructing a basis for each of these spaces and counting dimensions. In fact, we shall construct the basis of $F_{\mu',\mu}$ and prove that (C.1) is exact at the same time, by an induction procedure starting from the fact that $F_{0,\mu} = \mathscr{F}(M_\mu)$ and that the $e_t^{0,\mu} = \{t\}$ form a basis of $\mathscr{F}(M_\mu)$ provided that we have chosen one tableau t for each tabloid $\{t\}$. So we must devise a convenient procedure (for our proof) for selecting one tableau for each tabloid. For this purpose it will be convenient to have a slightly different description of the tabloids. Recall that a tabloid is a partition of the set $\{1,\ldots,n\}$ into subsets of size $\mu_1,\ldots,\mu_n$. Thus a tabloid can be thought of as a function from $1,\ldots,n$ to the labels of the subsets telling to which subset a particular number corresponds. Thus the function

$$g(1) = 2, \quad g(2) = 1, \quad g(3) = 1, \quad g(4) = 2$$

is the same as the tabloid

$$\left\{ \begin{matrix} 2 & 3 \\ 1 & 4 \end{matrix} \right\}$$

since 1 and 4 belong to the second subset and 2 and 3 to the first. We can write such a function by writing the value $g(i)$ over the integer i, so

$$
\begin{array}{cccc}
2 & 1 & 1 & 2 \\
1 & 2 & 3 & 4
\end{array}
$$

or, since the numbers in the second row are always in increasing order, we do not need to write them at all. So the above tabloid can be written simply as

$$2 \quad 1 \quad 1 \quad 2.$$

It is clear that any tabloid corresponding to the partition $(2,2)$ can be written as a sequence of two 1's and two 2's. Thus, for example, the sequence 1122 is the same as the tabloid $\left\{ \begin{array}{cc} 1 & 2 \\ 3 & 4 \end{array} \right\}$. More generally, a sequence of integers of length n will be called a μ sequence if it contains μ_1 1's, μ_2 2's, μ_3 3's, etc. From the above discussion it is clear that the set of μ sequences is the same as the set of μ tabloids. Thus, for example, the sequence 3 1 2 2 4 1 1 2 corresponds to the tabloid

$$
\left\{ \begin{array}{ccc}
2 & 6 & 7 \\
3 & 4 & 8 \\
1 & & \\
5 & &
\end{array} \right\}.
$$

We now describe a procedure for assigning a tableau to each μ sequence. We call a number occurring in a given μ sequence 'good' or 'bad' according to the following rules:

(i) All 1's are good.

(ii) An $i + 1$ is 'good' if the number of preceding good i's is greater than the number of preceding good $(i + 1)$'s. Thus listing the quality of each element below it in the sequence listed above we get

$$
\begin{array}{cccccccc}
3 & 1 & 2 & 2 & 4 & 1 & 1 & 2 \\
b & g & g & b & b & g & g & g \; .
\end{array}
$$

Now move along the sequence and put each good entry as far to the left and each bad entry as far to the right as possible in the appropriate row. Thus

$$
\begin{array}{ccc}
2 & 6 & 7 \\
3 & 8 & 4 \\
1 & & \\
5 & &
\end{array} \; .
$$

Notice that there are three good 1's and two good 2's, no good 3's or 4's. By 'as far as possible' we mean without disturbing the assignments already made. So the number 6, in the above example, which is the entry of the second good 1, goes to the right if the 2 on the first row.

Here is another example:

$$
\begin{array}{ccccccccccccccc}
1 & 2 & 3 & 4 & 5 & 6 & 7 & 8 & 9 & 10 & 11 & 12 & 13 & 14 & 15 \\
3 & 2 & 1 & 1 & 3 & 2 & 1 & 4 & 2 & 4 & 4 & 3 & 4 & 3 & 1 & 2 \\
b & b & g & g & b & g & g & b & g & b & g & g & g & g & g
\end{array}
$$

$$
\begin{array}{cccc}
3 & 4 & 7 & 14| \\
6 & 9 & 15 & \boxed{2} \\
11 & 13 & \overline{5} & 1 \\
\underline{12} & \overline{10} & 8 &
\end{array}
$$

There are four good 1's, three good 2's, two good 3's and one good 4. Notice that the numbers in the enclosed region are all increasing from left to right and from top down. (The numbers outside decrease from left to right in rows.) A moment's reflection shows that this is true in general. For any λ sequence, let $\lambda_1'' = \lambda_1''(s) = \lambda_1$ denote the number of good 1's, let $\lambda_2'' = \lambda_2''(s)$ denote the number of good 2's, etc. Then $\lambda_1'' \geqslant \lambda_2'' \geqslant \lambda_3'' \ldots$ (by the defining properties of good and bad) and the tableau we construct from s increases from left to right and down columns in the enclosed region of the partition pair (λ'', λ). (For example, if the second position in the third row is enclosed, it is filled by the second good 3. This means that an earlier number is placed to the left of it since it is the *second* good 3. It also means, by the defining property of *good*, that the second good 2 must have occurred earlier, so the number immediately above must also be smaller. It is clear that the same argument works for the kth good i, proving our contention.)

Let $S(\lambda', \lambda)$ denote the set of λ sequences with *at least* λ_i' good i's i.e. $\lambda_i''(s) \geqslant \lambda_i'$. Let $T(\lambda', \lambda)$ denote the corresponding set of tableaux. Thus every $t \varepsilon T(\lambda', \lambda)$ is 'standard inside λ'', i.e. increasing from left to right and from top down in the enclosed region given by λ'.

Our main task is to prove that the $\{e_t, t \varepsilon T(\lambda', \lambda)\}$ form a basis of $F_{\lambda', \lambda}$. The proof splits into two parts. By use of a partial order on tabloids, it will be relatively easy to show that the e_t are linearly independent. A much more delicate combinatorial argument, followed by an induction, will show that they actually span.

We order the λ tabloids as follows. We say that $\underline{t} < \underline{t}'$ if the number 1 occurs in a later row of $\underline{t}$ than of $\underline{t}'$. If 1 occurs in the same row of both, then $\underline{t} < \underline{t}'$ if the number 2 occurs in a later row of $\underline{t}$ than of $\underline{t}'$. If 2 occurs in the same row of both then we look at 3, etc. Of course if every number from 1 to n occurs in the same row of $\underline{t}$ and of $\underline{t}'$ then $\underline{t} = \underline{t}'$.

Let $\underline{t} \varepsilon T(\lambda', \lambda)$. Let $\sigma \varepsilon C_t^{\lambda', \lambda}$. Then outside of λ', all the entries of $\sigma \underline{t}$ agree with the entries of $\underline{t}$. If $\sigma \neq e$, then at least one column of $\sigma \underline{t}$ is not increasing from top down, so $\sigma \underline{t} < \underline{t}$. Thus all the tabloids $\underline{S}$ which enter into the expression for $e_t^{\lambda', \lambda}$ are $< t$ (except for t itself).

This implies that all the $e_t^{\lambda', \lambda}(t \varepsilon T(\lambda', \lambda))$ are linearly independent. Indeed, suppose that there were some linear relation

$$
\sum a_\alpha e_{t_\alpha}^{\lambda', \lambda} = 0 \qquad t_\alpha \varepsilon T(\lambda', \lambda).
$$

Let $\underline{t}_{\alpha_0}$ be the last $\underline{t}_\alpha$ that occurs in the above equation. Then all the tabloids $\underline{s}$ that enter into the equation are $< \underline{t}_{\alpha_0}$. The above equation can thus not hold unless $a_{\alpha_0} = 0$. Continuing we find that all the $a_\alpha = 0$.

The set $S(0, \lambda)$ consists of all sequences and $C_t^{0,\lambda} = \{e\}$ so $e_t = \underline{t}$. Thus the $e_t^{0,\lambda}$ span $F_{0,\lambda}$. Suppose we know that the $e_t^{\lambda',\lambda}$ span $F_{\lambda',\lambda}$, $t \varepsilon T(\lambda', \lambda)$.. In the sequence

$$0 \to F_{a(\lambda',\lambda)} \to F_{\lambda',\lambda} \xrightarrow{R} F_{r(\lambda',\lambda)} \to 0. \tag{C.1}$$

We have already verified that $Re_t^{\lambda',\lambda} = e_z^{r(\lambda',\lambda)}$, where z is obtained from t by moving all the unenclosed numbers in the cth row of t to the $(c - 1)$st row, where c is the first row which is not entirely enclosed. We will have proved that the $e_z^{r(\lambda',\lambda)}$ span *and* that the sequence is exact provided we can prove the following lemma.

Lemma

The following gives a one-to-one correspondence between $S(\lambda', \lambda) - S(a(\lambda', \lambda))$ *and* $S(r(\lambda', \lambda))$: change all bad c's to $(c - 1)$s.

Since we know that the $e_t^{0,\lambda}$ span, we can use the lemma inductively to conclude that $e_t^{\lambda',\lambda}$ span $F_{\lambda',\lambda}$ for all (λ', λ) and that the sequence $0 \to F_{a(\lambda',\lambda)} \to F_{\lambda',\lambda} \to F_{r(\lambda',\lambda)} \to 0$ is exact. Indeed, we know that R is surjective and $F_{a(\lambda',\lambda)} \subset \ker R$. We know that the $e_t^{\mu',\mu}$ are linearly independent. Hence

$$\dim F_{r(\lambda',\lambda)} + \dim F_{a(\lambda',\lambda)} \leqslant \dim F_{\lambda',\lambda}$$
$$\text{\tiny W} \qquad\qquad \text{\tiny W} \qquad\qquad \text{\tiny II}$$
$$\#S(r(\lambda', \lambda)) + \#S(a(\lambda', \lambda)) = \#S(\lambda', \lambda).$$

Our induction hypothesis is that we have equality on the right-hand column and the lemma asserts that we have equality on the bottom row. Hence all the inequalities must be equalities, proving simultaneously that (C.1) is exact and that the $e_t^{a(\lambda',\lambda)}$ and $e_t^{r(\lambda',\lambda)}$ span their respective spaces.

Before proving the lemma, let us illustrate how the procedure works for the case of $(2, 2)$ tabloids. Here are the $(2, 2)$ sequences listed in order and the corresponding tableaux:

$$2\ 2\ 1\ 1 \qquad \begin{array}{cc} 3 & 4 \\ 2 & 1 \end{array}$$

$$2\ 1\ 2\ 1 \qquad \begin{array}{cc} 2 & 4 \\ 3 & 1 \end{array}$$

$$2\ 1\ 1\ 2 \qquad \begin{array}{cc} 2 & 3 \\ 4 & 1 \end{array}$$

$$1\ 2\ 2\ 1 \qquad \begin{array}{cc} 1 & 4 \\ 2 & 3 \end{array}$$

$$1\ 2\ 1\ 2 \qquad \begin{array}{cc} 1 & 3 \\ 2 & 4 \end{array}$$

$$1\ 1\ 2\ 2 \qquad \begin{array}{cc} 1 & 2 \\ 3 & 4 \end{array}$$

The first element belongs to $S(2, 0; 2, 2) - S(2, 1; 2, 2)$ and under $R = R_{(2,0;2,2),(4,0)}$ we have

$$R(e^{((2,0)(2,2))}_{\substack{3\ 4\\2\ 1}}) = R\left(\left\{\begin{matrix}3 & 4\\2 & 1\end{matrix}\right\}\right) = \{1\ 2\ 3\ 4\} = \{3\ 4\ 2\ 1\}$$

spans the image, while the $e^{\lambda',\lambda}_t$ with $\lambda' = (2, 1)$, $\lambda = (2, 2)$ lie in the kernel. We know the kernel is five dimensional and that the five elements $e^{\lambda',\lambda}_t$ are linearly independent. Hence they form a basis for $F_{((2,1);(2,2))}$. Now look at the next sequence

$$0 \to F_{((2,2);(2,2))} \to F_{((2,1);(2,2))} \xrightarrow{R} F_{((3,1);(3,1))} \to 0.$$

We know that

$$Re^{\lambda',\lambda}_{\substack{2\ 4\\3\ 1}} = e_{\substack{2\ 4\ 1\\3}} \quad Re^{\lambda',\lambda}_{\substack{2\ 4\\4\ 1}} = e_{\substack{2\ 3\\4}}{}_1 Re^{\lambda',\lambda}_{\substack{1\\2\ 3}} = e_{\substack{1\ 4\ 3\\2}}.$$

On the right-hand side of these equations we have written e_t for $e^{((3,1);(3,1))}_t$. The elements e_t occurring on the right-hand side are linearly independent, and hence span the three-dimensional space $F_{((3,1);(3,1))}$. If we take $\lambda'' = (2, 2)$ then we know that the $e^{\lambda'',\lambda}_t$ lie in the kernel and are linearly independent for the last two t's in our list. This proves that the sequence is exact and that we have constructed a basis of $F_{((2,2);(2,2))}$.

Again, before proving the lemma, let us derive some of its consequences. First of all, since we can get to F_λ by a sequence of operations of type a, we conclude that

$$F_\lambda \text{ is an intersection of } \ker R \text{ for suitable } R\text{'s.} \tag{C.2}$$

Secondly, the basis we have constructed shows that the set of all λ sequences all of whose entries are good give a basis of $F_\lambda = F_{\lambda,\lambda}$. Now the tableau corresponding to such a sequence is standard, i.e. is increasing from left to right in rows and down columns. Conversely, it follows from the definitions that given any such standard tableau, the corresponding sequence has all of its elements good. Thus,

$$\text{the elements } e_t, t \text{ standard, form a basis of } F_\lambda. \tag{C.3}$$

Let u lie in F_λ and suppose that u is an integer combination of tabloids. Let t be the last tabloid that occurs in the expression for u, so that $u = at +$ expressions involving earlier tabloids, and where a is an integer. From (C.3) we know that u can be written as a linear combination of the e_s, where s is standard. Suppose that we have ordered the e_s according to the order of the standard tabloids, s. Let t' be the last standard tabloid that occurs in the expression of u as a linear combination of the e_s. So $u = a'e_{t'} + $ a linear combination of e_s with t' later than s. We claim that t' is later than all the other tabloids that occur in the right-hand side of this last expression. Indeed, for any standard tableau, s, the last tabloid that occurs in e_s is clearly s. Thus t' is later than all the other tabloids occurring in $e_{t'}$ and certainly later than all the tabloids occurring in the earlier e_s. But this implies that $t' = t$ and hence $a' = a$. Subtracting off ae_t from u, we get an element v in F_λ which is an integral combination of tabloids, and involves only earlier tabloids than those in the expression for u. Proceeding in this way we see that

any u in F_λ which is an integer combination of tabloids is an integer combination of the basis elements of (C.3).

Now for any $\pi \in S_n$, we have $\pi e_t = e_{\pi t}$ and so is an integer combination of tabloids. Hence

$$\text{in the basis given by (C.3) all of the matrix entries of } \rho_\lambda(\pi) \text{ are} \qquad \text{(C.4)}$$
$$\text{integers for all } \pi \in S_n.$$

From this follows the remarkable fact that

$$\text{all the entries in the character table of } S_n \text{ are integers.} \qquad \text{(C.5)}$$

C.2 Proof of the main combinatorial lemma

We recall the setup. We are considering a partition pair (λ', λ) and c is an integer such that $\lambda_i' = \lambda_i$ for all $i < c$ while $\lambda_c' < \lambda_c$. The set $S(\lambda', \lambda)$ is the set of all λ sequences with at least λ_1' good 1's, λ_2' good 2's etc. In particular, if $s \in S(\lambda', \lambda)$ all of the $(c-1)$'s in s are good. A sequence belonging to $S(\lambda', \lambda) - S(a(\lambda', \lambda))$ has exactly λ_c' good c's and $\lambda_c - \lambda_c'$ bad c's. We change each such sequence into a μ sequence where $r(\lambda', \lambda) = (\mu', \mu)$, by changing all the bad c's to $(c-1)$'s and leaving all the remaining entries untouched. This defines a map from $S(\lambda', \lambda) - S(a(\lambda', \lambda))$ into $S(0, \mu)$ – the set of μ sequences. The assertion of the lemma is that this defines a one-to-one map between $S(\lambda', \lambda) - S(a(\lambda', \lambda))$ and $S(r(\lambda', \lambda))$. The first thing that must be proved is that the image actually lies in $S(r(\lambda', \lambda))$. Since we have not changed anything outside of the $(c-1)$st and cth rows, what we must show is that the modified sequence has at least λ_{c-1} good $(c-1)$'s and that the remaining λ_c' c's are all good.

Consider what happens when we change a bad c to a $c-1$, one at a time from the right. Such a change does not affect the quality of any good c. Indeed, if the $c-1$ that is produced is bad, then the quality of no later $c-1$ is changed, and so neither is the quality of any c. If the $c-1$ produced is good, some later $c-1$ might become bad, but the total number of good $(c-1)$'s before each position does not decrease. Thus, after changing all the bad c's to $(c-1)$'s the remaining c's will be all good. Also, the total number of good $(c-1)$'s is at least equal to λ_{c-1}'. This proves that we do get a map from $S(\lambda', \lambda) - S(a(\lambda', \lambda))$ to $S(r(\lambda', \lambda))$.

We now prove that the map is one-to-one. Replace all the $(c-1)$'s by left-hand parentheses and all the c's by right-hand parentheses. Thus, the sequence

$$1\ 1\ 2\ 1\ 3\ 3\ 1\ 2\ 4\ 2\ 3\ 2\ 1$$

where $c = 3$ becomes

$$1\ 1\ (1))1(4(\ )(1.$$

All that matters are the parentheses, so we can simplify the above to

$$(\ \ ))((\ \ )($$

Write $p = \lambda_c - \lambda_c'$ and $q = \lambda_{c-1} + \lambda_c - 2\lambda_c'$, so that $p = 1$ and $q = 3$ in our example. We can write this as

$$\pi_0)\pi_1(\pi_2(\pi_3$$

where the π's are (possibly empty, possibly nested) closed parentheses systems. Under the map, this configuration is changed to

$$\pi_0(\pi_1(\pi_2(\pi_3,$$

that is, the one extra) at the beginning has been changed to a (. In general, the initial configuration will have a certain number, λ'_c, of closed parentheses systems, p right hand,), parentheses followed by $q - p$ left hand, (, parentheses. It will look like

$$\pi_0)\pi_1)\cdots\pi_{p-1})\pi_p(\pi_{p+1}(\pi_{p+2}\cdots(\pi_q$$

where the π_j are (possibly empty) closed parentheses systems. Our mapping changes this to

$$\pi_0(\pi_1(\cdots\pi_{p-1}(\pi_p(\pi_{p+1}(\cdots(\pi_q.$$

We now have a candidate for the inverse map from $S(r(\lambda', \lambda))$ to $S(\lambda', \lambda) - S(a(\lambda', \lambda))$. Change the first p extra left-hand parentheses to right-hand parentheses; i.e. change the $(c - 1)$'s at these positions in the sequence to c's. This is clearly an inverse map.

What must be checked is that all the $(c - 1)$'s are good. Since no position other than the $(c - 1)$st and cth is affected, we then clearly get an element of $S(\lambda', \lambda) - S(a(\lambda', \lambda))$. Since all 1's are good we may assume that $c > 2$.

Consider an element s in $S(r(\lambda', \lambda))$ and mark the positions corresponding to the first extra p left-hand parenthesis. In our example

$$1 \; 1 \; 2 \; 1 \; 3 \; \overset{\vee}{2} \; 1 \; 2 \; 4 \; 2 \; 3 \; 2 \; 1$$

Sublemma

At any position x in $s \in S(r(\lambda', \lambda))$ the number of unmarked $(c - 1)$'s before x is $\leqslant$ the number of good $(c - 1)$'s before x. The inequality is strict if x is a bad $c - 1$, which is unmarked.

Proof Suppose that x is a marked $c - 1$. The number of unmarked $(c - 1)$'s before x equals the number of c's before x by our procedure of marking. And this is $\leqslant$ the number of good $(c - 1)$'s before x since all the c's are good. This proves the lemma for the case when x is a marked $c - 1$, and the same proof works if we take x to be one stop past the end of the sequence, i.e. the proof shows that the total number of unmarked $(c - 1)$'s is $\leqslant$ the total number of good $(c - 1)$'s. Now start at the end of the sequence and work back. The only problem that can arise is when x is both a good and a marked $c - 1$. But we know the lemma to be true when x is a marked $c - 1$. Applying the sublemma to the position immediately after an unmarked bad $c - 1$ gives the strict inequality.

We can now prove the main lemma, i.e. prove that the sequence s^* obtained from s by changing all the marked $(c - 1)$'s to c's has the property that all the remaining (unmarked) $(c - 1)$'s are good. Since s belongs to $S(r(\lambda', \lambda))$ we know that s contains at most $p = \lambda_c - \lambda'_c$ bad $(c - 1)$'s. So for any $c - 1$ in s the number of preceding good $(c - 2)$'s is $>$ (the number of preceding $(c - 1)$'s) $- p$. Thus every $c - 1$ in s^* after the pth marked $c - 1$ in S^* is good, since we have removed p of the $(c - 1)$'s and changed them to c's. If a term x in s^* is a $c-1$ that occurs before the pth marked $c-1$, then x was an unmarked $c-1$ in s. Thus the number of $(c-1)$'s before x, in $s^* =$ the number of unmarked $(c-1)$'s in s before $x \leqslant$ the number of good $(c-1)$'s before x by the sublemma $\leqslant$ the number of good $(c-2)$'s before x, with strict inequality if x is a bad unmarked $c-1$. Thus the good unmarked $(c-1)$'s remain good, and all the bad $(c-1)$'s in s have either been changed to c's or have become good because some preceding good $c-1$ has been changed to a c. This completes the proof of the lemma, and with it the proof of the main theorem. We now know that the $e_t^{\lambda',\lambda}$ span $F_{\lambda',\lambda}$ and that the sequence

$$0 \to F_{a(\lambda',\lambda)} \to F_{\lambda',\lambda} \xrightarrow{R} F_{r(\lambda',\lambda)} \to 0$$

is exact, with all the consequences of this fact.

C.3   The Littlewood–Richardson rule and Young's rule

We now show how the preceding methods give proofs of two important formulas in the representation theory of S_n. Let F_μ be an irreducible module for S_m and F_ν an irreducible module for S_n. Then $F_\mu \otimes F_\nu$ is an irreducible representation space for $S_m \times S_n$ which we can regard as a subgroup of the group S_{m+n} in the obvious way. We can thus form the representation of S_{m+n} induced from the representation $F_\mu \otimes F_\nu$ of $S_m \times S_n$. The Littlewood–Richardson rule tells how to decompose this induced representation into irreducibles.

Consider the space E_n of formal integer combinations of proper diagrams associated to n, i.e. all expressions of the form $\sum a_\nu \nu$. (If we think of ν as labeling F_ν, then the expression $\sum a_\nu \nu$ with all the $a_\nu \geqslant 0$ labels the most general representation S_n. We can then think of E_n as the 'Grothendieck ring' or the ring of all 'virtual representations' of S_n.) Let (λ', λ) be a partition pair, where λ is a partition of some integer k. We shall associate an operator $O_{\lambda',\lambda} : E_n \to E_{n+k}$ to (λ', λ). Extending by linearity, it suffices to define $O_{\lambda',\lambda}\nu$. We do so as follows. Think of λ as consisting of λ_1 1's, λ_2 2's, etc. Apply the following steps:

(1) Start at $i=1$. Suppose we have added all the $(i-1)$'s to ν. Add all the i's to ν so no two i's lie in the same column and the ensuing figure has the shape of a proper diagram. Continue through all of λ.

(2) For each resulting diagram, read the added entries from right to left in successive rows. If the resulting sequence is not an $S(\lambda', \lambda)$ sequence, scratch out the result.

(3) Replace all the added integers in each term by X's and sum up the resulting diagrams.

Example   Suppose $\mu = (2, 1)$ and take $(\lambda', \lambda) = ((2, 1), (2, 1))$. Then step (1) gives

```
X̶ X̶ 1 1̶ 2       X X 1 1       X X 1 1       X X 1
X̶                X 2           X             X 1 2
```
2

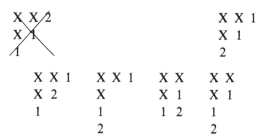

and we have crossed out the diagrams which are no good according to rule (2). Then rule (3) says that we end up with

$$
\begin{array}{ccc}
\text{X X X X} & \text{X X X X} & \text{X X X} \\
\text{X X} \quad + & \text{X} \quad + & \text{X X X} \quad + \\
 & \text{X} &
\end{array}
$$

$$
\begin{array}{cccc}
\text{X X X} & \text{X X X} & \text{X X X} & \text{X X} \\
\text{X X} \quad + & \text{X X} \quad + & \text{X} \quad + & \text{X X} \\
\text{X} & \text{X} & \text{X} & \text{X X} \\
 & & \text{X} &
\end{array}
$$

$$
+ \quad
\begin{array}{c}
\text{X X} \\
\text{X X} \\
\text{X} \\
\text{X}
\end{array}
$$

Thus

$$
O_{(2,1),(2,1)}(2,1) = (4,2) + (4,1,1) + (3,3) + 2(3,2,1)
$$
$$
+ (3,1,1,1) + (2,2,2) + (2,2,1,1).
$$

We shall denote the operator $O_{\lambda,\lambda}$ by O_λ. Then we can formulate the Littlewood–Richardson rule:

> the representation of S_{m+n} induced from $F_\mu \otimes F_\nu$ of $S_m \times S_n$ is given, as an element of E_{m+n}, by $O_\nu\mu$. $\hspace{1em}$ (C.6)

Notice that rule (2) has no effect if $\lambda' = 0$. All that remains are rules (1) and (3). Thus

$$
O_{(0,\lambda)} = O_{\lambda_k} \cdot O_{\lambda_{k-1}} \cdots \cdots O_{\lambda_1}, \tag{C.7}
$$

where $\lambda = (\lambda_1, \ldots, \lambda_k)$. We can formulate *Young's rule*:

$$
\mathscr{F}(M_\lambda) = O_{0,\lambda}(0) \text{ as an element of } E_m. \tag{C.8}
$$

In fact, we shall prove the more general rule

$$
F_{\lambda',\lambda} = O_{\lambda',\lambda}(0) \text{ as an element of } E_m. \tag{C.9}
$$

All of these results will follow from the following consequence of our basic combinatorial lemma

$$
O_{\lambda',\lambda} = O_{a(\lambda',\lambda)} + O_{r(\lambda',\lambda)} \tag{C.10}
$$

as elements of $\mathrm{Hom}(E_n, E_{n+m})$.

We now set out to prove (C.10). We have $\lambda'_j = \lambda_j$ for $j < c$, where c is the integer entering into the definition of a and r. Thus any sequence in $S(\lambda', \lambda)$ has all the λ_{c-1}'s good. Now let us see what happens when we try to change a bad c to a $c-1$. There are two problems which might occur with rule (1). We might worry that in adding the c's a bad c might be to the right of a good c in the same row. This cannot happen because a c immediately following a bad c must itself be bad. Thus we may change all the bad c's to $(c-1)$'s in any term of $O_{\lambda', \lambda}$ provided that we do not worry about the second part of rule (1) which says that no two $(c-1)$'s can lie in the same column. We must show that this causes no trouble either. That is, we must show that no bad c can lie directly below a $c-1$. Suppose that a c lies below a $c-1$. So we have a situation like

$$
\begin{array}{cccccc}
c-1 & 0 & 0 & 0 & 0 & 0 \\
c & & 0 & 0 & 0 &
\end{array}
$$

at some position. We wish to prove that the c is good. Moving to the right along the row with the c, we will encounter a certain number of c's, followed by either an empty space or some number larger than c. On the row immediately above, the numbers must be non-decreasing, so $\geq c-1$ and $< c$ as long as they lie above a c. So the picture looks like

$$
\begin{array}{cccccc}
c-1 & c-1 \cdots c-1 & 0 & 0 & 0 \\
c & c. & c & 0 \\
& & \uparrow &
\end{array}
$$

Now from the definition of good and bad it follows that at any position in any sequence, the number of preceding good $(c-1)$'s is always $\geq$ the number of preceding good c's. Let us apply this to the position of the rightmost $c-1$ sitting above the c, marked by an arrow. Since all the $c-1$'s are good, and we are reading from right to left and down rows, all the $(c-1)$'s are excess good $(c-1)$'s over good c's, hence all the c's occurring immediately below in the next row must be good. Hence any c which lies below a $c-1$ must be good. Thus, in the diagrams which occur in $O_{\lambda', \lambda \mu}$, we may change all bad c's to $(c-1)$'s and rules (1) and (2) will still be satisfied. But changing bad c's to $(c-1)$'s is precisely the bijection we described between $S(\lambda', \lambda) - S(a(\lambda', \lambda))$ and $S(r(\lambda', \lambda))$. Grouping together all the diagrams whose sequence belongs to $S(a(\lambda', \lambda))$ and changing the bad c's in the remaining diagrams to $(c-1)$'s proves (C.10).

If $\lambda' \neq \lambda$ then $a(\lambda', \lambda)$ (is either empty or) has more good entries, and repeated application of r also increases the number of good entries since all the 1's are good. Hence repeated application of r and a always ends up with some (μ, μ) as was illustrated on p. 331. Hence, repeated application of (C.10) shows that

$O_{\lambda', \lambda} = \sum a_\mu O_\mu$, where each μ is a proper partition of m and the a_μ are (C.11)
integers. The $a_\mu = 0$ unless $\mu \geq \lambda$. If λ is proper then $a_\lambda = 1$. Otherwise
$a_\lambda = 0$.

We may solve for $O_{(0, \lambda)}$ (with λ proper) in terms of O_μ with resulting coefficient $\alpha_\lambda = 1$. Thus solving (C.11)

$$we\ can\ write\ O_\mu = \sum b^\mu_\beta O_{(0, \beta)}\ with\ integer\ coefficients\ b^\mu_\beta \qquad \text{(C.12)}$$

We next claim that

$$O_\mu(0) = \mu \ \textit{for any proper partition } \mu. \tag{C.13}$$

Indeed, let v be any diagram that appears in the expression for $O_\mu(0)$. This means that we can fill out all the partitions of v with μ_1 1's, μ_2 2's etc., so that rules (1) and (2) hold. We must show that all the 1's occur in the first row of v, all the 2's in the second, etc. Indeed, since the numbers must be strictly increasing down columns, no i can appear in the kth row, where $k > i$. Suppose an i appeared in some jth row with $j < i$. Choose i to be the smallest integer for which this happens. No $i - 1$ can occur in a higher row than this j, by the minimality of i. Also no $i - 1$ can occur to the right of this i since the rows are not decreasing from left to right. But this means that reading from right to left and down rows, this i is not preceded by any $i - 1$ so it must be bad, contradicting rule (2) which requires, for $O_\mu = O_{\mu,\mu}$, that all the entries be good.

Notice that (C.13) is the special case of (C.9) when $\lambda' = \lambda$. We can now use the exact sequence (C.1) and induction to prove (C.9) in general. Indeed we may assume it is true for all (μ', μ) with $\mu > \lambda$, and hence in particular for $r(\lambda', \lambda)$ and we may assume that it is true for $a(\lambda', \lambda)$. But in the ring E_n the exact sequence (C.1) says that $F_{\lambda',\lambda} = F_{a(\lambda',\lambda)} + F_{r(\lambda',\lambda)}$ and, by (C.10), $O_{\lambda',\lambda}(0) = O_{a(\lambda',\lambda)}(0) + O_{r(\lambda',\lambda)}(0)$. So (C.9), and, in particular, Young's rule, (C.8) are true.

Now to the Littlewood–Richardson rule. By (C.13)

$$
\begin{aligned}
O_\mu(v) &= O_\mu O_v(0) \\
&= \sum b_\beta^\mu O_{(0,\beta)} b_\gamma^v O_{(0,\gamma)}(0) \text{ by } (C.12) \\
&= \sum b_\beta^\mu b_\gamma^v O_{\beta_k} O_{\beta_{k-1}} \cdots O_{\beta_1} O_{\gamma_l} O_{\gamma_{l-1}} \cdots O_{\gamma_1}(0)
\end{aligned}
$$

By Young's rule, (C.8), applied to S_{m+n},

$$O_{\beta_k} \cdots O_{\beta_1} O_{\gamma_l} \cdots O_{\gamma_1}(0) = \mathscr{F}(M_{\beta \times \gamma})$$

(where $\beta \times \gamma$ is the corresponding partition of $m + n$) which is the representation of S_{m+n} induced from the trivial representation of the subgroup $S_\beta \times S_\gamma \subset S_m \times S_n$ of S_{m+n}. By the transitivity of induction, this then implies that

$$O_{(0,\beta)} O_{(0,\gamma)}(0) = \mathscr{F}(M_\beta) \times \mathscr{F}(M_\gamma) \uparrow S_{m+n}$$

where we have used the symbol $\uparrow$ for induction from $S_m \times S_n$ to S_{m+n}. Then

$$
\begin{aligned}
O_\mu(v) &= \sum b_\beta^\mu b_\gamma^v F_{(0,\beta)} \times F_{(0,\gamma)} \uparrow S_{m+n} \\
&= [(\sum b_\beta^\mu F_{(0,\beta)}) \times (\sum b_\gamma^v F_{(0,\gamma)})] \uparrow S_{m+n} \\
&= [(\sum b_\beta^\mu O_{(0,\beta)}(0)) \times (\sum b_\gamma^v O_{(0,\gamma)}(0))] \uparrow S_{m+n} \\
&= [O_\mu(0) \times O_v(0)] \uparrow S_{m+n} \\
&= [F_\mu \times F_v] \uparrow S_{m+n}
\end{aligned}
$$

which is the Littlewood–Richardson rule.

C.4   The ring of virtual representations of all the S_n

In the preceding section we introduced the space E_n of virtual representations of the group S_n. We recall its definition. For any group G we can introduce the semigroup $R(G)$ where elements of $R(G)$ are equivalence classes of representations of G, and if

$$0 \to r_1 \to r \to r_2 \to 0$$

is an exact sequence of representations (in other words r_1 is a representation on an invariant subspace of r and r_2 is the corresponding quotient representation) we identify the equivalence class

$$(r) \text{ with } (r_1) + (r_2).$$

It is routine to check that this does indeed define an addition on the space of equivalence classes of representations. (In the case that all representations of G are completely reducible, for example if G is finite, such an exact sequence is the same as asserting that r is equivalent to the direct sum of r_1 and r_2. Thus, in the case of finite groups, the addition in $R(G)$ is just the transcription to equivalence classes of the notion of the direct sum of two representations.) Addition is commutative and associative, with the zero being the equivalence class of the zero-dimensional representation. From the semigroup $R(G)$ we construct the corresponding group $E(G)$ in the usual way by introducing negative elements. Thus $E(G)$ is the group (or $\mathbb{Z}$ module) of 'virtual representations' of G. In the preceding section we defined E_n as $E(S_n)$. Now for any group we could make $E(G)$ into a ring by using the tensor product to define a multiplication on $E(G)$. This is *not* what we are going to do, however. Instead, we are going to introduce a multiplication on the direct sum

$$E = \bigoplus_1^\infty E_n.$$

The multiplication is the one that we studied in the preceding section: given a representation r_1 of S_m and r_2 of S_n we form the representation $r_1 \otimes r_2$ of $S_m \times S_n$ which we then induce up to get a representation of S_{m+n}. Passing to equivalence classes, this defines a map of $R(S_m) \times R(S_n) \to R(S_{m+n})$ which is associative relative to sums, and hence extends to a multiplication $E_m \times E_n \to E_{m+n}$. It is routine to check that this makes E into a graded commutative ring. Let λ be a proper partition of m. We shall let $[\lambda]$ denote the element of E_n corresponding to the irreducible representation F_λ. Thus the $[\lambda]$ form a basis of E_m. The Littlewood–Richardson rule can thus be regarded as providing the multiplication table of $E_m \times E_n \to E_{m+n}$ relative to the bases given by the proper partitions.

Let λ be a proper partition of m. Then $[\lambda_1]$ is the element of E_{λ_1} corresponding to the trivial representation of S_{λ_1} and similarly for all the λ_i occurring in λ. Thus

$$[\lambda_1][\lambda_2]\cdots[\lambda_m]$$

is the element of E_m corresponding to the representation induced from the trivial representation of $S_{\lambda_1} \times \cdots \times S_{\lambda_m}$, which is just the representation of S_m on $\mathscr{F}(M_\lambda)$. Thus Young's rule tells us how to write this element in terms of our generators. In other words, it tells us how to compute the (integer) entries of the matrix $(m_{\lambda\mu})$, where

$$[\lambda_1][\lambda_2]\cdots[\lambda_m] = \sum_\mu m_{\lambda\mu}[\mu].$$

For example, for the group S_3 we have

$$[3] = [3]$$
$$[2][1] = [3] + [(2,1)]$$
$$[1]^3 = [3] + 2[(2,1)] + [1]$$

where, as usual, we have written $[1]^3$ for $[1][1][1]$. So the matrix $(m_{\lambda\mu})$ for S_3 is

$$\begin{array}{ccc} 1 & 0 & 0 \\ 1 & 1 & 0. \\ 1 & 2 & 1 \end{array}$$

For general m, let us order the proper Young diagrams lexicographically: let us say that μ is lexically earlier than λ if $\mu_1 > \lambda_1$, or, if they are equal, then $\mu_2 > \lambda_2$, or, if they are equal as well, then $\mu_3 > \lambda_3$, etc. This is a total ordering on the set of diagrams, and it is consistent with the partial order introduced in Chapter 2 in the sense that if $\mu < \lambda$ in the Chapter 2 partial order it is certainly earlier than λ in the lexical order. Now recall from Chapter 2 that in the decomposition of $\mathscr{F}(M_\lambda)$ into irreducibles, the representation F_λ occurs once, and the only F_μ that can occur are those for which $\mu < \lambda$ in the sense of Chapter 2. This means that if we use the lexical order to arrange the λ, the matrix $(m_{\lambda\mu})$ is lower triangular with 1's along the diagonal. (As we said, the individual entries below the diagonal are to be computed by Young's rule.) But this means that the matrix $(m_{\lambda\mu})$ is invertible over the integers. This inverse matrix can, of course, be readily computed recursively from $(m_{\lambda\mu})$, but this involves actually computing the $m_{\lambda\mu}$. In fact, there is a remarkable formula which computes this inverse directly which we will now explain.

First let us introduce a bit of notation. For any negative integer, j, we will let the symbol $[j]$ denote the value 0. We will also let the symbol $[0]$ denote the integer 1. We have now assigned a meaning to $[k]$ for all integers, k. For any proper partition λ consider the following matrix, whose entries are in the ring E:

$$([\lambda_i - i + j]).$$

In other words, the diagonal entries of this matrix are the elements $[\lambda_i]$ at the ith diagonal position. Then along each row the integer inside the [ ] increases by 1 as we move one step to the right (or decreases by 1 as we move one step to the left). Thus, for the partition $(2,1)$ we would get the matrix

$$\begin{pmatrix} [2] & [3] & [4] \\ [0] & [1] & [2] \\ [-2] & [-1] & [0] \end{pmatrix}$$

which according to our conventions is the same as the matrix

$$\begin{pmatrix} [2] & [3] & [4] \\ 1 & [1] & [2] \\ 0 & 0 & 1 \end{pmatrix}$$

Since the ring E is commutative, it makes perfectly good sense to talk of the determinant of any matrix with entries in E, the value of this determinant will then be an element of E. The assertion of the formula of 'determinantal form' is that

$$[\lambda] = \det([\lambda_i - i + j]). \tag{C.14}$$

(Notice that in computing this determinant, the rows corresponding to $\lambda_p = 0$ make only a trivial contribution of multiplying by 1. That is, in evaluating the determinant we need only consider the top left $k \times k$ matrix, where k is the largest integer such that $\lambda_k \neq 0$, i.e. the number of rows in the Young diagram.) Thus (C.14) asserts that

$$[(2, 1)] = \det\begin{pmatrix} [2] & [3] \\ 1 & [1] \end{pmatrix} = [2][1] - [3].$$

The proof of (C.14) involves applying the Littlewood–Richardson rule. Before embarking on the proof we make a number of remarks and definitions. First of all observe that the integers occurring in the [ ] of the *last* column of the matrix are precisely the hook lengths, h_i, of the *first* column of the Young diagram. (Recall the definition of hook length from Chapter 2.)

Now we are going to prove (C.14) by induction on k, the number of non-zero rows in the Young diagram. It is obviously true for $k = 0$ and $k = 1$. So we assume that the formula is true for diagrams having fewer than k rows, and consider the assertion for a diagram with k rows. We will expand the determinant in (C.14) by the last column. Omitting the last column and the ith row gives a matrix M_i whose diagonal entries are

$$[\lambda_1], \ldots, [\lambda_{i-1}], [\lambda_{i+1} - 1][\lambda_{i+2} - 1] \cdots [\lambda_k - 1]$$

and the expansion formula for determinants says that the right-hand side of (C.14) is equal to

$$(\det M_k)[h_{k1}] - (\det M_{k-1})[h_{k-1}] + \cdots \pm (\det M_1)[h_{11}]. \tag{C.15}$$

Each of the matrices M_i corresponds to a Young diagram, and it is instructive to visualize what this diagram is in terms of our original diagram λ. Thus M_k corresponds to the diagram obtained from λ by eliminating the last row. We can think of the diagram associated to M_{k-1} as follows: do not touch the first $k - 2$ rows of λ. On the $(k - 1)$st row eliminate all positions starting from the right up to and including the position over the kth row and eliminate the entire kth row.

```
X X X X X X X
X X X X X X
X X̶ X̶ X̶
X̶ X̶
```

This has the effect of giving a diagram with $k-1$ rows, whose first $k-2$ rows are the same as those of λ and whose $(k-1)$st row is the kth row of λ reduced by 1. For M_{k-2} start with the extreme right-hand position of the $(k-2)$nd row, eliminate all entries up to and including the position immediately above the end of the $(k-1)$st row and continue eliminating as before:

```
X X X X X X X
X X X X X X
X X X X
X X
```

This clearly leaves a diagram with $k-1$ rows, whose $(k-2)$nd row is 1 shorter than the $(k-1)$st row of λ and whose $(k-1)$st row is 1 shorter than the kth row of λ. Indeed, the prescription is now clear. M_i corresponds to the diagram obtained from λ by starting at the end of the ith row and eliminating the entire portion of the rim from that position down. Thus the four diagrams in the above example are

```
X X X X X X    X X X X X X X X
X X X          X X X
X              X

X X X X X X X X    X X X X X X X X
X X X X X X        X X X X X X
X                  X X X X
```
.

Let λ^i denote the diagram corresponding to M_i. By our induction hypothesis, $\det M_i = [\lambda^i]$. Thus our column expansion becomes

$$[\lambda^k][h_{k1}] - \cdots \pm [\lambda^1][h_{11}]. \tag{C.16}$$

We must evaluate each term $[\lambda^i][h_{i1}]$ by the Littlewood–Richardson rule which says that we must add h_{i1} 1's to the diagram λ^i so that we get a proper diagram and subject to the constraint that no two added 1's lie in the same column, and add up all the diagrams obtained by replacing the 1's by X's. Now any diagram that is obtained in this way from λ^i must have its rows at least as large as λ^i, and hence its first $i-1$ rows contain all the entries of the corresponding rows of λ. On the other hand, the $(i+1)$st row of such a diagram cannot be as large as the $(i+1)$st row of λ, since the last node in such a row would have to be filled by a 1 and we would also need a 1 to fill the corresponding position in the ith row of λ^i. Similarly, none of the later rows of any diagram obtained from λ^i can be as large as those of λ. Thus the diagram λ itself appears only in the terms coming from the last diagram λ^k, and is obtained by replacing all the terms we took off in constructing λ^k from λ. (It is immediate that the number of terms on the rim removed in forming λ^i from λ is exactly equal to the hook length h_{i1}.) Thus λ occurs once in the terms coming from the application of the Littlewood–Richardson rule to the summands in (C.15); in fact λ comes from the last summand (and is the last diagram to occur in terms of the total ordering). We claim that all the other terms cancel. Let us illustrate a piece of this cancellation by looking at the terms coming from λ^4 and λ^3 in the above example. From λ^4 we get λ and various other diagrams obtained by adding two 1's to λ^4. To λ^3 we must add five 1's in all ways so that no two 1's lie in the same column. Among the various possibilities, we could use up three of the 1's to get λ^4:

$$
\begin{array}{ccccccc}
X & X & X & X & X & X & X \\
X & X & X & X & X & \\
X & 1 & 1 & 1 & & \\
\end{array}
$$

and the remaning two 1's to obtain all the diagrams we got from λ^4 with the exception of λ, since this would involve two 1's in the same column. As these diagrams coming from λ^3 all have a minus sign in front of them, we have cancelled all the diagrams coming from λ^4 except λ. We still have some left over diagrams coming from λ^3, those for which we did not use the 1's to fill up the third row of λ. But these will all be cancelled by those diagrams coming from λ^2 where we have used some (three) of the $h_{2,1} = 8$ 1's to fill up the second row of λ. The general situation should now be clear. Divide the collection of diagrams coming from λ^i into two sets, $A(i)$ and $B(i)$, where $A(i)$ consists of those diagrams whose ith row is at least as long as the ith row of λ and $B(i)$ are the others. Then $A(i) = B(i + 1)$, but their elements occur with opposite signs (C.15). Thus all the terms cancel except for $A(k)$ which consists only of λ, and which appears with a plus sign. This completes the proof of the determinantal form (C.14).

C.5   Dimension formulas

The dimension of the direct sum of two representations is the sum of their dimensions. Thus the dimension function extends to define an additive function, call it d, on E_n and hence on E,

$$
d : E \to \mathbb{Z}.
$$

We can thus apply d to (C.14) to compute $d([\lambda]) = \dim F_\lambda$. On the right-hand side of (C.14) we get a sum of products $[\mu_1] \cdots [\mu_k]$ (with a plus or minus sign in front). A non-zero such product corresponds to the representation of S_n on $\mathscr{F}(M_\mu)$, and $\dim \mathscr{F}(M_\mu) = \# M_\mu = n!/(\mu_1! \cdots \mu_k!)$. A zero product will arise if one of the entries has some μ_j negative. So let us make the convention that $1/r! = 0$ if $r < 0$. It then follows immediately from (C.14) that

$$
\dim F_\lambda = n! \det \left(\frac{1}{(\lambda_i - i + j)!} \right). \tag{C.16}
$$

We have already remarked that the integers occurring in the last column of the matrix in (C.14) are precisely the hook lengths of the first column in the diagram.

Thus we can rewrite (C.16) as

$$(1/n!)\dim F_\lambda = \det \begin{pmatrix} \dfrac{1}{(h_{11}-k+1)!} & \cdots & \dfrac{1}{(h_{11}-1)!} & \dfrac{1}{h_{11}!} \\ \vdots & & \vdots & \vdots \\ \dfrac{1}{(h_{k1}-k+1)!} & \cdots & \dfrac{1}{(h_{k1}-1)!} & \dfrac{1}{h_{k1}!} \end{pmatrix}$$

$$= \frac{1}{h_{11}!\cdots h_{k1}!} \det \begin{pmatrix} h_{11}(h_{11}-1)\cdots(h_{11}-k+2)\cdots h_{11} & 1 \\ \vdots & \vdots \\ h_{k1}(h_{k1}-1)\cdots(h_{k1}-k+2)\cdots h_{k1} & 1 \end{pmatrix}. \tag{C.17}$$

Now, we can manipulate this determinant in several ways. By adding columns successively from right to left we get the Vandermonde determinant

$$\det \begin{pmatrix} h_{11}^{k-1} & \cdots & h_{11} & 1 \\ \vdots & & \vdots & \vdots \\ h_{k1}^{k-1} & \cdots & h_{k1} & 1 \end{pmatrix} = \prod_{i<j}(h_{i1}-h_{j1}) \tag{C.18}$$

so

$$\dim F_\lambda = \frac{n!\prod\limits_{i<j}(h_{i1}-h_{j1})}{\prod h_{i1}!}. \tag{C.19}$$

This expresses $\dim F_\lambda$ in terms of the first column hook lengths. We claim that further manipulation of the determinant in (C.17) will prove that

$$\dim F_\lambda = \frac{n!}{\prod(\text{hook lengths})}. \tag{C.20}$$

We will prove this by induction on the size of the diagram. Pulling out a factor of $(1/h_{c1})$ from each row, we can rewrite the middle expression in (C.17) as

$$\frac{1}{h_{11}!\cdots h_{k1}!}\det\begin{pmatrix} (h_{11}-1)\cdots(h_{i1}-k+1)\cdots & h_{11}-1 & 1 \\ (h_{k1}-1)\cdots(h_{i1}-k+1)\cdots & h_{k1}-1 & 1 \end{pmatrix}$$

$$= \frac{1}{h_{11}\cdots h_{k1}}\det\begin{pmatrix} \dfrac{1}{(h_{11}-k)!} & \cdots & \dfrac{1}{(h_{11}-1)!} \\ \dfrac{1}{(h_{k1}-k)!} & \cdots & \dfrac{1}{(h_{k1}-k)!} \end{pmatrix}.$$

By our induction hypothesis this last determinant is the reciprocal of the product of the hook lengths of the diagram obtained from λ by removing all the entries in the first column, in other words, it is the reciprocal of the product of all the hook lengths except those of the first column. Taking into account the factor in front of the determinant, which is the reciprocal of the product of the first column hook lengths, we complete the proof of (C.19).

C.6   The Murnaghan–Nakayama rule

In this section we show how the determinantal form (C.14) implies the skew hook rule for computing characters that we described in Chapter 2. We begin with a lemma. Let ρ be a cycle of length r and let π be any permutation of the remaining $n - r$ numbers. Let λ be any partition of n. Then $\chi^{M_\lambda}(\pi\rho) = \#$(of fixed points of $\pi\rho$ acting on M_λ). Now since π and ρ act on different sets of numbers, a tabloid x in M_λ is fixed by $\pi\rho$ if and only if it is fixed both by π and by ρ. Since ρ is a cycle, the only way it can leave x fixed is for all the numbers occurring in ρ to lie in the same row of x. Thus we can divide the set of fixed points of ρ into k subsets according to which of the k rows of λ contains all the numbers in ρ. For each of these subsets we need to count the number of fixed points of π, and each of these numbers is the character of π, relative to the representation on an M_μ, where μ is obtained from λ by replacing λ_i by $\lambda_i - r$. Thus

$$\chi^{[\lambda_1]\cdots[\lambda_k]}(\pi\rho) = \sum \chi^{[\lambda_1]\cdots[\lambda_{i-1}][\lambda_i - r][\lambda_{i+1}]\cdots[\lambda_k]}(\pi). \tag{C.22}$$

We can now substitute (C.21) into (C.14). The result is the following rule: to compute $\chi^{F_\lambda}(\pi\rho)$ consider the sum of the k determinants obtained from the right-hand side of (C.14) by subtracting r from each of the rows of the matrix, and evaluate the character of the resulting element on π and sum. We claim that this is entirely equivalent to the Murnaghan–Nakayama rule. Indeed, we claim that subtracting r from the ith row will yield a zero determinant if no skew hook can be removed from the ith row of the diagram, and will yield the diagram with the skew hook removed, multiplied by $(-1)^{\text{leg length}}$ if a skew hook of length r can be removed. To see this it is clearly sufficient to understand the computation for the first row. Let us consider several alternatives. Suppose that $r \leqslant \lambda_1 - \lambda_2$. The determinant obtained from that of λ by subtracting r from each of the entries of the first row is that of a proper diagram, with r entries removed from the top row of the diagram, and the diagram with the top r skew hook removed in this case is

$$\begin{array}{llllllll}
\text{X} & \text{X} & \text{X} & \text{X} & \cancel{\text{X}} & \cancel{\text{X}} & \cancel{\text{X}} & \cancel{\text{X}} \\
\text{X} & \text{X} & \text{X} & & & & & \\
\text{X} & & & & & & & r = 4.
\end{array}$$

Suppose that $r = \lambda_1 - \lambda_2 + 1$. Then subtracting r from the entries in the top row on the right of (C.14) will leave the top row equal to the second row, so the resulting determinant is zero. Also, in this case no skew hook of length r can be removed from the top row.

$$\begin{array}{llllllll}
\text{X} & \text{X} & \cancel{\text{X}} & \cancel{\text{X}} & \cancel{\text{X}} & \cancel{\text{X}} & \cancel{\text{X}} & \cancel{\text{X}} \\
\text{X} & \text{X} & \text{X} & & & & & \\
\text{X} & & & & & & & r = 6.
\end{array}$$

Suppose that $\lambda_2 = \cdots = \lambda_p > \lambda_{p+1}$. In other words, suppose that the next $p - 1$ rows, $p \geqslant 2$, are equal. Then if $\lambda_1 - \lambda_2 + 1 \leqslant r \leqslant \lambda_1 - \lambda_p + p - 1$ we get zero for the same reason: two rows of the matrix are equal, and similarly no skew hook of length r can be removed from the first row. Suppose that $r = \lambda_1 - \lambda_p + p$ and $r \leqslant \lambda_1 + k$, where k is the number of rows in the diagram. The first column entries of the resulting matrix will now be $[\lambda_1 - r]$, $[\lambda_2] \cdots [\lambda_p][\lambda_{p+1}] \cdots [\lambda_k]$. These are not in decreasing order, but can be made into decreasing order by moving the first row past the pth. This has the effect of multiplying the determinants by $(-1)^{p-1}$. On the diagrammatic level, this is precisely the effect of removing an r hook from the first row (and multiplying by the appropriate sign):

$$
\begin{array}{l}
\text{X X X X̶ X̶ X̶ X̶ X̶ X̶} \\
\text{X X X X} \\
\text{X X X X} \\
\text{X X X X} \\
\text{X} \qquad\qquad\qquad\qquad r = 9.
\end{array}
$$

From here we can proceed inductively if there are more rows. That is, subtracting a larger value of r can be done in stages; if $r = r_1 + r_2$ first subtract off r_1 and then r_2 if $r_1 = \lambda_1 - \lambda_2 + p$; if $\lambda_1 + k < r$ then the top row will have all zeros, and no skew hook can be removed. In all cases we have verified that the removal of r skew hooks from the ith row (and multiplying by $(-1)^{\text{leg length}}$) is equivalent to subtracting r from the ith row of the determinant. This proves the Murnaghan–Nakayama rule.

C.7   Characters of $Gl(V)$

In Chapter 5 we saw how to associate an irreducible representation, U_λ, of $Gl(V)$ to each Young diagram λ (whose number of rows does not exceed dim V). We also proved the following fact: regard $S_m \times S_n$ as a subgroup of S_{m+n} and decompose the irreducible representation F_λ of S_{m+n}, when restricted to $S_m \times S_n$ into irreducibles:

$$F_\lambda|_{S_m \times S_n} = \sum a^\lambda_{\mu\nu} F_\mu \otimes F_\nu \tag{C.22}$$

then

$$U_\mu \otimes U_\nu = \sum a^\lambda_{\mu\nu} U_\lambda. \tag{C.23}$$

Now by the Frobenius reciprocity theorem, (C.22) says that

$$(F_\mu \otimes F_\nu)\!\uparrow_{S_{m+n}} = \sum a^\lambda_{\mu\nu} F_\lambda. \tag{C.24}$$

We can therefore interpret (C.23) and (C.24) as follows. Make $R(Gl(V))$ into a ring by using the product induced from tensor products of representations. (Recall that $R(Gl(V))$ is the space of equivalence classes of representations of $Gl(V)$ made into an additive group by including the 'negative' or 'virtual' representations.) The assignment $F_\lambda \to U_\lambda$ defines a map of the generators, and hence of E into $R(G)$. (We map diagrams with more than dim V rows into zero.) Then (C.23) and (C.24) say that this map is a ring homomorphism. We know from Chapter 5 that this map is *onto* the ring $R(U(V))$. In some sense, we should regard E as the ring $R(U(\infty))$ – the representation ring of the infinite unitary group, where no diagram would get sent to zero. In any event we can now use the determinantal form to evaluate the characters of U_λ in terms of the characters of the representations on spaces corresponding to single row diagrams, which after all are just the representations of $Gl(V)$ on symmetric powers. Indeed, let us introduce the following notation: for each integer k, let $\langle k \rangle$ be the function defined on $Gl(V)$ by

$$\langle k \rangle(A) = 0 \quad \text{if} \quad k < 0,$$

$$\langle 0 \rangle(A) = 1$$

$$\langle 1 \rangle(A) = \alpha_1 + \cdots + \alpha_d, \text{ where the } \alpha_i \text{ are the eigenvalues of } A$$

$$\langle 2 \rangle(A) = \alpha_1^2 + \cdots + \alpha_d^2 + \alpha_1\alpha_2 + \alpha_1\alpha_3 + \cdots + \alpha_{d-1}\alpha_d$$

and, in general for k positive

$$\langle k \rangle(A) = \sum_{k i_1 \leqslant i_2 \leqslant \cdots \leqslant i_k \leqslant d} \alpha_{i_1} \cdots \alpha_{i_k}.$$

By introducing a basis on V relative to which A is in lower triangular form, and the corresponding bases on $S^k(V)$ it is easy to see that

$$\langle k \rangle = \chi_{S^k(V)}.$$

It now follows from our homomorphism of E into $R(Gl(V))$ and the determinantal form, (C.14), that

$$\chi_{U_\lambda} = \det(\langle \lambda_i - i + j \rangle). \tag{C.25}$$

In particular

$$\chi_{S^k(V)}(I) = \dim S^k(V) = \binom{d+k-1}{d-1}$$

so

$$\dim U_\lambda = \det\left(\binom{d + \lambda_i - i + j - 1}{d-1}\right). \tag{C.26}$$

We claim that a study of the determinant on the right will prove that

$$\dim U_\lambda = \frac{\prod(d+j-i)}{\prod(\text{hook lengths})}. \tag{C.27}$$

Indeed, writing out the binomial coefficients, the determinant becomes

$$\det\left(\frac{d(d+1)\cdots(d+\lambda_i-1+j-i)}{(\lambda_i+j-i)!}\right). \tag{C.28}$$

This is some polynomial in d, call it $P(d)$. The degree of this polynomial is $\lambda_1 + \cdots + \lambda_k$ and the leading coefficient is

$$\det\left(\frac{1}{(\lambda_i+j-i)!}\right)$$

which we have already evaluated in our proof of (C.20) as being equal to

$$\frac{1}{\prod(\text{hook lengths})}.$$

We must check that the numerator in (C.27) divides $P(d)$, or, what is the same, that if a power of $(d+r)$ divides the numerator in (C.27) then it also divides $P(d)$. So let us examine what powers of $(d+r)$ divide the numerator in (C.27).

Let us distinguish between the cases $r \geqslant 0$ and $r < 0$. For $r \geqslant 0$, the entries in the numerator of (C.27) giving rise to factors of $(d+r)$ come from positions in the diagram which are on the diagonal r units to the right (or above) the main diagonal of the diagram. The number of such factors which occur is equal to the number of rows which contain this diagonal term. If the first p rows contain entries from this diagonal, then an examination of (C.28) shows that the first p rows of the matrix contains common factors of $(d+r)$. This proves that the appropriate power of $(d+r)$ divides $P(d)$ for $r \geqslant 0$. The proof for $r < 0$ is a bit more tricky, and we follow James (1978). Let us first check what power of $(d+r)$ divides the numerator in (C.27). Again, it is the number of rows in the diagram that contain the entries from the diagonal $j - i = r$. This diagonal now lies to the left (and below) the main diagonal, so the first $-r$ rows do not contribute. Let q be the largest integer such that $\lambda_q \geqslant q + r$, then the entries giving $(d+r)$ will occur in the rows from $-r+1$ through q, so the power of $(d+r)$ that divides the numerator of (C.27) is $p = q + r$. We must show that the determinant on the right-hand side of (C.26) has a zero of order at least p at the value $d = r$, or, what amounts to the same thing, that the rank of the matrix occurring on the right is at most $k + r$, when we set $d = r$. To prove this, we prove the following fact inductively. For any integer $0 < s \leqslant k$ we may replace the columns $j = s, s+1, \ldots, k$ by the entries

$$\binom{d + \lambda_i + j - i - s}{d - s}$$

by performing elementary column operations. Indeed, for $s = 1$, we have our original matrix. So we may assume it true for s. But for any column to the right of s, subtracting the $(j-1)$st column from the jth gives, by identity of binomial coefficients,

$$\binom{\lambda_i + d + j - i - s}{d - s} - \binom{\lambda_i + d + j - 1 - s}{d - s} = \binom{\lambda_i + d + j - i - s - 1}{d - s - 1}$$

proving our assertion. Now take $s = r$ and $d = -r$. Then

$$\binom{\lambda_i - r + j - i - r}{r - r} = \binom{\lambda_i + j - i}{0} = \begin{cases} 1 & \text{if } \lambda_i + j - i \geqslant 0 \\ 0 & \text{if } \lambda_i + j - i < 0 \end{cases}$$

so the matrix in (C.26) (with $d = r$) (is column equivalent to) a matrix with a block of all 1's in the upper right-hand corner, the block having q rows and $k + 1 + r$ columns. Thus the rank of the matrix is at most $-r - 1 + (k - q + 1) = k - q - r$ so the nullity is $p = q + r$ at least, which is what we had to prove. This completes the proof of (C.27)

Appendix D

WIGNER'S THEOREM

ON QUANTUM MECHANICAL SYMMETRIES

In this appendix we present a proof of Wigner's theorem concerning quantum mechanical symmetry operators. We follow Bargmann's exposition of Wigner's original proof. We restate the problem here. A quantum mechanical state is a projection, P, onto a one-dimensional subspace. If φ is a unit vector in $\mathscr{H}$, then P_φ denotes the projection onto the one-dimensional space spanned by φ. If P and Q are states, we define their transition probability $P \cdot Q$ by

$$P \cdot Q = \operatorname{tr} PQ.$$

A *quantum mechanical map*, T, from the states of a Hilbert space $\mathscr{H}$ to the states of a Hilbert space $\mathscr{H}'$ is a map which assigns, to each state P of $\mathscr{H}$, a state TP of $\mathscr{H}'$ so that

$$(TP) \cdot (TQ) = P \cdot Q \tag{D.1}$$

for any pair of states, P and Q of $\mathscr{H}$. Then Wigner's theorem says:

Let T be a quantum mechanical map from the states of $\mathscr{H}$ to the states of $\mathscr{H}'$. Then there exists a map, U, from vectors of $\mathscr{H}$ to vectors of $\mathscr{H}'$ such that

$$P_{U\varphi} = TP_\varphi, \tag{D.2}$$

$$U(\xi + \eta) = U\xi + U\eta \quad \text{for any} \quad \xi, \eta \in \mathscr{H}, \tag{D.3}$$

$$\langle U\xi, U\eta \rangle = \kappa(\langle \xi, \eta \rangle) \tag{D.4}$$

where either $\kappa(\lambda) = \lambda$ or $\kappa(\lambda) = \bar{\lambda}$ for all complex numbers λ, and

$$U\lambda\xi = \kappa(\lambda)U\xi \quad \text{for the same function, } \kappa. \tag{D.5}$$

If $\dim \mathscr{H} > 1$ then the alternative $\kappa(\lambda) = \lambda$ or $\kappa(\lambda) = \bar{\lambda}$ is determined by T, indeed the expression

$$\Delta(P_1, P_2, P_3) = \langle \varphi_1, \varphi_2 \rangle \langle \varphi_2, \varphi_3 \rangle \langle \varphi_3, \varphi_1 \rangle$$

where $P_i = P_{\varphi_i}$ and $\|\varphi_i\| = 1$ does not depend on the choice of the unit vectors φ_1 and we have

$$\Delta(TP_1, TP_2, TP_3) = \kappa(\Delta(P_1, P_2, P_3)). \tag{D.6}$$

If $\dim \mathscr{H} > 1$ we can find three states P_1, P_2, P_3 for which $\Delta(P_1, P_2, P_3)$ is not real so that (D.6) determines κ. Finally, given any unit vector, $\varphi \in \mathcal{H}$, and any unit vector $\varphi' \in \mathcal{H}'$ with

$$P_{\varphi'} = TP_\varphi$$

we can always choose U so that

$$U\varphi = \varphi'.$$

Having made such a choice, U is then completely determined. In particular, U is determined by T up to multiplication by a complex scalar of absolute value 1.

In particular, if we assume that T is, in addition, a bijection of the states of $\mathcal{H}$ onto the states of $\mathcal{H}'$ we conclude that in the theorem the map U is either unitary or anti-unitary according as $\kappa(\lambda) = \lambda$ or $\kappa(\lambda) = \bar{\lambda}$. We thus get the version of Wigner's theorem that we stated in Section 3.6.

Proof. The fact that Δ is independent of the choice of representatives is pretty clear: changing representatives means multiplying each φ_i by some α_i with $|\alpha_i| = 1$, and each α_i occurs along with its complex conjugate as a factor in the definition of Δ, and hence Δ is unchanged. If $\dim \mathcal{H} \geq 2$, let φ and ψ be orthogonal unit vectors and set $\varphi_1 = \varphi$, $\varphi_2 = (\varphi - \psi)/2^{\frac{1}{2}}$, and $\varphi_3 = (\varphi + (1-i)\psi)/3^{\frac{1}{2}}$. The corresponding value of Δ is $i/6$ as follows from a straightforward computation. Thus, if $\dim \mathcal{H} > 1$ we can find a complex value of Δ so that κ is determined by T. If $\dim \mathcal{H} = 1$, then any map U chosen as indicated clearly works. This shows that κ is not determined in the one-dimensional case, and that there is no problem with proving the theorem for this case.

We must therefore show that U exists and is uniquely determined by the choice of $U\varphi$ under the assumption that $\dim \mathcal{H} \geq 2$. We adopt the following notation. For any vector, $\xi \in \mathcal{H}$, the requirements (D.2) and (D.5) show that $U\xi$ is determined up to scalar multiple by a complex number of absolute value 1. We let ξ' be any one of these candidates for $U\xi$. Thus, if

$$\xi = \|\xi\| \varphi$$

then ξ' denotes any vector of length $\|\xi\|$ in the line determined by TP_φ. Thus, for instance, we have the equality

$$|\langle \xi', \eta' \rangle| = |\langle \xi, \eta \rangle| \tag{D.7}$$

for any pair of vectors ξ and η.

Let $\varphi_1, \ldots, \varphi_m$ be an orthonormal set of vectors in $\mathcal{H}$. Then it follows from (D.1) that $\varphi'_1, \ldots, \varphi'_m$ is an orthonormal set in $\mathcal{H}'$. Suppose that

$$\xi = \sum \beta_p \varphi_p.$$

We claim that

$$\xi' = \sum \beta'_p \varphi'_p \quad \text{with} \quad |\beta'_p| = |\beta_p|. \tag{D.8}$$

Indeed, take $\beta'_p = \langle \xi', \varphi'_p \rangle$. Then it follows from (D.1) that $|\beta'_p| = |\beta_p|$, and it also follows from (D.1) that

$$\| \xi' - \sum \beta'_p \varphi'_p \|^2 = \| \xi' \|^2 - \sum |\langle \xi', \varphi'_p \rangle|^2$$
$$= \| \xi \|^2 - \sum |\langle \xi, \varphi_p \rangle|^2$$
$$= \| \xi - \sum \langle \xi, \varphi_p \rangle \varphi_p \|^2$$
$$= \| \xi - \sum \beta_p \varphi_p \|^2$$
$$= 0.$$

We now proceed to the construction of U. Let $C = \varphi^\perp$ be the subspace of $\mathcal{H}$ consisting of all vectors orthogonal to φ. Let η be any vector in C, and let $\zeta = \varphi + \eta$. It follows that from (D.8) that every ζ' is of the form $\zeta' = \alpha' \varphi' + \eta'$ with $|\alpha| = 1$, and with

$$\langle \varphi', \eta' \rangle = 0 \quad \text{and} \quad \| \eta' \| = \| \eta \|.$$

There is thus a unique ζ' with $\alpha' = 1$, i.e. a unique ζ' of the form

$$\zeta' = \varphi' + \eta'.$$

This defines a *unique* η' for each $\eta \in C$. We thus set

$$U\eta = \text{the above } \eta' \quad \text{for } \eta \in C,$$

and so we have some map U defined on C. Our strategy is to first show that U satisfies all the requirements of the theorem for vectors in C, and then to extend U to all of $\mathcal{H}$ and show that it fulfils the requirements of the theorem.

Let ξ and η be vectors in C. We know from (D.1) that

$$|\langle U\xi, U\eta \rangle| = |\langle \xi, \eta \rangle|.$$

We also know that

$$|1 + \langle U\xi, U\eta \rangle|^2 = |\langle \varphi' + U\xi, \varphi' + U\eta \rangle|^2 = |\langle \varphi + \xi, \varphi + \eta \rangle|^2 = |1 + \langle \xi, \eta \rangle|^2.$$

Now for any complex number, β, we have $|1 + \beta|^2 = 1 + |\beta|^2 + 2 \operatorname{Re} \beta$. It thus follows from the two preceding equations that

$$\operatorname{Re} \langle U\xi, U\eta \rangle = \operatorname{Re} \langle \xi, \eta \rangle \quad \text{for all } \xi, \eta \text{ in } C.$$

For any vector $\xi \in C$ we know that both $\lambda U\xi$ and $U(\lambda \xi)$ lie in the same line and have the same length. They thus differ by some complex number of absolute value 1, and we can write

$$U(\lambda \xi) = \theta_\xi(\lambda) U\xi$$

where

$$|\theta_\xi(\lambda)| = |\lambda|, \quad \operatorname{Re} \theta_\xi(\lambda) = \operatorname{Re} \lambda$$

and thus

$$\theta_\xi(\lambda) = \lambda \quad \text{if} \quad \lambda \text{ is real.}$$

We now propose to show that for any two vectors, ξ and η in C, that $\theta_\xi = \theta_\eta$, that

$\langle U\xi, U\eta \rangle = \theta_\xi(\langle \xi, \eta \rangle)$ and that $U(\xi + \eta) = U\xi + U\eta$. These assertions all involve just two vectors at a time. We can thus restrict attention to a subspace of C of dimension (at most) 2. Let ψ_1 and ψ_2 be two orthonormal vectors in C, and let C' be the subspace spanned by these two vectors. (In case $\mathscr{H}$ is two dimensional, so that C is one dimensional, let C' denote all of C, which is now spanned by a single unit vector, ψ_1.) Suppose that $\xi = \beta_1\psi_1 + \beta_2\psi_2$ (where we suppress the ψ_2 term if dim $\mathscr{H} = 1$ in this and in all that follows). Then $U\xi = \beta'_1 U\psi_1 + \beta'_2 U\psi_2$ with $|\beta'_1| = |\beta_1|$ and $|\beta'_2| = |\beta_2|$. Suppose that $\beta_1 \neq 0$. Then

$$\langle \beta_1\psi_1, \bar{\beta}_1^{-1}\psi_1 \rangle = \langle \xi, \bar{\beta}_1^{-1}\psi_1 \rangle = \beta_1^{-1}\beta_1 = 1$$

is real, and so $\langle U\xi, U\overline{\beta_1^{-1}}\psi_1 \rangle = \theta_{\psi_1}(\beta_1)\overline{\theta_{\psi_1}(\bar{\beta}_1^{-1})}$ but $U(\overline{\beta_1^{-1}}\psi_1) = \theta_{\psi_1}(\bar{\beta}_1^{-1})U\psi_1$ and so we conclude that

$$\overline{\theta_{\psi_1}(\bar{\beta}_1^{-1})} \cdot \beta'_1 = \overline{\theta_{\psi_1}(\bar{\beta}_1^{-1})}\theta_{\psi_1}(\beta_1)$$

implying that

$$\beta'_1 = \theta_{\psi_1}(\beta_1),$$

and this equation clearly holds also if $\beta_1 = 0$. Similarly,

$$\beta'_2 = \theta_{\psi_2}(\beta_2).$$

If we take $\xi = \psi_1 + \psi_2$ we have $U(\psi_1 + \psi_2) = \psi'_1 + \psi'_2$, while

$$U(\beta\xi) = \theta_\xi(\beta)U(\xi) = \theta_\xi(\beta)(\psi'_1 + \psi'_2) = \theta_{\psi_1}(\beta)\psi'_1 + \theta_{\psi_2}(\beta)\psi'_2$$

which proves that

$$\theta_{\psi_1}(\beta) = \theta_{\psi_2}(\beta)$$

for all β. Thus, setting $\theta(\beta) = \theta_{\psi_1}(\beta)$ we have

$$U(\beta_1\psi_1 + \beta_2\psi_2) = \theta(\beta_1)\psi_1 + \theta(\beta_2)\psi_2.$$

If we take $\beta = i$, then $|\theta(\beta)| = |i| = 1$ and $\mathrm{Re}\,\theta(i) = \mathrm{Re}\,i = 0$, so that

$$\text{either } \theta(i) = i \quad \text{or} \quad \theta(i) = -i.$$

Now $\mathrm{Re}\,\theta(\bar{\beta})\theta(\lambda) = \mathrm{Re}\,\langle U\lambda\psi_1, U\beta\psi_1 \rangle = \mathrm{Re}\,\langle \lambda\psi_1, \beta\psi_1 \rangle = \mathrm{Re}\,\bar{\beta}\lambda$. But $\mathrm{Re}\,\bar{i}\lambda = \mathrm{Im}\,\lambda$ so we conclude that

$$\text{either } \theta(\lambda) = \lambda \text{ for all } \lambda$$
$$\text{or} \quad \theta(\lambda) = \bar{\lambda} \text{ for all } \lambda .$$

For either of these possibilities we have $\theta(\alpha + \beta) = \theta(\alpha) + \theta(\beta)$, $\theta(\alpha\beta) = \theta(\alpha)\theta(\beta)$ and $\theta(\bar{\alpha}) = \overline{\theta(\alpha)}$. If we write

$$\xi = \alpha_1\psi_1 + \alpha_2\psi_2 \qquad \text{and} \quad \eta = \beta_1\psi_1 + \beta_2\psi_2$$

then

$$U\xi = \theta(\alpha_1)\psi'_1 + \theta(\alpha_2)\psi'_2 \quad \text{and} \quad U\eta = \theta(\beta_1)\psi'_1 + \theta(\beta_2)\psi'_2$$

and it follows that

$$U(\xi + \eta) = U\xi + U\eta, \quad U\lambda\xi = \theta(\lambda)U\xi$$

and

$$\langle U\xi, U\eta \rangle = \theta(\alpha_1)\overline{\theta(\beta_1)} + \theta(\alpha_2)\overline{\theta(\beta_2)} = \theta(\alpha_1\bar{\beta}_1 + \alpha_2\bar{\beta}_2) = \theta(\langle \xi, \eta \rangle).$$

We have thus verified (D.2)–(D.5) with $\kappa = \theta$ for vectors in C' and hence for all vectors in C.

We have already seen that the map U was specified by

$$U(\varphi + \xi) = U\varphi + U\xi = \varphi' + U\xi \quad \text{with} \quad P_{U(\varphi+\xi)} = TP_{\varphi+\xi},$$

where $\xi \in C$. The most general vector in $\mathscr{H}$ can be written as

$$\alpha\varphi + \eta \quad \text{with} \quad n \in C.$$

If $\alpha \neq 0$, set $\xi = (\alpha^{-1}\eta)$ and

$$U(\alpha\xi + \eta) = U[\alpha(\varphi + \xi)] = \kappa(\alpha)(\varphi' + U\xi), \quad \text{where } k = \theta.$$

We now have defined U for all vectors in $\mathscr{H}$. Equations (D.2), (D.3) and (D.5) clearly hold, and (D.4) is easily verified. This completes the proof of Wigner's theorem.

Appendix E

COMPACT GROUPS, HAAR MEASURE,
AND THE PETER–WEYL THEOREM

A *topological group* is a group, G, which is a topological space such that the maps $G \times G \to G$ sending $a, b \rightsquigarrow ab$ and $G \to G$ sending $a \rightsquigarrow a^{-1}$ are continuous. The group G is called *compact* if it is a topological group and the underlying topological structure of G is that of a compact space. Many of the theorems proved in Chapter 2 for finite groups are valid with only minor modifications for the case of compact groups. One of the main techniques, that of averaging over the group, carries over with practically no change at all. In order to average over the group we have to replace the sum by an integral, and the first main theorem that we shall prove is that there exists a unique notion of invariant integration on the group. That is, that there exists a unique rule assigning to each continuous function, f, its integral, which we denote by $\int_G f(a)\,\mathrm{d}a$, which satisfies the usual axioms for an integral, and in addition is left invariant, i.e.

$$\int_G f(b^{-1}a)\,\mathrm{d}a = \int_G f(a)\,\mathrm{d}a \quad \text{for all} \quad b \in G \tag{E.1}$$

and is normalized,

$$\int_G 1\,\mathrm{d}a = 1. \tag{E.2}$$

This integral is known as the Haar integral, and the corresponding measure is called Haar measure. Our first task will be to establish the existence and uniqueness of the Haar integral, and establish some of its properties. In particular, we shall prove that the Haar integral is also right invariant, i.e. satisfies

$$\int_G f(ab)\,\mathrm{d}a = \int_G f(a)\,\mathrm{d}a. \tag{E.3}$$

The most familiar example of a topological group is the group $Gl(n)$ of non-singular matrices (either the group of $n \times n$ real matrices, $Gl(n, \mathbb{R})$ or the group of $n \times n$ complex matrices, $Gl(n, \mathbb{C})$). The topology is just the usual topology obtained by regarding $Gl(n)$ as a subset of the space of all $n \times n$ matrices, which has the topology of $\mathbb{R}^{n^2}$ or $\mathbb{C}^{n^2}$. Any subgroup of $Gl(n)$ which is closed and bounded as a subset of the set of all matrices will then be a compact group. Thus the orthogonal group, $O(n)$, is a compact group, as is the rotation group $SO(n)$, of all orthogonal matrices with determinant 1. Similarly, the group $U(n)$ of all $n \times n$ unitary matrices is a compact group as is the special unitary group, of all unitary matrices of determinant 1. These will be among the most important of our examples.

In what follows, we will not use the full fact that our groups are compact, but only that they are *totally bounded*: a group G is totally bounded if there exists a fundamental family of neighborhoods of the origin, U_i, and, for each i, a finite number of elements $g_1, \ldots, g_{j(i)}$ such that the sets $g_1 U_i, \ldots, g_{j(i)} U_i$ cover G for each fixed i. Thus, if G is compact it is certainly totally bounded. Also, suppose that G possesses a metric, d, which is invariant under right and left translation, and that the diameter of G is finite with respect to this metric. Then we can take U_i to be the ball of radius $1/i$ about e. This gives another example of a totally bounded group. For the rest of this section, we will assume without mentioning it any further that G is a totally bounded group.

The existence of the Haar integral will follow from the following theorem.

Theorem E.1   (The mean ergodic theorem)
Let T be a linear transformation of a normed vector space, V, such that
  (i) There is some constant, c, such that

$$\| T^n v \| \leqslant c \| v \|$$

for all $v \in V$ and all positive integers, n, and
  (ii) For some fixed $w \in V$ the sequence

$$S_n w = (1/n)(w + Tw + \cdots + T^{n-1} w)$$

possesses a subsequence which converges to some element $\bar{w}$ (or, in fact, only converges weakly to $\bar{w}$, i.e. $f(S_{n_j} w - \bar{w}) \to 0$ for any continuous linear function, f).
Then $T\bar{w} = w$ and the sequence $S_n w$ converges to $\bar{w}$.

Proof   We first observe that the subspace $\overline{(I - T)V}$ consists of all $z \in V$ such that $S_n z \to 0$. Indeed, if $z = (I - T)v$, then

$$\| S_n z \| = \| (1/n)(T^n v - v) \| \leqslant (c/n) \| v \| + (1/n) \| v \| \to 0,$$

so that every vector in $(I - T)V$ satisfies $S_n z \to 0$. On the other hand, (i) implies that the space of z satisfying $S_n z \to 0$ is closed, so that every element of $\overline{(I - T)V}$ satisfies $S_n z \to 0$. Conversely, suppose that $S_n z \to 0$. Then for any positive ε we can find some n such that

$$\| z - (z - S_n z) \| < \varepsilon.$$

We claim that for any n, $z - S_n z \in (I - T)V$. Indeed,

$$z - S_n z = (1/n)\{(I - T)z + (I - T^2)x + \cdots + (I - T^{n-1})z\}$$
$$= (1/n)\{(I - T)z + (I - T)(I + T)z + \cdots + (I - T)$$
$$\cdot (I + T + \cdots + T^{n-2})z \in (I - T)V\}.$$

Now suppose that

$$S_{n_j} w \to \bar{w}.$$

Then

$$TS_{n_j}w - S_{n_j}w = (1/n)(T^{n_j}w - w) \to 0$$

so

$$T\bar{w} = \bar{w}.$$

Now

$$T^n w = T^n \bar{w} + T^n(w - \bar{w}) = \bar{w} + T^n(w - \bar{w})$$

so

$$S_n w = \bar{w} + S_n(w - \bar{w})$$

By assumption,

$$w - S_{n_j}w \to w - \bar{w}$$

and

$$w - S_{n_j}w \in (I - T)V.$$

Hence $w - \bar{w} \in \overline{(I - T)V}$ so that

$$S_n(w - \bar{w}) \to 0,$$

which is what we want to prove. (If we only assume that $S_{n_j}w \to \bar{w}$ weakly, the argument proceeds in much the same fashion but uses the Hahn–Banach theorem. We conclude, as before, that $f(Tw - \bar{w}) = 0$ holds for any continuous linear function, and hence $Tw = \bar{w}$. It also follows that $w - \bar{w} \in \overline{(I - T)V}$. Indeed, if not, there would exist some linear function, f, vanishing on $\overline{(I - T)V}$ with $f(w - \bar{w}) \neq 0$. But $w - S_{n_j}w \in (I - T)V$ and $f(w - S_{n_j}w) \to f(w - \bar{w})$, yielding a contradiction.) Notice that if T is a unitary operator on a Hilbert space, then for any w we can find a weakly convergent subsequence of $S_{n_j}w$ since the unit sphere of a Hilbert space is weakly compact. Thus for any unitary operator in a Hilbert space our theorem guarantees that $S_n w$ converges to an invariant element for any w. This is the usual form of the mean ergodic theorem.

We now apply the theorem to the construction of an invariant integral on a totally bounded group. Let V be the space of uniformly continuous functions on G with the norm taken to be the sup norm, i.e.

$$\|f\| = \sup_{g \in G} |f(g)|.$$

(If G is compact every continuous function is uniformly continuous and V is simply the space of continuous functions.) If we take all the g_k's that entered into the definition of uniform boundedness and arrange them into a sequence, we clearly get some sequence $\{g_k\}$ which is dense in V. Define the operator T on V by the formula

$$(Tf)(g) = \sum_k \frac{1}{2^k} f(g_k^{-1}g)$$

If $|f(a) - f(b)| < \varepsilon$ whenever $a^{-1}b \in U$, where U is some neighborhood of e, then

$$|Tf(a) - Tf(b)| = \sum \frac{1}{2^k} |f(g_k^{-1}a) - f(g_k^{-1}b)| < \varepsilon$$

so that T does map V into itself, and it is clear that $\|T^n f\| \leqslant \|f\|$ for all n and f. Now

$S_n f(g) = \sum a_j f(h_j g)$ for some sequence of elements, h_j in G and some sequence of real numbers, a_j, with $\sum a_j = 1$, $a_j \geqslant 0$. Thus, by the same sort of estimate we conclude that sequence of functions, $S_n f$ is equicontinuous, i.e. that for any $\varepsilon > 0$ we can find some fixed neighborhood, U, of e such that

$$|S_n f(a) - S_n f(b)| < \varepsilon \quad \text{for all } n \text{ if } a^{-1}b \in U.$$

Since G possesses a dense sequence of elements, this implies (by the usual selection type argument) that we can choose a convergent subsequence of $S_n f$. By Theorem E.1, we conclude that for any $f \in V$, the sequence $S_n f$ converges to some element $\bar{f}$ satisfying $T\bar{f} = \bar{f}$.

We claim that $\bar{f}$ is a constant, and intend to show that the map, $f \rightsquigarrow \bar{f}$ has all the properties of an integral. To prove that $\bar{f}$ is a constant, we may assume, by taking real and imaginary parts, that f is real valued. Let $M = \sup \bar{f}$. If $\bar{f}$ is not constant, then there will be some $h \in G$ with $\bar{f}(h) < M - 2\varepsilon$ for some $\varepsilon > 0$. Since $\bar{f}$ is uniformly continuous, we will then have $\bar{f}(g) < M - \varepsilon$ for all g such that $gh^{-1} \in U$, where U is some suitable neighborhood of the origin. By construction, a finite number of the $g_i U$ will cover all of G. Thus for any $g \in G$ we will have

$$gh^{-1} \in g_i U \quad \text{for some} \quad i < N$$

where N is some sufficiently large number, and hence

$$g_i^{-1}gh^{-1} \in U \quad \text{so that} \quad \bar{f}(g_i^{-1}g) < M - \varepsilon.$$

But then

$$\bar{f}(g) = T\bar{f}(g) = \sum \frac{1}{2^j} \bar{f}(g_j^{-1}g) \leqslant \sum_{j \neq i} \frac{1}{2^j} M + \frac{M - \varepsilon}{2^N} \leqslant M - \frac{\varepsilon}{2^N}$$

contradicting the definition of M.

Thus $\bar{f}$ is a constant. We denote the number $\bar{f}$ by $\int_G f(g)\,dg$ and note that

$$\text{the map } f \rightsquigarrow \int_G f(g)\,dg \text{ is linear,} \tag{E.4}$$

$$\int_G 1\,dg = 1 \tag{E.2}$$

and

$$\text{if } f(g) \geqslant 0 \quad \text{for all} \quad g \in G \text{ then } \int_G f(g)\,dg \geqslant 0. \tag{E.5}$$

Thus we do indeed get all the properties of an integral. Notice that if we set $f_h(g) = f(gh)$ for some $h \in G$, then $(Tf)_h = T(f_h)$ from which it follows that

$$\int_G f(gh)\,dg = \int_G f(g)\,dg. \tag{E.3}$$

It follows from the construction of $\int_G f(g)\,dg$ that for any fixed function, f, and any $\varepsilon > 0$, we can find a sequence of real numbers, $a_k \geqslant 0$, and a sequence of group elements,

h_k, such that

$$\sup_{g\in G}\left|\sum a_k f(h_k^{-1}g) - \int_G f(g)\,dg\right| < \varepsilon. \tag{E.6}$$

Now we could just as well have started our whole procedure with a map $\tilde{T}$ given by $\tilde{T}f(g) = \sum \frac{1}{2^j}f(gp_j)$ for a suitable dense sequence of elements, p_j. The arguments would have gone exactly as before, and we would end up with some integral, $\bar{\int}$, satisfying (E.4), (E.2), (E.5), and with (E.3) replaced by

$$\bar{\int}_G f(p^{-1}g)\,dg = \bar{\int}_G f(g)\,dg$$

and with (E.6) replaced by

$$\sup_{g\in G}\left|\sum b_k f(gq_k) - \bar{\int}_G f(g)\,dg\right| < \varepsilon. \tag{E.7}$$

That is, left multiplication is replaced everywhere by right multiplication. Now (E.6) and (E.7) imply that

$$\sup_{g\in G}\left|\sum a_j b_k f(g_j^{-1}gq_k) - \int_G f(g)\,dg\right| < \varepsilon$$

and

$$\sup_{g\in G}\left|\sum a_j b_k f(g_j^{-1}gq_k) - \bar{\int}_G f(g)\,dg\right| < \varepsilon$$

so that

$$\int_G f(g)\,dg = \bar{\int}_G f(g)\,dg,$$

and therefore $\int_G f(g)\,dg$ is also left invariant, i.e. (E.1) holds. Notice that (E.2), (E.3), (E.4) and (E.5) uniquely characterize the integral $\int$, as do (E.1), (E.2), (E.4) and (E.5). Indeed, if $\tilde{\int}$ is some other integral satisfying (E.2)–(E.5) we have

$$\int_G f(g)\,dg - \varepsilon \leqslant \sum b_j f(gq_j) \leqslant \int_G f(g)\,dg + \varepsilon$$

for any real valued function, f, and suitable b_j with $\sum b_j = 1$ and suitable group elements, q_j. Applying the integral $\tilde{\int}$ to this inequality yields

$$\left|\int_G f(g)\,dg - \tilde{\int}_G f(g)\,dg\right| \leqslant \varepsilon.$$

Since this holds for all f and for all ε we conclude that $\int = \tilde{\int}$. A similar argument shows that (E.1) works as well as (E.3). We have thus established the existence and uniqueness of the Haar integral. It is also useful to remark that it follows from our construction that if f is a uniformly continuous function satisfying $f \geqslant 0$, and if $f(h) > 0$ for some h, then $\int_G f(g)\,dg > 0$.

Using the Haar integral, much of the theory of *finite*-dimensional representations goes through with little change. For a topological group, a representation will mean a continuous homomorphism of G into the group of continuous linear maps, $\text{End}(V)$, of some topological vector space, V. Here $\text{End}(V)$ is endowed with one of its suitable topologies (and the choice of topology might affect the notion of representation). For finite-dimensional vector spaces over $\mathbb{C}$ (or $\mathbb{R}$) there is only one topology – the one we mentioned above for $Gl(n)$. Thus there is no ambiguity in discussing finite-dimensional representations of topological groups, in particular in discussing compact groups. For two continuous functions, f_1 and f_2, on the compact group, G (or two uniformly continuous functions on a totally bounded group) we define their scalar product by

$$(f_1, f_2) = \int_G f_1(g)\overline{f_2(g)}\,dg$$

(which coincides with the definition given in Chapter 2 if the group happens to be finite). We now briefly indicate the minor changes to be made in redoing the results of Section 2.3 for the case of compact groups: Schur's lemma needs no change. In Proposition 3.1 replace the definition of S by

$$S = \int_G r_a^2 S_0 (r_a^1)^{-1}\,da.$$

Replace (3.1), Section 2.3, by

$$\text{If } r^1 \not\sim r^2 \quad \text{then} \quad \int_G r_{kl}^2(a) r_{ij}^1(a)\,da = 0 \tag{E.8}$$

and replace (3.2) by

$$\int_G r_{kl}(a) r_{ij}(a^{-1})\,da = \frac{1}{n}\delta_{li}\delta_{kj}. \tag{E.9}$$

The method of averaging over the group (with sum replaced by integral) still applies to show that any finite-dimensional representation is equivalent to a unitary representation, and hence that every finite-dimensional representation is completely reducible. For finite-dimensional unitary representations (3.3), (3.4), and all of Section 2.4, hold without change. On the other hand, the regular representation, which we must define with slightly more care, will be infinite dimensional if G is not finite. For the purposes of the present section, we will take the regular representation to be the unitary representation of G on $L^2(G)$. That is, we consider the action of G on the space of continuous functions on G, but complete this space (relative to the scalar product, $(,)$), to obtain the Hilbert space $L^2(G)$, and (E.1) implies that G acts as unitary transformations. Unless G is finite, the space $L^2(G)$ will be infinite dimensional, and thus the character cannot be defined in the usual way as a function on G. (For instance, the value of the character at e would have to be infinite.) In Appendix G we will see that is possible to generalise the definition of trace, and hence of character, so that it makes sense for many of the interesting infinite-dimensional representations that arise in practice, but the character will then be a distribution rather than a function. For the moment, however, we have no analogs of the results of Sections 2.5 and 2.6 for general compact groups, but we will return to these formulas in Appendix G.

The assertions concerning induced representations need a certain amount of reformulation for the case of compact groups. The key tool is an analysis of the regular representation, i.e. the representation of G on $L^2(G)$. The essential facts are summarized in the celebrated Peter–Weyl theorem, to be stated below. Before stating the theorem, it is convenient to make a definition.

Definition E.1

A (uniformly) continuous function, f, is called a *representative function* if the space spanned by all its translates, gf is finite dimensional, where, as usual, $gf(a) = f(g^{-1}a)$, $g \in G$.

Suppose that f is a representative function, and let $g_1 f, \ldots, g_n f$ be a maximal set of linearly independent translates of f. Then for any $g \in G$

$$f(g^{-1}u) = gf(u) = \sum h_i(g)(g_i f)(u) = \sum h_i(g) f_i(u)$$

where the h_i are suitable continuous functions on G and where we have set $f_i = g_i f$. Conversely, if f is a function satisfying such an equation, it is clear that f is a representative function. Replacing g by g^{-1} in the above equation shows that the set of right translates, $f(\cdot u)$ also span a finite-dimensional space. In particular, if we set $\tilde{f}(a) = f(a^{-1})$ then $\tilde{f}$ is a representative function, and conversely. Let W be the space spanned by the translates of f, so that the f_i form a basis for W. By the above equation

$$gf_j = (gg_j)f = \sum h_i(gg_j) f_i$$

so that the values $h_i(gg_j)$ are precisely the matrix entries of the representation of G on W, relative to the basis f_i. But

$$\tilde{f}(g) = f(g^{-1}) = \sum h_i(gg_j) f_i(e).$$

Since the $f_i(e)$ are just constants in the above equation, we see that $\tilde{f}$ is a linear combination of the matrix elements of some finite-dimensional representation of G, and hence so is f. Conversely, if we start with some finite-dimensional representation of G, and let r_{ij} be its matrix elements relative to some basis, then it is clear that each of the r_{ij} is a representative function. Since every finite-dimensional representation can be written as a direct sum of irreducibles, we conclude that the space of representative functions consists precisely of linear combinations of matrix coefficients of finite-dimensional representations.

We are now in a position to state the Peter–Weyl theorem:

Theorem E.2

(1) The representation functions are dense in $L^2(G)$

(2) The space $L^2(G)$ decomposes into a Hilbert space direct sum of irreducible representations of G, each of which is finite dimensional.

(3) Every irreducible representation of G is finite dimensional.

(4) Each irreducible representation of G occurs in $L^2(G)$ with a multiplicity equal to its dimension.

(5) Any unitary representation of G on any Hilbert space decomposes into a Hilbert space direct sum of (finite-dimensional) irreducible representations.

We begin by proving a special case of (3):

Proposition E.1

Let W be a closed subspace of $L^2(G)$ which is irreducible in the sense that it possesses no proper closed subspaces which are invariant under G. Then W is finite dimensional.

Let f be a unit vector in $L^2(G)$ and consider the function k_f defined on $G \times G$ by the formula

$$k_f(a, b) = \int_G \overline{f(ga)} f(gb) \, dg.$$

Since $f \in L^2(G)$, it is easy to see that k_f is a continuous function on $G \times G$. On the other hand, let P_f denote projection onto the line spanned by f, so that $P_f v = (v, f)$ and define the operator K_f by

$$K_f = \int_G g^{-1} P_f g \, dg,$$

so that for any v, $w \in L^2(G)$, we have, by unitarity,

$$(K_f v, w) = \int_G (P_f gv, gw) \, dg$$

$$= \int_G (gv, f)(f, gw) \, dg.$$

Assume, for the moment, that v and w are continuous functions. Then this last expression can be written as

$$\int_G (gv, f)(f, gw) \, dg = \int_G \int_G v(g^{-1}a) \overline{f(a)} \, da \int_G f(b) \overline{w(g^{-1}b)} \, db \, dg$$

$$= \int_G \int_G v(a) \overline{f(ga)} \, da \int_G f(gb) \overline{w(b)} \, db \, dg$$

$$= \int_G \int_G k_f(a, b) v(a) \overline{w(b)} \, da \, db$$

$$= \left(\int_G k_f(a, \cdot) v(a) \, da, \, w \right)$$

so that

$$K_f v = \int_G k_f(a, \cdot) v(a) \, da.$$

Thus K_f is given as an integral operator with the continuous kernel k_f. In particular, K_f carries any bounded set in $L^2(G)$ into an equicontinuous set, and hence is a compact operator:

$$|K_f v(b_1) - K_f v(b_2)| \le \|v\| \sup_{a \in G} |k_f(a, b_1) - k_f(a, b_2)|.$$

Finally, since K_f was obtained from P_f by averaging over the group, it follows in the usual fashion that K_f commutes with all elements of G. If f lies in some invariant subspace, W, then it is clear that $K_f(L^2(G)) \subset W$. The operator K_f is self-adjoint and

$$(K_f f, f) = \int_G (P_f g f, g f) \, dg > 0$$

since $(P_f g f, g f) \ge 0$ for all g and $(P_f f, f) = 1$. Thus $K_f \ne 0$. Therefore K_f, being a compact operator, has an eigenspace corresponding to some non-zero eigenvalue, and this eigenspace must be finite dimensional (since K_f is compact) and invariant under G (since K_f commutes with the elements of G). Thus, if W is a non-trivial invariant subspace of $L^2(G)$, we have produced a non-trivial invariant subspace of W which is finite dimensional (and hence closed). If W is irreducible, this subspace must coincide with W, proving the proposition.

Now let U_i be a fundamental sequence of neighborhoods of e, and let $f_i \ge 0$ be a sequence of continuous functions with supp $f_i \subset U_i$ and $f_i(e) > 0$. By multiplying f_i by a suitable constant we may further assume that $\int |f_i(g)| \, dg = 1$. We set

$$R_i = R_{f_i} = \int_G f(a) a \, da$$

in the regular representation, and notice that

$$R_i v(b) = \int_G f_i(a) v(a^{-1} b) \, da = \int_G f_i(ab^{-1}) v(a^{-1}) \, da = \int_G f_i(ba^{-1}) v(a) \, da$$

so that R_i is an integral operator with continuous kernel $h_i(a, b) = f_i(ba^{-1})$ and hence is compact. On the other hand, the sequence $R_i v$ approaches v as $i \to \infty$. Indeed,

$$\|R_i v - v\| = \left\| \int_G f_i(a)(av - v) \, da \right\|$$

$$\le \int_{U_i} |f_i(a)| \, \|av - v\| \, da$$

which approaches zero since $\int_G |f_i(a)| \, da = 1$ and $av \to v$ as $a \to e$. Let us write

$$L^2(G) = H_{0,i} + \sum_j H_{j,i} \quad \text{(Hilbert space direct sum),}$$

where $H_{0,i}$ is the zero eigenspace of R_i and the $H_{j,i}$ are non-zero eigenspaces, and hence the $H_{j,i}$ are finite dimensional for $j \neq 0$. We claim that

$$L^2(G) \text{ is the closure of the subspace } \sum_i \sum_{j \geq 1} H_{j,i}.$$

Indeed, if v is orthogonal to this subspace, then $v \in \cap H_{0,i}$ so that $R_i v = 0$ for all i, implying that $v = 0$.

For each j and i, let $W_{j,i}$ denote the intersection of all the closed invariant subspaces of $L^2(G)$ which contain $H_{j,i}$. It, itself, is clearly a closed invariant subspace containing $H_{j,i}$ and hence is the minimal such subspace. We claim that any closed invariant subspace of $W_{j,i}$ must have a non-zero intersection with $H_{j,i}$. Indeed, suppose that U is a closed invariant subspace of $W_{j,i}$ whose intersection with $H_{j,i}$ were zero. Then the orthogonal complement of U in $W_{j,i}$ would be a closed invariant subspace containing $H_{j,i}$ and hence must coincide with $W_{j,i}$, implying that $U = 0$. Consider all the intersections $U \cap H_{j,i}$, as U ranges over the closed invariant subspaces. The spaces $U \cap H_{j,i}$ are all finite dimensional and non-zero. Let us pick such a subspace, $H_{j,i}^1$, which has minimum dimension, and let us set

$$W_{j,i}^1 = \cap U,$$

where U ranges over closed invariant subspaces satisfying $U \subset W_{i,j}$ and $U \cap H_{j,i} = H_{j,i}^1$. Thus $W_{j,i}^1$ is the smallest closed invariant subspace of $W_{j,i}$ whose intersection with $H_{j,i}$ is $H_{j,i}^1$. But this implies that $W_{j,i}^1$ is irreducible. Indeed, any proper closed subspace of $W_{j,i}^1$, would have to intersect $H_{j,i}$ in a proper subspace of $H_{j,i}^1$ contradicting our choice of $H_{j,i}^1$. Let us now replace $H_{j,i}$ by $H_{j,i} \cap (H_{j,i}^1)^\perp$ and $W_{j,i}$ by $W_{j,i} \cap (W_{j,i}^1)^\perp$. Proceeding as before, we will find a collection, $H_{j,i}^1, H_{j,i}^2, \ldots$ of mutually orthogonal Subspaces of $H_{j,i}$, and a collection $W_{j,i}^1, W_{j,i}^2, \ldots$ of irreducible mutually orthogonal subspaces with $W_{j,i}^k \cap H_{j,i} = H_{j,i}^k$. Since $H_{j,i}$ is finite dimensional, there will be only finitely many such subspaces, $H_{j,i}^k$ and $W_{j,i}^k$.

Suppose that $(W_{j,i}^k, W_{r,s}^l) \neq \{0\}$. Then the orthogonal projection onto $W_{j,i}^k$ is non-trivial, when restricted to $W_{r,s}^l$, and hence, by Schur's lemma, these two irreducible subspaces give equivalent representations. It follows from the expresson for R_i that it has the same eigenvalue on these two subspaces, so that $W_{r,s}^l$ has a non-trivial intersection with $H_{j,i}$ and hence $W_{r,s}^l \subset W_{j,i}$.

Let us relabel the $W_{j,i}$ as W_1, W_2, etc. and set

$$U_1 = W_1, \quad U_2 = W_2 \cap U_1^\perp, \ldots,$$

$$U_{j+1} = W_j \cap (U_1 \oplus \cdots \oplus U_j)^\perp, \ldots$$

so that

$$W_1 + \cdots + W_j = U_1 \oplus \cdots \oplus U_j \text{ (orthogonal direct sum).}$$

By the above remarks, each U_j is the direct sum of finitely many irreducibles (which we know to be finite dimensional) and, since $L^2(G) = \overline{\sum H_{ij}}$ we conclude that $L^2(G)$ is the Hilbert space direct sum of the U_j's. This proves assertion (2) of the Peter–Weyl theorem.

We have actually proved a little more, namely that there are only finitely many irreducible subspaces in the direct sum decomposition of a given type. Indeed, on equivalent irreducible representations, the operators R_i must have the same eigenvalues, and hence all such irreducibles correspond to the same $H_{j,i}$, and thus there cannot be an infinite number of mutually orthogonal irreducible subspaces of the same type.

Before proceeding with the rest of the proof of the Peter–Weyl theorem, it will be of some use to us to isolate the preceding argument and state the conclusions in a broader context which will have other applications for us. Let G be a continuous group with left invariant measure $\int dg$, and let r be a unitary representation of G. If f is any continuous function with compact support defined on G, we can form the operator R_f. We say that the representation is *completely continuous* if each of the operators R_f is a compact operator. An examination of the above argument shows that we have proved the following proposition.

Proposition E.2
Let r be a completely continuous representation of the topological group G on the Hilbert space, H. Then we can decompose H into a Hilbert space direct sum $H = \oplus W_i$ of irreducible subspaces, W_i where there are only finitely many irreducible subspaces of any given equivalence class of irreducible representations.

Notice that if W is a closed invariant subspace of a completely continuous representation, then the restriction of the representation to W is again completely continuous.

We now return to the Peter–Weyl theorem. If G is any compact group, we can map $L^2(G)$ into $L^2(G \times G)$ by sending f into θf, where $\theta f(a,b) = f(ab^{-1})$. (The image consists of those functions in $L^2(G \times G)$ which are invariant under right translation by elements of the diagonal subgroup, i.e. the functions h on $G \times G$ which satisfy $h(ag, bg) = h(a, b)$ for all $g \in G$.) Notice that

$$\|\theta f\|^2 = \int_G \int_G |f(ab^{-1})|^2 \, da \, db = \int_G |f(a)|^2 \, da = \|f\|^2$$

so that the map θ is unitary. If we consider G as the subgroup $G \times \{e\}$ of $G \times G$ then θ is a G morphism. Finally, since left translation commutes with right translation, the image of θ is an invariant subspace of $L^2(G \times G)$, for the regular representation of $G \times G$. Thus, by restriction, we can regard $G \times G$ as acting on $L^2(G)$, and, by the representation of $G \times G$ on $L^2(G)$ is completely continuous. By Proposition E.2 we know that $L^2(G)$ decomposes under $G \times G$ into a direct sum of irreducibles with each type occurring a finite number of times, and by Proposition E.1 (applied to $G \times G$, where $L^2(G)$ is a subspace of $L^2(G \times G)$), we know that each irreducible is finite dimensional. We claim that every finite-dimensional irreducible representation of $G \times G$ occurs exactly once in this decomposition. Indeed let W_1 and W_2 be two irreducible subspaces of $L^2(G)$ under $G \times G$ which define equivalent representations. We claim that $W_1 = W_2$. Let $u_1, \ldots, u_n$ and $v_1, \ldots, v_n$ be bases of W_1 and W_2 such that $G \times G$ has the same matrix representation relative to these bases. Let $F(a, b) =$

$\sum u_i(a)v_i(b)$. Then, letting (r_{ij}) denote the matrix of the irreducible representation in question, we have

$$F(gah^{-1}, gbh^{-1}) = \sum u_j(gah^{-1})\overline{v_j(gbh^{-1})}$$

$$= \left(\sum_j r_{ij}(g,h)\overline{r_{ij}(g,h)}\right)\sum u_i(a)\overline{v_i(b)}$$

$$= F(a,b)$$

since the matrix (r_{ij}) is unitary. In particular,

$$\sum u_i(g)\overline{v_i(e)} = F(g,e) = F(e,g^{-1})$$

$$= \sum u_i(e)\overline{v_i(g^{-1})} = \sum u_i(e)\overline{\tilde{v}_i(g)}.$$

If we set $v = \sum u_i(e)v_i \in W_2$, the above equation asserts that $\overline{\tilde{v}} \in W_1$. Now the space of all functions of the form $\overline{\tilde{v}}$, $v \in W_2$ is clearly an invariant subspace for $G \times G$, and hence must coincide with W_1. If we take $W_1 = W_2$ in the above argument, we see that $\overline{\tilde{u}} \in W_1$ if and only if $u \in W_1$, so that any two equivalent irreducibles in the decomposition of $L^2(G)$ under $G \times G$ coincide.

We now propose to show that these irreducibles are isomorphic under $G \times G$ to the irreducible representations of $G \times G$ on $\mathrm{Hom}(V,V) = V \otimes V^*$, as V ranges over the finite-dimensional irreducible representations of G. Indeed, given $u \otimes v^* \in V \otimes V^*$ we obtain a function $f_{u \otimes v^*}$ on G defined by $f_{u \otimes v^*}(g) = v^*(gu)$, and it is clear that this extends to a map of $V \otimes V^* \to L^2(G)$. The group $G \times G$ acts on $V \otimes V^*$ and

$$f_{(g,h)(u \otimes v^*)}(a) = f_{gu \otimes h^{*-1}v^*}(a) = v^*(h^{-1}agu) = f_{u \otimes v^*}(h^{-1}ag)$$

so that the map is equivariant for the action of $G \times G$. Since G acts irreducibly on V, and hence also on V^*, we can conclude that $G \times G$ acts irreducibly on $V \otimes V^*$. Indeed, if we examine the proof of the assertion on p. 66, we see that the proof works without change for the case of compact groups. Thus each $V \otimes V^*$ does occur exactly once in the decomposition of $L^2(G)$ under $G \times G$. If we knew that every irreducible finite-dimensional representation of $G \times G$ must be a tensor product of irreducibles, we could conclude that every irreducible subspace in $L^2(G)$ must be equivalent to a representation of $G \times G$ on $\mathrm{Hom}(V', V) = V \otimes V'^*$. By Schur's lemma, we could conclude that $V \sim V'$ if we found a non-zero element in this space which was invariant under all $(g,g) \in G \times G$. Let us establish the existence of such a non-zero G-invariant vector in any $G \times G$ irreducible subspace, W, of $L^2(G)$. Let $w_1, \ldots, w_n$ be an orthonormal basis of W. Let $B(a,b) = \overline{w_1(ab^{-1})}w_1 + \cdots + \overline{w_n(ab^{-1})}w_n$, so that $B: G \times G \to W$, and

$$B(ha, gb) = \overline{w_1(hab^{-1}g^{-1})}w_1 + \cdots + \overline{w_n(hab^{-1}g^{-1})}w_n$$

$$= (h,h)B(a,b).$$

Since B does not map all of $G \times G$ into zero, it follows that $B(e,e) = 0$. But it also follows directly from the above equation and the definition of B that $B(e,e)$ is invariant under all (g, g). We still must prove that irreducibles of $G \times G$ are tensor products of irreducibles of G. This follows from proposition E.3.

Proposition E.3.

Let r be an irreducible unitary representation of the topological group $G_1 \times G_2$ on a finite-dimensional vector space W, where G_1 and G_2 are topological groups and $G_1 \times G_2$ is given the product topology. Then $r \sim r_1 \otimes r_2$, where r_i is an irreducible unitary representation of G_i, $i = 1, 2$.

Proof Consider the restriction of r to the group G_1, considered as the subgroup $G_1 \times \{e\}$ of $G_1 \times G_2$. The space W decomposes into a direct sum of irreducibles, $W = V_1 \oplus \cdots \oplus V_k$ under G_1. We claim that all the V_i are equivalent as representation spaces of G_1. Indeed, let W_1 be the sum of G_1-invariant subspaces, U, such that $\mathrm{Hom}_G(V_1, U) = \{0\}$, so that W_1 is the maximal such subspace. It is clear that W_1 is G_1 invariant, and we claim that it is also G_2 invariant. Indeed, $r_{(e,h)}$ commutes with all $r_{(g,e)}$ and so $r_{(e,h)} T \in \mathrm{Hom}_{G_1}(V_1, r_{(e,h)} U)$ if and only if $T \in \mathrm{Hom}_{G_1}(V_1, U)$. Thus W_1 is $G_1 \times G_2$ invariant, and since $V_1 \cap W_1 = \{0\}$, we conclude that $W_1 = \{0\}$. Thus all the V_i are equivalent. Let $U_2 = \mathrm{Hom}_{G_1}(V_1, W)$. By Schur's lemma we know that $\dim U_2 = k$, and by the above considerations we know that G_2 acts on U_2 by sending $T \sim \to r_{(e,h)} T$. We have the obvious evaluation map of $V_1 \otimes U_2 \to W$ sending $v \otimes T \sim \to Tv$, $v \in V_1$ and $T \in U_2$. The group G_1 acts on V_1 and the group G_2 acts on U_2 so that $G_1 \times G_2$ acts on $V_1 \otimes U_2$. It is easy to check that the evaluation map is equivariant with respect to the action of $G_1 \times G_2$ on both sides, and is non-trivial. It is therefore surjective, and the representation of G_2 on U_2 is irreducible, proving the proposition.

Let us now see where we stand in the proof of the Peter–Weyl theorem. It is easy to check that the functions $f_{u \otimes v^*}$ are precisely the matrix elements for the representation of G on $V \otimes V^*$, when we take u and v to be basis elements of V. Thus the irreducible subspaces of $L^2(G)$ consist precisely of the representative functions for the various irreducible representations, and thus we have proved (1). We have also proved (2). We have proved that each finite-dimensional irreducible representation occurs in $L^2(G)$ with a multiplicity equal to its dimension, which is the assertion or (4), provided that we know (3), i.e. that each irreducible representation is finite dimensional. Since (3) is a consequence of (5), it suffices for us to prove (5).

Proof of (5) For any pair of continuous functions, f_1 and f_2, on G we define their convolution, $f_1 * f_2$ by the formula

$$(f_1 * f_2)(a) = \int_G f(ag^{-1}) f_2(g) \, \mathrm{d}g.$$

If r is any unitary representation of G and we set

$$R_f = \int_G f(a) r_a \, \mathrm{d}a$$

then

$$R_{f_1}R_{f_2} = \int_G\int_G f_1(a)f_2(b)r_{ab}\,\mathrm{d}a\,\mathrm{d}b = \int_G\int_G f_1(ab^{-1})f_2(b)\,\mathrm{d}b\,r_a\,\mathrm{d}a$$

and so

$$R_{f_1}R_{f_2} = R_{f_1*f_2}. \tag{E.10}$$

If r^1 and r^2 are inequivalent irreducible representations we have for the matrix elements

$$r_{ij}^1 * r_{kl}^2(g) = \Sigma r_{ij}^1(g)\int_G r_{ij}^1(b^{-1})r_{kl}^2(b)\,\mathrm{d}b = 0$$

so that

$$r_{ij}^1 * r_{kl}^2 = 0 \text{ if } r^1 \text{ and } r^2 \text{ are inequivalent irreducible finite-dimensional} \tag{E.11}$$
representations,

and a similar argument shows that

$$r_{ij} * r_{kl} = (1/n)\delta_{jk}r_{il} \text{ if } r \text{ is an irreducible representation of degree } n. \tag{E.12}$$

Taking the trace of (E.11) and (E.13) gives, for irreducible characters

$$\chi^i * \chi^j = \begin{matrix} 0 & \text{if} & \chi^i \neq \chi^j \\ (1/\chi^i(e))\chi^i & \text{if} & \chi^i = \chi^j \end{matrix} \tag{E.13}$$

where χ^j and χ^j are irreducible characters.

Combining (E.10) with (E.13) shows once again that

$$P_i = \chi^i(e)R_{\bar\chi}i$$

is a projection operator for any unitary representation, r. We first propose to show that the sum of all these projections, as χ^i ranges over all the finite-dimensional irreducible characters, is the identity operator. The following amounts to the same thing: we want to show that if v is a vector in the representation space such that $P_iv = 0$ for all i, then $v = 0$. Indeed, let us define the function f by

$$f(g) = \langle v, r_g v \rangle.$$

Then

$$\overline{f * \chi^i}(a) = \int_G \langle v, \chi^i(b)r_{ab^{-1}}v \rangle\,\mathrm{d}b = (1/\chi^i(e))\langle v, r_a P_i v \rangle = 0.$$

Now the function f is continuous, and hence has an expansion in terms of the (orthogonal system of functions given by) the matrix coefficients, r_{ij}^k, as r^k ranges over the irreducible representations of G: if we set

$$c_{ij}^k(f) = (f, r_{ij}^k) \tag{E.14}$$

then

$$f = \sum n_{\bar k}^{\frac{1}{2}}c_{ij}^k(f)r_{ij}^k, \quad \|f\|^2 = \sum_k n_k\left(\sum_{i,j}|c_{ij}^k(f)|^2\right). \tag{E.15}$$

This is true for any $f \in L^2(G)$ by virtue of that portion of the Peter–Weyl theorem that we have already proved. Now

$$\operatorname{tr} R_f^k R_f^{k*} = \sum_{i,j} \int_G f(a) r_{ij}^k(a) \, \mathrm{d}a \int \bar{f}(b) \bar{r}_{ij}(b) \, \mathrm{d}b$$

$$= \sum |c_{ij}^k(\bar{f})|^2$$

and thus, since $\| f \|^2 = \| \bar{f} \|^2$ we can rewrite the second equation in (E.15) as

$$\| f \|^2 = \sum n_k \operatorname{tr} R_f^k R_f^{k*}. \tag{E.16}$$

Now for any representation and any continuous function, f, we have

$$R_{f^*} = R_{\bar{f}}, \quad \text{where } \bar{f}(g) = \overline{f(g^{-1})} \tag{E.17}$$

as can be easily verified. We can thus rewrite (E.16) as

$$\| f \|^2 = \sum n_k \operatorname{tr} R_{\bar{f}*f}^k. \tag{E.18}$$

Equations (E.14)–(E.18) are valid for all functions f. Let us now apply (E.18) to the function f given by $f(g) = \langle v, r_g v \rangle$. We know that

$$f * \chi^{-k} = 0$$

and therefore

$$\bar{f} * f * \chi^k = 0.$$

But

$$(\bar{f} * f * \chi^k)(e) = \operatorname{tr} R_{\bar{f}*f}^k,$$

and we conclude from (E.18) that $f = 0$. Since the function f is continuous, we conclude that $f(e) = 0$, which is just the assertion $v = 0$.

We have thus proved that no v is orthogonal to all the spaces $P_i H$, where H is the Hilbert space of the representation and P_i ranges over all the projections associated with the finite-dimensional irreducible representations. To complete the proof of the Peter–Weyl theorem, it suffices to show that if $P_i v = v$, then v lies in a finite-dimensional invariant subspace. Let W be the space spanned by all the vectors $r_a v$, $a \in G$. We will show that W is finite dimensional, and, in fact, has dimensional at most n_i^2. For this it suffices to show that any collection of more than n_i^2 vectors of the form $r_{a_j} v$ must be linearly dependent, or that the matrix whose entries are $\langle r_{a_i} v, r_{a_j} v \rangle$ is singular. This will certainly be the case if we can show that functions f_j given by

$$f_j(g) = \langle v, r_{ga_j} v \rangle = f(ga_j), \quad f(b) = \langle v, r_b v \rangle$$

are linearly dependent. But,

$$n_i(f * \bar{\chi}^i)(b) = \langle v, r_b P_i v \rangle = f(b)$$

and so

$$f(ga) = n_i \int_G f(gab^{-1}) \chi_i(b) \, \mathrm{d}b$$

$$= n_i \int_G f(ba^{-1}g^{-1}) \chi_i(b^{-1}) \, \mathrm{d}b$$

$$= n_i \int_G f(b) \chi_i(b^{-1}ga) \, \mathrm{d}b$$

is a superposition of the matrix function $r^i_{jl}(g)$, and there are only n^2_i linearly independent such functions. This completes the proof of the Peter–Weyl theorem.

There are a number of immediate corollaries and minor improvements of the Peter–Weyl theorem which are worth recording. We recall that a function, f, is called a central function if it satisfies $f(ab) = f(ba)$, i.e. if it is constant on conjugacy classes. For a central function, f, the operator $C_f = \int_G f(a)r_{a^{-1}}\,da$ satisfies

$$C_f r_b = \int_G f(a)r_{a^{-1}b}\,da = \int_G f(ba)r_{a^{-1}}\,da = \int_G f(ab)r_{a^{-1}}\,da = r_b C_f.$$

By Schur's lemma, if r is irreducible then C_f is a scalar operator and

$$\operatorname{tr} C_f = (f, \chi), \text{ so } C_f = (1/\chi(e))(f, \chi)I.$$

Writing this out as an equation for the matrix entries with $\chi = \chi^k$ gives

$$(f, r^k_{ij}) = \frac{1}{n_k}\delta_{ij}(f, \chi^k).$$

If we now apply (E.14) and (E.15) we get

$$\text{if } f \text{ is a central function then } f = \sum_k (f, \chi^k)\chi^k. \tag{E.19}$$

The equality in (E.15) and (E.19) are in the sense of $L^2(G)$, the series converge in the norm of the Hilbert space. If f is a continuous function, it need not be true that the series converge in the sup norm. Nevertheless, the representative functions are dense in the sup norm: This is the content of the next proposition, which asserts that any continuous function can be uniformly approximated by representative functions. But, as we mentioned, the specific 'Fourier series' given by the right-hand side in (E.15) or (E.19) need not converge in the uniform (sup) norm.

Proposition E.4
Given any continuous function, f, on the compact group, G, and given any $\varepsilon > 0$, there exists a representative function, q, such that

$$\sup_{g \in G} |f(g) - q(g)| < \varepsilon.$$

Proof If h is any continuous function on G and $p \in V \otimes V^*$ is a representative function, then, letting r denote the regular representation of G on $L^2(G)$, we have

$$R_h p \in V \otimes V^*.$$

On the other hand, if v denotes any element of $L^2(G)$, we have

$$(R_h v)(a) = \int_G h(g)v(g^{-1}a)\,dg = \int_G h(ag^{-1})v(g)\,dg = h * v.$$

Suppose we choose some neighborhood, U, of e such that

$$|f(x) - f(y)| < \varepsilon \quad \text{for } xy^{-1} \in U.$$

Let us choose h to be a non-negative continuous function with support in U, and with $\int_G h(g)\,dg = 1$. Then it is easy to see that

$$|(R_h f)(a) - f(a)| = |(h * f)(a) - f(a)| < \frac{\varepsilon}{2}.$$

If v is continuous we have

$$|h * v(a)| \leqslant \|h\| \, \|v\|.$$

Now choose a representative function, p, such that

$$\|f - p\| \leqslant \frac{\varepsilon}{2\|h\|}.$$

Then

$$|f(a) - h * p(a)| \leqslant |f(a) - h * f(a)| + |h * f(a) - h * p(a)| \leqslant \frac{\varepsilon}{2} + \frac{\varepsilon}{2},$$

proving the proposition with $q = h * p$.

As an immediate corollary we obtain

Proposition E.5
A compact group is commutative if and only if all its irreducible representations are one dimensional.

Proof If G is compact, then every irreducible representation must be finite dimensional. If G is not commutative, we can find a and b in G with $ab \neq ba$, and hence a continuous function f, with $f(ab) \neq f(ba)$. If all irreducible representations were one dimensional, all representative functions would be central, so that f could not be approximated by representative functions.

Let us give some examples of non-finite compact groups. The most familiar example is the group $C_\infty = O^+(2)$ of rotations in the plane. Its elements consist of rotations through angle θ, where θ is determined mod 2π. In other words, the group is $\mathbb{R}/2\pi\mathbb{Z}$. If we use $\theta \pmod{2\pi}$ to parametrize the group then it is clear that the Haar measure is $(1/2\pi)\,d\theta$. The group is commutative so that all its irreducible representations are one dimensional. Any one-dimensional represention is given by a $\mathbb{C}$-valued continuous function, φ, satisfying $\varphi(\theta_1 + \theta_2) = \varphi(\theta_1)\varphi(\theta_2)$ and $\varphi(\theta + 2\pi) = \varphi(\theta)$. The only such functions are the exponential functions, χ^n:

$$\chi^n(\theta) = \exp in\theta.$$

The orthogonality relations between the characters reduce to the usual orthogonality relations between exponentials. The coefficients in (E.14) are the usual Fourier coefficients. The right-hand side of the first equation in (E.15) is the standard Fourier series expansion of f, and (E.15) asserts that the Fourier series converges in L^2, with the second assertion in (E.15) being the Plancharel formula for Fourier series. Proposition E.4 is then (a version of) the Weirstrass approximation theorem. The equation $\chi^m \cdot \chi^n = \chi^{m+n}$ tells us all there is to know about the tensor product of two representations.

As our second group, consider the orthogonal group in two dimensions, $O(2) = D_\infty$. It contains C_∞ as a normal commutative subgroup of index 2, and $O(2)$ is the semidirect product of C_∞ with $\mathbb{Z}_2$: let τ denote any reflection in $O(2)$, and let us denote rotation through angle θ by ρ_θ. Then every element of the group is either of the form ρ_θ (if it is a rotation) or of the form $\tau\rho_\theta$ (if it is a reflection), and we have the relations

$$\tau^2 = 1, \quad \tau\rho_\theta\tau = \rho_{-\theta}.$$

If f is any continuous function on $O(2)$ it is easy to check that

$$\int_G f(g)\,dg = \frac{1}{4\pi}\int_0^{2\pi} f(\rho_\theta)\,d\theta + \frac{1}{4\pi}\int_0^{2\pi} f(\tau\rho_\theta)\,d\theta$$

is invariant under left (and right) translations and hence gives the formula for the invariant integral. Let us construct irreducible representations of $O(2)$ by analogy with our construction of the irreducible representations of the finite dihedral groups. We get two one-dimensional representations coming from the representations of $\mathbb{Z}_2$, regarded as the quotient group $O(2)/O^+(2)$. These have characters

$$\varphi^1 : \varphi^1(g) = 1$$

and

$$\varphi^2 : \varphi(\rho_\theta) = 1, \quad \varphi(\tau\rho_\theta) = -1.$$

We also have the two-dimensional irreducible representations, r^k, $k = 1, 2, \ldots$, given by

$$r^k_{\rho_\theta} = \begin{pmatrix} e^{ik\theta} & 0 \\ 0 & e^{-ik\theta} \end{pmatrix}$$

$$r^k_{\tau\rho_\theta} = \begin{pmatrix} 0 & e^{-ik\theta} \\ e^{ik\theta} & 0 \end{pmatrix}$$

with character, χ^k, given by

$$\chi^k(\rho_\theta) = 2\cos k\theta \quad \chi^k(\tau\rho_\theta) = 0.$$

We claim that these are all the irreducible representations. To prove this it suffices to show that the characters listed above form an orthonormal basis for the space of central function in $L^2(G)$. Now if a function, f, is central, then $f(\rho_{-\theta}\tau\rho_\theta) = f(\tau\rho_{2\theta}) = f(\tau)$, so that f is a constant on elements of the form $\tau\rho_\theta$, while $f(\tau\rho_\theta\tau) = f(\rho_{-\theta})$ so that f is an even function of θ. By a linear combination of φ^1 and φ^2 we can obtain all the constant functions of $\tau\rho_\theta$ and the cosines form an orthonormal basis for the even functions, proving that we have indeed found all the irreducible representations of D_∞. We have

$$\varphi_1 \cdot \chi = \chi \quad \text{for any character, } \chi$$

$$\varphi_2 \cdot \varphi_2 = \varphi_1, \quad \varphi_2 \cdot \chi^k = \chi^k$$

and

$$\chi^k \cdot \chi^l = \chi^{k+l} + \chi^{k-l} \ (k \neq l), \quad \chi^k \cdot \chi^k = \chi^{2k} + \varphi_1 + \varphi_2,$$

since $4\cos k\theta \cos l\theta = 2\cos(k+l)\theta + 2\cos(k-l)\theta$. This gives us the decomposition of the tensor product of two representations.

An entirely different kind of compact, or rather totally bounded, group is provided by the theory of almost periodic functions, which plays an important role in celestial mechanics. Let G be any group (not necessarily carrying a topology). A function, f, on G is called *almost periodic* if the family of functions, f_g, on $G \times G$, where

$$f_g(a, b) = f(agb)$$

is totally bounded relative to the uniform norm on $G \times G$. The set of all almost periodic functions forms a vector space, and each almost periodic function induces a notion of distance on G by setting

$$d_f(a, b) = \sup_{g,h \in G} |f(gah) - f(gbh)|.$$

It is clear that

$$d_f(a, b) = d_f(aa', ba') = d_f(a'a, a'b) \quad \text{for any } a' \in G,$$

i.e. d_f is both right and left invariant. Also

$$d_f(a_1 a_2, b_1 b_2) \leqslant d_f(a_1 a_2, a_1 b_2) + d_f(a_1 b_2, b_1 b_2) = d_f(a_2, b_2) + d_f(a_1, b_1),$$
$$d_f(a^{-1}, b^{-1}) = d_f(aa^{-1}b, ab^{-1}b) = d_f(b, a) = d_f(a, b)$$

and

$$|f(a) - f(b)| \leqslant d_f(a, b).$$

If $F = \{f_1, \ldots, f_n\}$ is any finite collection of almost periodic functions, and we set

$$d_F(a, b) = \max_i d_f(a, b),$$

then it is clear that d_F also satisfies the above relations, i.e.

$$d_F(a, b) = d_F(a'a, a'b) = d_F(aa', ba') \tag{E.20}$$

$$d_F(a_1 a_2, b_1 b_2) \leqslant d_F(a_1, b_1) + d_F(a_2, b_2) \tag{E.21}$$

$$d_F(a^{-1}, b^{-1}) = d_F(a, b) \tag{E.22}$$

and

$$|f(a) - f(b)| \leqslant d_F(a, b) \quad \text{if } f \in F. \tag{E.23}$$

If we let $H_F = \{a \mid d_F(a, e) = 0\}$ then it follows from (E.20) and (E.21) that H_F is a normal subgroup of G. Let G_F denote the quotient group, $G_F = G/H_F$. It follows from (E.20)–(E.22) that d_F induces a right and left invariant metric on G_F, which we shall continue to denote by d_F. It follows from (E.23) that any $f \in F$ induces a continuous function on G_F which we shall continue to denote by f. It is clear that if F and F' are two finite sets of almost periodic functions with $F \subset F'$ then

$$d_F(a, b) \leqslant d_{F'}(a, b)$$

and therefore $H_{F'} \subset H_F$ so that we get an induced map of $G_F \to G_{F'}$ which is continuous. If f is any almost periodic function, then the assertion that f_a is uniformly bounded means that for any $\varepsilon > 0$ we can find a finite set, $\{a_1, \ldots, a_k\}$ of elements if G such that

$$\min_i d_f(a, a_i) < \varepsilon \quad a \in G.$$

It follows that the same is true if we replace d_f by d_F, and thus the group G_F is uniformly bounded. (We could, if we like, complete it relative to the metric d_F and obtain a compact group, $\bar{G}_F$). We can now apply our theorem concerning the existence and uniqueness of Haar measure to conclude the following result due to von-Neumann.

Proposition E.6
Let A denote the space of almost periodic functions on the group G. There is a unique linear functional, μ, on A which satisfies

$$\mu(1) = 1$$

$$\mu(f) \geqslant 0 \quad \text{if } f \geqslant 0$$

and

$$\mu(f_g) = \mu(f), \quad \text{where } f_g(a) = f(g^{-1}a).$$

We can use the functional μ to introduce a scalar product, $(f_1, f_2) = \mu(f_1 \bar{f}_2)$ on the space of almost periodic functions, A, and complete it to get a Hilbert space, talk about almost periodic representative functions, apply the Peter–Weyl theorem and so on. We refer the reader to any standard modern text on almost periodic functions for the details. As an illustration of the scope of the results proved so far, we examine the classical case of almost periodic functions on the (commutative) group $\mathbb{R}$. For any real number, λ, the exponential function $\chi_\lambda(x) = \exp i\lambda x$ is clearly almost periodic, and satisfies

$$(\chi_\lambda)_y(x) = (\exp - i\lambda y)\chi_\lambda(x)$$

Thus $\mu(\chi_\lambda) = (\exp i\lambda y)\mu(x_\lambda)$ and therefore

$$\mu(\chi_\lambda) = \begin{matrix} 0 & \text{if } \lambda \neq 0 \\ 1 & \text{if } \lambda = 0 \end{matrix}$$

Notice that this same set of equalities holds for

$$v(\varphi) = \lim_{T \to \infty} \frac{1}{T} \int_0^T \varphi(t)\, dt$$

with $\phi = \chi_\lambda$.

Now let f be any almost periodic function on $\mathbb{R}$, and let $F = \{f\}$. It is then clear that the $d_f(a, b) = 0$ if and only if $f_a = f_b$, where $f_a(x) = f_b(x)$ and so we can identify the group G_F with the set of all f_a and the map of $\mathbb{R} \to \mathbb{R}_F$ sends a into f_a. The group $\mathbb{R}_F$ is commutative, and each (uniformly) continuous character on $\mathbb{R}_F$ induces a continuous character on $\mathbb{R}$ which must then be of the form χ_λ. Now suppose that the function f, in addition to being almost periodic, is also continuous as a function on $\mathbb{R}$. Then the function on $\mathbb{R}_F$ sending f_a into $f_a(0) = f(a)$ is a continuous function on $\mathbb{R}_F$. Hence, by Proposition E.4, we can find, for any $\varepsilon > 0$, characters $\chi_{\lambda_1}, \ldots, \chi_{\lambda_k}$ and constants

$c_1, \ldots, c_k$ so that

$$\sup_{a \in \mathbb{R}} | f(a) - \sum c_j \exp i\lambda_j a | < \varepsilon.$$

This is the content of the celebrated 'Bohr approximation theorem' which asserts that any continuous almost periodic function on $\mathbb{R}$ can be uniformly approximated by a linear combination of exponentials. Notice, among other things, that this implies that $v(f)$ converges for any continuous almost periodic function and that, since $v = \mu$ for exponentials,

$$\mu(f) = \lim \frac{1}{T} \int_0^T f(t) \, dt$$

for any continuous almost periodic function. For any λ we can form the scalar product

$$a_f(\lambda) = (f, \chi_\lambda) = \lim \frac{1}{T} \int_0^T f(t)(\exp - i\lambda t) \, dt$$

and, by the second equation in (E.15), we know that

$$\lim 1/T \int_0^T |f(t)|^2 \, dt = \sum_\lambda |a_f(\lambda)|^2.$$

In particular, there are only countably many values of λ for which the 'Fourier coefficients' $a_f(\lambda)$ do not vanish, and these λ's are called the 'frequencies' of the almost periodic function, f. The series $\sum a_f(\lambda_j) \chi_{\lambda_j}$ converges to f in the L^2 sense (relative to the mean value, μ) by (E.15). It is not difficult to show that in the Bohr approximation theorem we need only use those λ which are frequencies of f, but, of course, the actual 'Fourier series' need not converge. We leave the details to the reader.

We now return to the general theory of compact groups and close this section with a brief discussion of homogeneous vector bundles and induced representations.

Let G be a topological group and M a topological space. When we speak of an action of G on M, we now demand the additional requirement that the map $G \times M \to M$ be continuous. In the definition of a vector bundle, E over a topological space, M, we require, in addition to the previous requirements, that E be a topological space and that the projection $\pi : E \to M$ be continuous. We can let $C(E)$ denote the space of continuous sections of E and let $C_0(E)$ denote the space of continuous sections of compact support. It is clear that any $g \in G$ acts as a linear transformation on $C(E)$ and $C_0(E)$, sending the section s, into the section gs, where, as usual, $(gs)(x) = g(s(x))$. It is also clear that $(g_1 g_2)s = g_1(g_2 s)$. In the main we will be interested in the situation where the fibers, E_x, are finite-dimensional spaces, although much of what we shall say is valid in greater generality.

Let H be a closed subgroup of G. Then G/H is a topological space on which G acts transitively. If r is a representation of H on some finite-dimensional vector space, V, then we can form the topological space $E = G \underset{H}{\times} V$. As before, E consists of equivalence classes of pairs (g, v), where $(gh, v) = (g, r_h v)$. The topology on E is the quotient topology inherited from the product topology on $G \times V$. A section $s : G/H \to E$ gives a function,

f_s, from G to V by the equation $s(gH) = [(g, f_s(g))]$, and f_s satisfies the identity

$$f(gh) = r_h f(g) \quad \text{for any } h \in H. \tag{E.24}$$

Conversely, any function from G to V which satisfies (E.24) clearly determines a section of E over G/H. It is clear that s is continuous if and only if f_s is continuous.

Let E be a continuous vector bundle over a topological space, M, and suppose that each of the fibers, E_x, is a Hilbert space under a scalar product. $\langle \ , \ \rangle_x$, with the property that for any two continuous sections, s_1 and s_2, the function on M given by $x \rightsquigarrow \langle s_1(x), s_2(x) \rangle_x$ is continuous. We then say that E is a Hermitian vector bundle. If a group G acts on E so that the map $g : E_x \to E_{gx}$ is unitary for all g and x, then we say that E is a homogeneous Hermitian vector bundle for G. Suppose that E is a Hermitian vector bundle over M and that we have a notion of integration over M, i.e. we are given a measure on M. Then we can introduce a scalar product on $C_0(E)$ by setting

$$(s_1, s_2) = \int_M \langle s_1(x), s_2(x) \rangle_x \, dx$$

The completion of the space $C_0(E)$ relative to this scalar product is called $L^2(E)$. If E is a homogeneous Hermitian vector bundle for G and G preserves the measure on M then we clearly get a unitary representation of G on $L^2(E)$. An analysis of the representation of G on $L^2(E)$ is somewhat complicated, although most of the theorems concerning the Mackey decomposition can be reformulated so as to hold in this context. In fact, they were originally proved by Mackey in even greater generality. We shall not go into this point here. Even if the group G is compact, the situation is a bit complicated, because the action of G on $L^2(E)$ need not be completely continuous. However, if G acts transitively on M, i.e. if $M = G/H$, where H is a closed subgroup of the compact group, G, and E is the vector bundle induced from some unitary representation of H on a Hilbert space V, the situation is quite simple. First of all, there is a well defined integral on M which is invariant under G. Indeed, any function, f, on M gives rise to a function, $\hat{f}$, on G defined by

$$\hat{f}(g) = f(gH),$$

and we set

$$\int_M f(x) dx = \int_G \hat{f}(g) dg.$$

If s_1 and s_2 are continuous sections of E and we take

$$f(x) = \langle s_1(x), s_2(x) \rangle_x,$$

then it is easy to see that

$$\hat{f}(g) = \langle f_{s_1}(g), f_{s_2}(g) \rangle$$

where the scalar product on the right is the scalar product on V, and where, as above, f_s denotes the V-valued function on G corresponding to the section s. We can thus regard $L^2(E)$ as a subspace of $L^2(G, V)$, where $L^2(G, V)$ is the L^2 completion of the space of V-

valued functions on G, namely $L^2(E)$ is the subspace consisting of those $\varphi \in L^2(G)$ which satisfy

$$\varphi(gh) = r_h\varphi(g) \quad \text{for } h \in H.$$

It then follows directly that the representation of G on $L^2(E)$ is completely reducible, so that Proposition E.2 applies. We conclude that the induced representation, i.e. the representation of G on $L^2(E)$, decomposes into a direct sum of finite-dimensional irreducible subspaces, each occurring with finite multiplicity. In Appendix G we shall show how to introduce the notion of a character for the representation of G on $L^2(E)$, and we will be able to conclude that the Frobenius reciprocity theorem holds for induced representations of compact groups. (Actually, in Appendix G, we will be discussing a more restrictive class of groups – the Lie groups, to be defined below. However, an examination of our argument will show it to be valid in the more general case considered here.) A direct proof of the Frobenius reciprocity theorem is also quite easy, but we will not present it here.

Appendix F

A HISTORY OF
19TH CENTURY SPECTROSCOPY

The fact that metals glow when heated, and that the color is indicative of the temperature ('red hot', 'white hot') has undoubtedly been known from the earliest period of metal working. The fact that a flame becomes intensely yellow when table salt is added to the burning substance, or that a sufficiently hot fire becomes green in the presence of copper has also doubtless been recognized since antiquity. During the first half of the 19th century, observations of the solar spectrum (Wollaston, 1802; Fraunhofer, 1814) and the spectra of flames and electric arcs (Fraunhofer, 1814; and the work of Brewster, Talbot, Wheatstone, Angstrom and others in the period 1822–1855) culminated in the principles of spectrum analysis as formulated and applied by Bunsen and Kirchhoff (1859–1860). These principles seized the imagination of the scientific world because they revealed a method of investigating the chemical nature of substances independently of their distances from the laboratory: a new science was thus created, inasmuch as chemical analysis could be applied to the sun and other stellar bodies. But the beautiful simplicity of the first experiments, pointing apparently to the conclusion that each element had its characteristic and invariable spectrum, whether in the free state or in chemical combination, was soon found to be afflicted by complications. A bewildering variety of data accumulated on the spectra of elements and of compounds. In 1882 Schuster formulated the extraordinarily prescient suggestion that the main function of the study of spectra would be to obtain information about the structure of atoms and molecules and the nature of the forces that bind them together, and coined the name 'spectroscopy' for this new science. However, it was only with Balmer's formula (1885) for the hydrogen atomic spectrum and its extension by Rydberg and others (1890–1897) to the spectra of other elements that some order was introduced into the knowledge of atomic spectra. Bohr's model of the atom (1913) using Planck's quantum theory began the development of the understanding of atomic structure and atomic spectra, which culminated in quantum mechanical explanation (1926–1930). The theory of molecular spectra was being developed at the same time by Wigner and others. Wigner and Weyl emphasized the role of group theory in atomic spectra, while Wigner's fundamental paper on group theory and molecular structure (1930) laid the cornerstone for the use of molecular spectra in the determination of molecular symmetry and structure. Thus, Schuster's objectives were fulfilled. We now sketch the story in somewhat greater detail.

In a sense, our story can be said to begin with Newton's famous experiments on color.

In 1666, while Newton was a scholar at Trinity College, Cambridge, he obtained a triangular prism

> 'to try therewith the celebrated Phaenomena of Colours. ... having darkened my chamber, and made a small hole in my window shuts, to let in a convenient quantity of the Sun's light, I placed my Prisme at his entrance, that it might be thereby refracted to the opposite wall. It was at first a very pleasing divertisement, to view the vivid and intense colours produced thereby; but after a while applying myself to consider them more circumspectly, I became surprised to see them in *oblong* form, which according to the received laws of Refraction, I expected should have been *circular*.'

He was forced by the plague to leave Cambridge, and presented his memoir on the subject of color to the Royal Society in 1671. In it he writes

> 'Colours are not *Qualifications of light* derived from Refractions, of Reflections of natural Bodies (as 'tis generally believed) but *Original and connate properties* which in divers rays are divers. Some rays are disposed to exhibit a red colour and no other: some a yellow and no other, some a green and no other, and so of the rest. Nor are there only Rays proper and particular to the more eminent colours, but even to all the intermediate gradations. To the same degree of Refrangibility ever belongs the same colour, and to the same colour ever belongs the same degree of Refrangibility.
> The species of colour, and degree of Refrangibility proper to any particular sort of Rays, is not mutable by Refraction, nor by Reflection from natural bodies, nor by any other cause that I could yet observe. When any one sort of Rays hath been well parted from those of other kinds, it hath afterwards obstinately retained its colour, notwithstanding my utmost endeavours to change it.'

Thus, according to Newton, white light is in some sense a mixture of light of various colors, these colors being intrinsic properties of the rays themselves. Newton offered various hypotheses as to the nature of light, relating it both to corpuscles and to vibrations of the aether. He writes, in 1672, that the corpuscles corresponding to different colors, would, like sonorous bodies of different pitch, excite different types of vibrations of the aether, and if, by any means, the aether vibrations are separated from one another,

> 'the largest beget a Sensation of a *Red* colour, the least or shortest of a deep *Violet*, and the intermediate ones of intermediate colours; much after the manner that bodies, according to their several sizes, shapes and motions, excite vibrations in the Air of various bignesses, which according to those bignesses, make several Tones in Sound.'

Thus, Newton not only considers color as an intrinsic property of the light rays, he relates color to periodic phenomena associated with these rays.

The band of colors that appeared after the white light had been split up by a prism Newton called a *spectrum*.

The theory of light was greatly advanced during the next century and a half, principally by the work of Huygens, Young, Fresnel and Hamilton. It is not our purpose to describe these advances here. The next event in our story is a discovery by the English scientist, William H. Wollaston, in 1802. Wollaston (1766–1828) started out as a physician, but was unsucccessful in his medical career, and turned to original research. He was somewhat of a recluse, living alone and conducting most of his researches in the utmost secrecy. He was, however, active in the affairs of the Royal Society, being elected a Fellow in 1793 and made Secretary in 1806. His main research was in chemistry, especially of the metals associated with platinum. He apparently made a fortune out of a secret process for working platinum on a commercial scale. He did original work in many diverse fields, such as optics, electricity, acoustics, mineralogy, astronomy, physiology, botany and the theory of art. For instance, he established the identity of the currents of frictional electricity and those of Volta, by showing that they have the same chemical effect of decomposing metallic salts, asserting that voltaic electricity is 'less intense but produced in much larger quantity'. He noticed the existence of an oscillating current in certain electrical discharges. In 1821, after Oersted had shown that a magnetic needle is deflected by an electric current, he attempted, in the laboratory of the Royal Institution, in the presence of Humphry Davy, to convert that deflection into a continuous rotation, and also to obtain the reciprocal effect of the magnet on the current. He failed in both attempts, but Faraday, who overheard a portion of his conversation with Davy, succeeded, and this got Wollaston into an unsuccessful priority fight with Faraday.

In 1802, Wollaston was interested in investigating the number of 'primary colors' to be found in the solar spectrum. Wollaston found that when daylight was viewed through a glass prism held close to the eye, only four colors were visible: red, yellowish green, blue and violet. The boundaries of these colors were marked by certain dark lines. However, in addition to the dark lines 'marking the boundary' between colors, other distinct dark lines were found, seven in all. Wollaston's principal interest was in the spectral colors, and not the dark lines, which did not attract much attention. Wollaston also repeated the same experiment using candlelight and 'electric light'. The blue light of the lower part of the candle split into five images. The first was broad and red, terminating in a bright yellow line, the next two were green, and the last two blue. Again, Wollaston's observations on the flame spectra did not seem to elicit much response.

In this same article, Wollaston mentions the discovery that silver chloride is blackened not only by visible light, but also in the region beyond the violet, indicating the presence of ultra-violet rays. (This discovery was made independently by the German scientist J.W. Ritter (1776–1810) in 1801, but not published until 1803.) In 1800, the distinguished astronomer, Sir William Herschel, had discovered that there was a steady increase in the sun's heating power from the violet to the red end of the spectrum, and that the effects were even more intense beyond the visible red end. Thus the infra-red rays were also discovered at the beginning of the 19th century. The question of whether the infra-red ray, the so-called 'radiant heat', was of the same nature of light remained a matter of some controversy until the work of Kirchhoff towards the middle of the century and, finally, Maxwell's electromagnetic theory in the last quarter of the century explained visible light as only a small portion of a continuous range of frequencies of electromagnetic radiation. (Herschel himself originally accepted the idea that 'radiant heat' was a kind of light, but later rejected it.)

For 12 years, Wollaston's observations on the dark lines in the solar spectrum and the nature of flame spectra did not attract much attention. They were rediscovered, with much greater precision, by the German optician and physicist, Joseph von Fraunhofer (1787–1826) and published in 1814. Fraunhofer, the son of a glazier, was apprenticed at the age of 12 to a glass polisher and looking-glass maker. Two years later, he nearly lost his life when the house in which he was living collapsed. Maximilian Joseph, Elector (later King) of Bavaria, was present when Fraunhofer was extricated from the ruins, and gave him 18 ducats. He used part of this money to release himself from the last six months of his apprenticeship, and with the rest he bought a glass polishing machine. He then employed himself in the making of eyeglasses and metal engraving, devoting his spare time to the study of mathematics and optics. In 1809 he founded, with two partners, an optical institute, of which he became sole proprietor in 1818. He was responsible for many inventions in the field of optics, and his institute was manufacturing the most refined optical instruments of the period. In the course of his work, Fraunhofer was interested in finding a source of light of a 'pure' color, in order to be able to measure accurately the dispersive powers of various glasses. He found that colored flames yielded complex spectra when passed through a prism, but that all flames possessed in common a bright band in the yellow–orange region. He then decided to study sunlight, to see if the same bright band was present in its spectrum. Much to his surprise, instead of finding bright bands, he found that the spectrum of sunlight consisted of a continuous bright region traversed by many dark lines. In particular, he discovered that the bright band of the flame spectra consisted of two close bright lines, and that these seemed to coincide with two particular dark lines in the solar spectrum. Fraunhofer constructed a map of some 576 lines of varying degrees of darkness in the sun's spectrum, and labeled the more prominent of them by the alphabetic letters *A* to *H*. The double lines which apparently coincided with the bright lines of flames he labeled *D*. He then examined the spectra of some of the fixed stars, and discovered that in some of them many dark lines appeared that were absent from the solar spectrum, while dark lines which occurred in the sun's spectrum sometimes did not appear in the spectrum of a fixed star. This proved that the dark lines were not due to the earth's atmosphere, but were rather due to the sun (or star) itself. The question immediately arose as to the nature of the dark lines. Either for some strange reason the sun is producing radiation in all frequencies except those corresponding to the dark lines, or the sun is producing radiation of all frequencies, part of which is absorbed at the frequencies of the dark lines. Indeed, early on, Herschel suggested that the lines are absorption lines, saying that 'it is no impossible supposition that the deficient rays in the light of the sun and stars may be absorbed in passing through their own atmospheres'.

Indeed, Sir David Brewster (1781–1868), reporting to the British Association in 1831, describes experiments in which he passed lamplight through nitrous oxide gas, and discovered that the spectrum was crossed by hundreds of dark lines and bands. Thus, the phenomenon of absorption of light was actually observed in the laboratory. Brewster was aware of the importance of his experiment, and describes it as

'extending so widely the reasources of the practical optician and lying so close to the roots of atomical science, that I am persuaded it will open up a field of research, which will exhaust the labours of philosophers for centuries to come'.

On the other hand, Brewster was a strong opponent of the wave theory of light, and used the phenomenon of absorption as a strong argument against the wave theory. Arguing by analogy with the propagation of sound, he writes, in 1833,

'we can scarcely conceive an elastic medium so singularly constituted as to exhibit such extraordinary effects. We might readily understand how a medium could transmit sounds of a low pitch: but it is incomprehensible how any medium could transmit two sounds of nearly adjacent pitches, and yet obstruct a sound of intermediate pitch'.

In fact, the explanation of the Fraunhofer lines as absorption by the solar atmosphere was disputed as a result of an examination of the solar eclipse of 1836. It was argued that if the sun's atmosphere is the cause of the dark lines, the phenomenon should be enhanced if we observe light reaching us from the sun's 'edges' instead of traversing the atmosphere directly. But no such difference was observed. Thus, the acceptance of the wave theory of light on the one hand, which seemed to defy the phenomenon of absorption (which was actually observed in the laboratory), and the absence of any eclipse effects on the other, called into question the explanation of the Fraunhofer lines as absorption lines. Indeed, the question was considered open for the next 20 years, until the work of Stokes, Foucault and Kirchhoff, to be described below.

Getting back to Fraunhofer, we should mention another major contribution that he made, and that is to the theory of diffraction and the construction of the first diffraction gratings, around 1823. He gave a theoretical explanation of diffraction in terms of the wave theory of light, and indicated how diffraction gratings could be used to measure the wavelength of light. He realized that it would now be possible to formulate spectral information in terms of wavelengths, so providing an absolute standard, instead of using prism angles where the angles quoted would depend on the nature of the equipment used. Fraunhofer also constructed the first diffraction gratings, first by using wires wrapped around a flat frame, and later by ruling various plated surfaces. However, it was not until the last third of the century that diffraction gratings of sufficient precision were developed and used in the description of spectra. Even

Kirchhoff, in his classical investigations with Bunsen, and in his map of the solar spectrum, uses prismatic settings. It was Angstrom, in 1868, who gave the first systematic spectral measurements with a diffraction grating.

During the first half of the century, the idea that the flame spectra of various chemicals might be used to detect the presence of these chemicals was gradually being developed. The leading proponent of this idea was William Henry Fox Talbot (1800–77), the British mathematician who is perhaps best known for his work in the development of photography. Fox Talbot had discovered some of the essentials of the photographic process simultaneously with, or perhaps shortly earlier than, Daguerre. He devised the so-called calotype or talbotype reproducing process, and, after the discovery of the collodon process by Archer in 1851, devised a method of instantaneous photography. At one time he held practically all of the important patent rights in photography in England. (There is some dispute among historians whether, as a result, he did not do more harm than good to the progress of photography for a period of 20 years.) Fox Talbot was also very active in archaeology, and played a key role in the decipherment of Akkadian cuneiform. (He worked at the British Museum with Norris, the decipherer of Elamite and the Secretary of the Royal Asiatic Society. In one famous episode, he suggested that Norris give the same text to several Assyriologists to decipher without revealing that it was the identical text. A copy of a recently discovered cuneiform inscription stemming from three clay cylinders of the Assyrian King, Tiglath-Pileser I (1113–1074 BCE) were sent to four leading Assyriologists, Oppert, Hincks, Rawlinson and Fox Talbot, who were asked to translate it and send their results to the Society. The fact that all four translations agreed on most essential points led to public confidence in the science of Assyriology.)

In 1834 Fox Talbot suggests that one could apply spectrum analysis in the observation of flames to discern the presence of various elements. Thus, while by visual observation one could not distinguish the red flames of lithium from those of strontium, by use of a prism,

> 'The strontia flame exhibits a great number of red rays well separated from each other by dark intervals, not to mention an orange, and a very definite bright blue ray. The lithia exhibits one single red ray. Hence, I hesitate not to say that optical analysis can distinguish the minutest portions of those two substances from each other with as much certainty, if not more than any other known method'.

Two years later he writes that

> '... an extensive course of experiments should be made on the spectra of chemical flames, accompanied with accurate measurements of the bright and dark lines... The definite rays emitted by certain substances as, for example, the yellow rays of the salts of soda, possess a fixed and invariable character, which is analogous in some measures to the fixed proportion in which all bodies combine according to the atomic theory. It may be expected, therefore, that optical researches, carefully conducted, may throw some additional light upon chemistry.'

Fox Talbot's program was hindered for the next 25 years by the fact that the double *D* line in the yellow seemed to be present in all flame spectra. This difficulty was overcome by the Scottish chemist. William Swan, in 1857. Swan made use of the 'colorless flame' which had just been invented by Bunsen. Swan describes his experiments as follows:

> 'One tenth of a grain of common salt, carefully weighed in a balance indicating 1/100 of a grain, was dissolved in 5000 grains of distilled water. Two perfectly similar slips of platinum foil were then carefully ignited by the Bunsen lamp, until they nearly ceased to tinge the flame with yellow light; for to obtain the total absence of yellow light is apparently impossible. One of the slips was dipped into the solution of salt, and the other into distilled water, the quantity of the solution of salt adhering to the slip being considerably less than 1/20 grain, and both slips were held over the lamp until the water had evaporated. They were then simultaneously introduced into the opposite sides of the flame; when the slip which had been dipped into the solution of salt, invariably communicated to a considerable portion of the flame a bright yellow light, easily distinguishable from that caused by the slip which had been dipped into pure water. It is thus proved that a portion of chloride of sodium, weighing less than 1/1,000,000 of a grain is able to tinge a flame with bright yellow light.'

Thus the bright yellow lines were indicative of sodium salt impurities. For pure substances, the flame spectra should be characteristic of the materials, and the use of spectrum analysis could provide a tool of fantastic sensitivity to detect the presence of various metals.

The full development of the necessary experimental techniques was achieved two years later by Bunsen and Kirchhoff in their classic collaboration. Bunsen and Kirchhoff prepared salts of extraordinary purity by elaborate chemical techniques. They examined the flame spectra of the chlorides, bromides, iodides, hydrated oxides, sulphates and carbonates of potassium, sodium, lithium, strontium, calcium and barium. They concluded that the observed spectrum was not only characteristic of the chemical being investigated, but was actually characteristic of the *metal*. Thus, a metal gives the same characteristic spectrum no matter how it was combined chemically. They wrote that

> 'The different bodies with which the metals employed were combined, the variety of the chemical processes occurring in several, and the wide differences of temperature which these flames exhibit, produce no effect upon the position of the bright lines in the spectrum which are characteristic of each metal.'

Thus, spectrum analysis could be used to detect the presence of metallic elements! In fact, by the use of spectrum analysis, Bunsen and Kirchhoff discovered two new metals: caesium and rubidium. They noticed some unknown lines in the spectra of certain salts they were investigating. Bunsen then set out to obtain the substance or substances to which these were due. To this end he evaporated 40 tons of Durkheim mineral water to get about 17 grams of the mixed chlorides of the two substances. With about one-third of that quantity of caesium chloride he was able to prepare the most important compounds of the element and determine their characteristics, even making goniometric measurement of their crystals.

The possibility of the discovery of new elements captured the chemical imagination, and set many chemists on the road for the search of further elements, in the hope of achieving personal fame. This led to a rapid development of techniques and experiments leading very shortly to complications in the idea that spectra are characteristic of only the elements. But before describing these developments we should discuss an even more striking idea due to Kirchhoff, and that is his explanation of the Fraunhofer lines, and the consequent possible use of spectrum analysis to study the chemistry of the stars. This part of the story starts a few years earlier, some time prior to the summer of 1852, in a conversation that George Stokes had with William Thomson (Kelvin) at Cambridge. Stokes discussed the coincidence of the double dark *D* line in the solar spectrum with the bright yellow double line in many flames, and the fact that when salt is thrown onto an alcohol flame the flame turns bright yellow and consists almost solely of these two double lines. Stokes felt that light from any source was due, in the final analysis, to a 'vibratory movement' among the 'ultimate molecules' of the luminous body. Many years later, Stokes recounts the discussion:

> 'In conversation with Thomson I explained the connection of the bright and dark lines by the analogy of a set of piano strings tuned to the same note, which if struck would give out that note, and would also be ready to sound out, to take it up in fact, if it were sounded in air. This would imply absorbtion of the aerial vibrations. I told Thomson I believed there was vapour of sodium in the sun's atmosphere.'

Thomson was enthusiastic about the idea and began teaching it in his classes as early as the academic year 1852–3. In a letter to Thomson in February, 1854, Stokes mentions his own observations that the bright *D* lines were absent from a candle flame when the wick was snuffed clean so as not to project into the luminous envelope, and from an alcohol flame when the spirit was burned in a watch glass; which indicated that the substance responsible for the *D* lines was contained not in the alcohol but in the wick. Furthermore, he writes that he does not know of any other pure substance which gave these lines and that it

> 'would be extremely difficult to prove, except in the case of gases or substances volatile at a not very high temperature, that the bright line *D*, if observed in a flame, was not due to soda, such an infinitesimal quantity of soda would be competent to produce it.'

Thus Stokes anticipated the experimental discovery of Swan. But Stokes was not aware of any experiment which justified his explanation. He was not aware of any vapor whose absorption lines coincide with the *D* lines of sodium. In 1855 Stokes had dinner with Leon Foucault, who had come to London to receive the Copley medal from the Royal Society. At dinner, Foucault told Stokes of an experiment he had performed six years earlier. The results had been announced, but had remained relatively unnoticed. Foucault had wondered whether the double yellow line in the carbon electric arc emission spectrum coincided with the solar *D* lines. As he did not have an instrument to measure the wavelengths, he decided to pass the sunlight through the arc itself, in order to view the spectra superimposed on one another. Not only were the lines exactly coincident, but the dark solar lines appeared even darker when passed through the arc. Thus, Foucault concluded,

> 'the arc offers a medium which emits, on its own account, the *D* rays, and which, at the same time, absorbs them when these come from somewhere else.'

Thus, Stokes had experimental confirmation of his idea that a substance would tend to absorb radiation at exactly the same frequencies at which it tends to emit radiation.

Kirchhoff was led to exactly the same conclusion by an independent investigation and by arguments which were more general and based on thermodynamic considerations. At the turn of the century, Pierre Prevost (1751–1839), a Swiss physicist, remarked that a red-hot body becomes cooler by emitting radiation. Suppose we imagine an arrangement of a number of bodies so placed that they entirely intercept each other's radiation. If such a system achieves a steady state in which all the bodies remain at some definite temperature, each body will continue to radiate as if the others were not present, but each body must obviously receive as much energy as it radiates. This became known as 'Prevost's law of exchanges'. In 1858, Balfour Stewart observed that a plate of rock salt is much less diathermous for rays of radiant heat emitted by heated rock salt than it is for rays emitted by other hot substances. This led him to formulate the conclusion that the radiating power of any substance was equal to its absorbing power for every kind of radiant heat. In 1859, Kirchhoff gave an independent proof of this assertion, extending to all radiation, identifying radiant heat and visible radiation as different forms of the same phenomenon. The gist of Kirchhoff's argument is as follows. Let us imagine a hollow box whose walls are maintained at a constant temperature. The radiation inside it is in thermal equilibrium with the walls. If we imagine the radiation as consisting of some kind of dynamical system to which the laws of thermodynamics apply, we can talk of the energy present in the radiation and of the temperature of the radiation; and we would assign to it the same temperature as the temperature of the walls. This is true for any small element of volume in the cavity, so that we have homogeneous radiation throughout the cavity. The state within the cavity consists of radiation in equilibrium, depending solely on the temperature and independent of the particular physical or chemical processes giving rise to emission or absorption at the walls. We can consider radiation of various frequencies, v, and discuss the amount of energy present in radiation of frequency v per unit volume, at a given temperature, T. The first main assertion is that this energy is function of v and T, but independent of the nature of the walls of the cavity. Indeed, suppose that we have two hollow boxes, A and B, whose walls are different, and are maintained at the same temperature. Suppose that A has more energy than B at frequencies near v. We could then connect A and B by a small tube containing a color filter which is transparent to frequencies near v, but opaque to all other frequencies. Since the energy (and hence the intensity) of the radiation is greater in A than in B, radiation would flow from A to B, causing A to cool off and B to heat up, causing a temperature difference between A and B to arise without any work being done on the system, contradicting the second law of thermodynamics. We thus obtain a universal function, $F(v, T)$, representing the amount of steady state energy present in radiation at frequency v and temperature T. Now consider a wall made of some substance which absorbs radiation at a certain rate, i.e. let $a = a(v, T)$ be the fraction of the radiation absorbed (per unit area and unit time in homogeneous radiation) so that $aF(v, T)$ represents the amount of energy absorbed and converted into heat. On the other hand, the substance is also emitting radiation at frequency v at a certain rate, $e(v, T)$. In order for equilibrium to be maintained, the energy emitted must equal the amount absorbed, so that we must have

$$\frac{e}{a} = F(v, T).$$

This is the law of Balfour Stewart and Kirchhoff. In particular, if we have a perfectly 'black' body, one which absorbs all the radiation impinging on it so that $a = 1$, its emissivity is given by $F(v, T)$. A black body can be constructed as follows. Suppose we make a box and put a very small hole in it. Any light entering the hole will be absorbed by the walls, possibly after many reflections. One can thus hope to empirically measure the function F. Starting with the emissivity, F, of a perfectly black body, one can also deduce the law of Balfour Stewart and Kirchhoff as follows. Suppose we consider the exchange of energy between a surface element of some body A and a perfectly black body B. We may assume that the sizes and geometrical arrangements are such that F units of radiation pass per unit time from B to A. The energy passing from A to the black body will consist of two parts: the emitted radiation, e, and the reflected part, r. Since the bodies are in equilibrium, we must have $F = e + r$. If we consider a system in which all the rays are traversed in the opposite direction, the r is the part of the radiation that is reflected at A to fall somewhere on the black body, so that $r = F - aF$, where a is the fraction absorbed. Thus $F = e + F - aF$, or $e/a = F$, which is the law of Balfour Stewart and Kirchhoff. Kirchhoff was immediately aware of the significance of his result for deducing the presence of various elements in the atmospheres of the sun and the stars. He describes an experiment that he performed, similar to the one performed by Foucault and described above, and goes on to conclude that

> '.. the dark lines of the solar spectrum which are not evoked by the atmosphere of the earth, exist in consequence of the presence in the incandescent atmosphere of the sun, of those substances which in the spectrum of a flame produce bright lines at the same place. We may assume that the bright lines agreeing with the D in the spectrum of a flame always arise from sodium contained in it; the dark lines D in the solar spectrum allow us, therefore, to conclude that there exists sodium in the sun's atmosphere. Brewster has found bright lines in the spectrum of saltpeter at the place of Fraunhofer's lines A, a, B; these lines point to the existence of potassium in the sun's atmosphere.

From my observation, according to which no dark lines in the solar spectrum answer to the red lines of lithium, it would follow with probability, that in the atmosphere of the sun, lithium is either absent or is present in comparatively small quantity.'

Thus, by rather different kinds of arguments, Stokes and Kirchhoff were led to the same conclusion – that the emission lines and the absorption lines of a given element coincide. It is perhaps worthwhile to get ahead of ourselves at this juncture to get a glimpse of what some of the consequences of these two approaches would be over the next half century of physics. Kirchhoff's analysis implied the existence of the function F, and it thus became a problem of the first magnitude to determine, both theoretically and experimentally, what the function F actually is. The first step was taken by Stefan who showed in 1879 that the total radiation per unit volume, that is $\int F(v, T)\,dv$ was empirically determined to be proportional to the fourth power of the temperature. Five years later this was established theoretically by Boltzmann, using thermodynamical arguments. In 1893, Wien, again using thermodynamic arguments, established that F has the form

$$F(v, T)\,dv = \lambda^{-5}\Phi(\lambda T)\,d\lambda,$$

where $\lambda = c/v$ is the wavelength, and Φ is a function of a single variable. In 1896 he established that Φ has the form $\Phi(x) = C\exp - b/x$, where C and b are constants, so that

$$F(v, T)\,dv = C\lambda^{-5}(\exp - b/\lambda T)\,d\lambda$$

and this formula was found to hold for high frequencies, i.e. for short wavelengths, but not for low frequencies, i.e. long wave lengths. On the other hand, using arguments of statistical mechanics, and using arguments far more basic and rigorous than Wien, Rayleigh, in 1900, established the formula

$$F(v, T)\,dv = 8\pi kT\lambda^{-4}\,d\lambda$$

where

$$k = 1.38 \times 10^{-16}\,\text{erg deg}^{-1}$$

is Boltzmann's constant. Rayleigh's law worked reasonably well for low frequencies, but not for large frequencies, where, instead of dying out, as predicted by Wien's law and as observed, it diverged to infinity according to the fourth power of v. (This failure of Rayleigh's law Ehrenfest called the 'ultra-violet catastrophe'.) In 1900, Planck found the correct law by a combination of electrodynamic and thermodynamic considerations, and by attempting to interpolate between the Wien and Rayleigh laws. He found that

$$F(v, T) = \frac{av^3}{(\exp lv/T) - 1}$$

where a and l are constants. This formula agrees with Wien's for high frequencies and with Rayleigh's for low frequencies. Later on in the same year, he gave a theoretical interpretation of the formula, based on the assumption that the walls of the black body are Hertzian oscillators of various frequencies, v, and each could emit radiation only in integer multiples of a basic unit

$$\varepsilon = hv.$$

His formula then becomes

$$F(v, T) = \frac{8\pi hv^3}{[(\exp hv/kT) - 1]c^3}$$

where c is the speed of light, k is Boltzmann's constant, and h is a new universal constant, now known as Planck's constant:

$$h = 6\cdot625 \times 10^{-27} \text{ erg s.}$$

Thus, Kirchhoff's arguments led in a fairly straight path (when combined with the ideas of statistical mechanics) to the development of the quantum theory.

Whereas Kirchhoff argued on general principles, Stokes had proposed a somewhat mechanical model for the identity of emission and absorbtion spectra of a substance. The emission and absorption of radiation are due to the 'vibrations' of the 'ultimate molecules' of the substance. Now Stokes was a leading British mathematician of the 19th century, and, along with Maxwell and Kelvin, one of the three leading mathematical physicists. He was well aware of the fact that when objects vibrate their frequencies of vibration will, in general, vary in a rather complicated way depending on the nature of the objects. However, in propounding his ideas to the public, he used the analogy of the vibrating string (probably since the phenomenon of a piano string taking up vibrations from the air was a well known effect). Vibrating *strings* have the property that their frequencies are all integer multiples, 'harmonics', of a given fundamental frequency. The result was that instead of searching for the mechanism of 'vibration' of the 'ultimate molecules' (which was impossible anyway granted the state of knowledge concerning the nature of the 'ultimate molecules') people searched for harmonic relationships in the spectra, trying to express the observed frequencies as integer multiples of certain basic ones. The development of this line of thought (based, after all, on a popularized misunderstanding of a wrong idea) also led, by remarkable twists and turns, to major discoveries as to the nature of atoms and molecules, as we shall now sketch as part of the general story.

Let us return to the 1860s. Although the central dogma in Kirchhoff and Bunsen's program of spectrum analysis was the notion that each metal exhibited its characteristic spectrum independent of the nature of any chemical combination in which it may be found, they were aware of the possibility that compounds, too, might exhibit characteristic spectral behavior. Thus, they mention that the absorption lines of nitrous oxide differ from those of nitrogen and of oxygen, and suggest that similar behavior might be found in metallic compounds. Shortly thereafter, characteristic spectra of some metallic salts were found by the German chemist, Mitscherlich, then by the French chemist, Diacon. Thus, in 1862, Mitscherlich found that barium chloride exhibits two bright bands, rather than the lines characteristic of barium. Thus band spectra were associated with compounds while line spectra were associated with elements. However, over the next few years, it became apparent that various elements, such as oxygen and nitrogen, also exhibit both band and line spectra. By 1875, the view began to be accepted that band spectra were associated with molecules and line spectra with atoms. Thus, the French chemist, Salet, writes

> 'As for the theory of double spectra, it is easy to imagine one. As heat has ordinarily the effect of simplifying chemical compounds and reducing them to their elements; as, on the other hand, elementary channelled spectra are observed at the lowest temperatures and are very similar to those of compound bodies, one can suppose that they are produced by aggregations of homogenous atoms. This assumption will appear natural above all to chemists, for whom for a long time chlorine has been the chloride of chlorine (Cl_2), sulphur the sulphide of sulphur (S_2) etc. But these molecules are themselves capable of being resolved into atoms. It is to this allotropy of elevated temperatures that the variation of spectra seems due.'

The trouble was that certain substances exhibited not merely a band spectrum and a line spectrum, but several spectra: different line spectra and sometimes continuous spectra as well. The key explanation for this fact, although not accepted by most of his contemporaries, was suggested by the astronomer, Sir Norman Lockyer. Apparently influenced by the current evolutionary theory in biology, he outlined a theory of evolution of the elements. According to him, the various observed spectra of one and the same substance were due to the decomposition of the substance (due to high temperature or electrical tension) into its constituent parts. (This is not far from the current view which associates the continuous spectra to disassociation of molecules into atoms or atoms into ions and electrons, and the various differing line spectra of a given element to ions.) Although the details of Lockyer's theory were not substantiated, the psychological groundwork was laid for the consideration of the possible existence of subatomic particles.

In the meanwhile, the search for numerical regularities in the spectrum of a given element, and the relationship between the spectra of similar elements, was begun. In 1871, Stoney suggested that it would be useful to plot the spectra in terms of wave numbers, i.e. reciprocal wavelengths, in order to find simple numerical relations between them. He was, of course influenced by the string analogy of Stokes, and thought that he had found 'harmonic' relationships in various spectra, i.e. that various lines in a spectrum have wave numbers which are small integer multiples of a 'fundamental' wave number, much as the frequencies of a vibrating string are whole number multiples of the fundamental. And, during the 1870s, a number of papers appeared on 'spectral harmonics', listing such kinds of relationships. In 1881, this whole line of activity was demolished by a careful study done by Arthur Schuster, who showed that, in the main, all the relationships that were exhibited were not more than could be expected from purely random data on the basis of pure chance. In 1882, Schuster wrote that, instead of looking for such whole integer relationships, one should hope to analyse the spectrum in order to be able to understand the mechanism that produces the lines; that the main role of the spectra in the future would not be to identify the presence or absence of various elements, but to obtain information about the very constitution of atoms and molecules. Thus the name of the subject should no longer be spectrum analysis but rather spectroscopy. He writes

> 'It is the ambitious object of spectroscopy to study the vibrations of atoms and molecules in order to obtain what information we can about the nature of the forces which bind them together. But we must not too soon expect the discovery of any grand and very general law, for the constitution of what we call a molecule is no doubt a very complicated one, and the difficulty of the problem is so great that were it not for the primary importance of the result which we may finally hope to obtain, all but the most sanguine might be well discouraged to engage in an enquiry which, even after many years of work, may turn out to have been fruitless. We know a great deal more about the forces which produce the vibrations of sound than about those which produce vibrations of light. To find out the different tunes sent out by a vibrating system is a problem which may or may not be solvable in certain special cases, but it would baffle the most skillful mathematician to solve the inverse problem and to find out the shape of a bell by means of the sounds which it is capable of giving out. And this is the problem which ultimately spectroscopy hopes to solve in the case of light. In the meantime we must welcome with delight even the smallest step in the desired direction.

(It should be remarked that the physical understanding of the structure of atoms and molecules on the basis of spectroscopy and other tools came quite rapidly in the years following Schuster's formulation of his program. The mathematical problem that he posed, that of determining the geometry of an object from its vibrations, is a mathematical problem which has only recently seen substantial progress towards its solution.)

In a series of fundamental contributions from 1879–83, the Cambridge scientists Liveing and Dewar began to notice some patterns in the spectra of the alkali metals in the visible and ultra-violet range. Certain lines seem to fall into natural groupings or 'series' according to appearance and wave number patterns. They write

> 'We have already, in describing the visible spectra of the alkali metals and that of magnesium, called attention to probable harmonic relations between the lines. This relation manifests itself in three ways – first, by the repetition of similar groups of lines, secondly by a law of sequence in distance, producing a diminishing distance between successive repetitions of the same group as they decrease in wave length; and thirdly, a law of sequence as regards quality, an alternation of sharper and more diffuse groups, with a gradually increasing diffuseness and diminishing intensity of all the related groups as the wave length diminishes. The first relationship has long since been noticed in the case of the sodium lines which recur in pairs, and we have noticed that the potassium lines between the extreme red and the violet pairs are repetitions of a quadruple group.... We now

record a second harmonic series of potassium lines which appear to be
pairs, and the violet pair, and possibly the red pair too belong to this
series. Lithium shows a second harmonic series of single lines high up
on the scale. Calcium gives a long series of well marked triplets... The
alternations of the sharper and more diffuse groups are generally
apparent and are very marked in the case of calcium and zinc.'

Thus, the concept of spectral series was introduced, and, in 1883, Hartley noticed that
the separation between the component lines in pairs or triplets in a given series was
approximately constant if measured in wave numbers. In this way, one could begin to
assign various lines to series on the basis of objectively measurable quantities instead of
appearance. In 1885, Cornu noticed that some of the lines in aluminum and thallium
correspond with the hydrogen lines, after a change of origin and of scale is made.

   In that same year there occurred one of the most remarkable events in the history of
science. A 60 year old Swiss school teacher, who, as far as is known, had not had any
previous experience with spectra, or any theoretical justification of what he was doing,
or any previous scientific publication, came to some conclusions after playing with the
four numbers giving the wavelengths of the hydrogen spectrum. According to
Angstrom's carefully observed measurements, these wavelengths, in terms of 10^{-8} cm
are given as

H_α	(C line)	6562.10	(red)
H_β	(F line)	4860.74	(green)
H_γ	(near G line)	4340.10	(blue)
H_δ	(h line)	4101.2	(violet).

Balmer observed that if we let $b = 3645.6$, then the numbers $(9/5)b$, $(4/3)b$, $(25/21)b$ and
$(9/8)b$ take on the value 6562.08, 4860.8, 4340 and 4101.3, a discrepancy of less than
1/40 000 of a wavelength with the Angstrom measurements.

   Now Balmer observed that the numbers 9/5, 4/3, 25/21 and 9/8 can be written as
$3^2/(3^2 - 4)$, $4^2/(4^2 - 4)$, $5^2/(5^2 - 4)$ and $6^2/(6^2 - 4)$. This suggested that the general
formula for the hydrogen lines should be $[m^2/(m^2 - 4)]b$, and, in particular, there
should be a fifth line, corresponding to $m = 7$, at $(49/45)b = 3969.65$, which is just at the
edge of the visible spectrum. Balmer was not aware of such a line, but indeed a line at
this position had recently been found. When Balmer was informed of the existence of
such a line, he was convinced that he had the right formula and had his result published.
Actually, nine more lines had recently been discovered in the ultra-violet and they all
agreed with Balmer's formula, with m ranging from 8 to 16 with discrepancies of at most
one part in a thousand. Within a few years, 29 lines were observed in all, and with a
slight modification in the value of b (replacing the original value of b by 3646.13) they all
fit amazingly well with Balmer's formula, yielding a discrepency of, at most, one part in
thirty thousand, a truly remarkable result. There was no question that Balmer had the
right formula, although there was no understanding whatsoever of what could lie
behind such a formula. Balmer's discovery had a major impact, and the search was on
for formulas for the spectra of the other elements.

The main contributors to the elucidation of the structure of the line spectra over the 20 year period following Balmer's discovery were Rydberg, Schuster, Kayser and Runge, and Ritz. The key work was done by Rydberg, whose first step was to elucidate the notion of series which had previously been introduced by Liveing and Dewar, by not only using the physical appearance of the lines but also numerical relations involving the wave numbers. In this way, he distinguished three types of series, the 'sharp' series, in which the lines were comparatively sharp, a 'principal' series lying mostly in the ultra-violet, but whose first lines lying in the visible spectrum were usually the most intense, and a 'diffuse' series. (A fourth series, lying usually in the infrared, was introduced in 1908 by Bergmann, and has become known as the 'fundamental' series, even though there is nothing fundamental about this series.) Using wave numbers (number of waves per centimeter, so with units cm^{-1}), Rydberg discovered that all lines in a given series have wave numbers of the form

$$v_n = v_\infty - \frac{R}{(n + \mu)^2}, \qquad n = 2, 3, 4, \ldots,$$

where v_∞, the wave number of the limit of the series of lines, and the constant, μ, depend on the particular series and element under investigation, but the constant $R = 109721.6$ is independent of the series and of the element. (Setting $\mu = 0$ gives Balmer's formula for hydrogen.) Thus, for example, lithium has three main series, the sharp, principal and diffuse, whose wave numbers are given by

$$v_n^s = 28601.6 - \frac{109721.6}{(n + 0.5951)^2}$$

$$v_n^p = 43487.7 - \frac{109721.6}{(n + 0.9596)^2}$$

$$v_n^d = 28598.5 - \frac{109721.6}{(n + 0.9974)^2}.$$

Rydberg observed that if we set $n = 1$ in the right-hand fraction occurring in the expression for v_n^s we obtain

$$\frac{109721.6}{(1.5951)^2} = 43123.7$$

which is approximately equal to $43487.7 = v_\infty^p$, the limit of the principal series. Similarly, if n is set equal to one in the running fraction for the principal series, we obtain

$$\frac{109721.6}{(1.9596)^2} = 28573.1$$

which is approximately the limit of both the sharp series and the diffuse series. Rydberg concluded that the small differences were due to experimental error, and that, indeed, the limits of the sharp and diffuse series should coincide and be equal to the value obtained by setting $n = 1$ in the running term for the principal series. Also, Rydberg and Schuster independently observed that $v^p_\infty - v^s_\infty = v^p_1 = -v^s_1$. All this led to the conclusion that the wave numbers for the different series could be written as *differences*:

$$v^s_n = \frac{R}{(1+P)^2} - \frac{R}{(n+S)^2}$$

$$v^p_n = \frac{R}{(1+S)^2} - \frac{R}{(n+P)}$$

$$v^d_n = \frac{R}{(1+P)^2} - \frac{R}{(n+D)^2}.$$

Soon after the discovery of the fundamental series in 1908 by Bergmann, Runge pointed out that its wave numbers could also be written as differences:

$$v^f_n = \frac{R}{(2+D)^2} - \frac{R}{(n+F)^2}.$$

Ritz introduced the shorthand notation for the above formulas:

$$v^s_n = 1P - nS, \quad v^d_n = 1P - nD$$
$$v^p_n = 1S - nP \quad \text{and} \quad v^f_n = 2D - nF.$$

These formulas led Rydberg and Ritz to suggest that the numbers 1 and 2 occurring in the above formulas should be allowed to be replaced by an arbitrary positive integer to give a new series; thus, for example, $3P - nS$, $4S - nP$, etc. This suggestion was known as the *Ritz combination principle*. Applied to the hydrogen atom, it suggested that in addition to the Balmer series, $R(1/2^2 - 1/n^2)$, $n = 3, 4, 5, \ldots$ one should have an arbitrary series of the form $R(1/m^2 - 1/n^2)$, $n = m + 1, m + 2, \ldots$. In the course of time various series of this kind were discovered in the hydrogen spectrum: the Lyman series ($m = 1$) by Lyman (1906); the Paschen series ($m = 3$) (1908); the Brackett series ($m = 4$) (1923); the Pfund series ($m = 5$) (1924); the Humphreys series ($m = 6$) (1953).

In an expression such as $mS - nP$, the term nP is only approximately given as $R/(n + P)^2$; the actual term in the denominator was found to be somewhat more complicated, and various expressions were offered for it. Also, the Rydberg constant, R, was found to vary slightly from element to element. Furthermore, various series of lines appeared as doublets or triplets, so that an expression such as ^3P was introduced to indicate a triplet, and $^3P_0$, $^3P_1$, and $^3P_2$ a numerical value corresponding to the three lines of the triplet, with n^3P_1, for example, standing for a term whose approximate value was given as $R/(n + {}^3P_1)^2$. The various spectral lines could then be given as differences of the form $m^3P_2 - n^3D_1$, for example. The important point was that the wave numbers of the *spectral lines were analyzed as differences* of various expressions such as n^3P_1 called *terms*. The description of the spectrum as differences of terms was known as term analysis, and it became a matter of importance to attach physical significance to the terms themselves rather than to the lines. Before turning to this development, we should mention a very pregnant observation of Lord Rayleigh, in 1896, the full import of which could not be appreciated for another 30 years. Rayleigh pointed out that if the explanation of the spectral lines was to lie in a mechanical system oscillating about equilibrium, as suggested by Stokes, one should expect formulas relating the squares of the frequencies, rather than the frequencies themselves, as is the case in Balmer's formula and the discoveries of Rydberg and Schuster. (The fact that acoustical oscillations involve the squares of the frequencies is, of course, due to the fact that the wave equation is second order in space and time. The idea that one should consider a partial differential equation which is first order in time and second order in the space variables, the Schrodinger equation, was a step that would not be taken until 1926.)

We recall that the search for formulas for spectral lines was given a great amount of impetus by Stokes's suggestion that the emission or absorption of radiation was due to the vibrations of the 'ultimate molecules' of a substance. For line spectra, this meant that each atom would be vibrating at all the frequencies observed in its line spectrum. The first key step to get away from this idea was again taken by Schuster. Although J. J. Thomson (who was Schuster's student) did not publish the results of his conclusive experiments demonstrating the existence of the electron until 1897, the idea of a fundamental unit of charge had been around for a long time due to electrochemistry and the idea of its possible existence as a 'subatomic' particle (even though this seemed to many as a contradiction in terms) was in the air. In 1895, Schuster pointed out the consequences for spectroscopy of the existence of such a subatomic particle:

> 'Most of us, I believe, now accept a definite atomic charge of electricity, and if each charge is imagined to be capable of moving along the surface of an atom, it would represent two degrees of freedom. If a molecule is capable of sending out a homogeneous vibration, it means that there must be a definite position of equilibrium of the 'electron'. If there are several such positions, the vibrations may take place in several such periods. Any one molecule may perform for a certain time a simple periodic oscillation about one position of equilibrium, and owing to some impact the electron may be knocked over into a new position. The vibrations under these circumstances would not be quite homogeneous, but if the electron oscillates about any one position sufficiently long to perform a few thousand oscillations, we would hardly notice the want of homogeneity. Each electron at a given time would only send out vibrations which in our instruments would appear as homogeneous. Each molecule would thus succesively give rise to a number of spectral rays, and at any one time the electrons in the different molecules would, by the law of probability, be distributed over all possible positions of equilibrium so that we should always see all the vibrations which any one molecule of the gas is capable of giving out. The probability of an electron oscillating about one of its positions of equilibrium need not be the same in all cases. Hence a line may be weak not because the vibration has a smaller amplitude, but because fewer molecules give rise to it.'

In 1897 Zeeman discovered that the presence of a magnetic field caused the splitting of certain spectral lines. He communicated this result to Lorentz, who explained the effect in terms of a classical theory of the electron, predicting that the original line should split into three and also predicted the separation between the split lines. On closer examination Zeeman found that Lorentz's predictions were correct, giving strong evidence for the existence of the electron. This apparently stimulated J. J. Thomson to publish the results of his experiments on cathode rays which gave conclusive evidence of the existence of the electron and also gave an accurate measurement of e/m, where e is the electronic charge and m is the electronic mass. (However, within a few years Lorentz's explanation of the Zeeman effect was refuted, when many discoveries were made of lines which split into more than three components in a magnetic field – the so-called 'anomalous' Zeeman effect.)

Over the next 14 years, various models were proposed. For instance, Ritz proposed a magnetic model for the atom in which the electrons were spinning cylinders generating magnetic fields, and the atom was held together by magnetic forces. Various 'astronomical' models with electrons moving as planets were proposed, all suffering from the defect that since the electrons were in motion, they must be constantly radiating, and hence the atom must be continuously losing energy. The most accepted model of the period was Thomson's 'plum pudding' model, in which the atom was made of uniformly distributed positive electricity in which a number of electrons were embedded. In 1911, Rutherford published his famous paper in which he described his experiments bombarding matter with alpha particles, demonstrating the fact that the atom consisted of a positive nucleus occupying relatively small volume, surrounded by electrons. This set the stage for the Bohr model of the atom.

As we have already mentioned, Schuster had proposed that each atom contributes just one line to the spectrum and that the line is connected with the motion of a single electron. The Rutherford atom had electrons surrounding a central nucleus, and the electrons were probably in motion. Indeed, in his article on spectroscopy in the 1911 edition of the Encyclopaedia Britannica, Schuster writes, concerning possible physical explanations of the line spectra

> 'on the whole it seems probable that the system of moving electrons, which according to a modern theory constitute the atom, is not directly concerned in thermal radiation which would rather be due to a few more loosely connected electrons hanging on to the atom. The difficulty that a number of spectroscopic lines seem to involve at least an equal number of electrons may be got over by imagining that the atom may present several positions of equilibrium to the electron which it occupies in turn. A collision may be able to throw the electrons from one of these positions to another. According to this view the different lines are given out by different molecules, and we should have to take averages over a number of molecules to obtain the complete spectrum, just as we now take averages of energy to obtain the temperature.'

(There can be no doubt that Schuster's ideas were well known in Rutherford's laboratory at Manchester at the time of Bohr's visit there in 1913 when he wrote his fundamental paper on the structure of the atom. Schuster brought Rutherford to Manchester. Indeed, Schuster, who was professor at Manchester, was greatly impressed by Rutherford's work on alpha particles, done at McGill University in Montreal. Schuster visited Rutherford in 1906, and later wrote him offering to resign his chair at Manchester if Rutherford would take his place, and at the same time offered to finance personally a readership in mathematical physics. Rutherford came to Manchester in 1907 to take Schuster's chair and Bohr took up the readership in mathematical physics in 1913. Schuster was still quite active in the physics department through this whole period.)

The quantum hypothesis of Planck, and Einstein's theory of the photoelectric effect, in which light of frequency v is made up of photons, each of energy hv, were well known to the community of physicists. What remained was to connect these ideas to spectroscopy. The first step in this direction was taken in conjunction with molecular spectroscopy.

In 1886, Deslandres, a junior colleague of Cornu at the Ecole Polytechnique, used the new Rowland concave grating to examine the band spectra of molecules. He observed that the bands really consist of many lines when examined under high resolution. These lines are spaced regularly. Over the next few years it was observed that in the infra-red region, the spectra of diatomic molecules have an even simpler appearance. In the far infra-red, there is a series of lines whose frequencies are more or less separated by constant differences, Δv. In the near infra-red there are bands, but each band consists of a family of lines whose frequencies are also separated by the same difference, Δv (except that one line in the middle is usually missing). Now although nothing was known about the structure of atoms, it was assumed that molecules were made up of atoms, in particular diatomic molecules of two atoms. Thus, molecules had two types of 'degrees of freedom' not available to atoms, rotations of the molecule about the center of mass of the constituent atoms, and vibration of the atoms about their equilibrium position. These degrees of freedom played an important role in the current theories of the specific heats of gases. In 1912, Niels Bjerrum applied these ideas to the infra-red spectra, relating the lines in the far infra-red to molecular rotations, and the near infra-red bands to the combination of vibrations and rotations. Following the ideas of Planck and Einstein, and a suggestion of Nernst that molecular rotations should be quantized, he assumed that the rotational energy must be a multiple of hv, where v is the number of revolutions per second. Now if I is the moment of inertia of the rotating molecule, its rotational energy is

$$\tfrac{1}{2}I(2\pi v)^2$$

which must be a multiple of hv. Combining the rotational and vibrational energies gave, according to a principle due to Rayleigh, a family of oscillations whose frequencies are of the form

$$v_0 \pm v_n$$

where v_n is the nth rotational frequency. In 1913, Ehrenfest, in discussing quantum statistics, showed that angular momentum should be quantized as multiples of $\tfrac{1}{2}hv$, and not hv as suggested by Nernst and Bjerrum. Thus the nth rotational frequency is determined by the equation

$$\tfrac{1}{2}I(2\pi v_n)^2 = nhv_n$$

so

$$v_n = \frac{nh}{4\pi^2 I}$$

and thus the angular momentum, $2\pi Jv$, is quantized according to the law

$$2\pi Jv = n\hbar \quad (\text{where } \hbar = h/2\pi).$$

Now in the Bjerrum theory, the vibration–rotation bands of the infra-red spectrum were explained classically, on the basis of oscillators whose frequencies are $v_0 \pm v_n$. But how about the rotation lines in the far infra-red, at energies corresponding to the absence of vibration? By what mechanism does the molecule absorb or emit radiation precisely at the frequencies v_n, corresponding to the quantized levels of angular momentum? The explanation must be that by absorbing a quantum of energy hv_n, the molecule increases its rotational energy by this amount, and a molecule emits the quantum hv_n when its rotational energy drops by this amount. In other words, the observed lines in the infra-red spectrum of diatomic molecules must correspond to *differences* in the levels of rotational energy. The idea of discrete levels of energy was used by Nicholson in 1911 for the line spectra observed in nebulae and in the solar corona. He postulated the existence of certain elements 'nebulum' and 'protoflourine' and proposed atomic models for them, and suggested that the variable portion of their energy of an atom involved only its angular momentum which is quantized. In discussing the lines of the solar corona he writes

> 'The quantum theory has apparently not been put forward as an explanation of 'series' spectra consisting of a large number of related lines given by a comparatively simple atom. Yet, in the belief of the writer, it furnishes the true explanation in certain cases, and we are led to suppose that lines of a series may not emanate from the same atom, but from atoms whose internal angular momenta have, by radiation or otherwise, run down by various discrete amounts from a standard value. For example, in this view there are various kinds of hydrogen atom, identical in chemical properties and even in weight, but different in their internal motions.'

However, the notion that all the lines in the atomic spectrum corresponded to differences in energy levels of the atom was due to Bohr in his epoch making paper of 1913, 'On the constitution of atoms and molecules'. He writes that his principal assumptions are

> '(1) That the dynamical dquilibrium of the systems in the stationary states can be discussed by help of the ordinary mechanics, while the passing of the systems between different stationary states cannot be treated on that basis.

(2) That the latter process is followed by the emission of a
homogeneous radiation, for which the relation between the frequency
and the amount of energy emitted is the one given by Planck's theory.'

For the hydrogen atom he assumes that the electron moves in circular orbits about the
positively charged nucleus. Let m be the mass of the electron and ε its electric charge.
Let Z be the charge on the nucleus, so that for the hydrogen atom $Z = 1$; the analysis
applies equally well for any value of Z – thus for the He^+ ion, etc. The force attracting
the electron to the nucleus is then

$$\frac{Z\varepsilon^2}{r^2}$$

according to Coulomb's law, where r is the radius of the atom. Suppose that the
electron moves with velocity v, so that its angular momentum is

$$mvr.$$

The acceleration of the electron is v^2/r, and so, according to Newton's law,

$$\frac{mv^2}{r} = \frac{Z\varepsilon^2}{r^2}$$

or

$$mv^2 = \frac{Z\varepsilon^2}{r}.$$

Multiplying this equation by mr^2 and using the quantization of angular momentum,
$mvr = nh$, we get

$$n^2h^2 = Z\varepsilon^2 mr,$$

yielding the discrete values for r,

$$r = \frac{n^2h^2}{Zm\varepsilon^2}.$$

The potential energy is $-Z\varepsilon^2/r$ and the kinetic energy is $\frac{1}{2}mv^2 = \frac{1}{2}Z\varepsilon^2/r$ as derived
above. Thus the total energy is $-\frac{1}{2}Z\varepsilon^2/r$. Substituting the above values of r gives, as the
nth energy level,

$$E_n = -\frac{m\varepsilon^4 Z^2}{2h^2n^2}.$$

Applying the rule

$$h\nu = E_n - E_m$$

gives

$$\nu = K\left(\frac{1}{m^2} - \frac{1}{n^2}\right)$$

with

$$K = \frac{2\pi^2 m\varepsilon^4}{h^3}.$$

This is precisely the form of Balmer's formula for the hydrogen spectrum if we take $n = 2$. As Bohr points out, if we put $n = 3$ we get the series observed by Paschen and previously suspected by Ritz. Furthermore, Bohr writes, if we set $n = 1$ or $n = 4, 5, \ldots$

> 'we get series respectively in the extreme ultra-violet and the extreme ultra-red, which are not observed, but the existence of which may be expected.
>   The agreement in question is quantitative as well as qualitative. Putting

$$e = 4.7 \cdot 10^{-10}, \quad \frac{e}{m} = 5.31 \cdot 10^{17} \text{ and } h = 6.5 \cdot 10^{-27}$$

we get

$$\frac{2\pi^2 m \varepsilon^4}{h^3} = 3.1 \cdot 10^{15}$$

The observed value... is

$$3.290 \cdot 10^{15}$$

> 'The agreement between the theoretical and observed values is inside the uncertainty due to the experimental errors in the constants entering in the expression for the theoretical value.'

A little later on in the same paper he writes about other atomic spectra:

> 'According to Rydberg's theory – with the generalization given by Ritz – the frequency corresponding to the lines of the spectrum of an element can be expressed by

$$v = F_r(\tau_1) - F_s(\tau_2)$$

> where τ_1 and τ_2 are entire numbers, and $F_1, F_2, F_3, \ldots$ are functions of τ which approximately are equal to $K/(\tau + a_1)^2$ $K/(\tau + a_2)^2, \ldots K$ is a universal constant, equal to the factor outside the bracket in the formula... for the spectrum of hydrogen. The different series appear if we put τ_1 or τ_2 equal to a fixed number and let the other vary.
>   'The circumstance that the frequency can be written as the difference between two functions of entire numbers suggests an origin of the lines in the spectra in question similar to the one we have assumed for hydrogen; *i.e.* that the lines correspond to a radiation emitted during the passing of the system between two different stationary states. For systems containing more than one electron the detailed discussion may be very complicated, as there will be many different configurations of the electrons which can be taken into consideration as stationary states. This may account for the different sets of series in the line spectra emitted from the substances in question. Here I shall only try to show how, by help of the theory, it can be simply explained that the constant K entering into Rydberg's formula is the same for all substances.
>   'Let us assume that the spectrum in question corresponds to the

radiation emitted during the binding of an electron; and let us further assume that the system including the electron considered is neutral. The force on the electron, when at a great distance apart from the nucleus and the electrons previously bound will be very nearly the same as in the above case of the binding of an electron by the hydrogen nucleus. The energy corresponding to one of the stationary states will therefore for τ great be very nearly equal to the [of the hydrogen atom]... For τ great we consequently get

$$\lim(\tau^2 F_1(\tau)) = \lim(\tau^2 F_2(\tau)) = \cdots = \frac{2\pi^2 m \varepsilon^4}{h^3}$$

in conformity with Rydberg's theory.'

The course of research for the next 15 years was set: to discover the stationary states of the more complicated atoms, to understand why not *all* differences between stationary states show in the spectrum, i.e. the presence of 'selection rules', to find a quantum explanation of the Zeeman effect, and of the Stark effect (the splitting of the spectral lines in the presence of an electric field). We will not describe this part of the story in detail, referring to Chapters VI, VII and IX of Whittaker's book, *The history of aether and electricity* Vol. II. Suffice it to say that by the end of the period the entire physical conception of the universe had been revolutionized, with the development of quantum mechanics principally by Heisenberg, Born, Jordan, Schrödinger and Dirac. The detailed applications of these ideas to the spectroscopy of the atom and the molecule is the object of study of Chapter 3.

There is one experimental advance developed in the 1920s which we should describe here and that is the *Raman effect*. When a parallel beam of light goes through a transparent substance, a small fraction of the light is scattered in all directions. If the incident light is monochromatic, most of the scattered light has exactly the same frequency as that of the incident radiation. In 1923, in developing a theory of the scattering of radiaton, Smekal conjectured that some of the scattered waves should have frequencies differing from that of the incident radiation. No such phenomenon had been observed at the time of publication of Smekal's paper, but this effect was observed experimentally in 1928 by Sir Chandrasakara V. Raman, and has since been known as the Raman effect. It is found that if the incident light is monochromatic with frequency v, then in addition to the scattered light of frequency v, there is light of various frequencies $v = v'$ for various displacements, v'. (The intensity of the displaced frequencies is very small.) It is found that the amounts of the displacements, i.e. the values of the v', are independent of the incident frequency, v.

The basic theory of the absorption and emission of radiation and of the Raman effect for the spectroscopy of atoms and molecules was worked out in the late 1920s and early 1930s. We should mention that the first quantum mechanical treatment of the hydrogen atom was given by Pauli in 1925 on the basis of the matrix mechanics of Heisenberg, Born and Jordan, a full year before the appearance of the Schrödinger equation.) His treatment is (essentially) entirely group theoretical, and we have outlined it at the end of Chapter 4. The systematic application of group theoretical methods to the problems of atomic and molecular spectroscopy was developed by Wigner during this period, although the detailed execution of his ideas occupied a great many physicists and chemists over the next 20 years. In particular, see the monumental three volume work of Herzberg on the spectra of molecules. Within the past 30 years, there have been a number of technological developments which have greatly extended the experimental tools available in spectrum analysis. Thus, microwave technology has led to the measuremet of spectra in the microwave region (in particular for the rotational spectra of molecules), and commercially produced devices for the analysis of spectra in the infra-red region (vibrational spectroscopy) is a standard analytic tool in the chemistry laboratory. The vibrational spectra of about a quarter of a million chemical compounds have been investigated, and tens of thousands are tabulated in a form accessible to an electronic computer, so that spectrum analysis for the identification of chemical compounds have been mechanized. Thus Fox Talbot's program has been carried out in a fashion far exceeding his wildest expectations. In addition, the use of characteristic spectra of chemical groups is a standard tool (also programmable) for the structural analysis of molecules.

Atomic spectra, in particular line broadening and red shift effects, are a key tool in astronomy and astrophysics.

CHARACTERS AND FIXED POINT
FORMULAS FOR LIE GROUPS

In this appendix we will show how to extend a good bit of the formalism of Chapters 2 and 3 for the case of Lie groups. If G is a Lie group and M is a differentiable manifold, then when we talk of an action of G on M we require that the map $G \times M \to M$ giving the action be smooth. Similarly, in our definition of a vector bundle over a differentiable manifold we require that E be a differentiable manifold and that the projection, π, be smooth. The main result of this section will be to derive a formula which generalizes the Frobenius fixed point character formula, (2.3), Chapter 3. Since, as we have already remarked, the character of any infinite-dimensional representation cannot be a function, we must first introduce the class of objects to which our group characters belong. For the representations of interest to us, the characters turn out to be what are known as 'generalized functions' or 'distributions' and we must begin with the definition and study of certain properties of these objects. It will turn out that a study of the functorial properties of these objects, that is how they behave under smooth maps, together with some standard theorems about smooth maps themselves, will allow us to go quite far in generalizing the results of Section 3.2. We will thus begin this appendix with a rather lengthy detour into the geometrical theory of generalized functions. It should be pointed out that the methods of this appendix are quite different in spirit from those of the bulk of this book in that the notion of orthogonality does not play an important role.

Let M be a smooth manifold. We let $C^\infty(M)$ denote the space of all smooth (i.e. C^∞) functions on M. The space $C^\infty(M)$ is a vector space, and it also carries a topology: the topology of uniform convergence together with each finite number of derivatives on each compact subset of M. (Here derivatives are relative to some local coordinate system. For any compact subset, K, we cover K by a finite number of coordinate charts and require that we have convergence, together with derivatives up to a given order on each set $U \cap K$ for U belonging to the cover. The topology so obtained is seen to be independent of the various choices.)

We let $C_0^\infty(M)$ denote the space of smooth functions with compact support. It also is a vector space, it is a subspace of $C^\infty(M)$. However, we put a stronger topology on $C_0^\infty(M)$ than the one inherited from $C^\infty(M)$. For u_α to converge in $C_0^\infty(M)$ we require, in addition to convergence in $C^\infty(M)$ that there be some fixed compact set, K, such that all the u_α have their support lying in K.

A *distribution* on M is defined to be a continuous linear function on the space $C_0^\infty(M)$.

If γ is a distribution, we define its support as follows: a point $x \in M$ does *not* belong to supp γ if there is some neighborhood, U, of x such that $\langle \gamma, u \rangle = 0$ for any function, u, with supp $u \subset U$. Here $\langle \gamma, u \rangle$ denotes the value of γ on the function u. Thus supp γ is a closed set. Suppose that γ is a distribution with compact support. Then we can extend γ so as to be defined on all of $C^\infty(M)$: choose some smooth function, φ, of compact support such that $\phi = 1$ on some neighborhood of supp γ. Then define $\langle \gamma, u \rangle = \langle \gamma, \phi u \rangle$ for any smooth function, u. It is easy to see that this definition is independent of the choice of ϕ.

Let $f : M \to N$ be a smooth map, where M and N are smooth manifolds. Then f induces a continuous linear map, f^*, from $C^\infty(N) \to C^\infty(M)$ given by

$$f^* u = u \circ f.$$

If γ is a distribution of compact support on M, then we define the distribution, $f_* \gamma$, of compact support on N by the formula

$$\langle f_* \gamma, u \rangle = \langle \gamma, f^* u \rangle. \tag{G.1}$$

The map f_* is the adjoint of the map f^*. It defines a continuous linear map (relative to the weak topology) of the space of distributions of compact support on M to distributions of compact support on M. A smooth map, f, is called *proper*, if $f^{-1}(K)$ is compact for any compact subset, K, of N. If f is proper, then $f^* u$ will have compact support if u has compact support. Then we can use (G.1) to define $f_* \gamma$ for any distribution, γ. In short, under smooth maps, smooth functions 'pull back' and distributions of compact support 'push forward'. Under proper smooth maps, functions of compact support pull back to have compact support and all distributions push forward.

The maps f^* and f_* are functorial in the sense that if $f_1 : M_1 \to M_2$ and $f_2 : M_2 \to M_3$ are smooth maps, then

$$(f_2 \circ f_1)^* = f_1^* \circ f_2^* \quad \text{and} \quad (f_2 \circ f_1)_* = f_{2*} \circ f_{1*}. \tag{G.2}$$

A particularly nice class of distributions on a manifold, M, are the smooth *densities*. We recall that a density, ρ, on a manifold is an object which, in terms of any local coordinate chart, U, is given by a function, ρ_U, and, when we change to some other coordinate chart, U', the transition rule is

$$\rho_U = \rho_{U'} |d\underset{\sim}{y}/d\underset{\sim}{x}|$$

where $\underset{\sim}{y}$ denotes the coordinates on U' and $\underset{\sim}{x}$ the coordinates on U and where $d\underset{\sim}{y}/d\underset{\sim}{x}$ denotes the Jacobian determinant. This, of course, is how an integrand changes in the change of variables formula for an integral. We say that ρ vanishes at some point if ρ_U vanishes at that point, relative to some, and then hence to any choice of local coordinate chart. Thus we can define the support of a density as the closure of the set of points where it does not vanish. Similarly, we say that a density is smooth if each of the functions ρ_U is smooth, and it suffices to verify this condition for some coordinate cover. If ρ is a smooth density and if f is a smooth function, then $f\rho$ is again a smooth density.

The densities are objects that can be integrated over manifolds and thus each smooth density defines a linear functional on $C_0^\infty(M)$ according to the rule

$$\langle \rho, u \rangle = \int_M u\rho. \tag{G.3}$$

For this reason we will refer to a distribution as a 'generalized density'. We will denote the space of smooth densities on M by $C^\infty(\square M)$ and we will denote the space of generalized densities by $C^{-\infty}(\square M)$. If ρ is a density of compact support, then ρ has compact support as a distribution. Conversely, if ρ is smooth and has compact support as a distribution, then ρ has compact support, as can be easily proved. We will denote the space of smooth densities of compact support by $C_0^\infty(\square M)$ and the space of generalized densities of compact support by $C_0^{-\infty}(\square M)$. If ρ is a smooth density of compact support and u is a smooth function, then we can consider (G.3) as making u into a continuous linear function of ρ (where we put the obvious topology on the linear space $C_0^\infty(\square M)$. A *generalized function* is any continuous linear function on $C_0^\infty(\square M)$. We denote the space of generalized functions by $C^{-\infty}(M)$. We can define the support of a generalized function in a manner similar to the definition we gave above of the support of a generalized density. We denote the space of generalized functions of compact support by $C_0^{-\infty}(M)$.

Let ρ be a smooth density on M. We claim that ρ can be regarded as a section of a smooth line bundle over M. As this point of view will be of importance to us, we pause briefly to explain it.

Let V be a finite-dimensional vector space, with $\dim V = n$. By a volume element of V we mean a rule, α, which assigns to each n vectors, $v_1, \ldots, v_n$, a number $\alpha(v_1, \ldots, v_n)$ and satisfies the condition

$$\alpha(Av_1, \ldots, Av_n) = |\det A| \alpha(v_1, \ldots, v_n)$$

for any linear transformation, A. Any volume element, α, is clearly determined by its values on a basis, $v_1, \ldots, v_n$, and hence the set of all volume elements on V forms a one-dimensional vector space which we shall denote by $\square V$. If W is a subspace of V, and $w_1, \ldots, w_k$ is a basis of W, and $\bar{w}_{k+1}, \ldots, \bar{w}_n$ is a basis of V/W, then we can find a basis, $w_1, \ldots, w_k, w_{k+1}, \ldots, w_n$, such that the w_i project onto $\bar{w}_i$ for $i = k+1, \ldots, n$. Furthermore, any two such differ by a matrix whose determinant is one. Thus, if α_W is a volume element on W and $\alpha_{V/W}$ is a volume element on V/W then we get a volume element on V by setting

$$\alpha_V(w_1, \ldots, w_n) = \alpha_W(w_1, \ldots, w_k)\alpha_{V/W}(\bar{w}_{k+1}, \ldots, \bar{w}_n). \tag{G.4}$$

It is easy to check that α_V satisfies the identity for volume elements for those linear transformations A, such that $AW \subset W$, which is the consistency condition necessary for the above equation to make sense. We then extend α_V for all vectors $v_1, \ldots, v_n$ by the identity for volume elements. Conversely, if we start with α_V then (G.4) defines α_W and $\alpha_{V/W}$ up to the replacement $\alpha_W \sim \to c\alpha_W$ and $\alpha_{V/W} \sim \to c^{-1}\alpha_{V/W}$, where c is any non-zero constant. In short, if W is a subspace of V then we have the natural identification

$$\square V = \square W \otimes \square (V/W). \tag{G.5}$$

Let M be a differentiable manifold and let TM denote its tangent bundle. That is, TM_x is the tangent space to M at x for any $x \in M$. Then $\square(TM)_x$ is a one-dimensional vector space attached to each x, and it is easy to see that these vector spaces fit together to form a smooth line bundle over M, which we denote by $\square TM$. Let ρ be a smooth section of this line bundle. If $x^1, \ldots, x^n$ are local coordinates on a coordinate chart, U, of M, then we get n vector fields, $(\partial/\partial x^1), \ldots, (\partial/\partial x^n)$ on U, and we define the function ρ_U by

$$\rho_U = \rho((\partial/\partial x^1), \ldots, (\partial/\partial x^n)).$$

It is easy to see that the transition law for a density is satisfied, and thus ρ can be regarded as a density on M. Conversely, starting with a smooth density, we get a section of $\square TM$. Thus the space of smooth densities can be identified with the space of smooth sections of the line bundle, $\square TM$.

Let M and N be smooth manifolds. A smooth map, $f : M \to N$ is called a *submersion* if its differential, $df_x : TM_x \to TN_{f(x)}$ is surjective for all $x \in M$. Then the implicit function theorem guarantees that $f^{-1}(y)$ is a submanifold of M for each $y \in N$. We denote this manifold by M^y. Then, for each $x \in M^y$ the map df_x induces an isomorphism

$$\frac{TM_x}{(TM^y)_x} \sim TN_y,$$

since the kernel of the map $df_x : TM_x \to TN_y$ is precisely $(TM^y)_x$. We can now use (G.5) to write

$$\square TM_x \sim \square TN_{f(x)} \otimes \square TM_x^y.$$

In this way, a density ρ on M can be written as $\rho = \rho_N \otimes \rho^f$, where ρ_N is a density on N and, for each $y \in M$, we get a density ρ_y^f on $f^{-1}(y)$. If ρ has compact support, then each ρ_y^f has compact support and we can integrate it over $f^{-1}(y)$. In this way, by the process of 'integration over the fiber' we pass from a smooth density on M to a smooth density on N. Suppose that (z, y) are local coordinates on M with y local coordinates on N, so that $f(z, y) = y$, and thus z can be used as local coordinates on each fiber. If ρ is supported in such a neighborhood then we can write

$$\rho = \rho(z, y) \mathrm{d}z\, \mathrm{d}y$$

(where, of course, $z = (z_1, \ldots, z_k), \mathrm{d}z = \mathrm{d}z_1 \cdots \mathrm{d}z_k, \mathrm{d}y = \mathrm{d}y_1 \cdots \mathrm{d}y_l$). Then integration over the fiber assigns to ρ the density

$$\left(\int \rho(z, y) \mathrm{d}z \right) \mathrm{d}y$$

on N. If u is any smooth function on N with support in such a neighborhood, then we can write f^*u as $u(y)$. Then, by (G.1),

$$\langle f_* \rho, u \rangle = \langle \rho, f^* u \rangle$$

$$= \int \rho(z, y) u(y) \mathrm{d}z \, \mathrm{d}y$$

$$= \int \left(\int \rho(z, y) \mathrm{d}z \right) u(y) \mathrm{d}y$$

or,

$$f_* \rho = \int \rho(z, y) \mathrm{d}z.$$

By applying a partition of unity, we conclude that the same is true for any smooth density of compact support, i.e. that

> if $f: M \to N$ is a smooth submersion and ρ is a smooth density of compact support, then $f_* \rho$ is again a smooth density of compact support and is obtained by integrating ρ over the fibers of the submersion. In particular, f induces a continuous map f_*: $C_0^\infty(\square M) \to C_0^\infty(\square N)$ and hence, by (G.1), a map of $C^{-\infty}(N) \to C^{-\infty}(M)$ which we denote by f^*, and which extends the map $f^*: C^\infty(N) \to C^\infty(M)$.

As an example of the pull back of generalized densities, let us consider the case where $N = \mathbb{R}^l$ and we consider the generalized function, δ, on $\mathbb{R}^l$, where

$$\langle v \mathrm{d}y, \delta \rangle = v(0)$$

for any $v \in C^\infty(\mathbb{R}^l)$, where $\mathrm{d}y$ is Euclidean (Lebesgue) measure. Here we assume that $f: M \to \mathbb{R}^l$ is a submersion, and let $W = f^{-1}(0)$. Since f is a submersion, W is a submanifold. If $y_1, \dots, y_l$ denote Euclidean coordinates on $\mathbb{R}^l$, then we can find coordinates $z_1, \dots, z_k, y_1, \dots, y_l$, about any point of W. If $\rho = u \, \mathrm{d}z \, \mathrm{d}y$ is a density on M which is supported in a neighborhood where these coordinates are defined, then

$$\langle \rho, f^* \delta \rangle = \int u(z, 0) \mathrm{d}z.$$

Suppose that we had introduced some other local coordinate system $z_1, \dots, z_k$, $x_1, \dots, x_l$, in terms of which W is still given by the equations $x_i = 0$ (locally). Then $\mathrm{d}z \, \mathrm{d}y = |\det(\partial y / \partial x)| \mathrm{d}z \, \mathrm{d}x$. If, in terms of these coordinates, $\rho = w \, \mathrm{d}z \, \mathrm{d}x$, then we have

$$w(z, x) = u(z, y(z, x)) |y'(z, x)|,$$

where we have written y' for the Jacobian determinant $\det(\partial y / \partial x)$. In terms of these coordinates, we have

$$\langle \rho, f^* \delta \rangle = \int_W w(z, 0) |y'(z, 0)|^{-1} \mathrm{d}z.$$

In particular, we can consider the case where $\dim M = l = \dim N$, so that $k = 0$. In this case $W = f^{-1}(0)$ consists of a discrete set of points. Let $x = (x_1,...,x_l)$ denote local coordinates about each of these points, p, and let δ_p be the delta function given by each of these points. Then the above formula becomes.

$$f^*\delta = \sum_{f(p)=0} |y'(p)|^{-1} \delta_p.$$

If k is not necessarily zero, we have seen that the local expression for $f^*\delta$ is of the form $s \otimes |dx|^{-1}$, where $x = (x_1,...,x_k)$ are normal coordinates to W; i.e. if we replace x by other normal coordinates, $\bar{x}$, we must replace s by $\bar{s} = s|\det(\partial x/\partial \bar{x})|$. We can give a slightly more invariant interpretation to such an object as follows. Let W be a properly immersed submanifold of M. At each point $p \in W$ we have the identification $\square T_p M \approx \square T_p W \otimes \square N_p W$, where $N_p W = T_p M/T_p W$ denotes the normal space to W. Suppose that σ is a smooth section of the bundle $\square^{-1} N_p W$ along W. If ρ is a smooth density of compact support on M, then at each point p of W we can write $\rho_p = \rho_{1p} \otimes \rho_{2p}$, where $\rho_{1p} \in \square T_p W$ and $\rho_{2p} \in \square N_p W$. We can multiply ρ_{2p} with σ_p to get a number, and so obtain an element of $\square T_p W$. In this way we can multiply ρ and σ to obtain a smooth density on W which we can denote by $\rho\sigma$. Since ρ has compact support and W is properly immersed, $\rho\sigma$ has compact support on W, and can be integrated. We can therefore define

$$\langle \sigma, \rho \rangle = \int_W \rho\sigma.$$

In this way, a smooth section of $\square^{-1}(NW)$ defines a generalized function on M (supported along W). Such a generalized function will be called a δ function along W, or a $W - \delta$ function. Thus, if $f: M \to \mathbb{R}^l$ is a submersion, the generalized function $f^*\delta$ is a $W - \delta$ function, where $W = f^{-1}(0)$ which is given in terms of coordinates, y, on $\mathbb{R}^l$ and local normal coordinates, x, along W by the local expression

$$f^*\delta(x, z) = |y'|^{-1} \delta(x) \otimes dx^{-1}, \tag{G.6}$$

where the expression on the right means that when applied to $w(z, x)\, dz\, dx$ it gives $\int |y'(z, 0)|^{-1} w(z, 0)\, dz$.

We have seen how to pull back generalized functions under submersions. Suppose that X, Y, and Z are smooth manifolds, with $f: X \to Y$ and $g: Y \to Z$ both submersions. Then $gf: X \to Z$ is clearly a submersion, and we clearly have

$$(gf)^* = f^*g^*$$

when applied to generalized functions on Z. This suggests a way for defining pull backs for certain generalized sections even in the case where f is not necessarily a submersion. Indeed, suppose that $w = hg^*v$ is a generalized function on Y, where v is a generalized function on Z and where h is a smooth function on Y. Suppose that the map $gf: X \to Z$ is a submersion (at least on the inverse image of the support of h) even though f itself need not be a submersion. Then we would define f^*w as $(f^*h)((gf)^*v)$. One should check that this definition does not depend on the explicit description of the generalized function, w, as $w = hg^*v$. This can be done in a number of ways, the most general being a characterization of f^*w in terms of a limit process. We shall confine attention to δ functions associated to submanifolds, and in this case we can give an explicit definition of f^*w which does not depend on the local representation. Indeed, let w be a δ function associated to the submanifold $W = g^{-1}(0)$, where $g: Y \to \mathbb{R}^l$ is a submersion. Let $f: X \to Y$ be a smooth map. Then the composite map, $g \circ f$ will have 0 as a regular value (i.e. be a submersion on some neighborhood of $f^{-1}W$) if and only if for each $x \in f^{-1}W$, the image of $\mathrm{d}f_x: TX_x \to TY_{f(x)}$ projects surjectively onto $NW_{f(x)} = TY_{f(x)}/TW_{f(x)}$. Under these circumstances we say that the map f is *transversal* to the submanifold W, and it is clear that $f^{-1}W$ is a submanifold of codimension l in X, and that we have an isomorphism, given by $\mathrm{d}f_x$, of $N(f^{-1}W)_x$ with $N(W)_{f(x)}$. In particular, we get an isomorphism of $\square N(f^{-1}W)_x$ with $\square W_{f(x)}$. If w is a section of $\square NW$ over W, then this isomorphism allows us to define the section f^*w of $\square N(f^{-1})$ by setting $(f^*w)(x) = w(f(x))$, where we have identified $\square N(f^{-1}W)_x$ with $\square NW_{f(x)}$ via the above isomorphism. If $y_1, \ldots, y_l$ are local, normal coordinates to W, and if $s_1, \ldots, s_p, x_1, \ldots, x_l$, are local coordinates on X with the xs normal coordinates on $f^{-1}w$, then we can write w locally as $w = a \otimes \mathrm{d}y^{-1}$ in which case the local expression for f^*w is clearly

$$f^*w = |y'(s,0)|^{-1} a(f(s,0)) \delta(x) \otimes \mathrm{d}x^{-1} \tag{G.7}$$

where, as usual, y' denotes the Jacobian determinant of the y's with respect to the x's. Notice that our definition of f^*w is consistent with the previous definition of $f^*\delta$. Notice also the definition makes sense for a δ function associated to any properly immersed submanifold. Finally, observe that if $f: X \to Y$ and $g: Y \to Z$ are smooth maps, and if v is any δ function associated to a submanifold, S, of Z, and if g is transversal to S and f transversal to $g^{-1}(S)$, then the map $g \circ f$ is transversal to S and the formula

$$(g \circ f)^*v = f^*g^*v$$

holds.

Let us now examine the push forward of a δ function. Of course, we push forward generalized densities and not generalized functions, so we must define the notion of a δ density. Let w be a δ function associated with the submanifold, W, of X, and let ρ be a smooth density on X. Then $w\rho$ defines a generalized density on X by the formula

$$\langle w\rho, u \rangle = \langle u\rho, w \rangle$$

for any $u \in C_o^\infty(X)$. Such a $w\rho$ is called a δ density associated with W. Notice that at each point, x, of W, the density ρ defines an element of $\square TX_x = \square TW_x \otimes \square NW_x$, while w defines an element of $\square^{-1}NW_x$. Their product defines an element of $\square TW_x$, and thus $w\rho$ defines a smooth density on W which we shall denote by $\sigma_{w\rho}$. Let $\iota: W \to X$ denote the immersion of W into X, so that ι^*u denotes the restriction of u to W, if u is any smooth function on X. Then it is clear from the above definition of the δ density, $w\rho$, that

$$\langle w\rho, u \rangle = \langle \sigma_{w\rho}, \iota^*u \rangle = \int_W (\iota^*u)\sigma_{w\rho},$$

for any $u \in C_0^\infty(X)$, thus

$$w\rho = \iota_* \sigma_{w\rho}. \tag{G.8}$$

If $f: X \to Y$ is a proper map, it follows from (G.3) that

$$f_*(w\rho) = f_* \iota_* \sigma_{w\rho} = (f\iota)_* \sigma_{w\rho}. \tag{G.9}$$

If the map $f\iota$ is a submersion, then we can conclude that the generalized density $f_*(w\rho)$ is actually smooth. For example, suppose that $f: X \to Y$ is a submersion and that W intersects each fiber transversally. Then $f\iota$ is a submersion, and we conclude that any δ density along W pushes forward to a smooth density on Y given by (G.9).

So far we have been considering (generalized) functions and densities. On a product manifold, $Z \times X$, it is convenient to consider objects which transform as functions in one of the variables and as densities in the other, for instance, an object whose local expression would be $k(w, x)\mathrm{d}x$. If k is smooth, we shall call such an object a smooth section of $\square X$ over $Z \times X$. Similarly, one can talk of generalized sections of $\square X$ over $Z \times X$. (A little later on we shall define the notion of generalized section of any vector bundle over a manifold.) If κ is such a generalized section, it defines a linear transformation $K: C_0^\infty(X) \to C^{-\infty}(Z)$ by the formula

$$\langle Ku, \rho \rangle = \langle \kappa, \rho u \rangle \quad \text{for any} \quad \rho \in C_0^\infty(\square Z)$$

where ρu is the obvious smooth section of compact support of $\square Z$ over $Z \times X$. We say that κ is the *kernel* of K. (The Schwartz kernel theorem, which we shall not need, asserts that any continuous linear map, $K: C_0^\infty(X) \to C^{-\infty}(Z)$ has a kernel.) If κ is smooth, and is given by $\kappa = k\mathrm{d}x$, where k is a smooth function on $Z \times X$ and $\mathrm{d}x$ a smooth density on X, then it is easy to check that $Ku \in C^\infty(Z)$ and is given by

$$(Ku)(w) = \int_X k(z, x)u(x)\mathrm{d}x.$$

We can write the same formula in the general case if we interpret integration over X as the push forward π_{Z*}, where π_Z denotes the projection of $Z \times X$ onto Z:

$$Ku = \pi_{Z*}(\kappa u).$$

Suppose we are given a smooth map $h: Z \to X$. This defines a continuous linear map $h^*: C_0^\infty(X) \to C^\infty(Z)$, and we can look for the kernel corresponding to this map. We claim that the kernel exists and is in fact a δ section of $\square X$ associated to the submanifold of $Z \times X$ consisting of graph h. To prove this fact, it is clearly sufficient to verify that h^* is given by such a δ section when operating on smooth functions u of small support. Let $x_1, \ldots, x_l$ be local coordinates on some neighborhood, U, of X, and let $z_1, \ldots, z_k$ be local coordinates on some open subset of $h^{-1}U \subset W$. Then $x_1 - h_1(z_1, \ldots, z_k), \ldots, x_l - h_l(z_1, \ldots, z_k)$ are local normal coordinates to graph h. We shall write these coordinates as $x - h(z)$. Then $h^*u(z) = u(h(z))$, which we can write symbolically as

$$h^*u(z) = \int_X \delta(x - h(z))u(x)\mathrm{d}x,$$

or, more precisely, as

$$h^*u = \pi_{Z*}u\kappa$$

where κ is the δ section of $\square X$ associated to graph h whose local expression is given by

$$\kappa = [(x - h(z))^*\delta]dx, \tag{G.10}$$

with $x - h(z)$ the locally defined map of $Z \times X \to \mathbb{R}^l$.

Suppose that $Z = X$ is compact, and $\kappa = kdx$ is a smooth kernel. Then $C_0^\infty(X) = C^\infty(X)$, and the map $K: C^\infty(X) \to C^\infty(X)$ is compact, and has a trace which is easily seen to be given by

$$\operatorname{tr} K = \int_X k(x, x)dx.$$

We can write this as

$$\operatorname{tr} K = \pi_*\Delta^*\kappa, \tag{G.11}$$

where $\Delta: X \to X \times X$ is the diagonal map, $\Delta(x) = (x, x)$, and π maps X onto a point, $\pi: X \to \text{pt.}$, so that π_* is just integration over X. Now the right-hand side of (G.6) might be defined for certain κ which are not necessarily smooth, in which case we shall take the right-hand side to be the *definition* of $\operatorname{tr} K$. The map π_* is always well defined since X is compact. The problem is that Δ^* need not be defined. If κ is a δ section of $\square X$ associated with a submanifold W, of $X \times X$, and if the map Δ is transversal to W, then we know that $\Delta^*\kappa$, and hence $\operatorname{tr} K$ is defined by (G.6). Let us apply this remark to the case where $W = \text{graph } h$ for some map, $h: X \to X$. What is the condition that Δ be transversal to graph h? A point, $p \in X$ will belong to Δ^{-1} (graph h) if and only if $(p, p) \in \text{graph } h$, i.e. if and only if $h(p) = p$. At such points, the vectors of the form (ξ, ξ), $\xi \in TX_p$ which form the image of TX_p under $d\Delta_p$, and the vectors of the form $(\xi, dh_p\xi)$, which span the tangent space to graph h at the point (p, p) must span all of the tangent space to $X \times X$. This is easily seen to be equivalent to the condition that the map $(\text{id} - dh_p)$ of TX_p into itself be invertible. A fixed point of h satisfying this condition is called a *regular* fixed point. (Notice that a regular fixed point is necessarily isolated.) A map, all of whose fixed points are regular, is called a *Lefschetz* map. We have shown that (G.11) gives a definition of $\operatorname{tr} h^*$ if h is a Lefschetz map. Let us compute $\operatorname{tr} h^*$. Since X is compact and the fixed points are isolated, there are only finitely many fixed points. Let $x = (x_1, \ldots, x_l)$ be local coordinates about one of the fixed points, p, so that $x - h(x)$ gives normal coordinates to graph h near (p, p). Then by (G.10) and (G.7) we see that the local expression for $\Delta^*\kappa$ near each fixed point is

$$\frac{\delta(x)dx}{|\det(\text{id} - dh_p)|}$$

so that integrating over X gives

$$\operatorname{tr} h^* = \sum_{h(p) = p} \frac{1}{|\det(\text{id} - dh_p)|}. \tag{G.12}$$

Now suppose that G is a Lie group which acts on the compact manifold, X, so that we are given a smooth map.

$$f: G \times X \to X.$$

We assume that the action is locally transitive, which means

$$df_{(e,x)}: TG_e \times TX_x \to TX_x$$

maps $TG_e \times \{0\}$ surjectively into TX_x. From the fact that f defines a group action, it is clear that this implies that

$$df_{(g,x)}: TG_g \times TX_x \to TX_{gx}$$

(where, as usual, we write gx for $f(g, x)$) maps $TG_g \times \{0\}$ surjectively onto TX_{gx}.

Consider the submanifold, graph $f \subset G \times X \times X$, consisting of all points of the form $(g, x, f(g, x))$. We define the diagonal map, $\Delta: G \times X \to G \times X \times X$ by the formula

$$\Delta(g, x) = (g, x, x)$$

and observe that the transitivity implies that Δ is transversal to graph f. Thus Δ^{-1} (graph f) is a submanifold, W, of $G \times X$ which we shall call the isotropy bundle. As a set

$$W = \{(g, x) | g \cdot x = x\}.$$

The tangent space to W at $(g, x) \in W$ consists of those (η, ξ) such that $df_{(g,x)}(\eta, \xi) = \xi$. The transitivity implies that for any ξ we can find an η such that $df_{(g,x)}\eta = \xi$. Thus $f_{|W}: W \to X$ is a submersion.

The map $f^*: C^\infty(X) \to C^\infty(G \times X)$ has a kernel, κ, which is a δ section of $\square X$ along graph f. We can form

$$\Delta^*\kappa$$

which is a δ section of $\square X$ along W. We can then form

$$\chi = \tilde{\pi}_{G*}\Delta^*\kappa; \quad \tilde{\pi}_G(g, x) = g^{-1} \tag{G.13}$$

which is a well defined generalized function on G. We want to think of χ as the character of the action of G on $C^\infty(X)$, and we now examine in what sense this interpretation is valid. Suppose, for the moment, that κ was a smooth kernel, $k(g, x, y)dy$. (Of course this could never be the case for a kernel arising form f^* unless X was discrete.) Then the right-hand side of (G.8) would be the function $\chi(g) = \int_X k(g^{-1}, x, x) dx = \operatorname{tr} K_g$, where K_g is the operator whose kernel is $k_g(x, y) dy = k(g^{-1}, x, y)dy$. So in a very formal sense, (G.8) gives the trace of the action of g on $C^\infty(X)$. Of course such a trace will not exist in the ordinary sense, since $C^\infty(X)$ is not finite dimensional (unless X is discrete) and g will not be of trace class (since, for example, e acts as the identity). On the other hand, let ρ be a smooth density of compact support on G, and let us evaluate $\langle \chi, \rho \rangle$. Let us write (to conform to standard usage) $\rho = udg$, where u is a C^∞ function on G of compact support and dg is Haar measure. Then, if κ were a smooth kernel, we could write

$$\langle \chi, u \, dg \rangle = \int_G \int_X u(g) k(g^{-1}, x, x) \, dx \, dg$$

$$= \operatorname{tr} K_u$$

where $K_u \colon C^\infty(x) \to C^\infty(x)$ has the kernel

$$\kappa_u = \left[\int_G u(g) k(g^{-1}, x, y) \, dg \right] dy.$$

Now in our case the operator K_u is indeed compact; it is given by

$$(K_u v)(x) = \int_G u(g) v(g^{-1} \cdot x) \, dg \tag{G.14}$$

and has kernel

$$\kappa_u = \pi_{X \times X *}(u\kappa),$$

where

$$\pi_{X \times X} \colon G \times X \times X \to X \times X,$$

$$\pi_{X \times X}(g, x, y) = (x, y).$$

The transitivity of f implies that graph f is transversal to the fibers of $\pi_{X \times X}$ and hence, since $u\kappa$ is a δ section along graph f, that κ_u is a smooth kernel on $X \times X$ and hence that K_u is compact. (Indeed, the tangent space to graph f at any point consists of all vectors of the form $(\eta, \xi, df\eta + df\xi)$ and transitivity implies that the last two components are arbitrary, which is precisely the transitivity requirement.)

Thus K_u is indeed a compact operator and

$$\langle \chi, u \, dg \rangle = \pi'_* u\lambda \text{ (where } \pi' \colon G \to \text{pt. so } \pi'_* \text{ is integration over } G)$$

$$= \pi'_* \tilde{\pi}_{G*} \Delta^*(u\kappa)$$

$$= \pi_* \Delta^*(\pi_{X \times X *} \Delta^*) \text{ (where } \pi_* \text{ is integration over } X),$$

the interchange of the operators being justified since they act on different variables. We have thus proved:

For each $u \in C_0^\infty(G)$ the operator K_u given by (G.9) is compact and

$$\langle \chi, u \, dg \rangle = \operatorname{tr} K_u. \tag{G.15}$$

Now χ is only a generalized function. We might ask ourselves whether χ is actually a smooth function near certain points, g, of G. (A generalized function is smooth near some point if, when acting on all smooth densities supported in some small neighborhood of the point, it coincides with the action of a smooth function.) Since $\Delta^* \kappa$ is a δ section along W, we conclude from the discussion about push forward of δ sections that χ will be smooth near g if $\tilde{\pi}_G \iota \colon W \to G$ is a submersion on a neighborhood of

G, i.e. if g is a regular value for this map, and we can evaluate by applying (G.4). So we must investigate which g are regular values. We claim that g is a regular value for $\tilde{\pi}_G \iota : W \to G$ if and only if the map

$$f_{g^{-1}} : X \to X, \quad f_{g^{-1}}(x) = f(g^{-1}, x) = g^{-1} \cdot x$$

is a Lefschetz map. Indeed, g is a regular value means that at all points $(g^{-1}, x) \in W$, the tangent space to W maps onto the tangent space to G. The tangent space to W consists of all (η, ξ) such that

$$\mathrm{d}f_{(g^{-1}, x)}(\eta, \xi) = \xi$$

We can write this equation as

$$\mathrm{d}f_{(g^{-1}, x)}\eta = \xi - \mathrm{d}f_{g^{-1}}\xi.$$

By transitivity, the left-hand side is all of TX_x; hence we can solve this equation for all η if and only if the map $\mathrm{id} - \mathrm{d}f_{g^{-1}}$ is surjective, i.e. an isomorphism. Since this has to hold at all fixed points of $f_{g^{-1}}$, we conclude that g is a regular value if and only if $f_{g^{-1}}$ is Lefschetz. Now (G.9) says that we can evaluate λ at such regular g by ordinary integration over the fibre, and this integration has already been carried out in (G.7) with h playing the role of $f_{g^{-1}}$. We conclude:

> if g is a point such that $f_{g^{-1}}$ is a Lefschetz map then the generalized
> function, χ, given by (G.8) is smooth near g and

$$\chi(g) = \sum_{g \cdot x = x} \frac{1}{|\det(\mathrm{id} - \mathrm{d}f_{g^{-1}})|}. \tag{G.16}$$

Formula (G.16) is the analogue of formula (5.1), Chapter 2, when a discrete set is replaced by a compact smooth manifold. We now discuss the analogue of the Frobenius fixed point formula, (G.12). All that is involved in getting such an analogue is a slight extension of the formalism, where we must consider 'generalized sections' of vector bundles instead of just generalized functions or generalized densities. We now describe the set up.

Let s be a smooth section of a vector bundle $E \to M$. Let E^* denote the dual vector bundle (whose fibers are just the dual spaces to the fibers of E) and let $E' = E^* \otimes \square \, TM$ obtained by tensoring E^* with the line bundle of densities. Then s defines a linear functional on $C_0^\infty(E')$, the space of smooth sections of compact support of E'. Indeed, if $t \otimes \rho$ is such a section, then s pairs with t to give a function, which we may denote by $\langle s, t \rangle$, and

$$\langle s, t \otimes \rho \rangle = \int_M \langle s, t \rangle \rho.$$

A *generalized section* of E is then any continuous linear functional on $C_0^\infty(E')$. If $E \to M$ is a smooth vector bundle, we get a smooth vector bundle $\mathrm{Hom}(E, E)$ over $M \times M$ whose fiber over (x, y) is $\mathrm{Hom}(E_x, E_y)$. A *kernel* is a generalized section of $\mathrm{Hom}(E, E) \otimes \square \, TM$, and it is clear that any kernel determines a continuous linear map of $C_0^\infty(E) \to C^{-\infty}(E)$, where $C^{-\infty}(E)$ denotes the space of generalized sections of E. If κ is a smooth kernel and K is the corresponding linear map, then K is again compact, and it is easy to see that $\operatorname{tr} K$ is now given by

$$\operatorname{tr} K = \int_M \operatorname{tr} k(x,x)\,\mathrm{d}x,$$

where $k(x,x) \in \operatorname{Hom}(E_x, E_x)$ and so its trace is defined. For a general kernel, we do essentially the same thing, replace (G.6) by

$$\operatorname{tr} K = \pi_* \operatorname{tr} \Delta^* \kappa, \tag{G.17}$$

provided that $\Lambda^* \kappa$ is defined. If $\Delta^* \kappa$ is defined, it will be a generalized section of the bundle $\operatorname{Hom}(E,E) \otimes \Box TM$ on the diagonal, where $\operatorname{Hom}(E,E)$ is now a bundle over M whose fiber over each point x is $\operatorname{Hom}(E_x, E_x)$. Now tr is a well defined bundle map from this bundle into the trivial bundle over M. If $b: E \to F$ is a bundle map between two vector bundles over M, and s is a generalized section of E, then it is clear how to define the generalized section bs of F. Since tr defines a bundle map from $\operatorname{Hom}(E,E) \otimes \Box TM$ to $\Box TM$, we see that $\operatorname{tr} \Delta^* \kappa$ is a generalized density on M, provided that $\Delta^* \kappa$ is well defined. In particular, suppose that h is an automorphism of the bundle E over M, and that the induced map of M into itself, which we shall also denote by h, is a Leschetz map. Then h induces a kernel which is a δ section, and an operator, h^*, where

$$(h^* s)(x) = r(h(x))^{-1} s(h(x)),$$

where $r(x): E_x \to E_{h(x)}$ is the linear map on the fibers given by h. We then obtain a formula similar to (G.12) except that the 1 occurring in the numerator must be replaced by $\operatorname{tr} r(x)^{-1}$. Suppose that the group G acts as a group of automorphisms of E in such a fashion that it acts transitively on M. The previous discussion again applies: we define the character of the induced action of G on the sections of E by

$$\chi = \tilde\pi_{G*} \operatorname{tr} \Delta^* \kappa, \tag{G.18}$$

where κ is the kernel on $G \times M \times M$ corresponding to the induced action of G on sections of E. For any smooth function u on G of compact support the operator K_u is compact and

$$\langle \chi, u\,\mathrm{d}g \rangle = \operatorname{tr} K_u. \tag{G.19}$$

If $a \in G$ is such that the map $f_{a^{-1}}: M \to M$ is Lefschetz, then χ is smooth near a and we have the Frobenius fixed point formula

$$\chi(g) = \sum_{ax=x} \frac{\operatorname{tr} a: E_x \to E_x}{|\det(\operatorname{id} - \mathrm{d}f_{a^{-1}})|}. \tag{G.20}$$

Suppose that $M = G/H$, where H is a closed subgroup, and that the vector bundle E comes from a representation of H whose character is σ. Then, as in the proof of (3.3), Chapter 3, we can write any x as gH, and x will be a fixed point of a if $g^{-1}ag \in H$, in which case the numerator is just $\sigma(g^{-1}ag)$. To compute the denominator, we observe that

$$f_{a^{-1}} = f_g f_{g^{-1}a^{-1}g} f_g$$

so that

$$\det(\mathrm{id} - \mathrm{d}f_{g^{-1}}) = \det(\mathrm{id} - f_h), \quad \text{where } h = g^{-1}a^{-1}g,$$

and we are computing $\mathrm{d}f_h$ acting on $TM_H = TG_e/TH_e$. Let ξ be an element of TG_e and let $(\exp t\xi)\cdot H$ be the curve on M that it generates. Then

$$h(\exp t\xi)\cdot H = [h(\exp t\xi)h^{-1}]\cdot H = (\exp t[\mathrm{Ad}\,h\xi])\cdot H$$

so, differentiating,

$$\mathrm{d}f_h = \mathrm{Ad}\,h \quad \text{on} \quad TG_e/TH_e.$$

Substituting into (G.19) gives

$$\chi(g) = \sum_{\substack{ax=x \\ x=gH}} \frac{\sigma(g^{-1}ag)}{|\det(\mathrm{id} - \mathrm{Ad}(g^{-1}a^{-1}g)_{TG_e/TH_e})|} \tag{G.21}$$

which is the Frobenius formula for induced characters.

Let us now look at the case where G is compact, to see if we can get the Frobenius reciprocity formula for induced characters. Let χ be the character induced from the representation of H with character, σ, and let ψ be the character of some irreducible (hence finite-dimensional) representation of G, so that ψ is a smooth function on G. We can evaluate

$$\langle \chi, \bar\psi \, \mathrm{d}g \rangle = \langle \tilde\pi_{G*} \, \mathrm{tr}\, \Delta^* \kappa, \bar\psi \, \mathrm{d}g \rangle = \langle \mathrm{tr}\, \Delta^* \kappa, (\tilde\pi_{G*}\bar\psi) \, \mathrm{d}g \rangle,$$

the last expression being an evaluation in $G \times M$ of a δ section of $\square M$ along W with a smooth section of $\square G$ to get a number. This last expression has a more transparent meaning if we write it as

$$\int\int \bar\psi(g)[\mathrm{tr}\,g : E_x \to E_x]\delta(x - g^{-1}\cdot x) \, \mathrm{d}g \, \mathrm{d}x.$$

According to (G.7) we can evaluate the expression as an honest integral over W. Now W is fibered over M, and so we can evaluate the integral by first integrating over the fibers and then over M. The fiber over $x = gH$ is $G_x = gHg^{-1}$, and the integral over the fiber is

$$\int \bar\psi(a)\sigma(g^{-1}ag) \, \mathrm{d}a = \int \bar\psi(gag^{-1})\sigma(a) \, \mathrm{d}a = \int_H \bar\psi(a)\sigma(a) \, \mathrm{d}a,$$

which is a constant, independent of x. Since the total volume of G and of H are both one, we conclude that the volume of M is also one, and conclude that

$$\langle \chi_\sigma, \bar\psi \, \mathrm{d}g \rangle_G = \int_H \bar\psi_{|H}(a)\sigma(a) \, \mathrm{d}a. \tag{G.22}$$

The right-hand side is the integer giving the intertwining of the two finite-dimensional representations of H; if σ is irreducible, it is just the number of times that σ occurs in the restriction of ψ to H. We can evaluate the left-hand side by writing it as the trace of the operator $K_{\bar{\psi}}$. In computing this trace, we will get the same answer whether we compute it on $L^2(E)$ or on $C^\infty(E)$. But on $L^2(E)$ we already know that $K_{\bar{\psi}} = (1/d)P_\psi$, where d is the dimension of the irreducible representation corresponding to ψ, i.e. $d = \psi(e)$, and P_ψ is the projection onto the sum of subspaces isomorphic to the ψ representation. Thus $\operatorname{tr} K_\psi$ gives precisely the number of times that ψ representations occur in the induced representation. This completes the proof of the Frobenius reciprocity theorem for compact Lie groups.

Let us now return to the more general case where we do not necessarily assume that G is compact, and examine the question of the nature of the singularities of the character, χ. Since χ is the push forward of a smooth section on W, its singularities are certainly no worse than those of the push forward of a measure, which is always a measure. We can ask whether the measure is absolutely continuous relative to Haar measure, i.e. whether the generalized function, χ, is actually locally integrable. For this we use the following criterion due to Dave Schaeffer:

> Let $F: X \to Y$ be a smooth map, and let $A_F \subset X$ be the set of critical points of f, i.e. the set of x at which df_x is not surjective. If A_F has measure zero, then for any smooth density ρ on X, the generalized density, $F_*\rho$ is an absolutely continuous measure on Y. If A_F is not of measure zero, then we can always find a smooth ρ with $F_*\rho$ not absolutely continuous.

The proof is quite easy. Let $C_F = F(A_F)$, and $C_{F,\rho} = F(A_F \cap \operatorname{supp} \rho)$. By Sard's theorem, we know that C_F has measure zero, and so $C_{F,\rho}$ is compact and of measure zero. We know that ρ is smooth on $Y - C_{F,\rho}$ since F is a submersion of A_F. To show that $F_*\rho$ is absolutely continuous, it therefore suffices to show that $F_*\rho$ assigns measure zero to $C_{F,\rho}$. But

$$\int_{C_{F,\rho}} F_*\rho = \int_{F^{-1}(C_{F,\rho})} \rho$$

$$= \int_{A_F} \rho + \int_{(X - A_F) \cap F^{-1}(C_{F,\rho})} \rho.$$

The first integral vanishes since A_F has measure zero, and the second integral vanishes since F is a submersion on $X \to A_F$. Conversely, if A_F does not have measure zero then we can find a smooth density such that the first integral is not zero.

Let us apply this criterion to the case of characters on analytic groups. (Actually all Lie groups are analytic. Rather than prove this fact, we simply observe that all the groups that arise as algebraic matrix groups that we shall consider are obviously analytic.) In this case, where G and H are analytic groups, the manifold W is analytic, and so the set of singular points for the projection $\tilde{\pi}_G$ is an analytic subvariety. In this case (assume for simplicity that W is connected) either the set of singular points is a proper analytic subvariety, and hence of measure zero, which will happen if there is at least one regular point, or $\tilde{\pi}_G$ is singular at all points, in which case χ is concentrated on a proper subset of G. To establish the existence of at least one regular point in a given case, it is convenient to make use of the computation we have already done showing that

$$df_h = \operatorname{Ad} h \quad \text{on} \quad TG_e/TH_e.$$

If we write

$$\operatorname{Ad}(\exp s\eta) = \operatorname{id} + sa\,d\eta + O(s^2)$$

the element $(\exp s\eta, H)$ will be regular if, and only if, $ad\eta$ on TG_e/TH_e is invertible (since $\operatorname{id} - df_{\exp s\eta}) \, sa\,d\eta + O(s^2))$ at least for small values of s. Thus,

> if there exists an element η in TH_e such that $ad\eta$ is invertible on TG_e/TH_e, then χ is locally integrable.

Let us give an illustration of this remark and of the fixed point formula for characters. Let us take $G = SL(n, \mathbb{R})$. Now any matrix in G can be written as the product ODT, where O is orthogonal, D is diagonal with positive entries, and T is upper triangular with ones on the diagonal. If we let $K = SO(n)$ and A the group of positive diagonal matrices and N the group of upper triangular matrices, we can rephrase the above matrix decomposition as $G = KAN$. We give the proof of this fact, which is a special case of what is known as the Iwasawa decomposition.

It is clear from linear algebra that any matrix can be written as the sum of an antisymmetric matrix, a diagonal matrix and an upper triangular matrix with zeros on the diagonal. Exponentiating gives the desired decomposition for matrices close to the identity. To get it globally, we use the following lemma about topological groups.

> Let G be a connected topological group with G_1 and G_2 subgroups with G_1 compact and G_2 connected. Suppose that there are neighborhoods, U_1 in G_1 and U_2 in G_2 such that $U_1 U_2$ is a neighborhood of e in G. Then $G = G_1 G_2$.

Proof For any a in G_1, we can find a neighborhood, W, of e in G_2, so that $aWa^{-1} \subset U_1 U_2$. Such a W works for all b close to a, and thus, since G_1 is compact, we can find a W that works for all a in G_1. Thus $WG_1 \subset G_1 G_2$, and so

$$(W \cdot W)G_1 = W(WG_1) \subset WG_1 G_2 \subset G_1 G_2$$

and so on, so $W^k G_1 \subset G_1 G_2$. Since G_2 is connected, the powers of W exhaust all of G_2 so that

$$G_2 G_1 \subset G_1 G_2.$$

Now

$$G_1(G_1 G_2) = G_1 G_2 \quad \text{and} \quad G_2(G_1 G_2)$$
$$= (G_2 G_1)G_2 \subset G_1 G_2, \quad \text{so} \, (G_1 G_2)(G_1 G_2) \subset G_1 G_2$$

and $(G_1 G_2)^{-1} = G_2 G_1 \subset G_1 G_2$. Thus $G_1 G_2$ is a subgroup and it is open, since it contains $U_1 U_2$. Since G is connected this implies that $G = G_1 G_2$. Of course, in general this product decomposition is not unique.

In the case of $SL(n, \mathbb{R})$ the product decomposition is easily seen to be unique: if $OAT = e$, then O is both orthogonal and triangular with positive entries along the diagonal and hence $O = e$.

Let us take $H = AN$ so that G/H is compact. For any diagonal matrix $D \in A$ with diagonal entries $\lambda_1, \ldots, \lambda_m$, the matrix Ad D on TG_e/TH_e has eigenvalues $\lambda_j \lambda_i^{-1}(j > i)$. Thus, if all the λ's are distinct, the transformation id $-$ Ad D is invertible, and so regular elements exist. For example, formula (G.21) applies to $\chi(D)$, where the denominator is $|\prod(1 - \lambda_j \lambda_j^{-1})|$. The formula, in this case, was first given by Gel'fand and Neumark.

FURTHER READING

CHAPTER 1

There are many good elementary books on group theory. For the general role of symmetry in art and science, there is the short, delightful book by **Hermann Weyl**, *Symmetry*, Princeton University Press (1952), written for a popular audience. Another, longer, book in the same vein is *Symmetry in Science and Art* by **A. V. Shubnikov and V. A. Koptsik**, Plenum Press, New York (1974), written at a more technical level. *The Fascination of Groups* by **F. J. Budden**, Cambridge University Press (1972) gives a long and loving look at many of the groups introduced in this chapter, with many exercises. The classic source for icosahedral symmetry is **Felix Klein** (1849–1925) *Lectures on the Icosahedron and the Solution of Equations of the Fifth Degree*, Dover Publications, New York (1956). A very readable source for crystallography is **F. C. Phillips**, *An Introduction to Crystallography*, Longman, London (1963). See also several books by **M. J. Buerger** including *Elementary Crystallography, an Introduction to the Fundamental Geometrical Features of Crystals*, Wiley, New York (1963) and *Introduction to Crystal Geometry*, McGraw-Hill, New York (1971). An excellent and detailed description and classification of the space groups can be found in *Die Bewengungsgruppe der Kristalographie* by **J. J. Burckhardt**, Birkhauser, Basel (1966). A more modern source, in English, is *Space Groups for Solid State Scientists* by **G. Burns and A. M. Glaser**, Academic Press (1990). Many of the groups in this chapter are examples of finite subgroups of the orthogonal groups generated by reflections. This important class of groups plays a central role in several areas of mathematics. A beautiful treatment of this subject is given in *Reflection Groups and Coxeter Groups* by **J. E. Humphreys**, Cambridge University Press (1990). For the history of the origins of group theory, see *The Genesis of the Abstract Group Concept* by **H. Wussing**, MIT Press (1984). For the history of the classification of the 230 space groups, see the article by **J. J. Burckhardt** in *Arch. Hist. Exact Sci.* **4** (1967), p. 235 (in German).

CHAPTER 2

The best succinct introduction to the basic facts about representations of finite groups can be found in **J. P. Serre**, *Linear Representations of Finite Groups*, Springer-Verlag, Heidelberg (1977). The comprehensive source is **C. W. Curtis and**

I. Reiner, *Representation Theory of Finite Groups and Associative Algebras*, Interscience Publishers, New York (1962). See also *Introduction to Group Theory with Applications* by **G. Burns**, Academic Press, New York (1977) and the first six chapters in *Representation Theory, a First Course* by **W. Fulton and J. Harris**, Springer-Verlag, Heidelberg (1991). The representation theory of the symmetric group has a rich literature of its own. The approach presented here and in Appendix C follows a 1977 paper of James. For this treatment, see **G. D. James**, *The Representation Theory of the Symmetric Groups*, Springer Lecture Notes 682, Springer-Verlag, Heidelberg (1978) and **G. D. James and A. Kerber**, *The Representation Theory of the Symmetric Group*, Encyclopedia of Mathematics and its Applications vol. 16, Addison-Wesley, Reading, MA (1981). There is also the Hopf algebra approach: see **A. V. Zelevinsky**, *Representations of Finite Classical Groups*, Springer Lecture Notes 869, Springer-Verlag, Heidelberg (1981); and Chapter 4 in *Quantum Groups* by **S. Shnider and S. Sternberg**, International Press, Hong Kong (1994). For a beautiful treatment from the combinatorial point of view, see **B. E. Sagan**, *The Symmetric Group: Representations, Combinatorial Algorithms and Symmetric Functions*, Wadsworth and Brooks/Cole Advanced Books and Software, Pacific Grove, CA (1991). A classic is **I. G. Macdonald**, *Symmetric Functions and Hall Polynominals*, Clarendon Press, Oxford (1979). In this book we only briefly touch on projective representations, a topic created by Schur at the beginning of this century. For this subject, see two books by **G. Karpilovsky**, *Projective Representations of Finite Groups*, Marcel Dekker, New York (1985) and *The Schur Multiplier*, London Mathematical Society Monographs, New Series 2, Oxford Science Publications (1987), and also **P. N. Hoffman and J. F. Humphreys**, *Projective Representations of the Symmetric Groups, Q-Functions and Shifted Tableaux*.

CHAPTER 3

For a treatment of the use of spectroscopy in the determination of molecular structure, see **P. J. Wheatley**, *The Determination of Molecular Structure*, Oxford University Press (1968) and **C. A. Coulson**, *The Shape and Structure of Molecules*, revised by Roy McWeeny, Clarendon Press, Oxford (1982). See also **F. A. Cotton**, *Chemical Applications of Group Theory*, Wiley, New York (1990). The best introduction for mathematicians to quantum mechanics is still *The Mathematical Foundations of Quantum Mechanics* by **G. W. Mackey**, Benjamin, New York (1963). An alternative point of view, together with an excellent discussion of elementary particle physics going considerably beyond the treatment in this book, is *Quantum Mechanics and the Particles of Nature, an Outline for Mathematicians* by **A. Sudbery**, Cambridge University Press (1989). The old workhorses, *Quantum Mechanics* by **L. I. Schiff**, McGraw-Hill, New York (1968) and **H. Eyring, G. E. Kimball, and J. E. Walter**, *Quantum Chemistry*, Chapman and Hall, London (1944), are still very serviceable. The discussion of the irreducible representations of the Poincaré group and the related partial differential equations of mathematical physics (presented in the next chapter) follows the treatment in **D. J. Simms**, *Lie Groups and Quantum Mechanics*, Springer

Lecture Notes in Mathematics 52, Springer-Verlag, Heidelberg (1968). A thorough treatment of tensor products can be found in **W. Greub**, *Multilinear Algebra*, Springer-Verlag, Heidelberg (1978). The extension of the fixed point formulas in this chapter culminates in the Atiyah–Bott–Lefschetz fixed point formula. A self-contained, yet concise, treatment of this subject can be found in the notes by **J. Roe**, *Elliptic Operators, Topology and Asymptotic Methods*, Pitman Research Notes in Mathematics Series 179, Longman Scientific & Technical, Burnt Mill, Harlow (1988).

CHAPTER 4

A comprehensive text on representations of the compact Lie groups is **D. P. Zhelobenko**, *Compact Lie Groups and Their Representations*, Translations of Mathematical Monographs 40, The American Mathematical Society, Providence, RI (1973). For the chemical applications discussed in the text, see **C. A. Coulson**, *Coulson's Valence*, updated by Roy McWeeny, Oxford University Press (1979) and also the text by Eyring *et al.* mentioned above. For group theory and nuclear structure, see *Nuclear Shell Theory* by **A. de Shalit and I. Talmi**, Academic Press, New York (1963) and **A. Bohr and B. R. Mottelson**, Benjamin, New York (1963). The classic text on Lie algebras is *Lie Algebras* by **N. Jacobson**, Interscience, New York (1962). Additional excellent texts are: **J. P. Serre**, *Lie Algebras and Lie Groups*, Benjamin, New York (1965) and his *Complex Semi-Simple Lie Algebras*, Springer-Verlag, Heidelberg (1987); **J. E. Humphreys**, *Introduction to Lie Algebras and Representation Theory*, Springer-Verlag, Heidelberg (1980); **A. W. Knapp**, *Lie Groups, Lie Algebras and Cohomology*, Mathematical Notes 34, Princeton University Press (1988); and the book by Fulton and Harris mentioned above.

CHAPTER 5

The classic book on the relation between the symmetric and general linear groups (and much else) is **H. Weyl**, *The Classical Groups*, Princeton University Press (1946). For a modern treatment, see the book by Fulton and Harris. There are books at various levels of technicality on elementary particle physics. My favorite book written for the general scientific audience is **Y. Ne'eman and Y. Kirsch**, *The Particle Hunters*, Cambridge University Press (1986). See also **J. C. Polkinghorne**, *The Particle Play*, Freeman, London (1979). At a more technical level, see **I. S. Hughes**, *Elementary Particles*, Cambridge University Press (1985). The book by Sudbery mentioned above includes an account of the successful Weinberg–Salam model of electroweak unification (which we do not discuss in the text) and of the more speculative $SU(5)$ grand unification models. Another text that includes a discussion of grand unification from the group theoretical point of view (and many of the other topics covered in the present text) is *A Course on the Application of Group Theory to Quantum Mechanics* by **I. V. Schensted**, NEO Press, Peaks Island, ME (1976). A brief sketch and model of quantum field theory can be found in the last chapter of the book by Sudbery. Two very readable texts that do Feynman diagram computations in elementary particle physics without going into the full details of quantum field theory are *Quarks*

& *Leptons: An Introduction Course in Modern Particle Physics* by **F. Halzen and A. D. Martin**, Wiley, New York (1984) and *Introduction to Elementary Particles* by **D. Griffiths**, Wiley, New York (1987). For a history of the acceptance of the quark theory and the role of sociological factors in modern physics, see *Constructing Quarks: a Sociological History of Particle Physics* by **A. Pickering**, University of Chicago Press (1984).

INDEX

Printed in the United States
By Bookmasters